Signals and Communication Technology

For further volumes:
http://www.springer.com/series/4748

Avishy Y. Carmi · Lyudmila S. Mihaylova
Simon J. Godsill
Editors

Compressed Sensing & Sparse Filtering

Springer

Editors
Avishy Y. Carmi
Department of Mechanical and Aerospace Engineering
Nanyang Technological University
Singapore
Singapore

Simon J. Godsill
Department of Engineering
University of Cambridge
Cambridge
UK

Lyudmila S. Mihaylova
School of Computing and Communications
Lancaster University
Lancaster
UK

ISSN 1860-4862
ISBN 978-3-642-38397-7 ISBN 978-3-642-38398-4 (eBook)
DOI 10.1007/978-3-642-38398-4
Springer Heidelberg New York Dordrecht London

Library of Congress Control Number: 2013939576

Printed on acid-free paper

Springer is part of Springer Science+Business Media (www.springer.com)

Preface

Contemporary signal processing technologies are frequently required to cope with undersampled, rare, missing or even conflicting measurements. In some cases the amount of available data can be below a threshold which seemingly inhibits plausible inference. Often, in such cases, conventional inference methods fall short of providing reliable solutions. As normally signals of interest can be discerned into only a few fundamental components in some mathematical domain, dedicated inference techniques seek to find solutions of the lowest complexity, a concept which has proved to be extremely useful when dealing with limited data.

This book is aimed at presenting concepts, methods and algorithms able to cope with undersampled and limited data. One such trend that recently gained popularity and to some extent revolutionised signal processing is compressed sensing. Compressed sensing builds upon the observation that many signals in nature are nearly sparse (or compressible, as they are normally referred to) in some domain, and consequently they can be reconstructed to within high accuracy from far fewer observations than traditionally held to be necessary.

Apart from compressed sensing this book contains other related approaches. Each methodology has its own formalities for dealing with such problems. As an example, in the Bayesian approach, sparseness promoting priors such as Laplace and Cauchy are normally used for penalising improbable model variables, thus promoting low complexity solutions. Compressed sensing techniques and homotopy-type solutions, such as the LASSO, utilise l_1-norm penalties for obtaining sparse solutions using fewer observations than conventionally needed. The book emphasizes on the role of sparsity as a machinery for promoting low complexity representations and likewise, its connections to variable selection and dimensionality reduction in various engineering problems.

This book is intended for researchers, academics and practitioners with interest in various aspects and applications of sparse signal processing.

Book Outline

Each chapter in the present book forms a complete self-contained work that can be read independently of others. The reader who is not acquainted with the subject matter and in particular with compressed sensing is advised to read at least the first half of Chap. 1. A brief description of the content of each chapter is provided below.

- Chapter 1 is a concise exposition to the basic theory of compressed sensing. It assumes no prior knowledge of the subject and gradually builds the theory while elaborating on the basic results. The second half of this chapter is mostly concerned with the application of compressed sensing ideas to dynamic systems and sparse state estimation.
- Chapter 2 is concerned with the geometrical foundations of compressed sensing. The geometric point of view adopted by the author not only underlies many of the initial theoretical developments on which much of the theory of compressed sensing is built, but has also allowed ideas to be extended to much more general recovery problems and structures. A unifying framework is that of non-convex, low-dimensional constraint sets in which the signal to be recovered is assumed to reside. The sparse signal structure of traditional compressed sensing translates into a union of low-dimensional subspaces, each subspace being spanned by a small number of the coordinate axes. The union of subspaces interpretation is readily generalised and many other recovery problems can be seen to fall into this setting. For example, instead of vector data, in many problems, data are more naturally expressed in matrix form (for example a video is often best represented in a pixel by time matrix). A powerful constraint on matrices are constraints on the matrix rank. For example, in low-rank matrix recovery, the goal is to reconstruct a low-rank matrix given only a subset of its entries. Importantly, low-rank matrices also lie in a union of subspaces structure, although now, there are infinitely many subspaces (though each of these is finite dimensional). Many other examples of union of subspaces signal models appear in applications, including sparse wavelet-tree structures (which form a subset of the general sparse model) and finite rate of innovations models, where we can have infinitely many infinite dimensional subspaces. The chapter provides an introduction to these and related geometrical concepts and shows how they can be used to (a) develop algorithms to recover signals with given structures and (b) allow theoretical results that characterise the performance of these algorithmic approaches.
- Chapter 3 extends the basic theory of compressed sensing to the general case of exponential-family noise that includes Gaussian noise as a particular case; the underlying recovery problem is then formulated as l_1-regularized generalized linear model (GLM) regression. In this chapter it is further shown that, under standard restricted isometry property assumptions on the design matrix, l_1-minimization can provide stable recovery of a sparse signal in presence of

exponential-family noise. Sufficient conditions on the noise distribution are also provided that guarantee stable recovery.

- Chapter 4 provides a brief review of some of the state of the art in nuclear norm optimization algorithms. The nuclear norm of a matrix, as the tightest convex surrogate of the matrix rank, has fueled much of the recent research and has proved to be a powerful tool in many areas. In this chapter the authors propose a novel application of the nuclear norm to the linear model recovery problem, as well as a viable algorithm for solution of the recovery problem.
- Chapter 5 presents very recent developments in the area of non-negative tensor factorization which admit sparse representations. Specifically, it considers the approximate factorization of third and fourth order tensors into non-negative sums of types of outer-products of objects with one dimension less using the so-called t-product. A demonstration on an application in facial recognition shows the potential promise of the overall approach. This chapter also discusses a number of algorithmic options for solving the resulting optimization problems, and modification of such algorithms for increasing the underlying sparsity.
- Chapter 6 describes the application of compressed sensing and sub-Nyquist sampling to cognitive radio. Cognitive radio has become one of the most promising solutions for addressing the spectral under-utilization problem in wireless communication systems. This chapter pays a special attention to the use of sub-Nyquist sampling and compressed sensing techniques for realizing wideband spectrum sensing. In addition, an adaptive compressed sensing approach is described for wideband spectrum sensing.
- Chapter 7 presents a few identification algorithms for sparse nonlinear multi input multi output (MIMO) systems. The algorithms are potentially useful in a variety of application areas including digital transmission systems incorporating power amplifier(s) along with multiple antennas, cognitive processing, adaptive control of nonlinear multivariable systems, and multivariable biological systems. Sparsity is a key constraint imposed on the model. The presence of sparsity is often dictated by physical considerations as in wireless fading channel estimation. In other cases it appears as a pragmatic modelling approach that seeks to cope with the curse of dimensionality, particularly acute in nonlinear systems like Volterra type series. The authors discuss three identification approaches: conventional identification based on both input and output samples, semi-blind identification placing emphasis on minimal input resources and blind identification whereby only output samples are available plus a priori information on input characteristics. Based on this taxonomy a variety of algorithms, existing and new, are studied and evaluated by simulations.
- Chapter 8 is concerned with the optimization formulation of the Kalman filtering and smoothing problems. The authors use this perspective to develop a variety of extensions and applications. They consider various extensions of Kalman smoothing to systems with nonlinear process and measurement models, systems with linear and nonlinear inequality constraints, systems with outliers in the measurements or sudden changes in the state, and systems where the sparsity

of the state sequence must be accounted for. All extensions preserve the computational efficiency of the classic algorithms, and most of the extensions are illustrated with numerical examples, which are part of an open source Kalman smoothing Matlab/Octave package.

- Chapter 9 develops a novel Kalman filtering-based method for estimating the coefficients of sparse, or more broadly, compressible autoregressive models using fewer observations than normally required. The proposed algorithm facilitates sequential processing of observations and is shown to attain a good recovery performance, particularly under substantial deviations from ideal conditions. In the second half of this chapter, a few information-theoretic bounds are derived pertaining to the problem at hand. The obtained bounds establish the relation between the complexity of the autoregressive process and the attainable estimation accuracy through the use of a novel measure of complexity. This measure is suggested as a substitute to the generally incomputable restricted isometric property.
- Chapter 10 introduces selective gossip which is an algorithm that applies the idea of iterative information exchange to vectors of data. Instead of communicating the entire vector and wasting network resources, the derived approach adaptively focuses communication on the most significant entries of the vector. The authors prove that nodes running selective gossip asymptotically reach consensus on these significant entries, and they simultaneously reach an agreement on the indices of entries which are insignificant. The results demonstrate that selective gossip provides significant communication savings in terms of number of scalars transmitted. In the second part of this chapter, a distributed particle filter is derived employing selective gossip. It is then shown that distributed particle filters employing selective gossip provide comparable results to the centralized bootstrap particle filter while decreasing the communication overhead compared to using randomized gossip to distribute the filter computations.
- Chapter 11 describes a recent work on the design and analysis of recursive algorithms for causally reconstructing a time sequence of (approximately) sparse signals from a greatly reduced number of linear projection measurements. The signals are sparse in some transform domain referred to as the sparsity basis and their sparsity patterns (support set of the sparsity basis coefficients) can change with time. The term "recursive" implies that only the previous signal's estimate and the current measurements are used to get the current signal's estimate. The authors briefly summarize their exact reconstruction results for the noise-free case and likewise present error bounds and error stability results for the noisy case. Connections with related work are also discussed. A key example application where the underlying recovery problem occurs is dynamic magnetic resonance imaging (MRI) for real-time medical applications such as interventional radiology and MRI-guided surgery, or in functional MRI to track brain activation changes. Cross-sectional images of the brain, heart, larynx or other human organ images are piecewise smooth, and thus approximately sparse in the

wavelet domain. In a time sequence, their sparsity pattern changes with time, but quite slowly. The same is also often true for the nonzero signal values. This simple fact, which was first observed by the authors, is the key reason that the proposed recursive algorithms can achieve provably exact or accurate reconstruction from very few measurements.

- Chapter 12 considers the problem of reconstructing time-varying sparse signals in a sensor network with limited communication resources. In each time interval, the fusion centre transmits the predicted signal estimate and its corresponding error covariance to a selected subset of sensors. The selected sensors compute quantized innovations and transmit them to the fusion centre. The authors consider the situation where the signal is sparse, i.e. a large fraction of its components is zero-valued. Algorithms are presented for signal estimation in the described scenario, and their complexity is analysed. It is shown that the proposed algorithms maintain near-optimal performance even in the case where sensors transmit a single bit (i.e., the sign of innovation) to the fusion centre.
- Chapter 13 is concerned with the application of sparsity and compressed sensing ideas in imaging radars, also known as synthetic aperture radars (SARs). The authors provide a brief overview of how sparsity-driven imaging has recently been used in various radar imaging scenarios. They then focus on the problem of imaging from undersampled data, and point to recent work on the exploitation of compressed sensing theory in the context of radar imaging. This chapter considers and describes in detail the geometry and measurement model for multi-static radar imaging, where spatially distributed multiple transmitters and receivers are involved in data collection from the scene to be imaged. The mono-static case, where transmitters and receivers are collocated is treated as a special case. For both the mono-static and the multi-static scenarios the authors examine various ways and patterns of undersampling the data. These patterns reflect spectral and spatial diversity trade-offs. Characterization of the expected quality of the reconstructed images in these scenarios prior to actual data collection is a problem of central interest in task planning for multi-mode radars. Compressed sensing theory argues that the mutual coherence of the measurement probes is related to the reconstruction performance in imaging sparse scenes. With this motivation the authors propose a closely related, but more effective parameter they termed the t %-average mutual coherence as a sensing configuration quality measure and examine its ability to predict reconstruction quality in various monostatic and ultra-narrow band multi-static configurations.
- Chapter 14 shows how a sparse solution can be obtained for a range of problems in a Bayesian setting by using prior models on sparsity structure. As an example, a model to remove impulse and background noise from audio signals via their representation in time–frequency space using Gabor wavelets is presented. A range of prior models for the sparse structure of the signal in this space is introduced, including simple Bernoulli priors on each coefficient, Markov chains linking neighbouring coefficients in time or frequency and Markov random fields, imposing two-dimensional coherence on the coefficients. The effect of

each of these priors on the reconstruction of a corrupted audio signal is shown. Impulse removal is also covered, with similar sparsity priors being applied to the location of impulse noise in the audio signal. Inference is performed by sampling from the posterior distribution of the model variables using the Gibbs sampler.

- Chapter 15 presents the methods that are currently exploited for sparse optimization in speech. It also demonstrates how sparse representations can be constructed for classification and recognition tasks, and gives an overview of recent results that were obtained with sparse representations.

February 2013

Avishy Y. Carmi
Lyudmila S. Mihaylova
Simon J. Godsill

Contents

Chapter 1
Introduction to Compressed Sensing and Sparse Filtering

Avishy Y. Carmi, Lyudmila S. Mihaylova and Simon J. Godsill

Abstract Compressed sensing is a concept bearing far-reaching implications to signal acquisition and recovery which yet continues to penetrate various engineering and scientific domains. Presently, there is a wealth of theoretical results that extend the basic ideas of compressed sensing essentially making analogies to notions from other fields of mathematics. The objective of this chapter is to introduce the reader to the basic theory of compressed sensing as emanated in the first few works on the subject. The first part of this chapter is therefore a concise exposition to compressed sensing which requires no prior background. The second half of this chapter slightly extends the theory and discusses its applicability to filtering of dynamic sparse signals.

1.1 What is Compressed Sensing?

First and foremost, compressed sensing (CS) is a very useful concept when dealing with limited and redundant data. The basic idea is as simple and sensible as one might expect from a theory that has flourished exceptionally fast over the past six years. Let us try to summarize it as follows. The conventional paradigm of data processing normally involves acquisition and compression stages. The compression phase is carried out either explicitly via a dedicated algorithm or implicitly as part

A. Y. Carmi (✉)
Department of Mechanical and Aerospace Engineering, Nanyang Technological University, Singapore, Singapore
e-mail: acarmi@ntu.edu.sg

L. S. Mihaylova
School of Computing and Communications, Lancaster University, Lancaster, United Kingdom
e-mail: mila.mihaylova@lancaster.ac.uk

S. J. Godsill
Department of Engineering, University of Cambridge, Cambridge, United Kingdom
e-mail: sjg@eng.cam.ac.uk

A. Y. Carmi et al. (eds.), *Compressed Sensing & Sparse Filtering*,
Signals and Communication Technology, DOI: 10.1007/978-3-642-38398-4_1,

of the inference methodology. It essentially involves elimination of redundancies and insignificant parts within the data for the mere purpose of producing a concise representation of a mathematical object of interest (be it a continuous-time signal, a vector in an Euclidean space, a set, a matrix or a tensor). Data acquisition and compression are somewhat contradictory in their intentions: whereas in the acquisition stage we are interested in collecting sufficiently enough data for making inference, in the compression stage parts of it which are not useful for our purpose are being thrown out. The conventional paradigm is therefore a wasteful process and a question is raised whether we could acquire less data in the first place for obtaining a compressed representation of the sought-after mathematical object. Compressed sensing is an umbrella term for the methodologies and concepts involved in reconstructing compressed representations of mathematical objects using limited amount of data, typically much less than the objects' ambient dimension (i.e., the object dimension when it is uncompressed).

Organization of This Chapter

We begin with the study of a classical signal reconstruction problem. This example illuminates some of the basic ideas underlying compressed sensing. An extensive overview of the basic theory of compressed sensing is then given in Sect. 1.3. The second part of this chapter, starting in Sect. 1.4, extends some of the basic notions from the first part and is mainly concerned with the estimation of dynamic sparse signals. In addition, some applications of compressed sensing are discussed in Sect. 1.5. Finally, conclusions are offered in the last section.

1.2 Classical Example: Shannon-Nyquist Sampling

Perhaps the best way to demonstrate the concept of compressed sensing is to consider a concrete example. The Shannon-Nyquist sampling paradigm is a classical archetype for this purpose. Consider a signal y in the time domain. The Shannon-Nyquist theory provides the conditions for perfectly reconstructing the signal y from a set of discrete samples $\{Y_k\}_{k=1}^{N}$. For achieving this we need to acquire equally time-spaced samples at a rate exceeding twice the bandwidth of y. Thus, the Shannon-Nyquist sampling theory is valid assuming that: (1) we know the signal bandwidth in advance, (2) sampling should be carried out in a fashion guaranteeing equally spaced samples and at a sufficiently high sampling rate. In practice, these requirements allow us to faithfully represent the signal in the Fourier domain. These two representations, the signal in the time domain (y) and the obtained discrete representation (Y_k) are equivalent in the sense the we can perfectly recover one of them given the other.

Further we compute the discrete Fourier transform (DFT), X_k, of the signal Y_k (which is equivalent to y in the above mentioned sense). Thus,

$$X_k = (1/\sqrt{N}) \sum_{j=1}^{N} Y_j \exp\left(-2\pi(j-1)(k-1)i/N\right) \quad (1.1)$$

Letting $X = (X_k)_{k=1}^N \in \mathbb{R}^N$ and $Y = (Y_k)_{k=1}^N \in \mathbb{R}^N$, the relation between the discrete representation of y and its Fourier transform can be written compactly as $X = FY$ with F being the unitary DFT matrix. The converse procedure of reconstructing the original signal from its Fourier representation can then be described by

$$Y = \bar{F}X \longrightarrow y \quad (1.2)$$

where $\bar{F} = F^T$ is the inverse DFT matrix (which in this case equals the conjugate transpose of F).

At this point we note the following. Our original signal y may turn out to have only a few significant entries in terms of magnitude when it is considered in the Fourier domain. In other words, the vector X could possibly have only a few meaningful entries while all others nearly vanish. We refer to such an X as *compressible vector* or *sparse vector* if the insignificant entries exactly vanish. Such occasion indicates that the signal energy does not spread uniformly over the spectrum and hence it can be adequately approximated by a reduced representation. The reduced representation of y could not be immediately obtained in the time domain and we had to turn to an alternative domain in which it appears sparse or compressible.

If we could tell the locations of the most significant entries of X in terms of magnitude then we could obtain an almost identical representation of y using less samples in Y. This argument follows by considering a sparse version of X in which all insignificant entries are set to zero. Lets denote this vector by Z and let us look at the difference $\Delta Y = Y - \bar{Y} = \bar{F}(X - Z)$. This difference cannot be large, meaning that we can use Z to adequately approximate the original signal

$$Z \longrightarrow \bar{Y} \approx Y \longrightarrow y \quad (1.3)$$

But as we know the locations of non vanishing entries in Z we can use only a fraction of the samples in Y to construct Z. Thus,

$$Y^m = \bar{F}_{m\times m} Z^m = \bar{F}_{m\times N} Z \quad (1.4)$$

where $m < N$ is the number of non-vanishing entries of Z, and $Y^m \in \mathbb{R}^m$, $Z^m \in \mathbb{R}^m$, $F_{m\times m} \in \mathbb{R}^{m\times m}$, $\bar{F}_{m\times N} \in \mathbb{R}^{m\times N}$ are obtained by eliminating the columns/rows corresponding to vanishing entries in Z. To conclude, we note that at least theoretically we could reconstruct y by sampling $m < N$ points at specific locations, a premise which translates into a sampling rate that may go far below the Nyquist rate.

Compressed sensing do exactly that but without knowing the specific locations of the non-vanishing coefficients. Following our notation that means we are aimed to solve an underdetermined system of linear equations

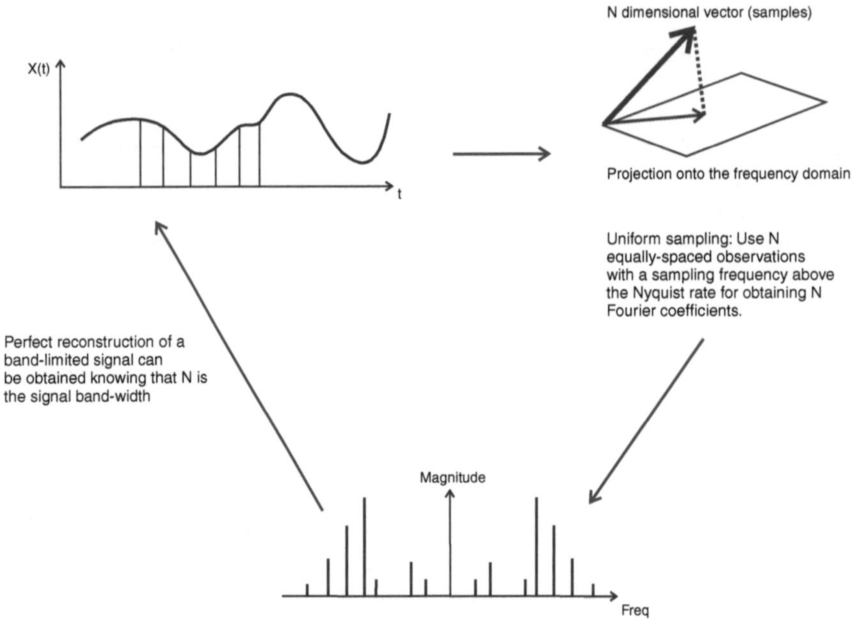

Fig. 1.1 Conventional (Shannon-Nyquist) sampling and reconstruction

$$Y^m = \bar{F}_{m \times N} Z \tag{1.5}$$

where the number of entries in Y^m is possibly much less than the dimension of the sparse vector Z. Once Z is obtained we could adequately reconstruct $\bar{Y} = \bar{F} Z$ from which y can be obtained. The problem (1.5) is generally NP-hard. Yet the interesting detail is that under certain conditions a unique and exact solution Z can be efficiently computed. An illustration of our argument is provided in Figs. 1.1 and 1.2.

1.3 Basic Theory of Compressed Sensing

Consider a signal represented by a vector in the n-dimensional Euclidean space $\chi \in \mathbb{R}^n$, which is sparse in some domain, i.e., it can be represented using a relatively small number of projections in some known, possibly orthonormal, basis, $\psi \in \mathbb{R}^{n \times n}$. Thus, we may write

$$\chi = \psi x = \sum_{i=1}^{n} x_i \psi_i = \sum_{x_j \in \text{supp}(x)} x_j \psi_j, \qquad \|x\|_0 < n, \tag{1.6}$$

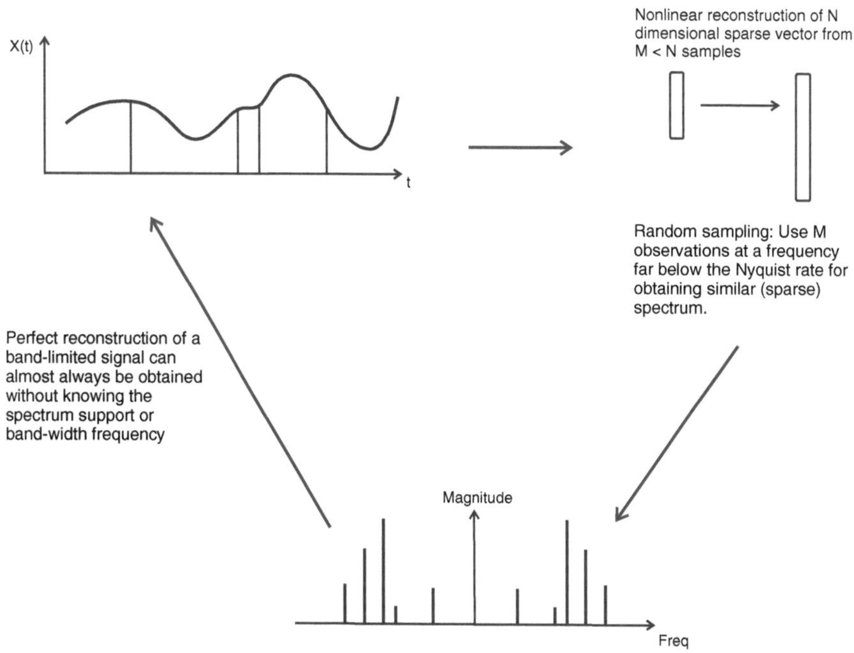

Fig. 1.2 Compressed sensing: random sub-sampling and efficient recovery

where supp(x) and $\|x\|_0$ (the zero norm) are the respective notations for the support of x and its dimension (i.e., the number of non-zero components of x), and ψ_i is the i-th column of the transpose ψ^T of the matrix ψ. The problem of compressed sensing considers the recovery of x (and therefore of χ) from a limited number, $m < n$, of incoherent and possibly noisy measurements (or, in other words, sensing a compressible signal from a limited number of incoherent measurement) [12]. The $\mathbb{R}^m$-valued measurements/observations vector y obeys a linear relation of the form

$$y = H'\chi = Hx \tag{1.7}$$

where $H \in \mathbb{R}^{m \times n}$, and $H = H'\psi$ is known as the *sensing matrix* or *dictionary*. In many practical applications, the observation vector y may be either inaccurate or contaminated by noise. In this case, which will be referred to as the *stochastic CS problem*, an additional noise term is added to the right-hand side of (1.7).

In general, if the sparseness degree of x, denoted by

$$s := \|x\|_0, \tag{1.8}$$

obeys $2s \leq \text{spark}(H)$, where the spark of a matrix is as defined below, then (1.7) has an exact solution.

Definition 1 *(The Spark of a Matrix).* The spark of a matrix H, denoted as spark(H), is the smallest number of columns in H that constitute a linearly dependent set.

If the solution of (1.7) is exact then it is also known to be the sparsest one, which formally translates into the problem

$$(\text{P0}) \quad \begin{cases} \min_x \|x\|_0 \\ \text{subject to } \|y - Hx\|_2^2 \leq \varepsilon \end{cases} \tag{1.9}$$

with $\varepsilon = 0$ (an estimate of x can be obtained in the stochastic case by letting ϵ be of order of the noise variance). The problem (P0) is known to be NP-hard, which implies that in practice, an optimizer cannot be computed efficiently.

1.3.1 A Convex Relaxation Approach

In the late 1990's, the l_1-norm was suggested as a sparseness-promoting term in the seminal works that introduced the LASSO operator [39] and the Basis Pursuit (BP) [18]. Recasting the sparse recovery problem (P0) using the l_1-norm provides a convex relaxation, making an efficient solution possible using a myriad of well-established optimization techniques. Commonly, there are two convex formulations that are proposed to replace (9.26): The quadratically-constrained linear program, which takes the form

$$(\text{P1}) \quad \begin{cases} \min_x \|x\|_1 \\ \text{subject to } \|y - Hx\|_2^2 \leq \varepsilon \end{cases} \tag{1.10}$$

or the quadratic program

$$\begin{cases} \min_x \|y - Hx\|_2^2 \\ \text{subject to } \|x\|_1 \leq \varepsilon' \end{cases} \tag{1.11}$$

The theoretical justification of replacing (P0) with a convex relaxation (P1) has been central in several follow-up works. One of the insights due to [21] is given below.

Theorem 1 ([21]). *Assume that $H = [H_1, \ldots, H_n]$ is composed of unit-length columns, $\|H_i\|_2 = 1$, $i = 1, \ldots, n$. Let also $\epsilon = 0$ in both (P0) and (P1). If in addition the sparsest solution x^s of (P0) obeys*

$$\|x^s\|_0 \leq \frac{\sqrt{2} - \frac{1}{2}}{M(H)} \tag{1.12}$$

where $M(H)$ denotes the maximum coherence among the columns of H, i.e., $M(H) = \max_{i,j} |H_i^T H_j|$, then the solution of (P1) is exact (i.e., the solution of (P1) is also that of (P0)).

In essence, Theorem 1 tells us that depending on the sparseness degree of the solution x^s and on the coherence of the matrix H, the exact solution of (P0) can be efficiently computed via solving the convex relaxation (P1). This is a very strong result which is in the heart of the basic theory of compressed sensing. A refinement of this result appears in [10–12]. Here the maximal coherence $M(H)$ is replaced by the notion of *restricted isometry property*, or RIP in short. This property of the sensing matrix is defined as follows.

Definition 2 *(Restricted Isometry Property of Order* **k** *).* Let $\delta_k \in (0, 1)$ be the smallest number such that

$$(1 - \delta_k)\|x\|_2^2 \leq \|Hx\|_2^2 \leq (1 + \delta_k)\|x\|_2^2 \tag{1.13}$$

for all k-sparse vectors $x \in \mathbb{R}^n$, i.e., vectors consisting of not more than k non-vanishing entries.

Roughly, the RIP implies that the columns of a matrix nearly form an orthonormal basis. More importantly, it is a necessary and sufficient condition guaranteeing efficient computation of the solution to the sparse recovery problem (P0). This result is summarized in the following theorems.

Theorem 2 ([11]). *If $\delta_{2s} < \sqrt{2} - 1$ then for all s-sparse vectors x, such that $y = Hx$, the solution of (P1) is exact, that is to say, the solution of (P1) is also the solution of (P0) for $\epsilon = 0$.*

The reconstruction accuracy of the sparse solution in the stochastic case where $\epsilon > 0$ is provided by the following complementary result.

Theorem 3 ([11]). *Suppose $y = Hx^s + e$ where e is a noise term obeying $\|e\|_2^2 \leq \epsilon$. If in addition $\delta_{2s} < \sqrt{2} - 1$ and x^s (the sought-after vector) is s-sparse then the solution $\hat{x}$ of (P1) obeys*

$$\|\hat{x} - x^s\|_2 \leq C_0 s^{-1/2} \|\hat{x} - \hat{x}^s\|_1 + C_1 \epsilon^{1/2} \tag{1.14}$$

where $\hat{x}^s$ is the best s-sparse approximation of $\hat{x}$, that is, a sparse vector comprising of not more than s most significant entries of $\hat{x}$. The coefficients C_0 and C_1 are independent of $\hat{x}$, x^s and e and are explicitly given in [11].

Note that whenever $\epsilon = 0$ then $\hat{x} = \hat{x}^s$ (i.e., the solution is sparse) and therefore Theorems 2 and 3 coincide.

Compressible Signals

In practice, the unknown signal $x = (x_i)_{i=1}^n \in \mathbb{R}^n$ may be *nearly* sparse, in the sense of having many relatively small components, which are not identically zero. Such representations, frequently encountered in real-world applications, are termed *compressible*. Most of the results in the CS literature naturally extend to the compressible case, assuming some behavior of the small nonzero components. Such a behavior is suggested in [10], where the compressible components sequence is assumed to decay according to the power law

$$|x_i| \leq \kappa i^{(-1/r)}, \qquad |x_i| \geq |x_{i+1}|, \tag{1.15}$$

where $\kappa > 0$ and $r > 0$ are the radius of a weak l_r-ball to which x is confined, and a decay factor, respectively. In this case, a measure of the signal sparseness degree, s, can be obtained as

$$\hat{s} = n - \text{card}\{i \mid 1 \leq i \leq n, \; |x_i| \leq \varepsilon''\} \tag{1.16}$$

for some sufficiently small $\varepsilon'' > 0$, where 'card' denotes the cardinality of a set.

1.3.2 Methods of Solution

The previous results allow us to solve a problem which at first glance seems intractable. This is made possible by resorting to alternative convex programs (1.10) and (1.11). These programs and alike have been studied vastly and nowadays there are myriad of existing optimization methods that are designed to efficiently solve them. We take this opportunity to mention a few notable methods that have been used for solving the CS problem.

The majority of CS methods can be broadly divided into three classes depending on their solution approach. These refer to: convex relaxations, non-convex local optimization techniques and greedy search algorithms. Convex relaxations are used in various methods such as LASSO [39], the Dantzig selector [9], basis pursuit and basis pursuit de-noising [18], and least angle regression [20]. Non-convex optimization approaches include Bayesian methodologies such as the relevance vector machine otherwise known as sparse Bayesian learning [40], Bayesian compressed sensing (BCS) [25], as well as stochastic search algorithms [23, 32, 35]. Notable greedy search algorithms are the matching pursuit (MP) [31], the orthogonal MP [36], and the orthogonal least squares [17], iterative hard thresholding (IHT) [7], the gradient projection [22] and gradient pursuit [6].

1.3.3 Construction of Sensing Matrices

The results of the previous sections rely on the characterization of the sensing matrix. Two properties of this matrix have been pointed out, namely, maximal coherence and the RIP. Although these properties are closely related, the RIP has become the standard notion when dealing with sensing matrices. As mentioned earlier, the RIP is both necessary and sufficient condition for efficient and adequate reconstruction of sparse and compressible signals. It is therefore of much interest to portray the kind of matrices that normally satisfy this property. Such matrices are sometimes refer to as *good CS matrices*, or simply *RIP matrices*.

One of the first results in CS considered an undersampled DFT matrix [12]. This is merely the same matrix $\bar{F}_{m\times N}$ used in our example in Sect. 1.2. This matrix is obtained in [12] by choosing rows uniformly at random from the square DFT matrix $\bar{F}$. The resulting matrix $\bar{F}_{m\times N}$ can be shown to satisfy the RIP with a probability approaching one assuming the number of rows (i.e., the number of observations) is $m \geq cs \log N$ where $c > 1$ is some constant and s is the underlying sparseness degree of the sought-after signal. The basic result concerning undersampled Fourier RIP matrices is summarized as follows.

Theorem 4 ([12]). *Assume that x is s-sparse and that we are given m Fourier coefficients (observations) with frequencies selected uniformly at random, i.e., we have $y = Hx$ where $H \in \mathbb{R}^{m\times n}$ is an undersampled DFT matrix. Suppose that the number of observations obeys*

$$m \geq cs \log n \tag{1.17}$$

Then solving (P1) reconstructs x exactly with overwhelming probability.

One may have noticed that the undersampled DFT matrix is constructed in a randomized fashion. Randomization is a key ingredient in the construction of most types of RIP matrices. The reason why this is so has to do with a phenomenon known as concentration of measure. Roughly, this property of a probability distribution implies that a significant part of the probability mass is concentrated near the mean. Concentration of measure is explained in more detail in the ensuing.

Apart from DFT matrices there are several other notable random constructions which are detailed below.

- *Gaussian ensembles.* Suppose that the entries of the sensing matrix $H \in \mathbb{R}^{m\times n}$ are randomly sampled from a zero-mean Gaussian distribution with variance $1/\sqrt{m}$. If in addition the sparseness degree of the sought-after vector x obeys

$$s = \mathcal{O}(m/\log(n/m)) \tag{1.18}$$

  then H obeys the RIP condition with probability $1 - \mathcal{O}(\exp(-\gamma n))$ for some $\gamma > 0$. In this case, the RIP constant $\delta_{2s} = 0.5$. This argument is based on known concentration of measure results of Gaussian matrices (see for example [38, 41] and references therein).

- *Bernoulli ensembles*. Suppose that the entries of H are randomly sampled from a Bernoulli distribution of the form

$$\Pr(H_{i,j}) = \begin{cases} 1/2, & \text{if } H_{i,j} = 1/\sqrt{m} \\ 1/2, & \text{if } H_{i,j} = -1/\sqrt{m} \end{cases} \tag{1.19}$$

  Then under the condition (1.18) the matrix H is RIP with probability $1 - \mathcal{O}(\exp(-\gamma n))$ for some $\gamma > 0$.
- *Incoherent ensembles*. Consider a matrix $H \in \mathbb{R}^{m \times n}$ which is obtained by choosing m rows uniformly at random from a $n \times n$ orthogonal matrix. The columns of H are then normalized such that $\|H_i\|_2 = 1$. If, additionally, the sparseness degree of the sought-after vector x obeys

$$s = \mathcal{O}\left(\frac{m}{M(H)^2 n \log^4 n}\right) \tag{1.20}$$

  where (as in Theorem 1) $M(H) = \max_{i,j} |H_i^T H_j|$ is the maximal coherence, then the matrix H is RIP with high probability.

The previous recovery statements can be recast taking into account the random nature of H. Following this rationale in the case of Theorems 2 and 3 yields the following complementary statements.

Theorem 5 *If H is any one of the mentioned random constructions then for all s-sparse vectors x, such that $y = Hx$, the solution of (P1) is almost always exact. In detail, if H is either Gaussian or Bernoulli ensembles then the solution of (P1) is exact with probability* $1 - \mathcal{O}(\exp(-\gamma n))$ *for some* $\gamma > 0$.

Theorem 6 *Suppose $y = Hx^s + e$ where e is a noise term obeying $\|e\|_2^2 \leq \epsilon$. If in addition H is any one of the mentioned random constructions and x^s (the sought-after vector) is s-sparse then with high probability the solution $\hat{x}$ of (P1) obeys*

$$\|\hat{x} - x^s\|_2 \leq C_0 s^{-1/2} \|\hat{x} - \hat{x}^s\|_1 + C_1 \epsilon^{1/2}$$

Concentration of Measure

Restrictions on the problem's dimensionality such as those in (1.17) and (1.18) are rather common in the theory of CS. In the case of random constructions these restrictions are imposed so as to guarantee well-behaved tail bounds of the underlying distribution. These tail bounds are collectively referred to as concentration of measure inequalities. For example, a widely used Gaussian tail bound is (see [5] p. 118)

$$\Pr\left(\left|\|Hx\|_2^2 - \|x\|_2^2\right| > c\|x\|_2^2\right) < s \cdot \exp\left\{-c^2 m/2\right\} \tag{1.21}$$

for some $c > 0$. The bound complementary to the above turns out to be a probabilistic statement on the RIP of H. Further letting $c^2 = 2(\bar{c}s)^{-1}\frac{\log(s/\alpha)}{\log n} \in (0, 1)$ where $\bar{c} > 1, \alpha \in (0, 1)$, it can be verified that

$$\Pr\left(\left|\|Hx\|_2^2 - \|x\|_2^2\right| \leq c\|x\|_2^2\right) \geq 1 - s \cdot \exp\left\{-c^2 m/2\right\} \geq 1 - \alpha \tag{1.22}$$

for some $m \geq \bar{c}s \log n$. From the definition of c^2 it can be immediately reocgnized that for having $c^2 < 1$ one must maintain a ratio s/n of the order of α. Having this in mind and recalling (1.22), we conclude that in such constructions the RIP becomes highly probable whenever

$$s/n \ll 1 \tag{1.23}$$

Deterministic RIP Constructions

Recently there have been a few notable attempts to construct RIP matrices in a deterministic fashion. These compositions essentially do away from random sampling and hence have been collectively termed deterministic RIP constructions. In the course of this, a few known types of structured matrices have been used such as Toeplitz, cyclic and generalized Vandermonde [8]. Other deterministic constructions utilize expander graphs [24]. At the moment, the best attainable dimensionality bound associated with deterministic RIP matrices is not comparable to the one achieved using random ensembles [19] (e.g., (1.18)).

1.4 Sparse Filtering and Dynamic Compressed Sensing

The basic CS framework is mainly concerned with parameter estimation, or time-invariant signals. An effort is yet being made for developing efficient CS techniques that would be able to perform in high dimensional non dynamic settings. Only recently, CS has been applied for the recovery of time-varying sparse signals (i.e., sparse random processes). There is no wonder why there is such unbalance between the two realms of non-dynamic and dynamic CS. The fundamentals of CS build upon convex optimization perspectives and as such it is conventionally assumed that the measurements are available in a batch form. This obviously restricts the theory to such signals for which the complexity does not considerably increase over time. Furthermore, the treatment of process dynamics, which are normally governed by probabilistic transition kernels, is not a straightforward task as far as optimization approaches are concerned.

In light of the above, a much more practical approach for treating dynamic sparse signals would be somehow based on state filtering methodologies. Followed by the pioneering works in [13] and [42], which show how the Kalman filter (KF) can be used in this respect, several dynamic CS schemes have been proposed over the past years. Thus, the work in [1] derives a l_1-regularized recursive least squares estimator. This type of estimator is capable of dealing with dynamic signals and support variations via the use of a "forgetting factor". In other works, the LASSO is amended for performing in dynamic settings with possible abrupt changes in the signal support [2, 4].

The KF algorithm constitutes a vital part in the works of [3, 16], and [29]. Indeed, the KF is elegant and simple and above all is the *linear* optimal minimum mean square error (MMSE) estimator irrespective of noise statistics. Despite its appealing features, rarely it is used in its standard formulation which is primarily designed for linear time-varying models. Modifying the KF structure and extending its capabilities have already become a common practice in many engineering and scientific fields. The resulting KF-based methods are vastly used for nonlinear filtering, constrained state estimation, distributed estimation, learning in neural networks, and fault-tolerant filtering.

The KF-based methodologies for dynamic CS can be divided into two broad classes: hybrid, and self-reliant. Whereas the former class refers to KF-based approaches involving the utilization of peripheral optimization schemes for handling sparseness and support variations, the latter class refers to methods that are entirely independent of any such scheme. Hybrid KF-based approaches refer to works such as [3, 16, 29, 42]. The only self-reliant KF method available to that end is the one of [13].

The self-reliant KF method in [13] benefits from ease of implementation. It avoids intervening in the KF process which thereby maintains the filtering statistics as adequate as possible. The key idea behind it is to apply the KF in constrained filtering settings using the so-called pseudo-measurement technique [28]. It may, however, exhibit an inferior performance when improperly tuned or when insufficient number of iterations had been carried out.

1.4.1 A Note on Compressed Sensing and Nonlinear Filtering

Only a small number of works in the CS literature attempt to extend the capabilities of the theory to nonlinear sensing problems (i.e., having nonlinear sensing functions). The reasons why this endeavour has not generated much interest may be two. Firstly, nonlinear sensing formulations seem to be much rare in such areas where CS notions and techniques are vastly employed, and secondly, many results in the theory cease being elegant and applicative in the nonlinear sensing realm, thus loosing their attractiveness. It is more likely the latter reason is the dominant one, as otherwise new application areas would naturally have been flourishing by now.

Nonlinear modeling of natural phenomena is common in many fields of engineering and science. This in turn renders nonlinear filtering one of the most challenging undertaking in a wide range of applications. An invaluable benefit could be achieved by carrying over the qualities of CS that would enable nonlinear filtering from undersampled and limited data.

1.4.2 Discrete-Time Sparse State Estimation

As we are primarily concerned with sparse, or more broadly with compressible dynamic systems, the following definitions are imperative.

Definition 3 *(Compressible Random Process).* An $\mathbb{R}^n$-valued random process $\{x_k\}_{k\geq 0}$ is said to be compressible if its instantaneous realisation at every $k \geq 0$ consists of only $s_k \ll n$ prominent entries in terms of their magnitudes.

Definition 4 *(Compressible System).* Consider a generalised discrete-time system of the form

$$x_{k+1} = f(x_k, w_k) \tag{1.24a}$$

$$z_{k+1} = g(x_{k+1}, v_{k+1}) \tag{1.24b}$$

where $x_k \in \mathbb{R}^n$, $z_k \in \mathbb{R}$ denote, respectively, the state and observation random processes. The smooth functions f and g are, respectively, the process and sensing mappings. The corresponding noises w_k and v_k are assumed to be statistically independent white sequences. The system governed by the equations (1.24) is said to be compressible, or otherwise, having compressible states, if the process $\{x_k\}_{k\geq 0}$ is compressible.

Consider the m-lifted observability mapping [34] associated with (1.24), where m denotes the number of observations needed for uniquely determining the state of the respective deterministic system. Thus,

$$D(x) := \begin{bmatrix} g(x) \\ g(f(x)) \\ g(f \circ f(x)) \\ \vdots \\ g(f^{(l)}(x)) \\ \vdots \end{bmatrix} \in \mathbb{R}^m \tag{1.25}$$

where $g(f^{(l)}(x)) := g(\underbrace{f \circ \cdots \circ f}_{l \text{ times}}(x))$. In our analysis throughout this chapter we assume that $D(x)$ is Lipschitz over the domain of interest, normally an open set in $\mathbb{R}^n$. This property which underlie differentiable structures essentially allows us to scrutinize the interplay between local and global behaviors of a function. When applied to the observability mapping $D(x)$, the Lipschitz condition conveys the extent to which a certain state is indistinguishable from its neighbors. Moreover, it portrays the mapping highest rate of change in the underlying neighborhood which thereby allows discerning the local isometric properties of the mapping. Both these features may respectively be viewed as the interpretation of the left-hand and right-hand sides of the Lipschitz inequality

$$\gamma_2 \le \frac{\|D(x+h) - D(x)\|_2^2}{\|h\|_2^2} \le \gamma_1, \quad \forall x, \ \forall h \neq 0 \tag{1.26}$$

where $\gamma_1, \gamma_2 \ge 0$. Note that, if in addition, $f(\cdot)$ and $g(\cdot)$ are linear, the condition (1.26) reduces to

$$\gamma_2 \le \left\| \frac{\partial D(x)}{\partial x} \bar{h} \right\|_2^2 \le \gamma_1, \quad \|\bar{h}\|_2 = 1 \tag{1.27}$$

which is merely the spectral range of the Gramian matrix $G = \left(\frac{\partial D(x)}{\partial x}\right)^T \frac{\partial D(x)}{\partial x}$ independent of the value of x. The generally non-square matrix $\partial D(x)/\partial x$ takes the role of conventional observability mapping for time-invariant systems.

Suppose for a moment that no process noise is present and that $m \ge n$ possibly noisy observations are gathered, $z_{1:m} = [z_1^T, \ldots, z_m^T]^T$. We wish to obtain a minimum squared error (MSE) estimate of x_0 (and subsequently of any $x_k, k = 1, 2, \ldots$) in the sense

$$\min_{\hat{x}_0} \|z_{1:m} - D(\hat{x}_0)\|_2^2 \tag{1.28}$$

For linear systems, the solution of (1.28) is no other than the well known least squares estimator. In this respect, the observability matrix, or alternatively the Gramian (which is independent of x), indicate whether the obtained solution is unique. For this to happen, both these matrices should not be rank deficient, i.e., they both should be full rank, which entails $\gamma_2 > 0$ in (1.26) and (1.27). Similarly, in the general case (1.24), global observability is ensured if and only if $\gamma_2 > 0$ for every x in the system's state space.

Definition 5 *(Global Observability Condition).* Let

$$\mathcal{O} = \left\{ x \,\middle|\, \frac{\|D(x+h) - D(x)\|_2^2}{\|h\|_2^2} = 0, \ \ \forall h \neq 0 \right\} \tag{1.29}$$

then the system (1.24) is globally observable if and only if $\mathcal{O} = \{\emptyset\}$. In that case,

$$\gamma_2 = \inf_{\substack{x \\ h \neq 0}} \left(\frac{\|D(x+h) - D(x)\|_2^2}{\|h\|_2^2} \right) > 0 \tag{1.30}$$

It can be readily verified that the smallest amount of observations that may allow a unique recovery in (1.28) is $m = n$. Surprisingly enough, this prerequisite can be relaxed, essentially allowing fewer observations than n, following the rationale underlying CS. Thus, by assuming that the system (1.24) is sparse (i.e., the insignificant entries in x vanish), a formulation alternative to (1.28) is given by the following program

$$\text{(P0)} \qquad \min_{\hat{x}} \|\hat{x}\|_0 \ \text{ subject to } \|z_{1:m} - D(\hat{x})\|_2 \le \epsilon \tag{1.31}$$

where, as before, $\|\cdot\|_0$ denotes the number of non vanishing entries in x. The tuning parameter ϵ normally has a magnitude comparable to the noise standard deviation. The solution of the program (P0) may be unique for $s \leq m < n$, where s is the actual number of non vanishing entries in x.

The solution of (P0) is known to be NP-hard and therefore cannot, in general, be computed in a finite time. As proposed by CS theory, a relaxation to (P0) is obtained by substituting the l_0 quasi-norm with the (convex) l_1 norm, that is

$$\text{(P1)} \qquad \min_{\hat{x}} \|\hat{x}\|_1 \quad \text{subject to } \|z_{1:m} - D(\hat{x})\|_2 \leq \epsilon \tag{1.32}$$

The seminal result in the theory of CS asserts that for a *linear system* with $\epsilon = 0$, the solution of the problem (P1) coincides with the exact one (P0), assuming x is s-sparse (i.e., having no more than s non vanishing entries), and $\partial D(x)/\partial x$ obeys

$$\left| \left\| \frac{\partial D(x)}{\partial x} \bar{h} \right\|_2^2 - 1 \right| \leq \delta_{2s}, \quad \|\bar{h}\|_2 = 1 \tag{1.33}$$

with $\delta_{2s} \leq \sqrt{2} - 1$ for every s-sparse $\bar{h}$. The above condition, which is otherwise known as the RIP, guarantees a moderately distorted (in the sense of distance preserving) projection of the high dimensional state space $\mathbb{R}^n$ onto the lower dimensional observation space $\mathbb{R}^m$. Formally, the RIP restricts the spectral range of any sub-matrix of $\partial D(x)/\partial x$ having m rows and not more than $2s$ columns. Hence,

$$1 - \delta_{2s} \leq \sigma_{min}(D_T), \sigma_{max}(D_T) \leq 1 + \delta_{2s} \tag{1.34}$$

for any $D_T \in \mathbb{R}^{m \times |T|}$, where T denotes a set of column indices with cardinality $|T| \leq 2s$, and $\sigma_{min}, \sigma_{max}$ stand for the smallest and largest singular values, respectively.

A complementary result to the above guarantees that in the presence of noise (i.e., $\epsilon > 0$), and while assuming the RIP constant δ_{2s} is bounded as before, the solution obtained by solving the convex relaxation (P1) is accurate to within magnitudes of the noise standard deviation.

1.4.3 Compressive Observability

Unsurprisingly, the effectiveness of compressible state filtering techniques depends on the properties of the state dynamics and the sensing equation. The notions of observability and estimability, which are common in standard systems, are applicable in this case, albeit not necessarily conveying the benefits brought forth by CS, such as the possible reduction in the required amount of observations. Our purpose in this part is to extend these notions for providing conditions under which CS techniques can be successfully applied. The following distinction is to be made: compressible systems which are deemed unobservable may be *compressively observable* (and the same holds for estimability). This in turn, opens up a wide range of possibilities for the design of effective sensing schemes guaranteeing compressive observability.

Assuming differentiability of the mapping $D(x)$ allows us to examine the observability locally at any given point in the state space. The formal definition provided below makes use of the notion of directional derivative.

Definition 6 *(Local Observability).* Let

$$d\mathcal{O}(x_0) = \left\{ h \,\middle|\, \lim_{\tau \to 0} \frac{\|D(x_0 + \tau h) - D(x_0)\|_2^2}{\|\tau h\|_2^2} = \|\nabla_{\bar{h}} D(x_0)\|_2^2 = 0, \quad \forall h \neq 0 \right\} \tag{1.35}$$

where $\nabla_{\bar{h}} D(x_0)$ denotes the directional derivative of $D(x)$ along $\bar{h} = h/\|h\|_2$, computed at x_0, that is $\left.\frac{\partial D(x)}{\partial x}\right|_{x=x_0} \bar{h}$. Then the system (1.24) is locally observable at x_0 if and only if $d\mathcal{O}(x_0) = \{\emptyset\}$.

We note that alternatively Definition 6 can be expressed by means of the local Gramian matrix, yielding an equivalent condition

$$\text{Rank}\,(G(x_0)) = n, \quad G(x_0) = \left(\left.\frac{\partial D(x)}{\partial x}\right|_{x=x_0} \right)^T \left.\frac{\partial D(x)}{\partial x}\right|_{x=x_0} \tag{1.36}$$

It can be easily verified that for the above condition to hold the mapping $D(x)$ should consist of at least n rows, or in other words, at least n observations are needed for the system to be locally observable irrespective of the instantaneous state. This requirement is radically changed, however, taking into consideration the compressibility of the underlying system.

Compressible State Variations

The compressibility assumption implies that the directional derivative $\nabla_{\bar{h}} D(x_0)$ in (1.35) involves only compressible variations $\bar{h}$. In particular, if the state x_0 is s_0-sparse (i.e., consists of not more than s_0 non vanishing entries), then $\|\bar{h}\|_0 \leq 2s_0$. This readily follows upon noting that $h = x - x_0$, where the state x, which is likewise s_0-sparse, is in the neighborhood of x_0. A small dimensional illustration of this property is provided in Fig. 1.3, where 1-sparse compressible states in $\mathbb{R}^3$ (forming the vertices $[\pm 1, 0, 0]$, $[0, \pm 1, 0]$, $[0, 0, \pm 1]$ of a polytope) are connected via the variation vectors (the edges of the polytope). This subtle detail allows us to define a compressible analogue of the local observability condition.

Proposition 1 (Local Observability for Compressible Systems). *Let*

$$d\mathcal{O}(x_0) = \left\{ \bar{h} \,\middle|\, \left| \|\nabla_{\bar{h}} D(x_0)\|_2^2 - 1 \right| > \delta_{4s_0}, \quad \forall h \neq 0, \ \|\bar{h}\|_0 \leq 2s_0 \right\} \tag{1.37}$$

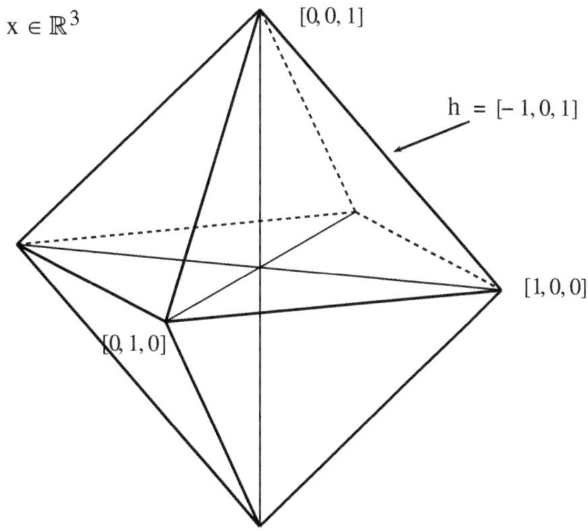

Fig. 1.3 Illustration of state variations in a compressible state space

Then the (compressible) system (1.24) *is said to be compressively observable at* x_0 *if and only if there exists such* $\delta_{4s_0} \in (0, 1)$ *for which* $d\mathcal{O}(x_0) = \{\emptyset\}$.

Proposition 1 can guarantee adequate recovery of x_0 using (considerably) less observations than conventionally needed (see Definition 6). This premise is directly related to the underlying concept in CS where an l_1 relaxation allows accurate and even exact reconstruction under certain restrictions on the compressibility level s_0, and on δ_{4s_0}.

Local Isometries and the Johnson-Lindenstrauss Lemma

In consideration of the conventions above, a compressible system may both be unobservable and compressively observable locally at any admissible (and compressible) state. This, however, would come into being for a certain class of systems for which the underlying condition in Proposition 1 holds. When the problem at hand reduces to the commonly studied example of linear parameter estimation, the conventional theory of CS provides the instrumental sensing matrices which satisfy Proposition 1 for reasonable values of δ_{2s_0} with overwhelming probability. In that case, the condition in Proposition 1 coincides with the acclaimed RIP, which is essentially a global feature of the system as it holds irrespective of the parameters themselves. Notwithstanding, Proposition 1 encompasses the broader class of time-varying and possibly nonlinear systems.

Several random sensing matrices which satisfy the RIP have been extensively studied in the theory of CS (see Sect. 1.3.3). These constructions rely on familiar concentration of measure results which are mainly pronounced in high dimensions. The prevalent compositions consist of Gaussian, Bernoulli, and Fourier matrices and recently also a class of deterministic matrices. As conveyed by Proposition 1 having an RIP Jacobian $\frac{\partial D(x)}{\partial x}\big|_{x=x_0}$ renders the system compressively observable at x_0.

Our notion of compressive observability becomes tangible by recalling the Johnson-Lindenstrauss (JL) lemma which is a statement about the existence of general Lipschitz low-distortion mappings [26].

Lemma 1 *(Johnson-Lindenstrauss).* Given some $\delta \in (0, 1)$, a set $\mathcal{X}$ of l points in $\mathbb{R}^n$ and a number $m_0 = \mathcal{O}(\ln(l)/\delta^2)$, there is a Lipschitz function $D : \mathbb{R}^n \rightarrow \mathbb{R}^m$ where $m > m_0$ such that

$$(1-\delta) \parallel x - \hat{x} \parallel_2 \leq \parallel D(x) - D(\hat{x}) \parallel_2 \leq (1+\delta) \parallel x - \hat{x} \parallel_2 \tag{1.38}$$

for all $x, \hat{x} \in \mathcal{X}$.

If we further consider a case where $x = \hat{x} + h$ with sufficiently small $\parallel h \parallel_2$, then by taking the first-order Taylor expansion of $D(x)$ around $\hat{x}$ it can be easily recognized that the JL Lemma reduces to approximately the RIP of the Jacobian $\partial D(x)/\partial x$ computed locally at $\hat{x}$, that is

$$(1-\delta) \parallel h \parallel_2 \leq \parallel \nabla_h D(\hat{x}) + o\left(\parallel h \parallel_2^2\right) \parallel_2 \leq (1+\delta) \parallel h \parallel_2 \tag{1.39}$$

In that sense the Lipschitz function that satisfies the JL relation (1.38) *locally* obeys the RIP at $\hat{x}$ for the perturbation vector h. The property (1.39) of the mapping $D(x)$ is termed *Local RIP* here. Similarly to the linear case, the level of the local RIP of $D(x)$ at $\hat{x}$ is determined according to the maximal sparseness degree s of the perturbation h for which (1.39) holds. Obviously, when considering the recovery of a sufficiently small and sparse h, CS can be applied where the Jacobian $\partial D(x)/\partial x$ assumes the role of the traditional sensing matrix. The general nonlinear program (1.32) would then reduce to a linear CS problem

$$\min \parallel h \parallel_1 \text{ subject to } \parallel z_{1:k} - D(\hat{x}) - \nabla_h D(\hat{x}) \parallel_2 \leq \epsilon \tag{1.40}$$

where the accuracy of recovery would be related to the local RIP constant δ_{4s}.

1.4.4 MMSE Estimation of Sparse Linear Gaussian Processes

Consider an $\mathbb{R}^n$-valued sparse random process $\{x_k\}_{k\geq 0}$ which evolves according to

$$x_{k+1} = Ax_k + w_k \tag{1.41}$$

where $A \in \mathbb{R}^{n \times n}$ is the state transition matrix and $\{w_k\}_{k\geq 0}$ is a zero-mean white Gaussian sequence with covariance $Q_k \succeq 0$. The initial distribution of x_0 is Gaussian with mean μ_0 and covariance P_0. The signal x_k is measured by the $\mathbb{R}^m$-valued random process

$$z_k = Hx_k + \zeta_k \tag{1.42}$$

where $\{\zeta_k\}_{k\geq 1}$ is a zero-mean white Gaussian sequence with covariance $R_k \succ 0$, and $H \in \mathbb{R}^{m\times n}$.

Letting $z_{1:k} := [z_1, \ldots, z_k]$, our problem is defined as follows. We are interested in finding a $z_{1:k}$-measurable estimator, $\hat{x}_k$, that is optimal in some sense. Often, the sought-after estimator is the one that minimizes the mean square error $E\left[\| x_k - \hat{x}_k \|_2^2\right]$. It is well-known that if the linear system (1.41), (1.42) is observable then the solution to this problem can be obtained using the KF. On the other hand, if the system is unobservable, then the regular KF algorithm is useless; if, for instance, $A = I_{n\times n}$ (the n by n unit matrix), then it may seem hopeless to reconstruct x_k from an under-determined system in which $m < n$ and $\text{rank}(H) < n$.

As mentioned earlier, in the deterministic case (i. e., when x is a parameter vector with $A = I_{n\times n}$), one may accurately recover x by solving the subset search problem [12]

$$\begin{aligned} &\min \| \hat{x} \|_0 \\ &\text{subject to} \sum_{i=1}^{k} (z_i - H\hat{x})^T R_i^{-1} (z_i - H\hat{x}) \leq \epsilon \end{aligned} \tag{1.43}$$

for a sufficiently small ϵ. Following a similar rationale, in the stochastic case the sought-after optimal estimator satisfies

$$\min \| \hat{x}_k \|_0 \ \text{subject to} \ E_{x_k|z_{1:k}}\left[\| x_k - \hat{x}_k \|_2^2\right] \leq \epsilon \tag{1.44}$$

where $E_{x_k|z_{1:k}}[\cdot]$ denotes the expectation of x_k conditioned on the observation history $z_{1:k}$. As the above subset search problems are generally NP-hard we resort to a closely-related convex relaxation of the form [13]

$$\min_{\hat{x}_k} E_{x_k|z_{1:k}}\left[\| x_k - \hat{x}_k \|_2^2\right] \ \text{subject to} \ \|\hat{x}_k\|_1 \leq \epsilon' \tag{1.45}$$

for some $\epsilon' > 0$.

The problem (1.45) can be addressed in the framework of constrained state filtering. In this respect, the l_1 constraint is imposed on the state estimate $\hat{x}_k$ at any given time point. When the underlying system is linear and Gaussian then a sub-optimal estimator can be obtained by modifying the standard Kalman filter algorithm (for further details the reader is referred to the Chapter "Compressive System Identification" and likewise to [3, 13, 16, 29, 42]).

1.5 Applications of Compressed Sensing

Compressed sensing has been extensively studied and applied in the following domains: medical image processing [37], compression [14], coding and machine learning including face recognition, detection and tracking of objects in video [30, 33, 43], sensor networks [27] and cognitive radio.

Especially video based object tracking is widely investigated with CS methods. The amount of data provided by video cameras in real time is enormous. In order to cope with this increased data flow, CS techniques are employed for background subtraction over a part of the video frame. Whereas traditional background subtraction techniques require that the full image is available, the CS-based background subtraction utilizes a reduced image size. The first CS-based background subtraction algorithm [15] performs background subtraction on compressive measurements of a scene, while retaining the ability to reconstruct the foreground. However, in this algorithm the measurement matrix is fixed. In [44] a technique is proposed that adaptively adjusts the number of compressive measurements. This leads to an adaptive scheme which is shown [43] to outperform the basic CS-based background subtraction algorithm [15].

In target tracking with video data the object template has a sparse representation. For instance, in [33] the target is modeled as a sparse representation of multiple predefined templates. The convex relaxation-based tracking algorithm needs to cope with the underlying complexity and hence different l_1 minimization techniques are used, e.g. the Orthogonal Matching Pursuit (OMP) [30] or the l_1-regularized least squares [33].

Compressed Sensing Repository

There are several web sites that provide both code and key papers in the areas of CS. These are listed below:

- A vast collection of paper and software can be found on the web site of Rice University http://dsp.rice.edu/cs.
- Compressive sensing - the big picture https://sites.google.com/site/igorcarron2/cs
- Sparse Optimization Toolbox containing optimization programs for sparse signal recovery, including MP, Basis Pursuit and constrained total variation pursuit. It can be downloaded from: http://www.mathworks.com/matlabcentral/fileexchange/16204
- SPARSELAB: SparseLab is a Matlab software package designed to solve sparse recovery problems. Available on: http://sparselab.stanford.edu/

1.6 Conclusions

This chapter is a concise exposition to the basic theory of compressed sensing. We assume no prior knowledge of the subject and gradually build the theory while elaborating on the basic results. The last part of this chapter is devoted to the application of compressed sensing ideas to dynamic systems and sparse state estimation.

References

1. Angelosante D, Bazerque JA, Giannakis GB (2010) Online adaptive estimation of sparse signals: where RLS meets the l_1-norm. IEEE Trans Sig Process 58:3436–3447
2. Angelosante D, Giannakis GB, Grossi E (2009) Compressed sensing of time-varying signals. In: Proceedings of the 16th international conference on digital signal processing, pp 1–8
3. Asif MS, Charles A, Romberg J, Rozell C (2011) Estimation and dynamic updating of time-varying signals with sparse variations. In: Proceedings of the international conference on acoustics, speech sig process (ICASSP), pp 3908–3911
4. Asif MS, Romberg J (2009) Dynamic updating for sparse time varying signals. In: Proceedings of the conference on information sciences and systems
5. Ball K (2002) Convex geometry and functional analysis. In: Handbook of Banach space geometry. Elsevier
6. Blumensath T, Davies ME (2008) Gradient pursuits. IEEE Trans Sig Process 56(6):2370–2382
7. Blumensath T, Yaghoobi M, Davies M (2007) Iterative hard thresholding and l_0 regularisation. In: Proceedings of the IEEE international conference on acoustics, speech and, signal processing, pp III-877–III-880
8. Calderbank R, Howard S, Jafarpour S (2010) Construction of a large class of deterministic sensing matrices that satisfy a statistical isometry property. IEEE J Sel Top Sig Process 4:358–374
9. Candes E, Tao T (2007) The Dantzig selector: statistical estimation when p is much larger than n. Ann Stat 35:2313–2351
10. Candes EJ (2006) Compressive sampling. In: Proceedings of the international congress of mathematicians, European Mathematical Society,Madrid, pp 1433–1452
11. Candes EJ (2008) The restricted isometry property and its implications for compressed sensing. C R Math 346:589–592
12. Candes EJ, Romberg J, Tao T (2006) Robust uncertainty principles: exact signal reconstruction from highly incomplete frequency information. IEEE Trans Inf Theory 52:489–509
13. Carmi A, Gurfil P, Kanevsky D (2010) Methods for sparse signal recovery using Kalman filtering with embedded pseudo-measurement norms and quasi-norms. IEEE Trans Signal Process 58(4):2405–2409
14. Carmi A, Kanevsky D, Ramabhadran B (2010) Bayesian compressive sensing for phonetic classification. In: Proceedings of the international conference on acoustics, speech and signal processing, pp 4370–4373
15. Cevher V, Sankaranarayanan A, Duarte M, Reddy D, Baraniuk R, Chellappa R (2008) Compressive Sensing for Background Subtraction
16. Charles A, Asif MS, Romberg J, Rozell C (2011) Sparsity penalties in dynamical system estimation. In: Proceedings from the conference on information sciences and systems, pp 1–6
17. Chen S, Billings SA, Luo W (1989) Orthogonal least squares methods and their application to non-linear system identification. Int J Control 50:1873–1896
18. Chen SS, Donoho DL, Saunders MA (1998) Atomic decomposition by basis pursuit. SIAM J Sci Comput 20(1):33–61

19. DeVore RA (2007) Deterministic constructions of compressed sensing matrices. J Complex 23:918–925
20. Efron B, Hastie T, Johnstone I, Tibshirani R (2004) Least angle regression. Ann Stat 32(2):407–499
21. Elad M, Bruckstein AM (2002) A generalized uncertainty principle and sparse representation in pairs of bases. IEEE Trans Inf Theory 48:2558–2567
22. Figueiredo MAT, Nowak RD, Wright SJ (December 2007) Gradient projection for sparse reconstruction: application to compressed sensing and other inverse problems. IEEE J Sel Top Sig Process 1:586–597
23. Geweke J (1996) Variable selection and model comparison in regression. In Bayesian Statistics 5, (Eds J. M. Bernardo, J. O. Berger, A. P. Dawid and A. F. M. Smith), Oxford University Press, pp 609–620
24. Jafarpour S, Xu W, Hassibi B, Calderbank R (2009) Efficient and robust compressed sensing using optimized expander graphs. IEEE Trans Inf Theory 55:4299–4308
25. Ji S, Xue Y, Carin L (2008) Bayesian compressive sensing. IEEE Trans Signal Process 56:2346–2356
26. Johnson W, Lindenstrauss J (1984) Extensions of lipschitz maps into a hilbert space. Contemp Math 26:189–206
27. Joshi S, Boyd S (2009) Sensor selection via convex optimization. IEEE Trans Signal Process 57(2):451–462
28. Julier SJ, LaViola JJ (2007) On Kalman filtering with nonlinear equality constraints. IEEE Trans Signal Process 55(6):2774–2784
29. Kalouptsidis N, Mileounis G, Babadi B, Tarokh V (2011) Adaptive algorithms for sparse system identification. Signal Process 91:1910–1919
30. Li H, Shen C, Shi Q (2011) Real-time visual tracking using compressive sensing. In: Proceedings of the IEEE conference on computer vision and pattern recognition (CVPR), pp 1305–1312
31. Mallat S, Zhang Z (1993) Matching pursuits with time-frequency dictionaries. IEEE Trans Signal Process 4:3397–3415
32. McCulloch RE, George EI (1997) Approaches for Bayesian variable selection. Stat Sinica 7:339–374
33. Mei X, Ling H (2011) Robust visual tracking and vehicle classification via sparse representation. IEEE Trans Pattern Anal Mach Intell (PAMI) 33(11):2259–2272
34. Moraal PE, Grizzle JW (1995) Observer design for nonlinear system with discrete-time measurements. IEEE Trans Autom Control 40:395–404
35. Olshausen BA, Millman K (2000) Learning sparse codes with a mixture-of-Gaussians prior. Advances in Neural Information Processing Systems (NIPS), pp 841–847
36. Pati YC, Rezifar R, Krishnaprasad PS (1993) Orthogonal matching pursuit: recursive function approximation with applications to wavelet decomposition. In: Proceedings of the 27th Asilomar conference on signals, systems and computers, vol 1, pp 40–44
37. Qiu C, Lu W, Vaswani N (2009) Real-time dynamic MR image reconstruction using Kalman filtered compressed sensing. In:Proceedings of the IEEE international conference on acoustics, speech and signal processing, pp 393–396
38. Szarek SJ (1991) Condition numbers of random matrices. J Complex 7:131–149
39. Tibshirani R (1996) Regression shrinkage and selection via the LASSO. J Roy Stat Soc Ser B, 58(1):267–288
40. Tipping ME (2001) Sparse Bayesian learning and the relevance vector machine. J Mach Learn Res 1:211–244
41. Tropp JA, Gilbert AC (2007) Signal recovery from random measurements via orthogonal matching pursuit. IEEE Trans Inf Theory 53:4655–4666
42. Vaswani N (2008) Kalman filtered compressed sensing. In: Proceedings of the international conference on image processing (ICIP), pp 893–896
43. Warnell G, Chellappa R (2012) Compressive sensing in visual tracking, recent developments in video surveillance. In: El-Alfy H (ed), InTech

44. Warnell G, Reddy D, Chellappa R (2012) Adaptive rate compressive sensing for background subtraction., In: Proceedings of the IEEE international conference on acoustics, speech, and, signal processing

Chapter 2
The Geometry of Compressed Sensing

Thomas Blumensath

Abstract Most developments in compressed sensing have revolved around the exploitation of signal structures that can be expressed and understood most easily using a geometrical interpretation. This geometric point of view not only underlies many of the initial theoretical developments on which much of the theory of compressed sensing is built, but has also allowed ideas to be extended to much more general recovery problems and structures. A unifying framework is that of non-convex, low-dimensional constraint sets in which the signal to be recovered is assumed to reside. The sparse signal structure of traditional compressed sensing translates into a union of low dimensional subspaces, each subspace being spanned by a small number of the coordinate axes. The union of subspaces interpretation is readily generalised and many other recovery problems can be seen to fall into this setting. For example, instead of vector data, in many problems, data is more naturally expressed in matrix form (for example a video is often best represented in a pixel by time matrix). A powerful constraint on matrices are constraints on the matrix rank. For example, in low-rank matrix recovery, the goal is to reconstruct a low-rank matrix given only a subset of its entries. Importantly, low-rank matrices also lie in a union of subspaces structure, although now, there are infinitely many subspaces (though each of these is finite dimensional). Many other examples of union of subspaces signal models appear in applications, including sparse wavelet-tree structures (which form a subset of the general sparse model) and finite rate of innovations models, where we can have infinitely many infinite dimensional subspaces. In this chapter, I will provide an introduction to these and related geometrical concepts and will show how they can be used to (a) develop algorithms to recover signals with given structures and (b) allow theoretical results that characterise the performance of these algorithmic approaches.

T. Blumensath (✉)
ISVR Signal Processing and Control Group, University of Southampton,
Southampton, UK
e-mail: thomas.blumensath@soton.ac.uk

A. Y. Carmi et al. (eds.), *Compressed Sensing & Sparse Filtering*,
Signals and Communication Technology, DOI: 10.1007/978-3-642-38398-4_2,

2.1 Introduction

How do we know something is there, if we haven't seen it, or, to use the cliche, how do we know that the falling tree still makes a sound even if there is no one to listen? This is far more than a purely philosophical question. It is at the heart of all of scientific discovery, indeed, one could say that all of science is ultimately a quest for rules that allow us to predict the unobserved. In science, this is done by observing certain aspects of nature which are then used to build models which in turn allow us to make predictions about things we have not yet seen.

Similar questions also arise in engineering. We live in a digital world where images, sounds and all kinds of other information are stored, transmitted and processed as finite collections of numbers. Whether it is your favourite TV show or the medical images acquired at your last hospital appointment, all are represented using zeros and ones on a computer. But how is this possible? Sound pressure varies continuously at your ear, so how can this continuously varying signal be described by a finite number of bits? In fact, the digital information stored on your favourite CD only describes the sound pressure measured at regular intervals. Similarly, a movie typically consists of (only) tens of images each second, yet the light intensity, originally measured by the camera, changes continuously with time. Digital movies and sound recordings are thus mere approximations of the original physical signal.

The question thus arises, "How much of the information is preserved in these approximations?" and "How do we infer what the signal was in the places we haven't seen?" that is, "How then do we interpret these approximations?" For example, our movie is only represented with a relatively small number of different images each second. Any changes that occur at a timescale that is faster than this, are not captured. In effect, to interpret the movie, we assume that such changes do not occur. Whilst this is not true in the movie example, our eyes are not able to resolve changes faster than those captured in a normal film. However, we are also all aware that this can lead to "errors". We have all experienced the illusion of a propeller on a plain or the wheel on a car that, when changing its speed appears to change direction. This "aliasing" is due to the fact that we don't interpret the data correctly, that is, our *model* (i.e. the assumption that changes are slow) is incorrect. As we don't know what happens to the propeller or wheel between frames, our brain makes the assumption that the propeller or wheel has moved the smaller of the two possible distances between the two observations.

The moral of this story is that we constantly have to make judgements about things that "happen where we have not looked" and that we do this using assumptions or *models*. Similar judgements have to be made by any algorithm that deals with measured continuous signals and a detailed theoretical understanding of these phenomena is thus fundamental to our ability to capture, process and reconstruct continuous physical phenomena.

The Shannon Nyquist Whitaker sampling theorem [1, 2] is the classical example of such a theoretical treatment of the problem. Consider a signal $x(t)$ that changes continuously with time t. This could be, for example, the sound pressure measured with a microphone. To represent $x(t)$ digitally, we sample it by taking equally spaced measurements $x(t_i)$ at time points t_i, where $t_i - t_{i-1} = \Delta_t$ is the constant sampling interval. Moving to bold face vector[1] notation, our representation of the original continuous signal $x(t)$ is now the vector $\mathbf{x}$ (which has either a finite number of entries if $x(t)$ was sampled over a finite interval, or could in theory be infinitely long). $\mathbf{x}$ is not yet a truly digital representation of $x(t)$, as each entry in the vector $\mathbf{x}$ is a real number, which also cannot be represented exactly in digital form. Nevertheless, for the purpose of this chapter, we will ignore this additional complication and assume that the effect of the additional errors introduced by the required quantisation of real numbers are negligibly small. Instead, the leitmotif here will be the interpretation of $\mathbf{x}$. Which properties must $x(t)$ possess so that it is fully described by the measurements in the vector $\mathbf{x}$?

We will see that there is an intimate interplay between (1) the way we measure a signal (e.g. the sampling interval in our Shannon sampling example), (2) the class of signals that we can describe exactly using the measurements $\mathbf{x}$ and (3) the way in which we can reconstruct $x(t)$ from the measurements $\mathbf{x}$. For continuous signals sampled at equally spaced intervals, the relationship between these three points is precisely what is captured by the Shannon Nyquist Whitiker sampling theorem, which states that:

Theorem 1 *If a continuous signal is sampled at equally spaced intervals and if the signal is band-limited with a bandwidth of less than half the sampling rate, then the signal can be reconstructed exactly using a linear reconstruction. Furthermore, the reconstruction filter only depends on the sampling rate and the frequency band occupied by the signal.*

The signal model here assumes $x(t)$ to be band-limited and, in order to be able to interpret or reconstruct $x(t)$, the frequency support must be known. If we sample at regular intervals and use a reconstruction that assumes the incorrect frequency support, or if the signal is not band-limited, then we will not be interpreting or reconstructing the signal correctly and aliasing will occur similar to the propeller or wheel example.

In this chapter, the sampling problem will be addressed in a more general setting. In particular, more general signal models will be considered. A more general theory will bring many advantages. For example, a sampling theory that allows a more general class of signal models allows us to design particular sampling schemes that are tailored to a specific problem. This, in turn, can lead to sampling approaches capable of, for example, sampling non-bandlimited signals or, sampling at a rate well below that required by the Nyquist rate. However, this can only be achieved

[1] In this chapter, we use two, somewhat different, meanings for the term vector. On the one hand, we call any one dimensional array of real or complex numbers a vector, this is the meaning used here. Below, we will introduce a more abstract definition of vectors as elements of some mathematical space. Which of these two definitions is appropriate at any one point in this chapter should be clear from the context.

if the sampling theory provides us with the tools to model and account for known signal structures.

There are several mathematical approaches to capture and model signal structure. Our view here will be predominantly geometrical. Similar to a sphere of radius 6,371 km which is a good model to use to describe my location on the earth's surface (up to small errors that would account for the fact that the earth is not completely spherical or that I might on occasion take a plain or visit an underground cave), similar geometrical models can be used to describe constraints on signals. In general, most signals such as sounds, images and movies can be thought of as living in some signal space, where we can define the distance between two signals or can measure angles. But similar to the assumption that I am not likely to be found anywhere in space but am restricted to the earths surface (after all, I am unlikely to spend any of my upcoming holidays on the moon), so are many types of signals only encountered in or close to a subset of the space they inhabit. For example, the assumption in Shannon's sampling theorem that signals are band-limited, translates into the geometric assumption that signals lie on a subspace (think of a subspace as the equivalent to an infinite piece of paper in our three dimensional world).

Many traditional sampling results are based on convex sets, such as subspaces. Whilst convex signal models lead to relatively simple sampling approaches, which are easily studied with current mathematical tools, non-convex models are significantly more flexible. However, the utility gained through the increased flexibility also leads to an escalation in the complexity of both the theoretical treatment of the sampling problem as well as their successful implementation. Non-convex signal models typically require non-linear reconstruction techniques, so that, for these models, an additional important aspect arises: the computational speed or complexity of signal reconstruction. In particular, many advanced signal models lead to NP-hard reconstruction problems. It thus becomes paramount to restrict sampling strategies for these signal models to a subset of linear operators that allow fast reconstruction.

The archetypal example here is compressed sensing [3–6]. Compressed sensing assumes signals to be sparse in some way. For finite dimensional signals that can be expressed as vectors, sparsity means that most of the entries in the vector that represent the signal are zero. It is important to note here that the sparse vector itself does not have to be the signal of interest. Instead, the sparse vector can equally well be a representation of a signal in some basis (wavelet and Fourier bases are popular examples). For finite dimensional signals, Fourier domain sparsity assumes a signal to be constructed from the mixture of a few sinusoids, where the frequencies of each of the sinusoids has to be taken from a fixed, finite-dimensional regularly spaced grid.

A related area that has gained more prominence recently is matrix completion [7, 8]. In the matrix completion problem, the signal of interest is a data-matrix, but, instead of measuring the data for each entry in the matrix, only a small subset of the matrix entries is filled with measurements initially. The task is then to estimate the missing entries using the measured entries only. This can again only be done if we assume the data to follow some known model. A popular model for matrix completion, which is related to the sparse model used in compressed sensing and which is found to describe many phenomena of interest, is a low rank matrix model. In these models, the full data matrix is assumed to have a rank which is significantly smaller than the maximum rank a matrix of the same dimensions could have. A popular example is the movie recommender system, where a matrix is constructed in which each entry contains a rating of a movie by a person. For each person in the system there is thus a row in the matrix and each film has an associated column. However, people are only able to watch and rate a small fraction of all movies, so that the missing entries have to be inferred from the few ratings made. Once the missing entries have been filled in, the system can then recommend movies to people on the system that they are likely to rate highly. A common assumption in these systems is that the full data matrix is of low rank, an assumption that is justified by an argument that stipulates that a persons preference in movies is primarily driven by a small number of underlying factors.

Compressed sensing and matrix factorisation can be seen as two particular instances of a more general class of constrained inverse problems [9]. In this chapter, the main ideas that define the class of problems we discuss will be that they (1) use non-convex constraints to model the signals we will be able to reconstruct and (2) pose computationally challenging reconstruction problems so that we will require efficient reconstruction methods. As promised in the chapter title, we here take a geometrical point of view, which will allow us to study important properties of non-convex signal models and their interplay with different efficient reconstruction methods.

2.2 Geometrical Signal Models

2.2.1 A Geometrical Primer

Before continuing our study of the geometry of data recovery problems, it makes sense to define and fix several mathematical concepts and notation.

Throughout this chapter, we will talk about *signals* which will be mathematical descriptions of physical phenomena such as sounds, images or movies. From a mathematical point of view, *signals* are functions and a function is a mapping that assigns a unique real or complex number to each set of functional *parameters*. The parameters of a function are taken from the reals or from a subset of the reals. For example, sound pressure can be described as a function that assigns a unique pressure to each point in time. Similarly, an image can be understood as a function that assigns a real number (describing the image intensity) for each location in the image plain. In contrast to the sound example, where the sound parameter ran over all possible time instances, for images it is typical to restrict the domain of the image parameters to intervals of real numbers. Another important class of functions are finite length vectors. For example, a ten dimensional vector can be understood as a collection of ten real or complex numbers. Such a vector is also a function, but here, the parameters are restricted to an interval of integers (i.e. 1, 2, ..., 10).

2.2.1.1 Vector Space

The material in this section can be found in any good textbook on analysis and functional analysis. Good, however rather technical, examples are [10] and [11].

A mathematical space is a collection of mathematical objects, such as numbers or functions, together with a set of properties. Properties can include, for example, additivity of elements (so that for any two elements, there is an element of the space that is the sum of the two elements). Other properties of mathematical objects that are important for a geometrical interpretation are length or size, distance between objects and angle between objects.

In this chapters, signals will be formally described as mathematical objects that live in a *vector space*. This means that they all have a certain set of universal properties common to all vector spaces. Formally, a linear vector space $\mathcal{V}$ over a Field $\mathcal{F}$ (which in this chapter will either be the real numbers ($\mathbb{R}$) or the complex numbers ($\mathbb{C}$)) is a selection of objects (called vectors) together with certain operations on these elements, which have the following set of properties:

1. The space has an addition operator $+$, so that for any two elements $\mathbf{x}_1, \mathbf{x}_2 \in \mathcal{V}$ the product $\mathbf{x}_1 + \mathbf{x}_2$ is also an element of the set $\mathcal{V}$.
2. The addition is commutative (i.e. $\mathbf{x}_1 + \mathbf{x}_2 = \mathbf{x}_2 + \mathbf{x}_1$) and associative (i.e. $(\mathbf{x}_1 + \mathbf{x}_2) + \mathbf{x}_3 = \mathbf{x}_1 + (\mathbf{x}_2 + \mathbf{x}_3)$).
3. There is a *zero* element $\mathbf{x}_0 \in \mathcal{V}$ for which $\mathbf{x} + \mathbf{x}_0 = \mathbf{x}$ holds for all $\mathbf{x} \in \mathcal{V}$. We will write the zero element as $\mathbf{0}$.
4. For all $\mathbf{x} \in \mathcal{V}$, there is an element $-\mathbf{x}$, such that $\mathbf{x} + (-\mathbf{x}) = \mathbf{0}$.
5. The space has a scalar multiplication operator $\cdot$, so that for all elements $\alpha, \beta \in \mathcal{F}$ and any $\mathbf{x} \in \mathcal{V}$ the element $\alpha \cdot \mathbf{x}$ is an element in $\mathcal{V}$ and, furthermore, $\alpha \cdot (\beta \cdot \mathbf{x}) = (\alpha\beta) \cdot \mathbf{x}$, $(\alpha + \beta) \cdot \mathbf{x} = \alpha \cdot \mathbf{x} + \beta \cdot \mathbf{x}$, $\alpha \cdot (\mathbf{x}_1 + \mathbf{x}_2) = \alpha \cdot \mathbf{x}_1 + \alpha \cdot \mathbf{x}_2$ and $1 \cdot \mathbf{x} = \mathbf{x}$.

Thus, vector spaces are collections of elements that can be added and subtracted and which can be multiplied by real or complex numbers (or more general by elements from its base field).

Banach Space

By equating real world signals with elements of a vector space, we can use vector addition and scalar multiplication to describe signal addition and scaling. The next useful concept we introduce is that of the *size* or length of a signal. Once we are able to talk about the size of a signal, then we can also talk about the *size of the difference* between two signals, which then enables us to formally define the distance or difference between two signals. The ability to talk about length and distance of signals is our first step in a geometrical interpretation of signal processing problems and is thus one of the most fundamental concepts discussed here.

The length of an element of a vector space $\mathcal{V}$ will be measured by a *norm*. We write $\|\mathbf{x}\|$ to denote the norm of the element $\mathbf{x}$. A norm is a non-negative function that assigns a real number to an element of a vector space and has the following properties:

1. The zero element $\mathbf{0}$ is the only element in the vector space that has a norm of zero, that is $\|\mathbf{x}\| = 0$ if and only if $\mathbf{x} = \mathbf{0}$.
2. The norm satisfies the triangle inequality $\|\mathbf{x}_1 + \mathbf{x}_2\| \leq \|\mathbf{x}\|_1 + \|\mathbf{x}_2\|$ for all $\mathbf{x}_1, \mathbf{x}_2 \in \mathcal{V}$.
3. The norm increases proportionally when scaling an element in the vector space, that is, $\|\alpha\mathbf{x}\| = |\alpha|\|\mathbf{x}\|$ for all $\mathbf{x} \in \mathcal{V}$ and $\alpha \in \mathcal{F}$.

The second of these properties is one of the fundamental properties that will allow us to use some of our geometrical intuition when discussing signal properties as it links the length of the sum of two vectors to the length of each vector individually. The geometrical picture is that of a triangle, where the length of any one side of the triangle (which is the same as the sum of the two other sides) is always shorter or at most as long as the sum of the lengths of each of the other two sides. Or, using another well known geometrical property, the length between two points is the straight line.

Thus, the concept of a norm not only tells us how 'large' an element in a vector space is, it also tells us how far apart different elements in the space are. From the properties of vector spaces, we know that the element $\mathbf{x} = \mathbf{x}_1 - \mathbf{x}_2$, that is the difference between the elements $\mathbf{x}_1$ and $\mathbf{x}_2$ is itself a vector. Therefore, if we have defined a norm on the vector space, then the norm $\|\mathbf{x}_1 - \mathbf{x}_2\|$ will be defined and will measures the *distance* between these two elements.

With the definition of distance comes another property, that of convergence of sequences of elements. Assume we have a collection infinitely many elements $\{\mathbf{x}_i\}$ (which do not have to be all different). This sequence is said to be Cauchy convergent if the distance $\|\mathbf{x}_m - \mathbf{x}_n\|$ can be made arbitrary small for all $n, m > N$ if we only choose N itself large enough. In other words, if we restrict our consideration to elements that are far enough from the beginning of the sequence, then any two elements will be arbitrarily close to each other.

Cauchy convergence might seem a little bit odd at first and another form of convergence might be more intuitive to the reader new to these idea. A sequence $\{\mathbf{x}_i\}$ is said to converge to a point $\mathbf{x}_{lim}$, if the distance $\|\mathbf{x}_i - \mathbf{x}_{lim}\|$ converges to zero in the limit. In this form of convergence, the sequence of elements will get arbitrarily close to a certain point, which we will here call $\mathbf{x}_{lim}$. The difference to Cauchy convergence is that, in the definition of Cauchy convergence, which is the more general of the two properties, whilst elements in the sequence are guaranteed to stay close to each other, there might not exist a single element within our vector space, which is a limit point, that is, to which the sequence will get arbitrarily close. However, rather than this being a property of the sequence itself, this is really a property of the space from which the elements of the sequence have been picked. In a sense, spaces in which there are sequences that are Cauchy convergent but which do not have a limit point are "incomplete". Thus a space where all Cauchy sequences converge are actually called complete spaces. As convergence is such a fundamental property, it is often useful to restrict discussions to complete spaces. Complete normed vector spaces thus have their own name, they are called *Banach Spaces*. All the spaces encountered in this chapter will be Banach spaces so that the concepts of Cauchy convergence and convergence will be identical.

Hilbert Space

A second geometrical concept which is as important as length, is that of the angle between two elements. Angles between vectors can be measured using inner products, which is a real or complex valued function of two elements of the vector space (written as $\langle \cdot, \cdot \rangle$) which satisfies the following two properties.

1. $\langle \mathbf{x}, \mathbf{x} \rangle \geq 0$ with $\langle \mathbf{x}, \mathbf{x} \rangle = 0$ if and only if $\mathbf{x} = \mathbf{0}$.
2. $\langle \mathbf{x}_1 + \mathbf{x}_2, \mathbf{x}_3 \rangle = \langle \mathbf{x}_1, \mathbf{x}_3 \rangle + \langle \mathbf{x}_2, \mathbf{x}_3 \rangle$.
3. $\langle \lambda \mathbf{x}_1, \mathbf{x}_2 \rangle = \lambda \langle \mathbf{x}_1, \mathbf{x}_2 \rangle$, where $\lambda \in \mathbb{C}$.
4. $\langle \mathbf{x}_1, \mathbf{x}_2 \rangle = \overline{\langle \mathbf{x}_2, \mathbf{x}_1 \rangle}$, where the bar $\bar{\cdot}$ indicates the complex conjugate.

Inner products can be used to 'induce' a norm, that is, they can be used to define a norm as follows

$$\| \cdot \| = \sqrt{\langle \cdot, \cdot \rangle}$$

Using the induced norm, inner products contain information on the angle between two elements. In fact, inner products combine information on angles and vector length, so that a quantity that has properties similar to the angle between to elements can be found by normalising the inner product

$$\frac{\langle \mathbf{x}_1, \mathbf{x}_2 \rangle}{\|\mathbf{x}_1\| \|\mathbf{x}_2\|}.$$

Thus, if $\mathbf{x}_1$ and $\mathbf{x}_2$ are the same vector, then their angle will be zero and the above normalised inner product is 1. Similarly, we will say that two vectors are at right angles or *orthogonal* if their inner product is 0.

With an induced norm there is an intimate link between norms and inner products. For example, the Pythagorean theorem holds

$$\|\mathbf{x}_1 + \mathbf{x}_2\|^2 = \|\mathbf{x}_1\|^2 + \|\mathbf{x}_2\|^2 \text{ if } \langle \mathbf{x}_1, \mathbf{x}_2 \rangle = 0,$$

which is a special case of the more general result that

$$\|\mathbf{x}_1 + \mathbf{x}_2\|^2 = \|\mathbf{x}_1\|^2 + \|\mathbf{x}_2\|^2 + 2\langle \mathbf{x}_1, \mathbf{x}_2 \rangle.$$

In addition, the following parallelogram law also holds

$$\|\mathbf{x}_1 + \mathbf{x}_2\|^2 + \|\mathbf{x}_1 - \mathbf{x}_2\|^2 = 2\|\mathbf{x}_1\|^2 + 2\|\mathbf{x}_2\|^2$$

and so does the following inequality

$$|\langle \mathbf{x}_1, \mathbf{x}_2 \rangle| \leq \|\mathbf{x}_1\| \|\mathbf{x}_2\|.$$

A vector space that has a norm that is induced by an inner product thus has very appealing geometrical properties. Such a space is called a Hilbert space if it is furthermore complete, that is, a Hilbert space is a complete inner product space with an induced norm.

Finite and Infinite Dimensional Spaces

We live in a three dimensional world, or mathematically speaking, in a three dimensional (thus finite dimensional) Hilbert space, yet many spaces of mathematical functions are actually infinite dimensional. In infinite dimensional spaces, some of our intuition still holds, yet, care has to be taken as there are also subtle differences. In essence, an infinite dimensional space is a space in which there are infinitely many vectors that are all orthogonal to each other. Orthogonality can be measured by the inner product, in fact, the inner product of orthogonal vectors is zero. In an infinite dimensional space, there are thus infinitely many vectors which all have a zero inner product with each other.

But infinity is even more subtle than this. In fact, there are infinities of different sizes. This might come as a surprise to some, yet the typical example are the sets of integers and the sets of real numbers. There are infinitely many integers, for any integer number I name, you will always be able to find a number that is larger. Real numbers on the other hand, not only contain all integers. There are infinitely many other real numbers that lie between any two distinct real numbers. It can indeed be shown that there will be 'more' real numbers than there are integers. When talking about infinities it is thus helpful to distinguish the infinity that is as large as the number of integers and infinities that are larger. Sets of infinitely many elements, that have as many elements as there are integers are said to be countable. The elements in a countable set can thus be labeled using integers (that is we could count them at least in theory). Sets that cannot be labeled with integers are called uncountably infinite. We will restrict the discussion here to Hilbert spaces that are at most countably infinite.

Basis

In a similar way in which we describe locations on earth (for example, using north-sout, east-west, and hight), it is useful to be able to find a way to describe the 'location' of vectors in a vector space. This will be done using a set of basis vectors (or basis directions). An important concept here is that any such description should ideally not contain replicated information; three parameters are enough to describe any location on earth and four parameters would only replicate some of this information. A similar concept holds in general vector spaces, even in infinitely large ones.

To capture the effect of replication of information, we use the concept of linear dependancy of a set of vectors. A set of vectors $\{\mathbf{x}_i\}$ is said to be linearly dependant if there are scalars λ_i (which are not all zero) such that $\sum_i \lambda_i \mathbf{x}_i = \mathbf{0}$ or $\sum_{i\neq j} \lambda_i \mathbf{x}_i = -\lambda_j \mathbf{x}_j$. Thus, if we use the vectors $\mathbf{x}_i$ to describe a vector $\mathbf{x}$ as $\mathbf{x} = \sum_i \alpha_i \mathbf{x}_i$, then we can always replace one of the vectors (say $\mathbf{x}_j$ with $\mathbf{x}_j = -\sum_{i\neq j} \lambda_i/\lambda_j \mathbf{x}_i$ so that the vector $\mathbf{x}$ is equally well described using one less vector. On the other hand, if there is no such set of scalars λ_i such that $\sum_i \lambda_i \mathbf{x}_i = \mathbf{0}$, then we say that the set of vectors $\{\mathbf{x}_i\}$ is linearly independent.

Any set of vectors $\{\mathbf{x}_i\}$, whether linearly dependant or not, can be used to describe certain vectors $\mathbf{x}$ as a linear combination $\mathbf{x} = \sum_i \alpha_i \mathbf{x}_i$. All those $\mathbf{x}$ which can be writen in this form for any given set $\{\mathbf{x}_i\}$ is called the linear span of the set $\{\mathbf{x}_i\}$, which is formally written as the set

$$\{\mathbf{x} = \sum_i \lambda_i \mathbf{x}_i, \text{ with } \lambda_i \in \mathcal{F}\}$$

where $\mathcal{F}$ is the field used in the definition of the vector space (e.g. $\mathcal{F}$ are the real or complex numbers).

A set $\{\mathbf{x}_i\}$ which is large enough to be able to describe *all* vectors in vector space and which furthermore is not too large so that its elements are linear independent is called a basis for the space.

We have already encountered the concept of orthogonality. A basis, in which any two elements are orthogonal is called an orthogonal basis. Furthermore, an orthogonal basis in which each element has unit length, is called an orthonormal basis. An important result in mathematics is the fact that every Hilbert space has an orthonormal basis. Furthermore, if the set of vectors in the basis is either finite or countably infinite, we say that the Hilbert space is separable.

We will here restrict our discussion to separable Hilbert spaces so that we can always find an at most countably infinite orthonormal basis set $\{\mathbf{x}_i\}$ that allows us to write any element of the Hilbert space as a linear combination

$$\mathbf{x} = \sum_i^{\infty} a_i \mathbf{x}_i. \tag{2.1}$$

2.2.1.2 Subspaces

A subset $\mathcal{S}$ of a vector space is called a linear subspace if any two elements $\mathbf{x}_1, \mathbf{x}_2 \in \mathcal{S}$ have the property that their linear combination $\lambda_1 \mathbf{x}_1 + \lambda_2 \mathbf{x}_2$ is also an element of the set $\mathcal{S}$. Here λ_1 and λ_2 are arbitrary scalars. The linear span of a set of vectors is a linear subspace.

2.2.1.3 Convex Sets

Closely related to linear subspaces are convex sets. A convex set is defined similarly to a linear subspace. A subset $\mathcal{S}$ of a vector space is called a convex subset if any two elements $\mathbf{x}_1, \mathbf{x}_2 \in \mathcal{S}$ have the property that their linear combination $\lambda_1\mathbf{x}_1 + \lambda_2\mathbf{x}_2$ is also an element of the set $\mathcal{S}$. However, the difference here is in the set of scalars allowed in the definition. Whilst in the definition of a linear subspace, λ_1 and λ_2 were allowed to be arbitrary scalars, for a set to be convex, we have the additional requirement that $\lambda_1, \lambda_2 \geq 0$ and that $\lambda_1 + \lambda_2 = 1$. It should thus be clear that a linear subspace is a convex set. A set that is not convex if there are $\mathbf{x}_1, \mathbf{x}_2 \in \mathcal{S}$ and $\lambda_1, \lambda_2 > 0$ with $\lambda_1 + \lambda_2 = 1$ for which the element $\lambda_1\mathbf{x}_1 + \lambda_2\mathbf{x}_2$ is *not* an element of the set $\mathcal{S}$ itself.

In a Hilbert space, in the same way in which we say that two vectors are orthogonal, we can also say that a vector $\mathbf{x}$ is orthogonal to a subset $\mathcal{S}$ if $\mathbf{x}$ is orthogonal to every element in $\mathcal{S}$. Similarly, if we have two subsets, these can be said to be orthogonal if every vector of one subset is orthogonal to every vector of the other subset. For example, the orthogonal complement of a subset is the set of all vectors that are orthogonal to the set. The orthogonal complement of any set is a closed[2] convex subspace.

For any closed convex subset $\mathcal{S}$ of a Hilbert space $\mathcal{H}$, it is allways possible to find a *best* approximation of any vector $\mathbf{x} \in \mathcal{H}$ by an element of the closed convex subset $\mathcal{S}$. That is, for any $\mathbf{x} \in \mathcal{H}$ there exists a $\mathbf{x}_0 \in \mathcal{S}$ such that

$$\|\mathbf{x} - \mathbf{x}_0\| = \inf_{\tilde{\mathbf{x}} \in \mathcal{S}} \|\mathbf{x} - \tilde{\mathbf{x}}\|.$$

We will call the element $\mathbf{x}_0$ the projection of $\mathbf{x}$ onto the closed convex subset $\mathcal{S}$.

This leads us to the important orthogonal projection theorem which states that for any closed linear subspace $\mathcal{S} \subset \mathcal{H}$ and any $\mathbf{x} \in \mathcal{H}$, we can always find a unique decomposition $\mathbf{x} = \mathbf{x}_{\mathcal{S}} + \mathbf{x}_{\overline{\mathcal{S}}}$, where $\mathbf{x}_{\mathcal{S}} \in \mathcal{S}$ and where $\mathbf{x}_{\overline{\mathcal{S}}}$ is orthogonal to $\mathcal{S}$. Furthermore, $\mathbf{x}_{\mathcal{S}} \in \mathcal{S}$ is the closest point in $\mathcal{S}$ to $\mathbf{x}$.

For any closed linear subspace $\mathcal{S}$, let $P_{\mathcal{S}}$ be the operator that maps and $\mathbf{x} \in \mathcal{H}$ to the element $\mathbf{x}_{\mathcal{S}}$ defined in the projection theorem. The operator P is self adjoint (that is $\langle P\mathbf{x}_1, \mathbf{x}_2\rangle = \langle \mathbf{x}_1, P\mathbf{x}_2\rangle$ for all $\mathbf{x}_1, \mathbf{x}_2 \in \mathcal{H}$), $P^2 = P$ and has an operator norm $\sup_{\mathbf{x}\neq\mathbf{0}} \|\Phi\mathbf{x}\|/\|\mathbf{x}\| = \|P\|$=1 whenever $P \neq 0$.

2.2.1.4 Unions of Simpler Geometrical Models

Having defined some of the basic geometric properties of Hilbert spaces, let us now return to the problem of signal modelling. Linear subspaces and closed convex sets have very appealing properties and these sets have long been used to define classes of signals which then allow us to find elements within these convex sets

---

[2] A subset $\mathcal{S} \subset \mathcal{H}$ is called closed if every sequence with elements in $\mathcal{S}$ that converges to an element of $\mathcal{H}$ has a limit in the subset $\mathcal{S}$ itself.

that can act as good representatives for a particular signal. Many classical signal processing ideas have been restricted to closed convex sets, yet, recent advances in our understanding of signal geometry have allowed us to extend similar ideas to more complex signal models, models that are no longer convex. This work has primarily looked at constraint sets that are the union over several (in many cases extremely large collections of) closed convex sets. In such a signal model, we are given a number of closed convex sets and assume that any signal lies within one of these sets, however, we are not sure in which set exactly we are to look for the signal.

Let us define these unions formally. Any union of closed and convex sets is defined as

$$\mathcal{S} = \bigcup_j \mathcal{S}_j : \text{ where the } \mathcal{S}_j \text{ are closed and convex subsets,} \tag{2.2}$$

Here each $\mathcal{S}_j$ can be any closed and convex subset of a larger Hilbert space and the union can be potentially over a countably infinite number of these sets. Of particular interest to us will be union models in which the $\mathcal{S}_j$ are closed subspaces.

An important example of a union of subspaces model is the sparse signal model in finite dimensions. Consider the Euclidean Hilbert space of dimension N whose elements we can represent using N element vectors. In a k-sparse model, the model subset $\mathcal{S}$ is the set of all vectors $\mathbf{x}$ that has no more then k non-zero entries. This model is in fact a union of subspace model. To see this, consider the support of a k-sparse vector, that is, consider the pattern of the location of the non-zero elements in this vector. If we add (or subtract) a k-sparse vector that has exactly the same support (that is, whose non-zero elements are in exactly the same location), then the sum (or difference) of these two vectors will again be k-sparse and will have the same support. Thus, the set of all k-sparse vectors which have the same support is a subspace. However, for any $k < N$, there will be many different support sets. In fact there will be $\binom{N}{k}$ such sets $\binom{N}{k}$, read N choose k, is the number of different ways in which we can choose k elements from a set of N elements). Thus, the set of all k-sparse vectors (irrespective of their support) is the union of $\binom{N}{k}$ subspaces. We also see that this set is non-convex as the sum of two k-sparse vectors with different support can potentially have up to $2k$ non-zero entries. In fact, the set of the sum of two (or three) k sparse vectors will be of importance later on, and we introduce some notation to specify these sets here.

In general, if $\mathbf{x} \in \mathcal{S}$ for some union $\mathcal{S}$, we will write

$$\mathcal{S} + \mathcal{S} = \{\mathbf{x} = \mathbf{x}_1 + \mathbf{x}_2 : \mathbf{x}_1, \mathbf{x}_2 \in \mathcal{S}\} \tag{2.3}$$

and

$$\mathcal{S} + \mathcal{S} + \mathcal{S} = \{\mathbf{x} = \mathbf{x}_1 + \mathbf{x}_2 + \mathbf{x}_3 : \mathbf{x}_1, \mathbf{x}_2, \mathbf{x}_3 \in \mathcal{S}\} \tag{2.4}$$

2.2.1.5 Operators on the Elements of a Space

One last fundamental notion that will be required throughout this chapter is that of an operator. In principle, an operator takes an element of one space and *transforms* it into the element of another space. We write $\mathbf{y} = \Phi(\mathbf{x})$, where $\mathbf{x}$ is an element of one space and $\mathbf{y}$ is the element of another space.

A linear operator has properties similar to a matrix. In particular it is linear, that is, for any two elements $\mathbf{x}_1$ and $\mathbf{x}_2$ from one space, it does not matter if we apply the operator to the sum of the two elements or if we apply the operator to each individual element and then sum the transformed elements. That is, $\Phi(\mathbf{x}_1 + \mathbf{x}_2) = \Phi(\mathbf{x}_1) + \Phi(\mathbf{x}_2)$. For linear operators, we generally write $\Phi\mathbf{x}$ instead of $\Phi(\mathbf{x})$. The parenthesis will be used primarily to indicate non-linear operators.

For linear operators, we can define a norm on an operator as follows

$$\|\Phi\| = \sup_{\mathbf{x}:\|\mathbf{x}\|\leq 1} \|\Phi\mathbf{x}\|, \tag{2.5}$$

that is, informally speaking, the operator norm is the maximum amount by which *any* vector can be lengthened when squeezed through the operator. Note that the operator norm as defined here depends on two vector norms, the norm of $\|\mathbf{x}\|$ and the norm of $\|\Phi\mathbf{x}\|$. In general, both of these norms can be arbitrary. In the case in which both $\mathbf{x}$ and $\Phi\mathbf{x}$ live in Hilbert spaces, then we assume that $\|\Phi\|$ is the norm defined using the Hilbert space norm.

An Operator is said to be invertible on a space $\mathcal{H}$ (or alternatively on a subset $\mathcal{S} \subset \mathcal{H}$) , for all $\mathbf{x} \in \mathcal{H}$ (or for all $\mathbf{x} \in \mathcal{S}$), if there exists an operator $\Phi\dagger$ such that $\mathbf{x} = \Phi^{\dagger}(\Phi(\mathbf{x}))$. If $\Phi\dagger$ is linear, then we say that Φ is linearly invertible on $\mathcal{H}$ (or on $\mathcal{S}$). An operator that is not invertible is said to be non-invertible.

If a linear operator between finite dimensional spaces is invertible, then the norm of $\|\Phi^{\dagger}\|$ is necessarily finite, however, in infinite dimensional spaces, it can happen that there are invertible linear operators whose norm is infinite. These operators are said to be ill-conditioned. Ill-conditioned operators would in theory allow us to recover $\mathbf{x}$ from $\mathbf{y} = \Phi\mathbf{x}$ uniquely, however, any small perturbation of $\mathbf{y}$ could potentially lead to an arbitrarily large change in the estimate of $\mathbf{x}$.

2.2.2 Examples and Sketch of Applications

So far, we have introduced an over-abundance of abstract mathematical ideas. Let us therefore take a step back here and discuss several important examples where geometrical ideas can help in the reconstruction of signals.

2.2.2.1 The Geometry of Shannon Sampling

The seminal work by Nyquist [1] and Shannon [2] is at the heart of much of traditional sampling theory. This theory deals with one instance of the signal recovery problem addressed throughout this book, although, we hardly think about it in this way any longer. The setting here is as follows, let **x** be a function over time with a domain spanning over the real numbers. For example, this might be the sound pressure produced by your favourite band. The aim is now to measure this sound pressure. Let us do this measurement by measuring the sound pressure intensity at infinitely many equally spaced intervals in time, so that our measurement **y** is an infinite sequence of real numbers and we again ask, how and when can we recover **x** from **y**. The Shannon sampling theorem answers exactly this question. In effect, if **x** is band-limited, then there is a simple linear reconstruction method that can recover **x** from **y** exactly. The band-width of the signal has to be less than half the inverse of the time interval between consecutive samples for this to work. Without going into too much detail (see for example [12] for a more detailed treatment), when we say that **x** is band-limited we mean that the Fourier transform of **x** (call this transform **X**) is a function whose support in restricted to a restricted frequency interval. This is our model. In fact, this is a subspace model. To see this, assume that you have two signals with the same frequency band-width. Adding these two signals (remember that the Fourier transform is linear) is the same as adding the Fourier transforms of the signals, so that the sum of any two band-limited signals is again band-limited. This is exactly our definition of a subspace. Thus, Shannon sampling uses a convex signal model and, as the model is convex, a simple reconstruction technique exists.

2.2.2.2 Sparse Signal Models in Euclidean Spaces

Instead of dealing with infinitely long sequences of numbers produced by 'proper' Shannon sampling, finite length approximations are the only practical approach to real problems. It is thus normal to assume that we can represent infinite dimensional signals using finite length vectors. In the same way, digitised images can be thought of as a collection of a finite number of real numbers. Let us therefore assume that our signal is well approximated using a vector in Euclidean space of dimension N.

If we were able to sample a signal using Shannon sampling ideas, then we would directly measure the elements in **x**. However, in many situations, we are unable to make enough measurements to use Shannon theory. For example, many measurement procedures are slow (e.g. in Magnetic Resonance Imaging, a patient has to lie in the scanner for several minutes to produce a single volumetric image), pose health risks (e.g. in X-ray computed tomography, X-ray dosage has to be limited to reduce exposure to ionising radiation) or are extremely expensive (certain hyperspectral imaging devices can come at a cost of thousands of dollars for a single pixel, so that traditional million pixel cameras with these elements would be prohibitively expensive).

Thus, we would want to reduce the number of measurements further and sample at a rate significantly below that described by Shannon theory. To do this is only possible if we use a much smaller signal set as our model. Single, very low-dimensional subspaces are not versatile enough to capture the diverse information present in most signals and images (if it were, we could just use Shannon theory), instead, more complex, low-dimensional, but non-convex models have to be used. One of the most powerful sets of models are sparse models. A sparse (Euclidean) vector $\mathbf{x}$ is a vector whose elements are zero apart from a small number of elements, which can have arbitrary magnitude. We say $\mathbf{x}$ is k-sparse if all but k of its elements are zero. Vectors with a fixed subset of non-zero elements lie in a single subspace, but in a sparse model, we allow all possible subsets of k elements to be non-zero, so that k-sparse vectors lie in the union of $\binom{N}{k}$ different subspaces.

Instead of sparsity in the canonical basis (e.g. in an image, instead of assuming that the image has many zero pixels), a great deal of flexibility is achieved if we allow sparsity in a different basis. In our three dimensional world, the canonical basis might be a description of locations in terms of north-south, east-west and up-down, yet we are free to use a coordinate transform to represent locations in another way. In my office, it might make more sense to specify locations in terms of their distance from the window, the side walls and the floor. As my office is not exactly aligned with the north-south axis (though luckily the floor is still level), the axis in the world coordinate system are rotations of the axis in my office representation. Exactly the same principle holds in the representation of signals. For example, we are not restricted to represent an image by specifying values for each pixel (the canonical space) but could instead specify two dimensional discrete wavelet coefficients to specify spatial frequencies of the image or we might represent the image using a 2-dimensional wavelet transform. These transforms often are nothing else than a rotation of the coordinate axis. In this case, they do not change the length of vectors, just their representation. In other cases, the new coordinate system might actually have axis that are not orthogonal in the original space or even have more coordinates than the original space. In these cases, we still assume that we can find a representation of any vector in our original space in the transformed space, but the length of elements in the two spaces might now differ. The importance of these transforms from our signal recovery perspective is that many signals have sparse or approximately sparse representations in some transformed domain. For example, images are often found to be sparse in a wavelet representation. Thus, using sparsity in transform domains greatly enhances our ability to use sparse models to describe structure in real signals.

When talking about sparsity in a different domain, we assume that there is a linear mapping that maps elements $\mathbf{x}$ of our signal space into the transformed domain. Call this mapping Ψ, so that $\mathbf{z} = \Psi\mathbf{x}$ is the reprensentation of $\mathbf{x}$ in the transformed domain. Importantly, we assume that there is a generalised inverse $\Psi^\dagger$ of Ψ, such that for all $\mathbf{x} \in \mathcal{H}$, $\mathbf{x} = \Psi^\dagger\mathbf{z} = \Psi^\dagger\Psi\mathbf{x}$.

2.2.2.3 Structured Sparse Models in Euclidean Space

Sparsity can be a powerful constraint and in many applications additional structure can be brought into play, further increasing the utility of sparse models. Structured sparse models are sparse models that only allow certain subsets of sparse support sets to be present. For example, a block-sparse vector is a sparse vector in which the non-zero coefficients are contained in pre-specified blocks. For example, if $\mathbf{x} \in \mathbb{C}^N$ and assume we have J blocks that partition $\mathbf{x}$, that is, if $B_j \subset \{1, \ldots, \mathcal{N}\}$, $j \in \{1, \ldots, J\}$ is the set of indices in block j, then we assume that the blocks (1) do not overlap (that is $B_i \bigcap B_j =$) and (2) every index of $\mathbf{x}$ is in at elast one block (that is $\bigcup_{j \in J} B_i = \{1, 2, 3, \ldots, N\}$). A signal that is k-block sparse is then defined as any $\mathbf{x}$ whose support is contained in no more than $k < J$ different sets B_j, that is

$$\mathrm{supp}(\mathbf{x}) \subset \bigcup_{\mathcal{J}} B_j : \mathcal{J} \subset \{1, 2, \cdots, J\},\ |\mathcal{J}| \leq k. \tag{2.6}$$

To define block-sparse signals here we imposed the restriction that the blocks do not overlap and that their union includes all elements of $\mathbf{x}$. In theory, we could drop these two restrictions, however, theoretical treatment of these more general models becomes much more difficult, and, in fact, this class of models would be so general that it would include all possible structured sparse models.

Another set of useful structured sparse models are tree-sparse models. Instead of partitioning the signal's support set into disjoint blocks, tree-sparse signals have non-zero coefficients that follow a tree structure in which all ancestors of a node are allowed to be non-zero whenever the node itself is non-zero. A sparse tree model is a model in which the tree is furthermore sparse, that is, the total number of non-zero elements is small. The simplest example, a one sparse tree, would only have a non-zero element at its root, whilst a two-sparse tree model would have as many possible support sets as there are children of the root, as such a model would have to include the root itself plus one of its children.

2.2.2.4 Low Rank Matrices

In many applications, data is best represented in matrix form. By specifying an appropriate inner product and norm for matrices, the Hilbert space formalism can also be applied to matrix problems so that geometrical ideas can be used to define subsets of matrices that can act as signal models. A powerful constraint here is the low-rank matrix model. The set of all M by N matrices of rank r that have the same column and row space (the space spanned by the matrix's row or column vectors) form a linear subspace, that is, we can add any two of these matrices and end up with another matrix of the same size and rank that has again the same column and row space (or, more precisely, whose row and column spaces are subspaces of the row and column spaces of the original matrices). However, a matrix with different column or row spaces does not lie in the same subspace and adding two matrices with different column or row spaces will result in a matrix that is likely to have a different rank from that of its two components. Thus, low-rank matrices lie in a non-convex subset of the space of all matrices.

2.2.2.5 Sparsity in Continuous Signals

Our last set of examples are again taken from infinite dimensional spaces, where continuous analogues to sparsity have been developed. In Shannon sampling, the sampling rate is directly related to the bandwidth of the signal we would like to sample. In several applications, this would lead to a prohibitive sampling rate so that again, additional signal structure has to be exploited. A signal model that is in some ways similar to the sparse model in Euclidean spaces is the analogue compressed sensing model first studied in [13] for known support and in [14] for unknown support. Here a continuous and band-limited[3] real valued time series $x(t)$ is assumed to have a Fourier transform $\mathcal{X}(f)$ whose support S is the union of K intervals of 'small' bandwidth B_K, i.e. $S \subset \bigcup_{k=1}^{K}[d_k,\ d_k + B_K]$, where the d_k are arbitrary scalars from the interval $[0,\ B_N - B_K]$. These signals can be understood as a continuous version of a sparse signal, but instead of having few non-zero "elements," only a small part of the functions support (say in the Fourier domain) is non-zero. As the support of the Fourier transform of a real valued function is symmetric, we here only consider the support in the positive interval $[0,\ B_N]$. If $KB_K < B_N$ then $\mathcal{X}(f)$ is zero for some frequencies f in $[0,\ B_N]$, mirroring sparsity in a vector. If we would fix the support S, then $\mathcal{X}(f)$ and therefore $x(t)$ would lie in a subspace of the space of all square integrable functions with bandwidth B_N. However, in a model where S is not fixed and where $KB_K < B_N$, there will be infinitely many distinct sets S satisfying this definition, so that $x(t)$ will lie in the union of infinitely many infinite dimensional subspaces. The set of all signals that have energy restricted to K bands with $KB_K < B_N$ thus is a non-convex set.

Another set of powerful models are so called Finite Rate of Innovations models. Consider again a real valued function of one variable $x(t)$. Such a function is said to have a finite rate of innovation [15] if it can be written as

$$x(t) = \sum_{n\in\mathbb{Z}} \sum_{r=0}^{R} c_{nr}\, g_r\left(\frac{t - t_n}{T}\right), \tag{2.7}$$

where $T, t_n \in \mathbb{R}$ and where the $g_r(\cdot)$ are either functions (or generalised functions/distributions such as the Dirac delta function). For such signals one can define a rate of innovation as follows

[3] That is, a signal whose Fourier transform $\mathcal{X}(f)$ is assumed to be zero apart from the set $S \subset [-B_N,\ B_N]$.

$$\rho = \lim_{\mathcal{T}\to\infty} \frac{1}{\mathcal{T}} C_x\left(-\frac{\mathcal{T}}{2}, \frac{\mathcal{T}}{2}\right), \tag{2.8}$$

where the function $C_x(t_a, t_b)$ is a counting function that counts the number of 'degrees of freedom' in the interval $[t_a, t_b]$, that is, $C_x(t_a, t_b)$ counts that number of parameters c_{nr} for which the functions g are centred within the interval $[t_a, t_b]$. For a function $x(t)$ to have a finite rate of innovation, it is obviously necessary that $\rho < \infty$. Extensions of these ideas to complex valued functions of several variables are also possible and make it possible to apply similar ideas to problems in image processing.

2.3 Linear Sampling Operators, Their Properties and How They Interact with Signal Constraint Sets

Having discussed several concepts and ideas that allow us to think about signal models using geometrical ideas, we now turn to the analysis of the sampling or measurement process itself. We introduced a set of powerful constraint sets above to allow us to deal with many problems in which we are unable to sample all relevant information, either due to corruption of signals or due to constraints on resources or fundamental physical properties of our measurement system. We will now try and develop an understanding of how the measurement system itself acts on these signal models.

Assume that our sampling system is linear, so that for any signal $\mathbf{x}$ we produce measurements $\mathbf{y} = \Phi\mathbf{x}$, where Φ is a linear sampling operator. There are two particular aspects of the sampling system Φ we should be concerned about. If we assume the signal follows a given model $\mathcal{S}$, then we want our measurement system to measure enough information to allows us to distinguish different signals from our model. It is thus natural to require that for any two $\mathbf{x}_1, \mathbf{x}_2 \in \mathcal{S}$ with $\mathbf{x}_1 \neq \mathbf{x}_2$ we have $\Phi\mathbf{x}_1 \neq \Phi\mathbf{x}_2$, so that any two distinct signals give distinct observations. In this case we should (at least in theory) be able to find the unique $\mathbf{x} \in \mathcal{S}$ that gave rise to an observation $\mathbf{y} = \mathcal{S}\mathbf{x}$.

The second fundamental requirement would be a certain robustness to noise. As nearly all measurements are noisy to some extend, if the measurement we have of a signal is slightly different from the measurement we would expect if there were no noise, then we would require that the signal that in a noiseless setting would give the actual measurement we observe is not too far from the true signal. More concretely, assume that we want to measure a signal $\mathbf{x}$, but observe the following noisy measurement $\mathbf{y} = \Phi\mathbf{x} + \mathbf{e}$ for some small noise term $\mathbf{e}$. Assume that there is a signal $\hat{\mathbf{x}}$ that also lies in our model and that satisfies $\mathbf{y} = \Phi\hat{\mathbf{x}} = \Phi\mathbf{x} + \mathbf{e}$. As it seems reasonable to assume that the true signal was $\hat{\mathbf{x}}$ given we observe $\mathbf{y}$ and don't know what $\mathbf{e}$ is, it would not be very useful that, for small $\mathbf{e}$, the difference between $\mathbf{x}$ and $\hat{\mathbf{x}}$ is large as we would then make large errors in our signal reconstruction, even under small noise perturbations.

2.3.1 A Geometrical Approach to Signal Recovery

Let us recall the signal recovery problem we would like to solve. A signal is measured and we would like to either ask specific questions about the signal or we would like to reconstruct the signal from the measurements. With our mathematical framework, both, the signal and the measurement will be represented as vectors which live in some vector space. In general, we say the signal $\mathbf{x}$ lives in a vector space $\mathcal{H}$ and the measurement $\mathbf{y}$ lives in a space $\mathcal{L}$. For most of this chapter, $\mathcal{H}$ and $\mathcal{L}$ will be Hilbert spaces, that is, it will make sense to talk about distance and angle between signals (or measurements). Each measurement is a transformation of a signal $\mathbf{x}$ into an observation $\mathbf{y}$. This transformation is done (mathematically speaking) by an operator $\Phi(\mathbf{x})$, which can either be linear or non-linear. For most of our discussion, we will restrict ourselves to linear operators, as these are easier to understand. However, we will also discuss how some of the ideas that hold for linear measurements can be applied to the setting where the measurements are slightly non-linear.

Nature does not follow the idealistic precision of a mathematical operator and any real measuring device will add at least some systematic or random noise to the measurements. We will thus use the following fundamental measurement equation that describes how any signal $\mathbf{x}$ is transformed into a particular measurement

$$\mathbf{y} = \Phi(\mathbf{x}) + \mathbf{e}, \tag{2.9}$$

where $\mathbf{e}$ is an unknown vector of measurement noise. This brings us to the fundamental problem of this book, given a measurement $\mathbf{y}$ and knowing enough about the measurement process to be able to describe Φ, how can we recover the original signal $\mathbf{x}$ and with what precision can we do this?

2.3.1.1 Lets Start Simple

In the simplest instance, if Φ is linear and invertible on the entire space, then we could simply estimate $\mathbf{x}$ as

$$\widehat{\mathbf{x}} = \Phi^{\dagger}\mathbf{y} = \Phi^{\dagger}\Phi\mathbf{x} + \Phi^{\dagger}\mathbf{e}. \tag{2.10}$$

How good this estimate is depends on how much the inverse $\Phi^{\dagger}$ amplifies the error $\mathbf{e}$. To see this, consider the difference between $\widehat{\mathbf{x}}$ and $\mathbf{x}$.

$$\|\widehat{\mathbf{x}} - \mathbf{x}\| = \|\Phi^{\dagger}\Phi\mathbf{x} + \Phi^{\dagger}\mathbf{e} - \mathbf{x}\| = \|\mathbf{x} + \Phi^{\dagger}\mathbf{e} - \mathbf{x}\| = \|\Phi^{\dagger}\mathbf{e}\|. \tag{2.11}$$

Relative to the size of $\mathbf{e}$, this error is thus

$$\frac{\|\widehat{\mathbf{x}} - \mathbf{x}\|}{\|\mathbf{e}\|} = \frac{\|\Phi^{\dagger}\mathbf{e}\|}{\|\mathbf{e}\|}, \tag{2.12}$$

which, by the definition of the operator norm, cannot be larger than the operator norm of $\Phi^\dagger$. This is related to the condition number of $\Phi^\dagger$, which is defined as the ration of the relative change in the size of $\mathbf{e}$ (i.e. $\|\Phi^\dagger\mathbf{e}\|/\|\mathbf{e}\|$) to the relative change in size of $\Phi\mathbf{x}$ (i.e. $\|\Phi^\dagger\Phi\mathbf{x}\|/\|\Phi\mathbf{x}\| = \|\mathbf{x}\|/\|\|\Phi\mathbf{x}\|$), which is easily seen to be the same as the ratio of the operator norms $\|\Phi\|/\|\Phi^\dagger\|$. Thus, if Φ is invertible and (in an infinite dimensional setting) Φ is well-conditioned, all we need to do to recover any signal from its measurement is to calculate the inverse of the operator and apply it. To guarantee that the reconstruction error is small, we need to make sure that the operator norm of the inverse is small. The inverse itself is linked to Φ itself, so that in designing a measurement system, if we can insure that it is linearly invertible, then all we need to do is ensure that the inverse operator has small norm or that the operator has a condition number close to 1.

Before moving on to more challenging signal recovery problems, it is worth thinking about the above recovery in terms of the geometry of the signal space. Any signal that lies within a certain distance (say d) from a point $\mathbf{c}$ is said to lie in a ball with centre $\mathbf{c}$ and radius d. Thus, the set of all error signals that have a length of less than ϵ say, lie in an ϵ ball (with centre at zero). In the above example, where Φ was linearly invertible on the entire signal space $\mathcal{H}$, the norm of $\Phi^\dagger$ (i.e. $\|\Phi^\dagger\|$) together with the size of the error $\mathbf{e}$ will then specify the radius around the point $\mathbf{x}$ in which the estimate $\widehat{\mathbf{x}}$ will lie. In the geometrical view of this chapter, we will not specify an explicit probabilistic model for the error $\mathbf{e}$. Instead, we assume that $\mathbf{e}$ is of restricted size $\|\mathbf{e}\| \leq \epsilon$. There is obviously a link between a probabilistic formulation and our geometrical point of view. For example, for an independent and identically distributed Gaussian noise term $\mathbf{e}$, with high probability, we know that the error will very likely be smaller than several (say 3) standard deviations. Similar probabilistic arguments, where we can assume an error bound *with high probability* can be made for other noise distributions as well.

2.3.1.2 More Complex, Yet Manageable

Now the case in which Φ is linear and invertible is trivial when compared to the much more challenging task of the stable recovery of $\mathbf{x}$ when Φ is non-invertible or ill-conditioned. If $\mathbf{x}$ is an element in some Hilbert space, but if there are at least two $\mathbf{x}_1 \neq \mathbf{x}_2$ such that $\mathbf{y} = \Phi\mathbf{x} = \Phi\mathbf{x}_1 = \Phi\mathbf{x}_2$, then there is no way in which we can choose among the two offending $\mathbf{x}_1$ and $\mathbf{x}_2$, given only the measurement $\mathbf{y}$. Typically, if Φ is linear and non-invertible, then, for each $\mathbf{y}$, there will be entire subspaces of elements $\mathbf{x}$ that would give exactly the same measurement $\mathbf{y}$. Non-invertible linear operators have the property that there are elements $\mathbf{x}_0 \neq \mathbf{0}$ such that $\Phi\mathbf{x}_0 = \mathbf{0}$. For such an element, if we take any other $\mathbf{x}_1$ such that $\mathbf{y} = \Phi\mathbf{x}_1$ and add $\mathbf{x}_0$ to $\mathbf{x}_1$ we get the same observation $\mathbf{y} = \Phi\mathbf{x}_1 = \Phi(\mathbf{x}_1 + \mathbf{x}_0)$ and we are in the above situation where we can't distinguish between $\mathbf{x}_1$ and $\mathbf{x}_2 = \mathbf{x}_1 + \mathbf{x}_0$. Furthermore, for linear operators Φ, if $\Phi\mathbf{x}_0 = \mathbf{0}$, then $\Phi\lambda\mathbf{x}_0 = \mathbf{0}$ for all scalars λ. Thus, the set of all $\mathbf{x}_0$ for which $\Phi\mathbf{x}_0 = \mathbf{0}$ is a subspace. This subspace is called the *null-space* of the linear operator Φ and will be denoted as $\mathcal{N}(\Phi)$.

In the case in which Φ is non-invertible, we can therefore only recover elements from $\mathcal{H}$ if we can restrict the search to a subset $\mathcal{S}$ of $\mathcal{H}$. For this restriction to work, we require that the measurement operator Φ is invertible at least on the subset $\mathcal{S}$. To repeat; what we mean by this is that for any two $\mathbf{x}_1, \mathbf{x}_2 \in \mathcal{S}$, with $\mathbf{x}_1 \neq \mathbf{x}_2$, we require that $\mathbf{y}_1 = \Phi\mathbf{x}_1 \neq \Phi\mathbf{x}_2 = \mathbf{y}_2$. Thus, if we have a signal model that restricts the class of signals we try to recover to a subset $\mathcal{S}$ of $\mathcal{H}$ and if Φ in invertible on the subset, then we are again able to recover $\mathbf{x} \in \mathcal{S}$, even in situations in which Φ is not invertible on all of $\mathcal{H}$.

The simplest constraint sets $\mathcal{S}$ are convex sets of which subspaces are particularly nice to deal with. For a subspace $\mathcal{S}$ it is easy to see that, if Φ is linear and invertible on $\mathcal{S}$, then the set $\Phi\mathcal{S} = \{\mathbf{y} = \Phi\mathbf{x} : \mathbf{x} \in \mathcal{S}\}$ is also a subspace. That is, for any two $\mathbf{x}_1, \mathbf{x}_2 \in \mathcal{S}$ the sum $\mathbf{x}_1 + \mathbf{x}_2$ is also in $\mathcal{S}$ and so $\Phi(\mathbf{x}_1 + \mathbf{x}_2) = \Phi\mathbf{x}_1 + \Phi\mathbf{x}_2$ will be in $\Phi\mathcal{S}$. To recover $\mathbf{x}$ from a noisy measurement $\mathbf{y} = \Phi\mathbf{x} + \epsilon$ we thus can project $\mathbf{y}$ onto the subspace $\mathcal{S}$ (call this projected element $\mathbf{y}_{\Phi\mathcal{S}}$ say) and then find an estimate $\widehat{\mathbf{x}}$ such that $\mathbf{y}_{\Phi\mathcal{S}} = \Phi\widehat{\mathbf{x}}$. In practice, as $\Phi\mathcal{S}$ is only defined implicitly, this might be a bit more involved than just described, however, conceptually, the steps of projection onto a subspace followed by the inversion on the subspace is appealing.

2.3.1.3 But Here Is the Problem

The same conceptual inversion can be carried out if $\mathcal{S}$ is any convex subset of $\mathcal{H}$ on which Φ is invertible. Even if Φ is no longer linear, similar ideas could be used. However, even if $\mathcal{S}$ is convex, if Φ is non-linear, then the set $\Phi(\mathcal{S})$ might no longer be convex. Thus, finding the equivalent of a projection onto the non-conves set $\Phi(\mathcal{S})$ is now far from trivial, even if we were able to invert $\Phi(\cdot)$ on the subset $\mathcal{S}$. A similar situation arises when Φ is linear but the constraint set $\mathcal{S}$ is non-convex to start with. In this case $\Phi\mathcal{S}$ is also non-convex in general and finding the closest element on $\Phi\mathcal{S}$ to an observation $\mathbf{y}$ is non-trivial. Furthermore, the search through the set $\mathcal{S}$ for the element that corresponds to an element in $\Phi\mathcal{S}$ is also tricky. These problems will be at the heart of this chapter.

Let us repeat the thought experiment in which we measure a signal $\mathbf{x}$ using a measurement operator Φ and where the observation is noisy. We have $\mathbf{y} = \Phi\mathbf{x} + \epsilon$ and we want to recover $\mathbf{x}$ from $\mathbf{y}$. Furthermore, the measurements are not conclusive in general that is, we are not able to distinguish all elements from the space $\mathcal{H}$ from their measurements. Thus, we use prior knowledge and devise a model that describes a subset of elements of $\mathcal{H}$ we expect to find. In the spirit of this chapter, this model comes in the form of a geometrical constraint set $\mathcal{S}$ in which we assume $\mathbf{x}$ to lie. Now, if our measurements have been designed appropriately for our model, then Φ will be invertible on $\mathcal{S}$, and our theoretical approach to reconstruct $\mathbf{x}$ from $\Phi\mathbf{x} + \epsilon$ general would be

1. Find a point in $\Phi\mathcal{S}$ that is closest to the observation $\mathbf{y}$. Call this point $\mathbf{y}_{\mathcal{S}}$.
2. As Φ is invertible on $\mathcal{S}$, find the point $\widehat{\mathbf{x}} \in \mathcal{S}$ for which $\mathbf{y}_{\mathcal{S}} = \Phi\widehat{\mathbf{x}}$.

For general sets $\mathcal{S}$ in general Hilbert spaces, there is no guarantee that there actually is a unique point in $\Phi\mathcal{S}$ that is closer to $\mathbf{y}$ than all other points. In this case, we would have to select arbitrarily from among the 'closest' points. For general sets $\mathcal{S}$, another problem that arises is that there might not even be a point that is closer to a given $\mathbf{y} \in \mathcal{H}$ than all the other points in $\mathcal{S}$. For example, let $\Phi\mathcal{S}$ be the set $\{\mathbf{y} = 1/n$, where n is a positive integer $\}$. If we were to observe any non-positive number (including zero), then there actually is no element in $\Phi\mathcal{S}$ that is the closest element. That is, for which ever element we choose from $\mathcal{S}$ (say we choose element $1/N$) there is always an infinite number of other elements which are closer to all non-positive numbers. In this case, we would have to be contempt in step (1) of our recovery scheme with the selection of a point in $\Phi\mathcal{S}$ that is nearly as close to $\mathbf{y}$ as possible.

The closest we can get to any one point $\mathbf{y}$ with any element in $\mathcal{S}$ is given by the infimum

$$\inf_{\mathbf{x}\in\mathcal{S}} \|\mathbf{y} - \Phi\mathbf{x}\|. \tag{2.13}$$

We have to take the infimum here instead of the minimum, as there might actually not be an element $\mathbf{x}$ that reaches this minimal distance. From the definition of the infimum and baring in mind that $\inf_{\mathbf{x}\in\mathcal{S}} \|\mathbf{y}-\Phi\mathbf{x}\|^2 < \infty$, we can derive the following lemma

Lemma 1 *Let $\mathcal{S}$ be a nonempty closed subset of a Hilbert space $\mathcal{H}$. Let Φ be an operator from $\mathcal{H}$ into a Hilbert space $\mathcal{L}$, then for all $\delta > 0$ and $\mathbf{y} \in \mathcal{L}$, there exist an element $\tilde{\mathbf{x}} \in \mathcal{S}$ for which*

$$\|\mathbf{y} - \Phi\tilde{\mathbf{x}}\| \leq \inf_{\mathbf{x}\in\mathcal{S}} \|\mathbf{y} - \Phi\mathbf{x}\| + \delta. \tag{2.14}$$

All this lemma is saying is that we can actually find an element in $\Phi\mathcal{S}$ that is up to an arbitrarily small distance as close to $\mathbf{y}$ as any other element in $\Phi\mathcal{S}$. Thus, we can talk about a relaxed form of projection, where, instead of finding the closest point in a set, we are contempt with a nearly closest point.

Thus consider the following mapping that for each $\mathbf{y}$ and a fixed and arbitrarily small δ returns a set of elements

$$m_{\mathcal{S}}^{\delta}(\mathbf{y}) = \{\tilde{\mathbf{y}} : \tilde{\mathbf{y}} \in \mathcal{S} \text{ and } \|\mathbf{y} - \tilde{\mathbf{y}}\| \leq \inf_{\mathbf{x}\in\mathcal{S}} \|\mathbf{y} - \Phi\mathbf{x}\| + \delta\}. \tag{2.15}$$

By the above lemma, the sets $m_{\mathcal{S}}^{\delta}(\mathbf{y})$ are non-empty for all $\delta > 0$. An operator that for each $\mathbf{y}$ returns a single element from the set $m_{\mathcal{S}}^{\delta}(\mathbf{y})$ will be said to be an δ-projection.

Thus, for each $\mathbf{y}$, we can find the δ-best $\mathbf{y}_{\mathcal{S}} \in \Phi\mathcal{S}$ and then search through $\mathcal{S}$ to find the unique $\widehat{\mathbf{x}}$ such that $\mathbf{y}_{\mathcal{S}} = \Phi\widehat{\mathbf{x}}$.

2.3.1.4 It Only Works If...

How far will $\widehat{\mathbf{x}}$ be from $\mathbf{x}$? To answer this question we need to introduce a further property of the operator Φ, namely a property that describes how much Φ 'stretches' or 'shrinks' elements. For example, if we have a vector $\mathbf{x}$ of length $\|\mathbf{x}\|$, once we have mapped this vector into the space $\mathcal{L}$, how does the length change? If Φ is linear, then we say that Φ is bounded if $\|\Phi\mathbf{x}\| \leq c\|\mathbf{x}\|$ holds for all $\mathbf{x} \in \mathcal{H}$ and for some fixed c, so that bounded linear operators can never 'stretch' vectors by more than the operator norm (which is finite for bounded operators). But how much can $\mathbf{x}$ be 'shrunk'? Remember that we are interested in problems in which Φ is ill-conditioned and non-invertible. For these problems, we have necessarily the tight lower bound $0 \leq \|\Phi\mathbf{x}\|$, that is, vectors in the null-space of Φ are mapped to zero vectors whilst for ill-conditioned Φ, vectors are potentially shrunk to arbitrarily small length. But this is exactly why we introduced the constraint set $\mathcal{S}$. Thus, instead of asking what happens to the length of all vectors in $\mathcal{H}$, we instead would like to know what happens to those vectors that live in our constraint set. Furthermore, as will become clear later, we are actually mainly interested in the diffeence between vectors, thus, we ask, what happens to the length of the difference of any two vectors $\mathbf{x}_1$ and $\mathbf{x}_2$ that lie in the subset $\mathcal{S}$. What is the maximum these differences are stretched and by how much might they be shrunk? More formally, we want to find the largest real number α and the smallest real number β such that

$$\alpha\|\mathbf{x}_1 - \mathbf{x}_2\| \leq \|\Phi(\mathbf{x}_1 - \mathbf{x}_2)\| \leq \beta\|\mathbf{x}_1 - \mathbf{x}_2\| \tag{2.16}$$

holds for all $\mathbf{x}_1, \mathbf{x}_2 \in \mathcal{S}$. We call the above inequality the bi-Lipschitz condition, with α and β being the bi-Lipschitz constants.

For once, if Φ is linear and if $\alpha > 0$, then Φ will actually be invertible on $\mathcal{S}$, that is, assume that $\mathbf{x}_1, \mathbf{x}_2 \in \mathcal{S}$ are different vectors, i.e. $\|\mathbf{x}_1 - \mathbf{x}_2\| > 0$, so that the lower bound in the bi-Lipschitz condition is non-zero. By the bi-Lipschitz condition this then implies that $\|\Phi(\mathbf{x}_1 - \mathbf{x}_2)\|$ will also be non-zero, which in turn requires that $\|\Phi\mathbf{x}_1 \neq \Phi\mathbf{x}_2\|$ so that Φ is *one to one* on $\mathcal{S}$ (that is, Φ maps distinct points in $\mathcal{S}$ into distinct points in $\mathcal{L}$.

However, a non-zero bound with $\alpha > 0$ actually tells us more. If we use our theoretical reconstruction technique, that is, we project $\mathbf{y}$ onto $\Phi\mathcal{S}$ (assuming for now that this projection exists, though a similar argument can be made for ϵ-projections) and then find the corresponding $\mathbf{x} \in \mathcal{S}$. Say $\mathbf{y}_{\mathcal{S}}$ is the projection and $\tilde{\mathbf{x}}$ is the corresponding element in $\mathcal{S}$ so that $\Phi\tilde{\mathbf{x}} = \mathbf{y}_{\mathcal{S}}$. How far will $\mathbf{x}$ be from $\tilde{\mathbf{x}}$? We have

$$\|\mathbf{x} - \tilde{\mathbf{x}}\| \leq \frac{1}{\alpha}\|\Phi\mathbf{x} - \Phi\tilde{\mathbf{x}}\| = \frac{1}{\alpha}\|\mathbf{y} - \mathbf{e} - \mathbf{y}_{\mathcal{S}}\| \leq \frac{1}{\alpha}\|\mathbf{y} - \mathbf{y}_{\mathcal{S}}\| + \frac{1}{\alpha}\|\mathbf{e}\| \leq \frac{2}{\alpha}\|\mathbf{e}\|,$$

where the second to last inequality is the triangle inequality (which is one of the properties of a norm) and where the last inequality is due to the fact that $\mathbf{y}_{\mathcal{S}}$ is the closest element in $\Phi\mathcal{S}$ to $\mathbf{y}$ and is thus closer to $\mathbf{y}$ than $\Phi\mathbf{x}$ itself. Thus $\|\Phi\mathbf{x} - \mathbf{y}\| = \|\mathbf{e}\| \geq \|\mathbf{y} - \mathbf{y}_{\mathcal{S}}\|$.

We thus have the following Lemma.

Lemma 2 *For any* $\mathbf{x} \in \mathcal{S}$*, let* $\mathbf{y} = \Phi\mathbf{x} + \mathbf{e}$*, where* Φ *satisfies the bi-Lipschitz condition with* $\alpha > 0$ *and let* $\mathbf{y}_{\Phi\mathcal{S}}$ *be the closest element in* $\Phi\mathcal{S}$ *to* $\mathbf{y}$*, then the error between* $\mathbf{x}$ *and* $\tilde{\mathbf{x}} \in \mathcal{S}$ *uniquely defined by* $\mathbf{y}_{\Phi\mathcal{S}} = \Phi\tilde{\mathbf{x}}$ *satisfies*

$$\|\mathbf{x} - \tilde{\mathbf{x}}\| \leq \frac{2}{\alpha}\|\mathbf{e}\|. \tag{2.17}$$

Therefore, if $\mathbf{x} \in \mathcal{S}$ and if Φ is linear and satisfies the bi-Lipschitz condition, then our theoretical reconstruction technique will recover a signal $\tilde{\mathbf{x}}$ that is no more than $\frac{2}{\alpha}\|\mathbf{e}\|$ away from the true signal $\mathbf{x}$. This is good news, we have just shown that, at least in theory, we should be able to recover any $\mathbf{x} \in \mathcal{S}$ as long as Φ is bi-Lipschitz on $\mathcal{S}$. The worst case accuracy of our recovered signal will then only depend on the amount of measurement noise $\mathbf{e}$ and on the inverse of the lower bi-Lipschitz constant α.

The same argument would hold for non-linear Φ if $\alpha\|\mathbf{x}_1 - \mathbf{x}_2\| \leq \|\Phi(\mathbf{x}_1) - \Phi(\mathbf{x}_2)\|$, so that a non-linear operator Φ with this condition also guarantees that the theoretical inverse is stable, that is, if we find that element $\mathbf{y}_\mathcal{S}$ in $\Phi\mathcal{S}$ closest to $\mathbf{y}$, then the $\tilde{\mathbf{x}}$ that satisfies $\mathbf{y}_\mathcal{S} = \Phi(\tilde{\mathbf{x}})$ will be close to $\mathbf{x}$.

2.3.1.5 All Models Are Wrong

$\mathcal{S}$ is a model for our signal and we assumed above that $\mathbf{x} \in \mathcal{S}$. However, as all models are wrong (at least in general), any errors in the model have to be taken into account in our discussion. Let us therefore consider what happens if $\mathbf{x}$ does not lie exactly in $\mathcal{S}$, but only 'near by'. To deal with this case in our framework, we will consider the projection of $\mathbf{x}$ onto $\mathcal{S}$. Again, as $\mathcal{S}$ can be a general non-convex set, this 'projection' is not guaranteed to exist and is definitely not required to be unique. The first problem can be dealt with in a similar way in which we dealt with the projection onto $\Phi\mathcal{S}$. We can either find an δ optimal point or, restrict discussions to sets $\mathcal{S}$ that allow us to find a closest point in $\mathcal{S}$ to all points $\mathbf{x} \in \mathcal{H}$. To simplify notation, we restrict ourselves here to the second case and assume there is such a closest point. However, this might not be unique. If there are more than one point that is closest to a point $\mathbf{x}$ we will thus assume that we choose one of these. We call this point $\mathbf{x}_\mathcal{S}$, so that $\mathbf{x}_\mathcal{S} \in \mathcal{S}$ and $\|\mathbf{x} - \mathbf{x}_\mathcal{S}\| \leq \inf_{\overline{\mathbf{x}} \in \mathcal{S}} \|\mathbf{x} - \overline{\mathbf{x}}\|$. What about error $\|\mathbf{x} - \mathbf{x}_\mathcal{S}\|$? To recover $\mathbf{x}$ from $\mathbf{y} = \Phi(\mathbf{x}) + \mathbf{e}$ we follow the same steps as before, we 'project' $\mathbf{y}$ onto $\Phi(\mathcal{S})$ and then find the corresponding $\overline{\mathbf{x}} \in \mathcal{S}$. How far is this estimate now from $\mathbf{x}$? Again, we require the stability condition $\alpha\|\mathbf{x}_1 - \mathbf{x}_2\| \leq \|\Phi(\mathbf{x}_1) - \Phi(\mathbf{x}_2)\|$ to hold, so that (this time using the non-linear notation)

$$\begin{aligned}\|\mathbf{x} - \tilde{\mathbf{x}}\| &= \|\mathbf{x} - \mathbf{x}_\mathcal{S} + \mathbf{x}_\mathcal{S} - \tilde{\mathbf{x}}\| \\ &\leq \|\mathbf{x} - \mathbf{x}_\mathcal{S}\| + \|\mathbf{x}_\mathcal{S} - \tilde{\mathbf{x}}\|\end{aligned}$$

$$\begin{aligned}
&\leq \frac{1}{\alpha}\|\Phi(\mathbf{x}_{\mathcal{S}}) - \Phi(\tilde{\mathbf{x}})\| + \|\mathbf{x} - \mathbf{x}_{\mathcal{S}}\| \\
&= \frac{1}{\alpha}\|\mathbf{y} - \tilde{\mathbf{e}} - \mathbf{y}_{\mathcal{S}}\| + \|\mathbf{x} - \mathbf{x}_{\mathcal{S}}\| \\
&\leq \frac{1}{\alpha}\|\mathbf{y} - \mathbf{y}_{\mathcal{S}}\| + \frac{1}{\alpha}\|\tilde{\mathbf{e}}\| + \|\mathbf{x} - \mathbf{x}_{\mathcal{S}}\| \\
&\leq \frac{1}{\alpha}\|\mathbf{e}\| + \frac{1}{\alpha}\|\tilde{\mathbf{e}}\| + \|\mathbf{x} - \mathbf{x}_{\mathcal{S}}\|, \\
&\leq \frac{2}{\alpha}\|\mathbf{e}\| + \frac{1}{\alpha}\|\Phi(\mathbf{x}) - \Phi(\mathbf{x}_{\mathcal{S}})\| + \|\mathbf{x} - \mathbf{x}_{\mathcal{S}}\|.
\end{aligned}$$

where $\tilde{\mathbf{e}} = \mathbf{e} + \Phi(\mathbf{x}) - \Phi(\mathbf{x}_{\mathcal{S}})$ and where the first inequality is again the triangle inequality. Thus, if our model is wrong, then our recovery lemma reads (spot the two small differences to the previous version, (1) $\mathbf{x}$ is no longer required to lie in $\mathcal{S}$ and (2) the distances of $\mathbf{x}$ from $\mathbf{x}_{\mathcal{S}}$ and of $\Phi(\mathbf{x})$ from $\Phi(\mathbf{x}_{\mathcal{S}})$ now join the error bound)

Lemma 3 *For any* $\mathbf{x}$, *let* $\mathbf{y} = \Phi\mathbf{x} + \mathbf{e}$, *where* Φ *satisfies the bi-Lipschitz condition with* $\alpha > 0$ *and let* $\mathbf{y}_{\Phi\mathcal{S}}$ *be the closest element in* $\Phi\mathcal{S}$ *to* $\mathbf{y}$, *then the error between* $\mathbf{x}$ *and* $\tilde{\mathbf{x}} \in' S$ *uniquely defined by* $\mathbf{y}_{\Phi\mathcal{S}} = \Phi\tilde{\mathbf{x}}$ *satisfies*

$$\|\mathbf{x} - \tilde{\mathbf{x}}\| \leq \frac{2}{\alpha}\|\mathbf{e}\| + \frac{1}{\alpha}\|\Phi(\mathbf{x}) - \Phi(\mathbf{x}_{\mathcal{S}})\| + \|\mathbf{x} - \mathbf{x}_{\mathcal{S}}\|. \tag{2.18}$$

Thus, even if $\mathbf{x}$ is no longer within our model, we can still use the model $\mathcal{S}$ to recover $\mathbf{x}$. All we loose in the accuracy of our reconstruction is then the additional error terms $\mathbf{x} - \mathbf{x}_{\mathcal{S}}$ an $\Phi(\mathbf{x}) - \Phi(\mathbf{x}_{\mathcal{S}})$. Thus, if $\mathbf{x}$ is close to $\mathcal{S}$, then we can still recover $\mathbf{x}$ with high accuracy.

We have thus demonstrated that it is possible to recover elements from $\mathcal{H}$ which are close to elements in $\mathcal{S}$ from noisey observations $\mathbf{y} = \Phi(\mathbf{x}) + \mathbf{e}$ whenever $\alpha\|\mathbf{x}_1 - \mathbf{x}_2\| \leq \|\Phi(\mathbf{x}_1) - \Phi(\mathbf{x}_2)\|$ holds for all $\mathbf{x}_1, \mathbf{x}_2 \in \mathcal{S}$. However, our approach to do this recovery required two steps, (1) find an element $\mathbf{y}_{\mathcal{S}}$ in $\Phi\mathcal{S}$ closest to $\mathbf{y}$ and (2) find that $\tilde{\mathbf{x}} \in \mathcal{S}$ such that $\Phi\tilde{\mathbf{x}} = \mathbf{y}_{\mathcal{S}}$. For many complex models $\mathcal{S}$, both of these steps are far from trivial. For several sets $\mathcal{S}$ that are of interest in many applications, we will thus study more practical methods to recover $\mathbf{x}$. Crucially, not only are these approaches computationally much more efficient than the approach described above, they will also be shown to have a similar worst case recovery error.

2.4 Geometry of Convex Relaxation

The first efficient approach we will discuss that can be used to recover data in certain data-recovery problems under non-convex constraints uses convexification of the constraint set. This is the traditional approach used in compressed sensing and its operation relies on some beautiful geometrical reasoning. Convexification based ideas have been developed predominantly for sparse problems, where there is a natural and powerful convex version of the constraint. Consider a real vector $\mathbf{x}$ and let $\|\mathbf{x}\|_0$ be the number of non-zero entries in the vector $\mathbf{x}$. If we want to optimise with the constraint that $\|\mathbf{x}\|_0$ is smaller than some specified integer, then we have a non-convex constraint. Similarly, if we would like to optimise $\mathbf{x}$ so that $\|\mathbf{x}\|_0$ is as small as possible, subject to some other constraint (for example $\mathbf{y} = \Phi\mathbf{x}$), then we are dealing with a non-convex cost function. To simplify these problems, we can replace the non-convex function $\|\mathbf{x}\|_0$ with the norm $\|\mathbf{x}\|_1$, i.e. with the ℓ_1 vector norm, which directly leads to convex problems that are much easier to solve numerically.

The question now is, under which conditions are the solutions to problems that use $\|\mathbf{x}\|_0$ equivalent or similar to the solutions solved with their convex version based on the norm $\|\mathbf{x}\|_1$? To study this problem, we will look at the geometry of the constraint set $\|\mathbf{x}\|_1 \leq 1$.

2.4.1 The Null-Space and Its Properties

Our treatment of the topic here is inspired by the work in [16, 18]. Consider the compressed sensing problem: minimise $\|\mathbf{x}\|_0$ such that $\mathbf{y} = \Phi\mathbf{x}$ and its convex counterpart: minimise $\|\mathbf{x}\|_1$ such that $\mathbf{y} = \Phi\mathbf{x}$. Let $\hat{\mathbf{x}}$ be the solution to the second one of these problems and let $\mathbf{x}_k$ be the best k-term approximation of the vector $\mathbf{x}$, thats is $\mathbf{x}_k$ satisfies $\|\mathbf{x}_k - \mathbf{x}\| = \min_{\tilde{\mathbf{x}}:\|\tilde{\mathbf{x}}\|_0 = k} \|\tilde{\mathbf{x}} - \mathbf{x}\|$.

The null-space of Φ will play a fundamental role in this section. Let $\mathbf{h}$ be a vector in this null-space, that is, we have $\Phi\mathbf{h} = \mathbf{0}$. We will also use the following measure that characterises how well vectors in this null-space align with the co-ordinate axis. The null-space property of Φ is defined as follows. Let C_k be the largest constant such that

$$C_k \sum_{i \in \mathcal{K}} |\mathbf{h}_i| \leq \sum_{i \notin \mathcal{K}} |\mathbf{h}_i|, \tag{2.19}$$

holds for all vectors $\mathbf{h}$ in the null-space of Φ and for all index sets $\mathcal{K}$ of size k or less. Importantly, if the above condition holds for all subsets of $\mathcal{K}$ elements of the vector $\mathbf{h}$, then it must also hold for the subset of the k largest elements. We can therefore write the above condition as

$$C_k \|\mathbf{h}_k\| \leq \|\mathbf{h} - \mathbf{h}_k\|, \tag{2.20}$$

where $\mathbf{h}_k$ is again the vector with the largest (in magnitude) k elements of $\mathbf{h}$ and zeros elsewhere. This condition is known as the null-space property of Φ and if it holds for $C_k \leq 1$, then we say that Φ satisfies the null-space property of order k.

2.4.2 The Null-Space Property for Signal Recovery

The null-space property directly implies a bound on the quality of the solution $\hat{\mathbf{x}}$ to the convex optimisation problem: minimise $\|\mathbf{x}\|_1$ such that $\mathbf{y} = \Phi\mathbf{x}$.

To see this, let $\mathbf{x}$ be any vector such that $\mathbf{y} = \Phi\mathbf{x}$ and let $\hat{\mathbf{x}}$ be the minimum of the optimisation problem so that $\|\hat{\mathbf{x}}\|_1 \leq \|\mathbf{x}\|_1$ and $\mathbf{y} = \Phi\hat{\mathbf{x}} = \Phi\mathbf{x}$. We want to bound the length of the error $\hat{\mathbf{x}} - \mathbf{x}$. To do this, we first note that the vector $\mathbf{h} = \hat{\mathbf{x}} - \mathbf{x}$ lies in the null-space of Φ. To see this we use the fact that $\mathbf{y} = \Phi\hat{\mathbf{x}} = \Phi\mathbf{x}$, so that $\mathbf{0} = \Phi\hat{\mathbf{x}} - \Phi\mathbf{x} = \Phi(\hat{\mathbf{x}} - \mathbf{x})$.

Note that the ℓ_1 norm has the property that for any vector $\mathbf{x}$, we have $\|\mathbf{x}\|_1 = \|\mathbf{x}_k\|_1 + \|\mathbf{x} - \mathbf{x}_k\|_1$. Furthermore, note that the null-space property implies that for all $\mathbf{h}$ that lie in the null-space of Φ

$$
\begin{aligned}
(C-1)(\|\mathbf{h}_k\|_1 + \|\mathbf{h} - \mathbf{h}_k\|_1) &= C\|\mathbf{h}_k\|_1 - \|\mathbf{h}_k\|_1 + C\|\mathbf{h} - \mathbf{h}_k\|_1 - \|\mathbf{h} - \mathbf{h}_k\|_1 \\
&\leq \|\mathbf{h} - \mathbf{h}_k\|_1 - \|\mathbf{h}_k\|_1 + C\|\mathbf{h} - \mathbf{h}_k\|_1 - C\|\mathbf{h}_k\|_1 \\
&= C + 1\,(\|\mathbf{h} - \mathbf{h}_k\|_1 - \|\mathbf{h}_k\|_1), \qquad (2.21)
\end{aligned}
$$

which we will use in the following form

$$
\|\mathbf{h}_k\|_1 + \|\mathbf{h} - \mathbf{h}_k\|_1 \leq \frac{C+1}{C-1}(\|\mathbf{h} - \mathbf{h}_k\|_1 - \|\mathbf{h}_k\|_1) \qquad (2.22)
$$

Using these two inequalities, we can then decompose and bound the ℓ_1 norm of the error $\mathbf{x} - \hat{\mathbf{x}}$.

$$
\begin{aligned}
\|\mathbf{x} - \hat{\mathbf{x}}\|_1 = \|\mathbf{h}\| &= \|\mathbf{h}_k\|_1 + \|\mathbf{h} - \mathbf{h}_k\|_1 \\
&\leq \frac{C+1}{C-1}(\|\mathbf{h} - \mathbf{h}_k\|_1 - \|\mathbf{h}_k\|_1) \\
&= \frac{C+1}{C-1}(\|\mathbf{h} - \mathbf{h}_k\|_1 - \|\mathbf{h}_k\|_1 - \|\mathbf{x} - \mathbf{x}_k\|_1 + \|\mathbf{x} - \mathbf{x}_k\|_1) \\
&\leq \frac{C+1}{C-1}(\|(\mathbf{h} - \mathbf{h}_k) + (\mathbf{x} - \mathbf{x}_k)\|_1 - \|\mathbf{h}_k\|_1 + \|\mathbf{x} - \mathbf{x}_k\|_1) \\
&= \frac{C+1}{C-1}(\|(\mathbf{h} - \mathbf{h}_k) + (\mathbf{x} - \mathbf{x}_k)\|_1 - \|\mathbf{h}_k\|_1 + \|\mathbf{x}_k\|_1 - \|\mathbf{x}_k\|_1 + \|\mathbf{x} - \mathbf{x}_k\|_1) \\
&\leq \frac{C+1}{C-1}(\|(\mathbf{h} - \mathbf{h}_k) + (\mathbf{x} - \mathbf{x}_k)\|_1 + \|\mathbf{h}_k + \mathbf{x}_k\|_1 - \|\mathbf{x}_k\|_1 + \|\mathbf{x} - \mathbf{x}_k\|_1) \\
&= \frac{C+1}{C-1}(\|\mathbf{x} + \mathbf{h}\|_1 - \|\mathbf{x}_k\|_1 + \|\mathbf{x} - \mathbf{x}_k\|_1) \\
&= \frac{C+1}{C-1}\left(\|\hat{\mathbf{x}}\|_1 - \|\mathbf{x}_k\|_1 + \|\mathbf{x} - \mathbf{x}_k\|_1\right) \\
&\leq \frac{C+1}{C-1}(\|\mathbf{x}\|_1 - \|\mathbf{x}_k\|_1 + \|\mathbf{x} - \mathbf{x}_k\|_1)
\end{aligned}
$$

$$= \frac{C+1}{C-1}(\|\mathbf{x}-\mathbf{x}_k\|_1 + \|\mathbf{x}-\mathbf{x}_k\|_1)$$
$$= 2\frac{C+1}{C-1}\|\mathbf{x}-\mathbf{x}_k\|_1 \tag{2.23}$$

Let us walk through this chain of equalities and inequalities at a more pedestrian speed. The first equality just re-states that the error $\mathbf{x}-\hat{\mathbf{x}}$ lies in the null-space of Φ. The second equality is then the first property above, whilst the first inequality is the second property in (2.22). The next equality simply adds and subtracts $\|\mathbf{x}-\mathbf{x}_k\|_1$, whilst in the following line we use the triangle inequality

$$\begin{aligned}\|(\mathbf{h}-\mathbf{h}_k)+(\mathbf{x}-\mathbf{x}_k)\|_1 &= \|(\mathbf{h}-\mathbf{h}_k)+(\mathbf{x}-\mathbf{x}_k)+(\mathbf{x}_k+\mathbf{h}_k)-(\mathbf{x}_k+\mathbf{h}_k)\|_1\\ &\le \|(\mathbf{h}-\mathbf{h}_k)+(\mathbf{x}-\mathbf{x}_k)+(\mathbf{x}_k+\mathbf{h}_k)\|_1 + \|(\mathbf{x}_k+\mathbf{h}_k)\|_1.\end{aligned}$$

We again add and subtract the same number, before making a second use of the triangle inequality. We then use the fact that the two vectors $(\mathbf{h}-\mathbf{h}_k)+(\mathbf{x}-\mathbf{x}_k)$ and $\mathbf{h}_k+\mathbf{x}_k$ have different support, so that we can again use property one. The next equality just uses the definition of $\hat{\mathbf{x}} = \mathbf{x}+\mathbf{h}$, whilst the last inequality uses the fact that $\|\hat{\mathbf{x}}\|_1 \le \|\mathbf{x}\|_1$ (remember, $\hat{\mathbf{x}}$ minimises the ℓ_1 norm among all $\mathbf{x}$ that satisfy $\mathbf{y} = \Phi\mathbf{x}$). We finish the argument by a final application of property one.

Interestingly, the requirement that the null-space property holds is not only sufficient for the above bound to hold (as we have just shown) but is also necessary in the following sense. If the null-space property is violated, then there exists a measurement matrix with this null-space so that the above bound is violated for some k [16]. Note however that this does not imply that the bound is violated necessarily for any particular measurement matrix Φ even if it has a null-space that violates the condition.

Note also that the result here is slightly different from that of the "ideal" algorithm of the previous section and is also different from the bounds we derive in the next section. Firstly, the null-spaced based results are not able to account for measurement errors. Secondly, the bound here is in terms of the ℓ_1 norm of the error $\mathbf{x}-\mathbf{x}_k$, that is, it tells us how well we can approximate vectors whose $N-k$ smallest coefficients have a small ℓ_1 norm. A theory based on ideas similar to the bi-Lipschitz condition on Φ can also be derived. This is done for example in [5, 17]. For example, in [17] we have the following result which is more similar to that in Lemma 3.

Theorem 2 *For any* $\mathbf{x}$*, assume* Φ *satisfies the bi-Lipschitz property*

$$(1-\gamma)\|\mathbf{x}_1+\mathbf{x}_2\|^2 \le \|\Phi(\mathbf{x}_1+\mathbf{x}_2)\|^2 \le (1+\gamma)\|\mathbf{x}_1+\mathbf{x}_2\|^2, \tag{2.24}$$

where $\gamma < \sqrt{2}-1$*. Given observations* $\mathbf{y} = \Phi\mathbf{x}+\mathbf{e}$*, the minimiser of the problem* $min_{\tilde{\mathbf{x}}}\|\tilde{\mathbf{x}}\|_1$ *subject to the constraint that* $\|\mathbf{y}-\Phi\tilde{\mathbf{x}}\| \le \|\mathbf{e}\|$ *recovers an estimate* $\hat{\mathbf{x}}$ *that satisfies*

$$\|\mathbf{x}-\hat{\mathbf{x}}\| \le_0 C\|\tilde{\mathbf{e}}\| + C_1\|\mathbf{x}_k-\mathbf{x}\|, \tag{2.25}$$

where $\widetilde{\mathbf{e}} = \Phi(\mathbf{x} - \mathbf{x}_k) + \mathbf{e}$ *and where* C_o *and* C_1 *are constants depending on* γ.

Instead of proving this result here (the interested reader is redirected to [17]), we instead return to the null-space property and study the geometrical implications this property has for the recovery of sparse vectors in somewhat more detail.

2.4.3 Random Null-Spaces and the Grassman Angle

To build a measurement system that would allow us to use ℓ_1 recovery with the tight error bounds derived in (2.23), we thus need to ensure that the measurement system satisfies the null-space property. One particularly powerful approach to construct measurement systems is through random construction methods and it can be shown that these systems often satisfy the required null-space properties. As the null-space property is fundamentally geometrical in nature, geometrical ideas can also be used to study and understand these construction techniques.

Instead of the careful construction of a matrix whose null-space satisfies the null-space property, it is significantly simpler to randomly choose a null-space and then construct a matrix that has the same null-space. In fact, this random construction is one of the only few known construction method that can build matrices that on the one hand satisfy the null-space property and on the other hand, are optimal in terms of the number of measurements. However, we must note that, if we use a random construction, then our desired property will only hold with high probability and is not absolutely guaranteed.

We will assume that the null-space is chosen randomly in such a way that its distribution is rotation invariant. With this we mean that, if $\mathbf{B}$ is a basis for a null-space of dimension N and if $\mathbf{U}$ is an orthonormal rotation matrix, then any rotation invariant distribution $p(\mathbf{B})$ must satisfy $p(\mathbf{B}) = p(\mathbf{UB})$. For example, if we choose the entries of the matrix $\Phi \in \mathbb{R}^{M \times N}$ to be drawn independently from a zero-mean unit variance normal distribution, and if $M < N$, then the distribution of the null-space of Φ will have this property.

The null-space property of a matrix Φ is related to the following property (see [16]).

Lemma 4 *Let* $\mathcal{K}$ *be a subset of* k *of the indices of a vector in* $\mathbb{R}^N$. *Then, the null-space property*

$$C\|\mathbf{h}\|_1 \leq \|\mathbf{h} - \mathbf{h}_k\|_1, \tag{2.26}$$

for all $\mathbf{h}$ *in the null-space is equivalent to the property that all vectors* $\mathbf{x}$ *supported on* $\mathcal{K}$ *satisfy*

$$\|\mathbf{x} + \mathbf{h}_k\|_1 + \|\frac{\mathbf{h} - \mathbf{h}_k}{C}\|_1 \geq \|\mathbf{x}\|_1 \tag{2.27}$$

for all $\mathbf{h}$ *in the null-space.*

We here use the notation $\mathbf{h}_k$ to refer to a version of the vector $\mathbf{h}$ in which all entries are set to 0 apart from those elements with indices in the set $\mathcal{K}$.

To derive a lower bound on the probability under which a randomly sampled subspace satisfies the null-space property, we can therefore derive an upper bound on the probability with which the above condition fails. That is, what is the probability that for any k-sparse vector $\mathbf{x}$ the condition in (2.27) will fail?

To answer this question, we first note that we can restrict our attention to vectors $\mathbf{x}$ that satisfy $\|\mathbf{x}\|_1 = 1$. This is because if (2.27) holds or fails for any $\mathbf{x}$, then it will also hold or fail for $c\mathbf{x}$ for any c.

Let us now look at the probability that a randomly chosen null-space violates (2.27) for a particular $\mathbf{x}$ with a given support set $\mathcal{K}$ and a particular sign patters. We will call this probability $P_{\mathcal{K}}$. To understand the geometric properties of $P_{\mathcal{K}}$ let us consider all vectors $\mathbf{x}$ which satisfy $\|\mathbf{x}\|_1 = 1$ and which have a support $\mathcal{K}$ with $|\mathcal{K}| = k$.

As we assume $\|\mathbf{x}\|_1 = 1$, the condition in (2.27) is related to the following geometrical object.

$$WB = \{\hat{\mathbf{x}} \in \mathbb{R}^N : \|\hat{\mathbf{x}}_k\|_1 + \|\frac{\hat{\mathbf{x}} - \hat{\mathbf{x}}_k}{C}\|_1 \leq 1\}. \tag{2.28}$$

We call this cross-polytope the weighted ℓ_1 ball. A sketch of WB, $\mathbf{x}$ and $\mathbf{h}$ is given in Fig. 2.1. The probability $P_{\mathcal{K}}$ is thus the probability that there exist a vector $\mathbf{h} \neq \mathbf{0}$ in the null-space of Φ so that for at least one k-sparse vector $\mathbf{x}$ with $\|\mathbf{x}\|_1 = 1$ and support $\mathcal{K}$, where $\text{sign}(\mathbf{x}_k)$ is fixed, we have

$$\|\mathbf{x} + \mathbf{h}_k\|_1 + \|\frac{\mathbf{h} - \mathbf{h}_k}{C}\|_1 < \|\mathbf{x}\|_1 = 1. \tag{2.29}$$

Note that all the k-sparse $\mathbf{x}$ we consider here (those with $\|\mathbf{x}\|_1 = 1$) lie on the surface of an ℓ_1 ball. Furthermore, because $\mathbf{x}$ is assumed to be k-sparse, $\mathbf{x}$ lies on a k-dimensional face. To get a geometrical intuition for this, think of a diamond (build by sticking two equal pyramids that have a square base and equal-lateral triangle sides together at the square surface). Such a diamond is a cross-polytope in three dimensions. Each of its eight triangular sides is a two dimensional face. Furthermore, the diamond has eight ridges, which are, in high-dimensional geometry language, one-dimensional faces. Finally, the six sharp corners are called zero-dimensional faces or, alternatively, vertices. To further build our geometrical intuition, assume that our co-ordinate axis are aligned with the diamond edges (or vertices). If these vertices lie at exactly the points $[100]$, $[-100]$, $[010]$, $[0-10]$, $[001]$ and $[00-1]$, then the diamond is the unit ℓ_1 ball in three dimensions. Importantly, note also that any 2-sparse vector with unit ℓ_1 norm will lie on one of the ridges (or two-dimensional faces). Now, once we fix the support set $\mathcal{K}$, then in our two dimensional example, x_k will lie in one of the three planes that align exactly with four of the eight ridges of the diamond. Exactly which four ridges depends on the support set $\mathcal{K}$. The weighted ℓ_1 ball in this three dimensional example would then be a stretched diamond in

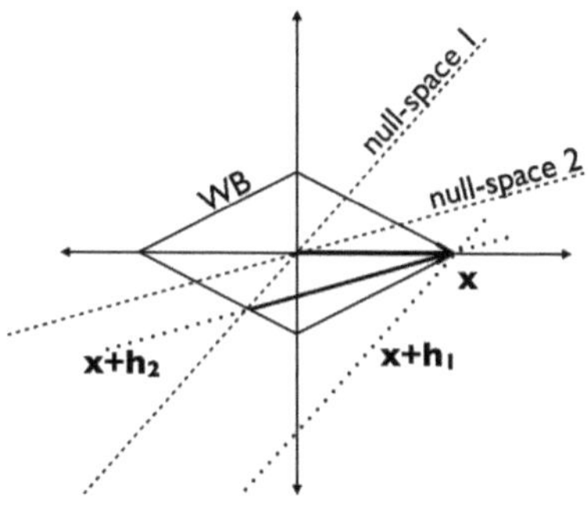

Fig. 2.1 Low dimensional sketch of the vectors, subspaces and ℓ_1 ball involved in the discussion of this section. For this simple two dimensional example, the null-spaces are chosen to be one-dimensional. A 1-sparse vectors $\mathbf{x}$ can be recovered if the null-space is null-space 1, as there is no null-space vector $\mathbf{h}_1$ such that $\mathbf{x} + \mathbf{h}_1$ lies within the cross-polytope WB (see dotted line labeled $\mathbf{x} + \mathbf{h}_1$). If, on the other hand, the null-space is null-space 2, then there exist null-space vectors $\mathbf{h}_2$ such that $\mathbf{x} + \mathbf{h}_2$ lies within WB (see the solid section of the dotted line labeled $\mathbf{x} + \mathbf{h}_2$) and $\mathbf{x}$ is not recoverable

which the two vertices that do not lie on the two dimensional subspace defined by the support set $\mathcal{K}$ are further away from the coordinate centre. Furthermore, we only consider one of the four possible sign patterns, which means that $\mathbf{x}_k$ is assumed to lie on only one of the four ridges.

The same principle holds in higher dimensions. Consider any particular k-sparse $\mathbf{x}$ that lies in the interior of a k-face of the "stretched" ℓ_1 ball (the k-face itself lies in the plane where no "stretching" has occurred). With interior of the face we here mean that we assume that $\mathbf{x}$ lies exactly within the k-face but not in any one of the $k-1$-faces, i.e. $\mathbf{x}$ has exactly k non-zero entries and not fewer. In our three dimensional pyramid for $k = 2$ this would mean that $\mathbf{x}$ lies on a ridge, but not exactly at a corner.

Let us now consider the probability that a randomly drawn subspace has vectors that satisfy

$$\|\mathbf{x} + \mathbf{h}_k\|_1 + \|\frac{\mathbf{h} - \mathbf{h}_k}{C}\|_1 < \|\mathbf{x}\|_1 = 1 \tag{2.30}$$

for at least one such $\mathbf{x}$, which as stated above is exactly the probability with which such a null-space would violate the null-space property for one of the possible sign patterns.

In our three dimensional example, the stretched cross-polytope is a stretched diamond and any 2-sparse vector $\mathbf{x}$ would be assumed to lie on one of the non-stretched ridges. Now if we were to draw a direction $\mathbf{d}$ at random (our null-space) and consider the affine subspace $\mathbf{x}+\mathbf{d}$, then what is the probability that this subspace does not go through the stretched diamond? Whilst in this three dimensional example with a two sparse vector, a one-dimensional subspace is really the only interesting scenario, in high dimensions, there is substantially more space and there is actually space to "attach" significantly higher dimensional subspaces onto low-dimensional faces of our diamond without the subspace actually cutting into the diamond itself.

Let us first try and think about the probability that a one-dimensional subspace does not intersect our stretched diamond if we attach it to a particular ridge. This probability is equivalent to a randomly drawn vector lying within a particular cone. To see this, take your diamond and shift it so that the point $\mathbf{x}$ lies at the centre of our coordinate system. In our three-dimensional example, all that we really care about in terms of the intersection of our subspace and the diamond is the intersection with the two *two*-faces that intersect at the ridge on which $\mathbf{x}$ lies. Now after shifting our diamond, our condition is violated as soon as a randomly drawn vector intersect any one of these two faces. And this happens with exactly the same probability with which a randomly drawn vector would lie within the cone generated by these two faces. Our problem is thus the same as one of specifying the probabilities with which randomly drawn subspaces lie within a cone specified by the faces of our cross-polytope that intersect with the face on which $\mathbf{x}$ lies.

Luckily, this is a problem that has been studied before. In fact, the probability that a randomly chosen low-dimensional subspace intersects with a skewed cross-polytope is equal to a geometric property known as the complementary Grassmann angle [19]. There even is a ready made formula available to calculate the complementary Grassman angle for any $(k-1)$ dimensional face [20].

$$P_{|\mathcal{K}|} = 2\sum_{i}\sum_{FACE_i} \beta(FACE_k, FACE_i)\gamma(FACE_i, WB). \tag{2.31}$$

The first sum is over all non-negative integers i and the second sum is over all (M+1+2i) dimensional faces $FACE_i$ of the skewed cross-polytope. Here $FACE_k$ is the k dimensional face on which we assume that $\mathbf{x}$ lies whilst WB is the entire cross-polytope itself. Both functions $\beta(\cdot,\cdot)$ and $\gamma(\cdot,\cdot)$ are functions of two faces of the cross-polytope (note that the entire polytope also counts as a face).

$\beta(FACE_1, FACE_2)$ is known as the internal angle. The internal angle is a geometrical property of the two faces. The angle is calculated by considering the following cone C. For all $\mathbf{x}$ in $FACE_1$, shift the polytope so that $\mathbf{x} = \mathbf{0}$ and let $C(\mathbf{x})$ be the cone of all vectors that leave $\mathbf{x}$ and intersect the face $FACE_2$. C is then the intersection of all cones $C = \bigcup_{\mathbf{x}\in FACE_1} C(\mathbf{x})$. The internal angle is now the proportion of the unit sphere of the same dimension as the cone, that is covered by the cone. The internal angle is zero if the two faces do not intersect and is unity if the faces are identical.

$\gamma(FACE_1, FACE_2)$ on the other hand is known as the external angle. The external angle is defined in a similar way, however, the cone is constructed differently by considering all outward normals to the hyperplanes that support the two faces. The external angle is again zero if the two faces do not intersect and is unity if the faces are identical.

The main effort now is the derivation for expressions that quantify these angles [16], but instead of going through this lengthy derivation here, or in fact, stating the expressions themselves, let us instead consider how these can be used to bound the probability we are interested in.

Now the above probability was for a given support set and a given sparsity pattern. However, we require the condition to hold for all support sets. To derive such a bound, let us first count the number of different support sets. For each support set, there are $2^{|\mathcal{K}|}$ different sign patterns. Furthermore, there are $\binom{n}{k}$ different support sets with k non-zero elements. We can therefore use a so called union bound to bound the probability of failure. A union bound uses the following simple probabilistic fact. If A and B are two events, then the probability that A or B holds (write $P(A \cup B)$) is always smaller or equal the probability that A holds ($P(A)$) added to the probability that B holds ($P(A)$). Thus

$$P(A \cup B) \leq P(A) + P(B). \tag{2.32}$$

If we apply this principle to the probabilities that (2.27) is violated for one of the support sets and one of the sign patterns, then we can bound this probability as

$$P(\textbf{Failure}) \leq \binom{n}{k} 2^k P_{\mathcal{K}}. \tag{2.33}$$

We thus see that the probability is bounded as a function of k and $P_{\mathcal{K}}$. $P_{\mathcal{K}}$ itself depends on the dimensions of the problem M and N as well as k. It also depends on C (as C specifies the amount of stretching in our weighted ℓ_1 ball). The main message is that, if we require a level of robustness (as defined by C and k) and want to observe a vector of length N, then we need to chose the number of observations large enough, so that the probability $\binom{n}{k}2^k P_{\mathcal{K}}$ is sufficiently small. In this case, a randomly chosen $N - M$ dimensional subspace will (with a probability bounded by $\binom{n}{k}2^k P_{\mathcal{K}}$) allow us to reconstruct our vector within the required precision. Unfortunately, closed form expressions are not available for the probabilistic bound derived here, however, numerical methods can be used to evaluate the required Grassmann angles for any required combination of C, M, N and k [16].

2.5 Geometry of Iterative Projection Algorithms

There are two main approaches to the solution of signal recovery problems under non-convex constraints. The first approach, discussed in the previous section, replaced the non-convex problem with a convex one, thus greatly simplifying it. In this section we look at an alternative, greedy methods. Greedy methods are iterative schemes that replace a non-convex optimisation problem with a sequence of simpler problems. The moniker 'greedy' here indicates that these methods 'greedily' grab a signal from the non-convex constraint set to satisfy these local optimisation constraints. Whilst there are many greedy algorithms, we here discuss a conceptually simple, yet extremely powerful approach that has similar performance guarantees to the convex relaxation based approaches discussed above, yet can also be used with many non-convex constraints for which there are no simple convex relaxations.

2.5.1 The Iterative Hard Thresholding Algorithm

The Iterative Hard Thresholding algorithm [21, 22], also known as the Iterative Projection or Projected Landweber Algorithm, is an iterative method that, as the name suggests, thresholds or projects an estimate iteratively. To see how this method works, let us consider again the optimisation problem we are trying to solve.

$$\min_{\mathbf{x}} \|\mathbf{y} - \Phi\mathbf{x}\|^2 \ : \ \mathbf{x} \in \mathcal{S}, \tag{2.34}$$

where $\mathcal{S}$ is a possibly non-convex constraint set.

Without any constraint, the simplest approach to tackle the above problem would be to use gradient optimisation (assuming that the gradient of $\|\mathbf{y} - \Phi\mathbf{x}\|^2$ exists). If $\mathbf{g}$ is the negative gradient of $\|\mathbf{y} - \Phi\mathbf{x}\|^2$ (or the Gâteaux derivative if $\mathbf{x}$ is a more general function), then this optimisation would update an estimate $\mathbf{x}^n$ using the iteration

$$\mathbf{x}^{n+1} = \mathbf{x}^n + \Omega\mathbf{g}, \tag{2.35}$$

where Ω is either a scalar step size, or more generally, a linear map to precondition and stabilise the problem. For example, Ω could be the inverse of $\Phi\Phi^T$ [23, 24] or the Hessian operator as in Newton's method. However, if Φ is non-invertible or ill-conditioned, then this optimisation will not lead to a unique and stable solution, which was the reason why the constraint set $\mathcal{S}$ was introduced in the first place. Thus, to utilise the constraint, we simply enforce the requirement that $\mathbf{x}^{n+1}$ lies in $\mathcal{S}$. To do this, the estimate $\mathbf{a} = \mathbf{x}^n + \mu\mathbf{g}$ has to be mapped to an element in $\mathcal{S}$ and, to keep the potential increase in our cost $\|\mathbf{y} - \Phi\mathbf{x}\|^2$ this mapping entails to a minimum, this mapping should not take us too far away from $\mathbf{a}$ itself. Thus, we would ideally like to find a point in $\mathcal{S}$ that is as close to $\mathbf{a}$ as possible. If we are able to calculate such a projection for the non-convex set $\mathcal{S}$, then we can use the Iterative Hard Thresholding algorithm.

$$\mathbf{x}^{n+1} = P_{\mathcal{S}}(\mathbf{x}^n + \Omega\mathbf{g}), \tag{2.36}$$

where $P_{\mathcal{S}}$ is this projection mapping.

This procedure might remind the reader of the approach we have discussed above, in which the reconstruction was done via a projection of $\mathbf{y}$ onto the closest element in $\Phi\mathcal{S}$. In principle, the projection $P_{\mathcal{S}}$ is defined in a similar way to the projection onto the set $\Phi\mathcal{S}$. Thus, if we are able to efficiently calculate the projection onto $\Phi\mathcal{S}$, then there would be no need to use the more complex Iterative Hard Thresholding algorithm. The point however is that, for many constraint sets $\mathcal{S}$ used in practice, calculating the projection $P_{\mathcal{S}}$ is significantly more efficient than to try and project onto the set $\Phi\mathcal{S}$. Several examples are given next.

2.5.2 Projections onto Non-convex Sets

Let us start by formalising again what we will mean when talking about projections onto the set $\mathcal{S} \subset \mathcal{H}$. A projection operator $P_{\mathcal{S}}$ will be any map that, for a given $\mathbf{x} \in \mathcal{H}$ returns a unique element $\mathbf{x}_{\mathcal{S}} \in \mathcal{S}$ such that

$$\|\mathbf{x} - \mathbf{x}_{\mathcal{S}}\| = \inf_{\tilde{\mathbf{x}} \in \mathcal{S}} \|\mathbf{x} - \tilde{\mathbf{x}}\|. \tag{2.37}$$

Again, in certain circumstances, there might not be any $\mathbf{x} \in \mathcal{S}$ for which this property holds with equality. In those cases, one can again relax the requirement on the projection and talk of ϵ projections as those mappings $P_{\mathcal{S}}$ that, for a given $\mathbf{x} \in \mathcal{H}$ returns a unique element $\mathbf{x}_{\mathcal{S}} \in \mathcal{S}$ such that

$$\|\mathbf{x} - \mathbf{x}_{\mathcal{S}}\| \leq \inf_{\tilde{\mathbf{x}} \in \mathcal{S}} \|\mathbf{x} - \tilde{\mathbf{x}}\| + \epsilon. \tag{2.38}$$

2.5.2.1 Sparsity

In Euclidean space, a sparse vector $\mathbf{x}$ is an element of $\mathbb{R}^N$ or $\mathbb{C}^N$ for which $x_i = 0$ for "many" of the indices $i \subset [1, 2, \ldots, N]$. A popular constraint set is then the set $\mathcal{S}_k$ of all vectors $\mathbf{x}$ in $\mathbb{R}^N$ (or in $\mathbb{C}^N$) that have no more than $k < N$ non-zero entries. As discussed above, this is a non-convex set and for general Φ finding the projection onto $\Phi\mathcal{S}_k$ is far from trivial, in fact this is a combinatorial search problem in general and we would have to look at each of the k-sparse subspaces in $\mathcal{S}_k$ in turn. However, projecting a vector $\mathbf{x}$ onto $\mathcal{S}_k$ itself is trivial, all one has to do is to identify the k largest (in magnitude) components $|x_i|$ and setting all other components to zero.[4]

2.5.2.2 Block-Sparsity

Like many other structured sparsity constraints, block-sparsity is not easy to deal with directly in the observation domain, that is, it is difficult to project onto $\Phi\mathcal{S}$. Yet again, projection onto $\mathcal{S}$ itself is trivial and is done in a similar way to the sparse case. The only difference now is that we have to calculate the length of $\mathbf{x}$ when restricted to each of the blocks. For example, if the individual blocks are labeled with indices j and if $\mathbf{x}_j$ is the sub-vectors of $\mathbf{x}$ containing only those elements in block j, then we calculate the length of each $\mathbf{x}_j$ and set all blocks to zero apart from those elements that are in the k largest blocks.

[4] We assume here that we use the norm $\sqrt{\sum x_i^2}$, though other Euclidean norms are treated with equal ease.

2.5.2.3 Tree-Sparsity

Tree sparse models are the other main structured sparsity model of interest. For a given Euclidean vector $\mathbf{x}$ and a pre-defined tree structure, finding the closest sparse vector that respects the tree structure is somewhat more complicated than the projection in the previous two examples. Luckily, there exist fast (yet in the worst case only approximate) algorithms that can be used. In particular, the *condensing sort and select algorithm* (CSSA) is relatively fast as it only requires a computational effort that, in many instances, is of the order of $\mathcal{O}(\mathcal{N} \log \mathcal{N})$ [25].

2.5.2.4 Sparsity in Other Bases

In the three examples above we have used constraint sets in which the signal model assumed sparsity in the canonical basis, that is, we thought about vectors as collections of N numbers and sparsity simply meant we were only allowed a few non-zero numbers. To do this, we have implicitly assumed that we write the vector as a collection of N real or complex numbers, that is, we assumed that we write our vectors in the traditional linear algebra notation as

$$\mathbf{x} = [x_1 x_2 x_3 \cdots x_N]^T. \tag{2.39}$$

Such a notation only specifies a vector if it is made with respect to some basis. Remember, it is best to think of a vector as a point in spaces, say the location of a particular flat in an apartment block, whose location you could specify as 3 floors up, corridor to the left and the third flat on the right, which could be written as [3 3 1]. However, other coordinate systems are possible and would lead to a different set of three numbers. The same is, as said before, also possible for our signal representation. If our signals $\mathbf{x}$ is a vector in Euclidean space, then we can write it as

$$\mathbf{x} = \sum_i a_i \mathbf{x}_i, \tag{2.40}$$

where the a_i are the numbers that specify the location and where the $\mathbf{x}_i$ are a particular basis. For example, for a sampled time series, we typically use what is called the canonical basis, where each basis vector is used to specify the signal intensity at each of the sample time-points. But, if we think about signals as points in space, then we are free to choose more convenient coordinate axis. This is particularly useful as sparsity is a property of collections of numbers (which are often informally called vectors, a sin we here freely commit too, but which, as we stressed above, are not to be confused with the definition of a vector as a point in space) and not of a point in space. A point in space can only be spares if we define the appropriate basis in which its representation is sparse and different signals are sparse in different bases. For example, many natural sounds are made up of a small number of harmonic components so that sounds are often fairly sparse using Fourier or other sinusoidal bases. Images, on the other hand, are often found to be sparse in representations based on wavelet bases.

It is easy to compute the projection of any signal onto any one of the basis vectors. This is done simply as

$$\langle \mathbf{x}, \mathbf{x}_i \rangle / \|\mathbf{x}_i\|, \tag{2.41}$$

where $\mathbf{x}_i$ is the basis vector we project onto. If all basis vectors are orthogonal, then we can also use this approach to project onto the linear subspace spanned by a subset of the basis vectors. Importantly, for orthogonal bases, the optimal choice of the coefficient for one basis vectror does not depend on the choice of the other basis vectors. This nice property does no longer hold if the basis is not orthogonal, however. Thus, finding a sparse approximation in any orthogonal basis is simple and can be computed by finding the representation of the signal in the basis, followed by a simple thresholding where only the elements are kept that have the largest magnitude. However, this simple approach is no longer possible in general when the basis is not orthogonal and the non-orthogonality will have to be taken into account.

2.5.2.5 Low Rank Matrices

As discussed above, data that comes in matrix form also allows the specification of powerful non-convex constraints. For matrices that are known to have a low rank we require a projection onto low rank matrices.Again, these projections are easy to calculate. The best approximation of a matrix with a matrix of rank k can be calculated using the Singular Value Decomposition of the matrix followed by thresholding of the singular values, such that only the largest k singular values are retained [26].

2.5.3 Convergence and Stable Recovery

The main question we should ask at this point is, "How good is the Iterative Hard Thresholding algorithm?" that is, if we are given an observation $\mathbf{y}$, where $\mathbf{y} = \Phi\mathbf{x}+\mathbf{e}$ and if we run the algorithm for a number of iterations, how close will our estimate $\hat{\mathbf{x}}$ be to the true, unknown signal $\mathbf{x}$?

An answer to this question is provided by the following theorem taken from [9].

Theorem 3 *Assume an arbitrary signal* $\mathbf{x}$ *in some Hilbert space* $\mathcal{H}$. *Assume you are given an observation* $\mathbf{y}$ *and a measurement operator* Φ *and you assume that* $\mathbf{y} = \Phi\mathbf{x} + \mathbf{e}$ *where* $\mathbf{e}$ *is an unknown error term. You furthermore know, from prior experience, that* $\mathbf{x}$ *lies close to a non-convex constraint-set* $\mathcal{S}$. *Then, if* Φ *satisfies the bi-Lipschitz condition on* $\mathcal{S}$ *with constants* $\beta/\alpha < 1.5$, *then the Iterative Hard Thresholding algorithm run with a step size* μ *that satisfies* $\beta \leq \frac{1}{\mu} < 1.5\alpha$ *and run for*

$$n^\star = \left\lceil 2\frac{\log(\delta\frac{\|\widetilde{\mathbf{e}}\|}{\|P_{\mathcal{S}}(\mathbf{x})\|})}{\log(2/(\mu\alpha)-2)}\right\rceil \tag{2.42}$$

iterations, will calculate a solution $\hat{\mathbf{x}}$ *satisfying*

$$\|\mathbf{x}-\hat{\mathbf{x}}\| \leq \left(\sqrt{c}+\delta\right)\|\widetilde{\mathbf{e}}\| + \|P_{\mathcal{S}}(\mathbf{x})-\mathbf{x}\| \tag{2.43}$$

where $c \leq \frac{4}{3\alpha-2\mu}$, $\widetilde{\mathbf{e}} = \Phi(\mathbf{x}-P_{\mathcal{S}}(\mathbf{x}))+\mathbf{e}$ *and* $\delta > 0$ *is arbitrary.*

There are several interesting observations to be made here. Let us start by looking at the number of iterations required by the theorem.

$$n^\star = \left\lceil 2\frac{\log(\delta\frac{\|\widetilde{\mathbf{e}}\|}{\|P_{\mathcal{S}}(\mathbf{x})\|})}{\log(2/(\mu\alpha)-2)}\right\rceil, \tag{2.44}$$

which depends on the ratio $\delta\frac{\|\widetilde{\mathbf{e}}\|}{\|P_{\mathcal{S}}(\mathbf{x})\|}$, which is a form of signal to noise ratio, however, the signal component here is $P_{\mathcal{S}}(\mathbf{x})$, that is, the projection of the true signal onto the closest element in $\mathcal{S}$. Similarly, the error term $\widetilde{\mathbf{e}} = \Phi(\mathbf{x}-P_{\mathcal{S}}(\mathbf{x}))+\mathbf{e}$, does not only account for the observation noise $\mathbf{e}$, but also for the distance between the true signal and the model $\mathbf{x}-P_{\mathcal{S}}(\mathbf{x})$). The flexibility in the choice of δ in the theorem allows us furthermore to trade the number of iterations with approximation accuracy. Importantly, δ influences the error bound linearly (halving δ will decrease the error bound dependance on $\widetilde{\mathbf{e}}$ with a constant proportion) but it feeds into the required iteration count within the logarithm, so that a linear change in the approximation error only requires a logarithmic increase in computation time.[5]

Let us also look closer at the approximation error itself. This is made up of two error terms, $\|\widetilde{\mathbf{e}}\|$ and $\|P_{\mathcal{S}}(\mathbf{x})-\mathbf{x}\|$. The second one of these terms $\|P_{\mathcal{S}}(\mathbf{x})-\mathbf{x}\|$, is the distance between the true signal $\mathbf{x}$ and its best approximation with an element from $\mathcal{S}$. Clearly, all our estimates are from the set $\mathcal{S}$, so we will be unable to get an approximation that is better than $\|P_{\mathcal{S}}(\mathbf{x})-\mathbf{x}\|$. The second terms, $\widetilde{\mathbf{e}} = \Phi(\mathbf{x}-P_{\mathcal{S}}(\mathbf{x}))+\mathbf{e}$, is made up of two error contributions, the observation noise $\mathbf{e}$ and the error $(\mathbf{x}-P_{\mathcal{S}}(\mathbf{x}))$ again, but this time, after being mapped into the observation space. The fraction of this error we actually have to suffer depends on the number of iterations we use (through δ) and the constant $c \leq \frac{4}{3\alpha-2\mu}$, which is bounded by μ and α. As μ ultimately depends on β, the constant c thus depends on the bi-Lipschitz properties of Φ on $\mathcal{S}$.

---

[5] Note that for $\delta < \frac{\|P_{\mathcal{S}}(\mathbf{x})\|}{\|\widetilde{\mathbf{e}}\|}$ and for μ and α as in the theorem, both, the numerator and the denominator in the iteration count are negative numbers, so that a decrease in delta leads to an increase in the required number of iterations. If we were to choose δ such that the numerator becomes positive, we would get a *negative* number of iterations. This has to be interpreted as meaning that we actually don't need to run the algorithm at all, as the associated estimate error is already achieved by the estimate $\hat{\mathbf{x}} = \mathbf{x}^0 = \mathbf{0}$.

2.5.4 The Proof

Proof We now show how the above theorem can be derived using the geometrical ideas developed throughout this chapter. The derivation here follows that in [9]. Our aim is to bound the distance between the true signal $\mathbf{x}$ and its estimate $\hat{\mathbf{x}}^n$ after iteration n. To do this, we start with the trusted triangle inequality to split this vector into two components, the error between $\mathbf{x}$ and $\mathbf{x}_{\mathcal{S}} = P_{\mathcal{S}}(\mathbf{x})$ and the error between $\mathbf{x}_{\mathcal{S}} = P_{\mathcal{S}}(\mathbf{x})$ and $\hat{\mathbf{x}}^n$. This gives the bound

$$\|\mathbf{x} - \hat{\mathbf{x}}^n\|_2 \leq \|\mathbf{x}_{\mathcal{S}} - \hat{\mathbf{x}}^n\|_2 + \|\mathbf{x}_{\mathcal{S}} - \mathbf{x}\|_2. \tag{2.45}$$

We see that the term $\|\mathbf{x}_{\mathcal{S}} - \mathbf{x}\|_2$ is already the last term in our error bound in the theorem and, as discussed before, we can't expect to do better than this, so we are done with this term and instead concentrate on the first term, the length of $\mathbf{x}_{\mathcal{S}} - \hat{\mathbf{x}}^n$ which we will bound further. Our aim here will be to bound $\|\mathbf{x}_{\mathcal{S}} - \hat{\mathbf{x}}^n\|_2$ in terms of the length of the error in the previous iteration plus some extra error terms independent of $\hat{\mathbf{x}}$.

Note that both $\mathbf{x}_{\mathcal{S}}$ and $\hat{\mathbf{x}}^n$ lie within the set $\mathcal{S}$, so that we can use the bi-Lipschitz condition for both of these vectors, in particular, we have

$$\|\mathbf{x}_{\mathcal{S}} - \hat{\mathbf{x}}^n\|_2^2 \leq \frac{1}{\alpha}\|\Phi(\mathbf{x}_{\mathcal{S}} - \hat{\mathbf{x}}^n)\|_2^2. \tag{2.46}$$

If we now use the definition $\widetilde{\mathbf{e}} = \Phi(\mathbf{x} - \mathbf{x}_{\mathcal{S}}) + \mathbf{e}$, we see that $\Phi\mathbf{x}_{\mathcal{S}} - \Phi\hat{\mathbf{x}}^n = \Phi\mathbf{x}_{\mathcal{S}} - \Phi\hat{\mathbf{x}}^n + \Phi\mathbf{x} - \Phi\mathbf{x}_{\mathcal{S}} + \mathbf{e} - \Phi(\mathbf{x} - \mathbf{x}_{\mathcal{S}}) + \mathbf{e} = (\mathbf{y} - \Phi\hat{\mathbf{x}}^n) - \widetilde{\mathbf{e}}$. We can thus express the square of the length of $\Phi(\mathbf{x} - \mathbf{x}_{\mathcal{S}}) + \mathbf{e}$ as the sum of the square of the length of $\mathbf{y} - \Phi\hat{\mathbf{x}}^n$ and $\widetilde{\mathbf{e}}$.

$$\begin{aligned}\|\Phi(\mathbf{x}_{\mathcal{S}} - \hat{\mathbf{x}}^n)\|_2^2 &= \|\mathbf{y} - \Phi\hat{\mathbf{x}}^n\|_2^2 + \|\widetilde{\mathbf{e}}\|_2^2 - 2\langle\widetilde{\mathbf{e}}, (\mathbf{y} - \Phi\hat{\mathbf{x}}^n)\rangle \\ &\leq \|\mathbf{y} - \Phi\hat{\mathbf{x}}^n\|_2^2 + \|\widetilde{\mathbf{e}}\|_2^2 + \|\widetilde{\mathbf{e}}\|_2^2 + \|\mathbf{y} - \Phi\hat{\mathbf{x}}^n\|_2^2 \\ &= 2\|\mathbf{y} - \Phi\hat{\mathbf{x}}^n\|_2^2 + 2\|\widetilde{\mathbf{e}}\|_2^2,\end{aligned} \tag{2.47}$$

with the last inequality derived through the inequalities

$$\begin{aligned}-2\langle\widetilde{\mathbf{e}}, (\mathbf{y} - \Phi\hat{\mathbf{x}}^n)\rangle &= -\|\widetilde{\mathbf{e}} + (\mathbf{y} - \Phi\hat{\mathbf{x}}^n)\|_2^2 + \|\widetilde{\mathbf{e}}\|_2^2 + \|(\mathbf{y} - \Phi\hat{\mathbf{x}}^n)\|_2^2 \\ &\leq \|\widetilde{\mathbf{e}}\|_2^2 + \|(\mathbf{y} - \Phi\hat{\mathbf{x}}^n)\|_2^2.\end{aligned} \tag{2.48}$$

We are now ready to bound the first term in (2.47). This is done using the abbreviation $\mathbf{g}^{n-1} = 2\Phi^T(\mathbf{y} - \Phi\hat{\mathbf{x}}^{n-1})$ and the inequality

$$\|\mathbf{y} - \Phi\hat{\mathbf{x}}^n\|_2^2 \leq (\mu^{-1} - \alpha)\|(\mathbf{x}_{\mathcal{S}} - \hat{\mathbf{x}}^{n-1})\|_2^2 + \|\widetilde{\mathbf{e}}\|_2^2 + (\beta - \mu^{-1})\|\hat{\mathbf{x}}^n - \hat{\mathbf{x}}^{n-1}\|_2^2, \tag{2.49}$$

which is due to the inequality

$$
\begin{aligned}
&\|\mathbf{y} - \Phi\hat{\mathbf{x}}^n\|_2^2 - \|\mathbf{y} - \Phi\hat{\mathbf{x}}^{n-1}\|_2^2 \\
&\leq -\langle(\mathbf{x}_{\mathcal{S}} - \hat{\mathbf{x}}^{n-1}), \mathbf{g}^{n-1}\rangle + \mu^{-1}\|\mathbf{x}_{\mathcal{S}} - \hat{\mathbf{x}}^{n-1}\|_2^2 + (\beta - \mu^{-1})\|\hat{\mathbf{x}}^n - \hat{\mathbf{x}}^{n-1}\|_2^2 \\
&\leq -\langle(\mathbf{x}_{\mathcal{S}} - \hat{\mathbf{x}}^{n-1}), \mathbf{g}^{n-1}\rangle + \|\Phi(\mathbf{x}_{\mathcal{S}} - \hat{\mathbf{x}}^{n-1})\|_2^2 \\
&+ (\mu^{-1} - \alpha)\|\mathbf{x}_{\mathcal{S}} - \hat{\mathbf{x}}^{n-1}\|_2^2 + (\beta - \mu^{-1})\|\hat{\mathbf{x}}^n - \hat{\mathbf{x}}^{n-1}\|_2^2 \\
&= \|\widetilde{\mathbf{e}}\|_2^2 - \|\mathbf{y} - \Phi\hat{\mathbf{x}}^{n-1}\|_2^2 + (\mu^{-1} - \alpha)\|(\mathbf{x}_{\mathcal{S}} - \hat{\mathbf{x}}^{n-1})\|_2^2 + (\beta - \mu^{-1})\|\hat{\mathbf{x}}^n - \hat{\mathbf{x}}^{n-1}\|_2^2.
\end{aligned}
\tag{2.50}
$$

Here, the second inequality is due to the non-symmetric RIP whilst the first inequality follows from the lemma [9]

Lemma 5 *If* $\hat{\mathbf{x}}^n = H_k(\hat{\mathbf{x}}^{n-1} + \mu\Phi^T(\mathbf{y} - \Phi\hat{\mathbf{x}}^{n-1}))$, *then*

$$
\begin{aligned}
&\|\mathbf{y} - \Phi\hat{\mathbf{x}}^n\|_2^2 - \|\mathbf{y} - \Phi\hat{\mathbf{x}}^{n-1}\|_2^2 \\
&\leq -\langle(\mathbf{x}_{\mathcal{S}} - \hat{\mathbf{x}}^{n-1}), \mathbf{g}^{n-1}\rangle + \mu^{-1}\|\mathbf{x}_{\mathcal{S}} - \hat{\mathbf{x}}^{n-1}\|_2^2 + (\beta - \mu^{-1})\|\hat{\mathbf{x}}^n - \hat{\mathbf{x}}^{n-1}\|_2^2
\end{aligned}
\tag{2.51}
$$

We can now combine the inequalities (2.45), (2.46) and (2.49). If $\beta \leq \mu^{-1}$, then we have

$$
\|\mathbf{x}_{\mathcal{S}} - \hat{\mathbf{x}}^n\|_2^2 \leq 2\left(\frac{1}{\mu\alpha} - 1\right)\|(\mathbf{x}_{\mathcal{S}} - \hat{\mathbf{x}}^{n-1})\|_2^2 + \frac{4}{\alpha}\|\widetilde{\mathbf{e}}\|_2^2. \tag{2.52}
$$

This is exactly the ound we were looking for as now the error between $\mathbf{x}_{\mathcal{S}}$ and the current estimate is smaller than a fraction of the difference between $\mathbf{x}_{\mathcal{S}}$ and the previous estimate (plus some additional noise term). Because we also have the restriction that $2(\frac{1}{\mu\alpha} - 1) < 1$, so that if we replace $\|(\mathbf{x}_{\mathcal{S}} - \hat{\mathbf{x}}^{n-1})\|_2^2$ with the bound in terms of $\|(\mathbf{x}_{\mathcal{S}} - \hat{\mathbf{x}}^{m-2})\|_2^2$ and then $\|(\mathbf{x}_{\mathcal{S}} - \hat{\mathbf{x}}^{n-2})\|_2^2$ with the bound in terms of $\|(\mathbf{x}_{\mathcal{S}} - \hat{\mathbf{x}}^{m-3})\|_2^2$ and so on until we end up with a bound in terms of $\|(\mathbf{x}_{\mathcal{S}} - \hat{\mathbf{x}}^0)\|_2^2$, where we assume that $\hat{\mathbf{x}}^0 = \mathbf{0}$, then we have

$$
\|\mathbf{x}_{\mathcal{S}} - \hat{\mathbf{x}}^n\|_2^2 \leq \left(2\left(\frac{1}{\mu\alpha} - 1\right)\right)^n \|\mathbf{x}_{\mathcal{S}}\|_2^2 + c\|\widetilde{\mathbf{e}}\|_2^2, \tag{2.53}
$$

with $c \leq \frac{4}{3\alpha - 2\mu^{-1}}$. These arguments then lead to the claim in the theorem. To see this, we first bound the distance of $\mathbf{x}$ from our estimate at iteration n

$$
\begin{aligned}
\|\mathbf{x} - \hat{\mathbf{x}}^n\|_2 &\leq \sqrt{\left(\frac{2}{\mu\alpha} - 2\right)^n \|\mathbf{x}_{\mathcal{S}}\|_2^2 + c\|\widetilde{\mathbf{e}}\|_2^2} + \|\mathbf{x}_{\mathcal{S}} - \mathbf{x}\|_2 \\
&\leq \left(\frac{2}{\mu\alpha} - 2\right)^{n/2} \|\mathbf{x}_{\mathcal{S}}\|_2 + c^{0.5}\|\widetilde{\mathbf{e}}\|_2 + \|\mathbf{x}_{\mathcal{S}} - \mathbf{x}\|_2,
\end{aligned}
\tag{2.54}
$$

which shows that after $n^\star = \left\lceil 2\frac{\log(\|\widetilde{\mathbf{e}}\|_2/\|\mathbf{x}_{\mathcal{S}}\|_2)}{\log(2/(\mu\alpha)-2)} \right\rceil$ iterations we have

$$\|\mathbf{x} - \mathbf{x}^{n^\star}\|_2 \leq (c^{0.5} + 1)\|\widetilde{\mathbf{e}}\|_2 + \|\mathbf{x}_{\mathcal{S}} - \mathbf{x}\|_2. \tag{2.55}$$

2.6 Extensions to Non-linear Observation Models

We are here interested in the development of a better understanding of what happens to the compressed sensing recovery problem when a signal is measured with some non-linear system. In particular, the hope is that, if the system is not too non-linear, then recovery should still be possible under similar assumptions to those made in linear compressed sensing. Considering non-linear measurements is not only of academic interest but has important implications for many real-world sampling systems, where measurement system can often not be designed to be perfectly linear. Assume therefore that our measurements are described by a nonlinear mapping $\Phi(\cdot)$ that maps elements of the normed vector spaces $\mathcal{H}$ into the normed vector spaces $\mathcal{B}$. The observation model is therefore

$$\mathbf{y} = \Phi(\mathbf{x}) + \mathbf{e}, \tag{2.56}$$

where $\mathbf{e} \in \mathcal{B}$ is an unknown but bounded error term.

In order to keep our development as general as possible, we will allow the error between $\mathbf{y}$ and $\Phi(\mathbf{x})$ to be measured with some general norm, that is, whilst we assume that $\mathbf{x}$ is an element from some Hilbert spaces $\mathcal{H}$, $\mathbf{y}$ will be allowed to lie in a more general Banach space $\mathcal{B}$ with norm $\|\cdot\|_{\mathcal{B}}$. Whilst we have not yet derived a full understanding of this recovery problem, some progress has been made. For example, we could show that the Iterative Hard Thresholding algorithm can also solve quite general non-convex optimisation problems under general Union of Subspaces non-convex constraints, given that a condition similar to the bi-Lipschitz property holds [28].

2.6.1 The Iterative Hard Thresholding Algorithm for Nonlinear Optimisation Under Non-convex Constraints

We start by treating the problem in a quite general framework where we want to optimise a non-convex function $f(\mathbf{x})$ under the constraint that $\mathbf{x}$ lies in a union of subspaces $\mathcal{S}$. This optimisation will be done using the Iterative Hard Thresholding method and to do this, we will need to specify an update direction. For example, we could assume that $f(\mathbf{x})$ is Fréchet differentiable with respect to $\mathbf{x}$. The Fréchet derivative is an extension of differentiation to function spaces and is defined as follows. A function is Fréchet differentiable if for each $\mathbf{x}_1$ there exist a linear functional $D_{\mathbf{x}_1}(\cdot)$ such that

$$\lim_{\mathbf{h}\to 0} \frac{f(\mathbf{x}_1 + \mathbf{h}) - f(\mathbf{x}_1) - D_{\mathbf{x}_1}(\mathbf{h})}{\|\mathbf{h}\|} = 0. \tag{2.57}$$

So not to have to deal with an abstract linear functional, we will use the Riesz representation theorem [29] which tells us that for each linear functional, we can find an equivalent inner product representation. Thus, we can always find a function ∇ so that we can write the functional $D_{\mathbf{x}_1}(\cdot)$ as an inner product

$$D_{\mathbf{x}_1}(\cdot) = \langle \nabla(\mathbf{x}_1), \cdot \rangle. \tag{2.58}$$

$\nabla(\mathbf{x}_1) \in \mathcal{H}$ is now an element of our function space.

In situations in which the space $\mathcal{H}$ is Euclidean, the Fréchet derivative is the differential of $f(\mathbf{x})$ at $\mathbf{x}_1$, in which case $\nabla(\mathbf{x}_1)$ is the gradient and $\langle \cdot, \cdot \rangle$ the Euclidean inner product. To simplify the discussion, we will therefore abuse terminology and call $\nabla(\mathbf{x}_1)$ the gradient even in more general Hilbert space settings.

Once we have specified the update direction $\nabla(\mathbf{x})$, we are in a good position to define an algorithmic strategy to optimise $f(\mathbf{x})$. In particular, the Iterative Hard Thresholding algorithm for non-linear optimisation problems can now be written as

$$\mathbf{x}^{n+1} = P_{\mathcal{S}}(\mathbf{x}^n - (\mu/2)\nabla(\mathbf{x}^n)), \tag{2.59}$$

where $\mathbf{x}^0 = \mathbf{0}$ and μ is a step size parameter chosen to satisfy the condition in theorem below.

2.6.2 Some Theoretic Considerations

Unfortunately, we can not expect the method to work for all constraint sets and for all problems. To specify those problems that can be recovered, we use the following generalisation of the bi-Lipschitz property called the *Restricted Strong Convexity Property* (RSGP) which, to our knowledge, was first introduced in [30]. The *Restricted Strong Convexity Constants* α and β are the largest respectively smallest constants for which

$$\alpha \leq \frac{f(\mathbf{x}_1) - f(\mathbf{x}_2) - Re\langle \nabla(\mathbf{x}_2), (\mathbf{x}_1 - \mathbf{x}_2)\rangle}{\|\mathbf{x}_1 - \mathbf{x}_2\|^2} \leq \beta, \tag{2.60}$$

holds for all $\mathbf{x}_1, \mathbf{x}_2$ for which $\mathbf{x}_1 - \mathbf{x}_2 \in \mathcal{S}+\mathcal{S}$, where the set $\mathcal{S}+\mathcal{S} = \{\mathbf{x} = \mathbf{x}_a + \mathbf{x}_b : \mathbf{x}_a, \mathbf{x}_b \in \mathcal{S}\}$.

Note that the bi-Lipschitz property is recovered if $f(\mathbf{x}) = \|\mathbf{y} - \Phi\mathbf{x}\|_2^2$, where Φ is linear. Also note that the main result in the next section requires the *Restricted Strong Convexity Property* to hold for all vectors $\mathbf{x}_1$ and $\mathbf{x}_2$, such that $\mathbf{x}_1 - \mathbf{x}_2 \in \mathcal{S}+\mathcal{S}+\mathcal{S}$, where the set $\mathcal{S}+\mathcal{S}+\mathcal{S} = \{\mathbf{y} = \mathbf{x}_1 + \mathbf{x}_2 + \mathbf{x}_3 : \mathbf{x}_1, \mathbf{x}_2, \mathbf{x}_3 \in \mathcal{S}\}$.

The performance result now mirrors that derived for the linear compressed sensing setting and states that for $f(\mathbf{x})$ which satisfy the Restricted Strong Convexity Property, the iterative hard thresholding algorithm can be used to find a vector $\mathbf{x} \in \mathcal{S}$ that is close to the true minimiser of $f(\mathbf{x})$. The formal theorem reads as follows.

Theorem 4 *Let $\mathcal{S}$ be a union of subspaces. Given the optimisation problem $f(\mathbf{x})$, where $f(\mathbf{x})$ is a positive function that satisfies the* Restricted Strict Convexity Property

$$\alpha \leq \frac{f(\mathbf{x}_1)f - f(\mathbf{x}_2) - Re\langle \nabla(\mathbf{x}_2), (\mathbf{x}_1 - \mathbf{x}_2)\rangle}{\|\mathbf{x}_1 - \mathbf{x}_2\|^2} \leq \beta, \tag{2.61}$$

for all $\mathbf{x}_1, \mathbf{x}_2 \in \mathcal{H}$ for which $\mathbf{x}_1 - \mathbf{x}_2 \in \mathcal{S}+\mathcal{S}+\mathcal{S}$ with constants $\beta \leq \frac{1}{\mu} \leq \frac{4}{3}\alpha$, then, after

$$n^\star = 2\frac{\ln\left(\delta\frac{f(x_{\mathcal{S}})}{\|\mathbf{x}_{\mathcal{S}}\|}\right)}{\ln 4(1-\mu\alpha)}, \tag{2.62}$$

iterations, the Iterative Hard Thresholding Algorithm calculates a solution $\mathbf{x}^{n^\star}$ that satisfies

$$\|\mathbf{x}^{n^\star} - \mathbf{x}\| \leq \left(2\sqrt{\frac{\mu}{1-c}} + \delta\right) f(\mathbf{x}_{\mathcal{S}}) + \|\mathbf{x} - \mathbf{x}_{\mathcal{S}}\| + \sqrt{\frac{2}{1-c}}\epsilon, \tag{2.63}$$

where $\mathbf{x}_{\mathcal{S}} = \mathrm{argmin}_{\mathbf{x}\in\mathcal{S}} f(\mathbf{x})$.

2.6.3 *Proof of Theorem 4*

Proof The proof, first presented in [28], is based around a subspace Γ, which is defined as the sum of no more than three subspaces of $\mathcal{S}$, such that $\mathbf{x}_{\mathcal{S}}, \mathbf{x}^n, \mathbf{x}^{n+1} \in \Gamma$. We also define P_Γ to be the orthogonal projection onto the subspace Γ and use the short hand notation $\mathbf{a}^n_\Gamma = P_\Gamma \mathbf{a}^n$ and $P_\Gamma \nabla(\mathbf{x}^n) = \nabla_\Gamma(\mathbf{x}^n)$.

As in [28], we start by establishing a few basic equalities, which follow from the fact that for all orthogonal projections P, we have $\langle P\mathbf{x}_1, P\mathbf{x}_2\rangle = \langle \mathbf{x}_1, P\mathbf{x}_2\rangle$. As both $\mathbf{x}_{\mathcal{S}}$ and $\mathbf{x}^n$ lie in Γ we have

$$\begin{aligned} Re\langle \nabla_\Gamma(\mathbf{x}^n), (\mathbf{x}_{\mathcal{S}} - \mathbf{x}^n)\rangle &= Re\langle P_\Gamma\nabla(\mathbf{x}^n), (\mathbf{x}_{\mathcal{S}} - \mathbf{x}^n)\rangle \\ &= Re\langle \nabla(\mathbf{x}^n), P_\Gamma(\mathbf{x}_{\mathcal{S}} - \mathbf{x}^n)\rangle \\ &= Re\langle \nabla(\mathbf{x}^n), (\mathbf{x}_{\mathcal{S}} - \mathbf{x}^n)\rangle \end{aligned} \tag{2.64}$$

and

$$\begin{aligned} \|\nabla_\Gamma(\mathbf{x}^n)\|^2 = \langle \nabla_\Gamma(\mathbf{x}^n), \nabla_\Gamma(\mathbf{x}^n)\rangle &= \langle P_\Gamma\nabla(\mathbf{x}^n), P_\Gamma\nabla(\mathbf{x}^n)\rangle \\ &= \langle \nabla(\mathbf{x}^n), P^*_\Gamma P_\Gamma\nabla(\mathbf{x}^n)\rangle \\ &= \langle \nabla(\mathbf{x}^n), \nabla_\Gamma(\mathbf{x}^n)\rangle, \end{aligned} \tag{2.65}$$

We will also make use of the following lemma.

Lemma 6 *Under the assumptions of the theorem,*

$$\|\frac{\mu}{2}\nabla_\Gamma(\mathbf{x}^n)\|^2 - \mu f(\mathbf{x}^n) \leq 0. \tag{2.66}$$

Proof The lemma can be established as follows. Using the *Restricted Strict Convexity Property* we have

$$\begin{aligned}
\|\frac{\mu}{2}\nabla_\Gamma(\mathbf{x}^n)\|^2 &= -\frac{\mu}{2}Re\langle\nabla(\mathbf{x}^n), -\frac{\mu}{2}\nabla_\Gamma(\mathbf{x}^n)\rangle \\
&\leq \frac{\mu}{2}\beta\|\frac{\mu}{2}\nabla_\Gamma(\mathbf{x}^n)\|^2 + \frac{\mu}{2}f(\mathbf{x}^n) - \frac{\mu}{2}f(\mathbf{x}^n - \frac{\mu}{2}\nabla_\Gamma(\mathbf{x}^n))) \\
&\leq \frac{\mu}{2}\beta\|\frac{\mu}{2}\nabla_\Gamma(\mathbf{x}^n)\|^2 + \frac{\mu}{2}f(\mathbf{x}^n).
\end{aligned} \tag{2.67}$$

Thus

$$(2-\mu\beta)\|\frac{\mu}{2}\nabla_\Gamma(\mathbf{x}^n)\|^2 \leq \mu f(\mathbf{x}^n), \tag{2.68}$$

which is the desired result as $\mu\beta \leq 1$ by assumption.

The main point of the theorem is to bound the distance between the current estimate $\mathbf{x}^{n+1}$ and the optimal estimate $\mathbf{x}_\mathcal{S}$. To derive this bound, let us write $\mathbf{a}^n_\Gamma = \mathbf{x}^n_\Gamma - \mu/2\nabla_\Gamma(\mathbf{x}^n)$. We then note that $\mathbf{x}^{n+1}$ is, up to ϵ the closest element in $\mathcal{S}$ to $\mathbf{a}^n_\Gamma$, so that

$$\begin{aligned}
\|\mathbf{x}^{n+1} - \mathbf{x}_\mathcal{S}\|^2 &\leq \Big(\|\mathbf{x}^{n+1} - \mathbf{a}^n_\Gamma\| + \|\mathbf{a}^n_\Gamma - \mathbf{x}_\mathcal{S}\|\Big)^2 \\
&\leq 4\|(\mathbf{a}^n_\Gamma - \mathbf{x}_\mathcal{S})\|^2 + 2\epsilon \\
&= 4\|\mathbf{x}^n - (\mu/2)\nabla_\Gamma(\mathbf{x}^n) - \mathbf{x}_\mathcal{S}\|^2 + 2\epsilon \\
&= 4\|(\mu/2)\nabla_\Gamma(\mathbf{x}^n) + (\mathbf{x}_\mathcal{S} - \mathbf{x}^n)\|^2 + 2\epsilon \\
&= \mu^2\|\nabla_\Gamma(\mathbf{x}^n)\|^2 + 4\|\mathbf{x}_\mathcal{S} - \mathbf{x}^n\|^2 + 4\mu Re\langle\nabla_\Gamma(\mathbf{x}^n), (\mathbf{x}_\mathcal{S} - \mathbf{x}^n)\rangle + 2\epsilon \\
&= \mu^2\|\nabla_\Gamma(\mathbf{x}^n)\|^2 + 4\|\mathbf{x}_\mathcal{S} - \mathbf{x}^n\|^2 + 4\mu Re\langle\nabla(\mathbf{x}^n), (\mathbf{x}_\mathcal{S} - \mathbf{x}^n)\rangle + 2\epsilon \\
&\leq 4\|\mathbf{x}_\mathcal{S} - \mathbf{x}^n\|^2 + \mu^2\|\nabla_\Gamma(\mathbf{x}^n)\|^2 \\
&\quad + 4\mu[-\alpha\|\mathbf{x}^n - \mathbf{x}_\mathcal{S}\|^2 + f(\mathbf{x}_\mathcal{S}) - f(\mathbf{x}^n)] + 2\epsilon \\
&= 4(1-\mu\alpha)\|\mathbf{x}_\mathcal{S} - \mathbf{x}^n\|^2 + 4\mu f(\mathbf{x}_\mathcal{S}) + 2\epsilon \\
&\quad + 4[\|(\mu/2)\nabla_\Gamma(\mathbf{x}^n)\|^2 - \mu(\mathbf{x}^n)] \\
&\leq 4(1-\mu\alpha)\|\mathbf{x}_\mathcal{S} - \mathbf{x}^n\|^2 + 4\mu f(\mathbf{x}_\mathcal{S}) + 2\epsilon.
\end{aligned} \tag{2.69}$$

Here, the second to last inequality is the RSCP and the last inequality is due to lemma 6.

We could thus bound the difference between the current estimate and $\mathbf{x}_\mathcal{S}$ in terms of the previous estimate and $\mathbf{x}_\mathcal{S}$ plus some error term.

$$\|\mathbf{x}^{n+1} - \mathbf{x}_{\mathcal{S}}\|^2 \leq 4(1 - \mu\alpha)\|\mathbf{x}_{\mathcal{S}} - \mathbf{x}^n\|^2 + 4\mu f(\mathbf{x}_{\mathcal{S}}) + 2\epsilon. \tag{2.70}$$

If we define the constant $c = 4(1 - \mu\alpha)$ and iterate the above expression (i.e. use the same bound to bound the last error with the one before that and so on), then we see that

$$\|\mathbf{x}^n - \mathbf{x}_{\mathcal{S}}\|^2 \leq c^n \|\mathbf{x}_{\mathcal{S}}\|^2 + \frac{4\mu}{1-c} f(\mathbf{x}_{\mathcal{S}}) + \frac{2}{1-c}\epsilon, \tag{2.71}$$

where the constant $\frac{1}{1-c}$ in front of the error term is a bound of the geometric series $\sum_n c^n$ due to the iterative procedure. Importantly, if $\frac{1}{\mu} < \frac{4}{3}\alpha$ we have $c = 4(1 - \mu\alpha) < 1$, so that c^n decreases with n. Taking the square root on both sides and noting that for positive a and b, $\sqrt{a^2 + b^2} \leq a + b$, we then have

$$\|\mathbf{x}^n - \mathbf{x}_{\mathcal{S}}\| \leq c^{n/2}\|\mathbf{x}_{\mathcal{S}}\| + 2\sqrt{\frac{\mu}{1-c}} f(\mathbf{x}_{\mathcal{S}}) + \sqrt{\frac{2}{1-c}}\epsilon. \tag{2.72}$$

The theorem now follows using the triangle inequality

$$\begin{aligned} \|\mathbf{x}^n - \mathbf{x}\| \leq& \|\mathbf{x}^n - \mathbf{x}_{\mathcal{S}}\| + \|\mathbf{x} - \mathbf{x}_{\mathcal{S}}\| \\ \leq& c^{n/2}\|\mathbf{x}_{\mathcal{S}}\| + 2\sqrt{\frac{\mu}{1-c}} f(\mathbf{x}_{\mathcal{S}}) + \sqrt{\frac{2}{1-c}}\epsilon \\ &+ \|\mathbf{x} - \mathbf{x}_{\mathcal{S}}\| \end{aligned} \tag{2.73}$$

and the iteration count is determined by setting

$$c^{n/2}\|\mathbf{x}_{\mathcal{S}}\| \leq \delta(\mathbf{x}_{\mathcal{S}}). \tag{2.74}$$

so that after

$$n = 2\frac{\ln\left(\delta \frac{f(\mathbf{x}_{\mathcal{S}})}{\|\mathbf{x}_{\mathcal{S}}\|}\right)}{\ln c}, \tag{2.75}$$

iterations

$$\|\mathbf{x}^n - \mathbf{x}\| \leq \left(2\sqrt{\frac{\mu}{1-c}} + \delta\right) f(\mathbf{x}_{\mathcal{S}}) + \|\mathbf{x} - \mathbf{x}_{\mathcal{S}}\| + \sqrt{\frac{2}{1-c}}\epsilon. \tag{2.76}$$

2.6.4 An Important Caveat

Whilst this is an important result that shows how the Iterative Hard Thresholding algorithm can be used for many non-linear optimisation problems, it does not directly translate into a simple application to Compressed Sensing under non-linear observations. It seems tempting to use $f(\mathbf{x}) = \|\mathbf{y} - \Phi(\mathbf{x})\|_{\mathcal{B}}^2$, where $\|\cdot\|_{\mathcal{B}}$ is some Banach space norm and where $\Phi(\cdot)$ is some non-linear function. If this $f(\mathbf{x})$ would satisfy the Restricted Strict Convexity property, then we could clearly use the algorithm to solve the non-linear compressed sensing problem in which we are given noisy observations

$$\mathbf{y} = \Phi(\mathbf{x}) + \mathbf{e}. \tag{2.77}$$

Unfortunately, whilst properties such as the restricted strict convexity property hold for certain non-linear functions such as those encountered in certain logistic regression problems [31], it is far from clear under which conditions on $f(\mathbf{x}) = \|\mathbf{y} - \Phi(\mathbf{x})\|_{\mathcal{B}}^2$ similar properties would hold.

Indeed, the following lemma shows that such a condition cannot be fulfilled in general for Hilbert spaces.

Lemma 7 *Assume $\mathcal{B}$ is a Hilbert space and assume $f(\mathbf{x})$ is convex on $\mathcal{S} + \mathcal{S}$ for all $\mathbf{y}$ (i.e. it Satisfies the Restricted Strict Convexity Property), then Φ is affine on all subspaces of $\mathcal{S} + \mathcal{S}$.*

Proof The proof is by contradiction. Assume Φ is not affine on any subspace of $\mathcal{S}+\mathcal{S}$. Thus, there is a subspace $S = \mathcal{S}_i + \mathcal{S}_j$, and $\mathbf{x}_n \in S$, such that for $\mathbf{x} = \sum_n \lambda_n \mathbf{x}_n$, where $\sum_n \lambda_n = 1$ and $0 \leq \lambda_n$, we have $\sum_n \Phi(\mathbf{x}_n) - \Phi(\mathbf{x}) \neq \mathbf{0}$. Now by assumption of strong convexity on S, we have (using $\mathbf{y}_n = \Phi(\mathbf{x}_n)$ and $-\bar{\mathbf{y}} = \mathbf{x}$)

$$\begin{aligned} \mathbf{0} &\leq \sum_n \lambda_n \|\mathbf{y} - \Phi(\mathbf{x}_n)\|^2 - \|\mathbf{y} - \Phi(\mathbf{x})\|^2 = \sum_n \lambda_n \|\mathbf{y} - \mathbf{y}_n\|^2 - \|\mathbf{y} - \bar{\mathbf{y}}\|^2 \\ &= 2\langle \mathbf{y}, \bar{\mathbf{y}} - \sum_n \lambda_n \mathbf{y}_n \rangle + \sum_n \lambda_n \|\mathbf{y}_n\|^2 - \|\bar{\mathbf{y}}\|^2. \end{aligned} \tag{2.78}$$

where the inequality is due to the assumption of convexity. But the above inequality cannot hold for all $\mathbf{y}$ (it fails for example for a multiple of $-(\bar{\mathbf{y}} - \sum_n \lambda_n \mathbf{y}_n)$). Thus Φ needs to be affine on the linear subsets of $\mathcal{S} + \mathcal{S}$.

2.6.5 An Alternative Approach

Fortunately, the above result does not prevent the existence of Φ for which $\|\mathbf{y} - \Phi(\mathbf{x})\|_{\mathcal{B}}^2$ has the Restricted Strict Convexity Property for at least some $\mathbf{y}$. Alternatively, one could also envisage an approach where the linearisation error is dealt with by considering a local linear approximation to $\Phi(\mathbf{x})$ of the form $\Phi(\mathbf{x}) = \Phi_{\mathbf{x}^\star}\mathbf{x} + g_{\mathbf{x}^\star}(\mathbf{x})$, where $\Phi_{\mathbf{x}^\star}$ is linear and satisfies a form of the linear bi-Lipschitz condition. In this case, one would need to bound the error $g_{\mathbf{x}^\star}(\mathbf{x})$. If this can indeed be done, then similar recovery results to those available in the linear case would seem feasible also for non-linear problems.

For example, we have [32]

Theorem 5 *Assume that* $\mathbf{y} = \Phi(\mathbf{x}) + \mathbf{e}$ *and that* $\Phi_{\mathbf{x}^\star}$ *is a linearisation of* $\Phi(\cdot)$ *at* $\mathbf{x}^\star$ *(i.e. the Jacobian of* $\Phi(\mathbf{x})$ *evaluated at* $\mathbf{x}^\star$*) so that the Iterative Hard Thresholding algorithm uses the iteration* $\mathbf{x}^{n+1} = P_{\mathcal{S}}(\mathbf{x}^n + \Phi^*_{\mathbf{x}^n}(\mathbf{y} - \Phi(\mathbf{x}^n))$. *Assume that* $\Phi_{\mathbf{x}^\star}$ *satisfies RIP*

$$\alpha\|\mathbf{x}_1 - \mathbf{x}_2\|_2^2 \leq \alpha\|\Phi_{\mathbf{x}^\star}(\mathbf{x}_1 - \mathbf{x}_2)\|_2^2 \leq \beta\|\mathbf{x}_1 - \mathbf{x}_2\|_2^2 \tag{2.79}$$

for all $\mathbf{x}_1, \mathbf{x}_2, \mathbf{x}^\star \in \mathcal{S}$. *Define* $\epsilon_{\mathcal{S}} = sup_{\mathbf{x}\in\mathcal{S}}\|\mathbf{y} - \Phi_{\mathbf{x}}\mathbf{x}_{\mathcal{S}}\|_2$ *and let* $\mathbf{e}^n_{\mathcal{S}} = \mathbf{y} - \Phi_{\mathbf{x}^n}\mathbf{x}_{\mathcal{S}}$, *then after*

$$k^\star = \left\lceil 2\frac{\ln(\delta\frac{\|\mathbf{e}_{\mathcal{S}}\|}{\|\mathbf{x}_{\mathcal{S}}\|})}{\ln(2/(\mu\alpha) - 2)} \right\rceil \tag{2.80}$$

iterations we have

$$\|\mathbf{x} - \mathbf{x}^{k^\star}\| \leq (c^{0.5} + \delta)\|\mathbf{e}_{\mathcal{S}}\| + \|\mathbf{x}_{\mathcal{S}} - \mathbf{x}\| + \sqrt{\frac{c\epsilon}{2\mu}}, \tag{2.81}$$

Proof The proof is similar to the linear case, with a few minor variations. In particular, we introduce the error term $\mathbf{e}^n_{\mathcal{S}} = \mathbf{y} - \Phi(\mathbf{x}^n) - \Phi_{\mathbf{x}^n}(\mathbf{x}_{\mathcal{S}} - \mathbf{x}^n)$ to bound $\|\mathbf{x}_{\mathcal{S}} - \mathbf{x}^{n+1}\|^2$ using the expression

$$\begin{aligned}
&\|\mathbf{x}_{\mathcal{S}} - \mathbf{x}^{n+1}\|^2 \\
&\leq \frac{1}{\alpha}\|\Phi_{\mathbf{x}^n}(\mathbf{x}_{\mathcal{S}} - \mathbf{x}^{n+1})\|^2 \\
&= \frac{1}{\alpha}\|\mathbf{y} - \Phi(\mathbf{x}^n) - \Phi_{\mathbf{x}^n}(\mathbf{x}^{n+1} - \mathbf{x}^n) - (\mathbf{y} - \Phi(\mathbf{x}^n) - \Phi_{\mathbf{x}^n}(\mathbf{x}_{\mathcal{S}} - \mathbf{x}^n))\|^2 \\
&= \frac{1}{\alpha}\Big(\|\mathbf{y} - \Phi(\mathbf{x}^n) - \Phi_{\mathbf{x}^n}(\mathbf{x}^{n+1} - \mathbf{x}^n)\|^2 + \|\mathbf{e}^n_{\mathcal{S}}\|^2 \\
&\quad -2\langle\mathbf{e}^n_{\mathcal{S}}, (\mathbf{y} - \Phi(\mathbf{x}^n) - \Phi_{\mathbf{x}^n}(\mathbf{x}^{n+1} - \mathbf{x}^n))\rangle\Big) \\
&\leq \frac{2}{\alpha}\|\mathbf{y} - \Phi(\mathbf{x}^n) - \Phi_{\mathbf{x}^n}(\mathbf{x}^{n+1} - \mathbf{x}^n)\|^2 + \frac{2}{\alpha}\|\mathbf{e}^n_{\mathcal{S}}\|^2.
\end{aligned} \tag{2.82}$$

Again using similar ideas to those of the linear proof, we use $\mathbf{g} = 2\Phi^*_{\mathbf{x}^n}(\mathbf{y} - \Phi(x^n))$ and expand

$$\begin{aligned}
&\|\mathbf{y} - \Phi(\mathbf{x}^n) - \Phi_{\mathbf{x}^n}(\mathbf{x}^{n+1} - \mathbf{x}^n)\|^2 - \|\mathbf{y} - \Phi(\mathbf{x}^n)\|^2 \\
&= -\langle(\mathbf{x}^{n+1} - \mathbf{x}^n), \mathbf{g}\rangle + \|\Phi_{\mathbf{x}^n}(\mathbf{x}^{n+1} - \mathbf{x}^n)\|^2 \\
&\leq -\frac{2}{\mu}\langle(\mathbf{x}^{n+1} - \mathbf{x}^n), \frac{\mu}{2}\mathbf{g}\rangle + \frac{1}{\mu}\|(\mathbf{x}^{n+1} - \mathbf{x}^n)\|^2 \\
&= \frac{1}{\mu}\left[\|\mathbf{x}^{n+1} - \mathbf{x}^n - \frac{\mu}{2}\mathbf{g}\|^2 - \frac{\mu}{2}\|\mathbf{g}\|^2\right] \\
&\leq \frac{1}{\mu}\left[\inf_{\mathbf{x}\in\mathcal{S}}\|\mathbf{x} - \mathbf{x}^n - \frac{\mu}{2}\mathbf{g}\|^2 + \epsilon - \frac{\mu}{2}\|\mathbf{g}\|^2\right]
\end{aligned}$$

$$
\begin{aligned}
&= \inf_{\mathbf{x}\in\mathcal{S}} \left[-\langle(\mathbf{x}-\mathbf{x}^n),\mathbf{g}\rangle + \frac{1}{\mu}\|(\mathbf{x}-\mathbf{x}^n)\|^2 + \frac{\epsilon}{\mu} \right] \\
&\leq -\langle(\mathbf{x}_\mathcal{S}-\mathbf{x}^n),\mathbf{g}\rangle + \frac{1}{\mu}\|(\mathbf{x}_\mathcal{S}-\mathbf{x}^n)\|^2 + \frac{\epsilon}{\mu} \\
&= -2\langle(\mathbf{x}_\mathcal{S}-\mathbf{x}^n), \Phi^*_{\mathbf{x}^n}(\mathbf{y}-\Phi(\mathbf{x}^n))\rangle + \frac{1}{\mu}\|\mathbf{x}_\mathcal{S}-\mathbf{x}^n\|^2 + \frac{\epsilon}{\mu} \\
&= -2\langle(\mathbf{x}_\mathcal{S}-\mathbf{x}^n), \Phi^*_{\mathbf{x}^n}(\mathbf{y}-\Phi(\mathbf{x}^n))\rangle + \alpha\|\mathbf{x}_\mathcal{S}-\mathbf{x}^n\|^2 \\
&\quad + (\frac{1}{\mu}-\alpha)\|\mathbf{x}_\mathcal{S}-\mathbf{x}^n\|^2 + \frac{\epsilon}{\mu} \\
&\leq -2\langle(\mathbf{x}_\mathcal{S}-\mathbf{x}^n), \Phi^*_{\mathbf{x}^n}(\mathbf{y}-\Phi(\mathbf{x}^n))\rangle + \|\Phi_{\mathbf{x}^n}(\mathbf{x}_\mathcal{S}-\mathbf{x}^n)\|^2 \\
&\quad + (\frac{1}{\mu}-\alpha)\|\mathbf{x}_\mathcal{S}-\mathbf{x}^n\|^2 + \frac{\epsilon}{\mu} \\
&= \|\mathbf{y}-\Phi(\mathbf{x}^n)-\Phi_{\mathbf{x}^n}(\mathbf{x}_\mathcal{S}-\mathbf{x}^n)\|^2 - \|\mathbf{y}-\Phi(\mathbf{x}^n)\|^2 \\
&\quad + (\frac{1}{\mu}-\alpha)\|\mathbf{x}_\mathcal{S}-\mathbf{x}^n\|^2 + \frac{\epsilon}{\mu} \\
&= \|\mathbf{e}^n_\mathcal{S}\|^2 - \|\mathbf{y}-\Phi(\mathbf{x}^n)\|^2 + (\frac{1}{\mu}-\alpha)\|(\mathbf{x}_\mathcal{S}-\mathbf{x}^n)\|^2 + \frac{\epsilon}{\mu},
\end{aligned}
\tag{2.83}
$$

where the first inequality is due to the bi-Lipschitz property and the choice of $\beta \leq \frac{1}{\mu}$ and the second inequality is the definition of $\mathbf{x}^{n+1} = P^\epsilon_\mathcal{S}(\mathbf{x}^n + \frac{\mu}{2}\mathbf{g})$. The third inequality is due to the fact that $\mathbf{x}_\mathcal{S} \in \mathcal{S}$ whilst the last inequality is the bi-Lipschitz property again.

This gives the bound

$$
\|\mathbf{y}-\Phi(\mathbf{x}^n)-\Phi_{\mathbf{x}^n}(\mathbf{x}^{n+1}-\mathbf{x}^n)\|^2 \leq (\frac{1}{\mu}-\alpha)\|(\mathbf{x}_\mathcal{S}-\mathbf{x}^n)\|^2 + \|\mathbf{e}^n_\mathcal{S}\|^2 + \frac{\epsilon}{\mu}, \tag{2.84}
$$

so that

$$
\|\mathbf{x}_\mathcal{S}-\mathbf{x}^{n+1}\|^2 \leq 2\left(\frac{1}{\mu\alpha}-1\right)\|(\mathbf{x}_\mathcal{S}-\mathbf{x}^n)\|^2 + \frac{4}{\alpha}\|\mathbf{e}^n_\mathcal{S}\|^2 + \frac{2\epsilon}{\mu\alpha}. \tag{2.85}
$$

This again expresses the distance of $\mathbf{x}^{n+1}$ from $\mathbf{x}_\mathcal{S}$ in terms of the distance of the estimate $\mathbf{x}^n$ calculated in the previous iteration.

The condition of the theorem $(2(\frac{1}{\mu\alpha}-1) < 1)$ again allows us to iterate this expression so that

$$
\|\mathbf{x}_\mathcal{S}-\mathbf{x}^k\|^2 \leq \left(2\left(\frac{1}{\mu\alpha}-1\right)\right)^k \|\mathbf{x}_\mathcal{S}\|^2 + c\epsilon_\mathcal{S} + \frac{c\epsilon}{2\mu}, \tag{2.86}
$$

where $c \leq \frac{4}{3\alpha-2\frac{1}{\mu}}$.

In conclusion, using the square root of (2.86), we have thus shown that

$$
\begin{aligned}
\|\mathbf{x}-\mathbf{x}^k\| &\le \sqrt{\hat{c}^k\|\mathbf{x}_{\mathcal{S}}\|^2 + c\|\mathbf{e}_{\mathcal{S}}\|^2 + \frac{c\epsilon}{2\mu}} + \|\mathbf{x}_{\mathcal{S}}-\mathbf{x}\| \\
&\le \hat{c}^{k/2}\|\mathbf{x}_{\mathcal{S}}\| + c^{0.5}\|\mathbf{e}_{\mathcal{S}}\| + \sqrt{\frac{c\epsilon}{2\mu}} + \|\mathbf{x}_{\mathcal{S}}-\mathbf{x}\|,
\end{aligned}
$$

where $\hat{c} = \frac{2}{\mu\alpha} - 2$. The theorem directly follows from this.

2.7 Conclusions

The use of geometrical ideas in signal processing can often lead to new insights and solutions. This is particularly true in the field of sampling. Sampling, the transition from the continuous world of physical phenomena to the discretised world of concrete computation, fundamentally relies on approximations and these, in turn, must be based on prior assumptions that incorporate models of the physical world. Geometric descriptions of these models have over the years proven exceedingly useful, culminating in the recent ascend of compressed sensing. Here, geometric considerations have led to significant advances in signal reconstruction and interpretation, particularly in settings, where complex prior constraints are to be imposed.

In this chapter, we have provided an introductory tour of some of the underlying mathematical concepts that make up modern geometry, focussing especially on those aspects relevant to modern sampling theory. Building on this mathematical framework, several aspects of sampling, and in particular compressed sensing, have been explored. For example, we have shown how geometric ideas can be used to extend the sparse signal models traditionally used in compressed sensing to much more general union of subspaces models, which are much more widely applicable. Geometric interpretations were further shown to be fundamental in the development and understanding of algorithmic signal reconstruction strategies that try to solve optimisation problems that are constraint by these models. But not only do these ideas allow us to construct efficient algorithms, geometric insights are also likely to play a major role in future developments, such as those discussed here in the context of non-linear sampling.

Acknowledgments This work was supported in part by the UKs Engineering and Physical Science Research Council grants EP/J005444/1 and D000246/1 and a Research Fellowship from the School of Mathematics at the University of Southampton.

References

1. Nyquist H (1928) Certain topics in telegraph transmission theory. Trans AIEE 47:617–644
2. Shannon CA, Weaver W (1949) The mathematical theory of communication. University of Illinois Press, Urbana

3. Donoho DL (2006) For most large underdetermined systems of linear equations the minimal 1-norm solution is also the sparsest solution. Commun Pure Appl Math 59(6):797–829
4. Candès E, Romberg J (2006) Quantitative robust uncertainty principles and optimally sparse decompositions. Found Comput Math 6(2):227–254
5. Candès E, Romberg J, Tao T (2006) Robust uncertainty principles: exact signal reconstruction from highly incomplete frequency information. IEEE Trans Inform Theory 52(2):489–509
6. Candès E, Romberg J, Tao T (2006) Stable signal recovery from incomplete and inaccurate measurements. Commun Pure Appl Math 59(8):1207–1223
7. Abernethy J, Bach F, Evgeniou T, Vert J-P (2006) Low-rank matrix factorization with attributes. arxiv:0611124v1
8. Recht B, Fazel M, Parrilo PA (2009) Guaranteed minimum-rank solution of linear matrix equations via nuclear norm minimization. Found Comput Math 9:717–772
9. Blumensath T (2010) Sampling and reconstructing signals from a union of linear subspaces. IEEE Trans Inf Theory 57(7):4660–4671
10. Rudin W (1976) Principles of mathematical analysis, 3rd edn. McGraw-Hill Higher Education, New York
11. Conway JB (1990) A course in functional analysis. Graduate texts in mathematics, 2nd edn. Springer, Berlin
12. Unser M (2000) Sampling-50 years after Shannon. Proc IEEE 88(4):569–587
13. Landau HJ (1967) Necessary density conditions for sampling and interpolation of certain entire functions. Acta Math 117:37–52
14. Mishali M, Eldar YC (2009) Blind multi-band signal reconstruction: compressed sensing for analog signals. IEEE Trans Signal Process 57(3):993–1009
15. Vetterli M, Marziliano P, Blu T (2002) Sampling signals with finite rate of innovation. IEEE Trans Signal Process 50(6):1417–1428
16. Xu W, Hassibi B (to appear) Compressive sensing over the grassmann manifold: a unified geometric framework. IEEE Trans Inf Theory
17. Cands EJ (2006) The restricted isometry property and its implications for compressed sensing. Compte Rendus de l'Academie des Sciences, Serie I(346):589–592
18. Donoho DL (2006) High-dimensional centrally symmetric polytopes with neighborliness proportional to dimension. Discrete Comput Geom 35(4):617–652
19. Gruenbaum B (1968) Grassmann angles of convex polytopes. Acta Math 121:293–302
20. Gruenbaum B (2003) Convex polytopes. Graduate texts in mathematics, vol 221, 2nd edn. Springer-Verlag, New York
21. Blumensath T, Davies M (2008) Iterative thresholding for sparse approximations. J Fourier Anal Appl 14(5):629–654
22. Blumensath T, Davies M (2009) Iterative hard thresholding for compressed sensing. Appl Comput Harmon Anal 27(3):265–274
23. Qui K, Dogandzic A (2010) ECME thresholding methods for sparse signal reconstruction. arXiv:1004.4880v3
24. Cevher V, (2011) On accelerated hard thresholding methods for sparse approximation. EPFL Technical Report, February 17, 2011
25. Baraniuk RG (1999) Optimal tree approximation with wavelets. Wavelet Appl Sig Image Process VII 3813:196–207
26. Goldfarb D, Ma S (2010) Convergence of fixed point continuation algorithms for matrix rank minimization. arXiv:09063499v3
27. Needell D, Tropp JA (2008) CoSaMP: iterative signal recovery from incomplete and inaccurate samples. Appl Comput Harmon Anal 26(3):301–321
28. Blumensath T (2010) Compressed sensing with nonlinear observations. Technical report. http://eprints.soton.ac.uk/164753
29. Rudin W (1966) Real and complex analysis, McGraw-Hill, New York
30. Negahban S, Ravikumar P, Wainwright MJ, Yu B (2009) A unified framework for the analysis of regularized M-estimators. Advances in neural information processing systems, Vancouver, Canada

31. Bahmani S, Raj B, Boufounos P (2012) Greedy sparsity-constrained optimization. arXiv:1203.5483v1
32. Blumensath T (2012) Compressed sensing with nonlinear observations and related nonlinear optimisation problems. arXiv:1205.1650v1

Chapter 3
Sparse Signal Recovery with Exponential-Family Noise

Irina Rish and Genady Grabarnik

Abstract The problem of sparse signal recovery from a relatively small number of noisy measurements has been studied extensively in the recent compressed sensing literature. Typically, the signal reconstruction problem is formulated as l_1-regularized *linear* regression. From a statistical point of view, this problem is equivalent to maximum a posteriori probability (MAP) parameter estimation with Laplace prior on the vector of parameters (i.e., signal) and linear measurements disturbed by *Gaussian noise*. Classical results in compressed sensing (e.g., [7]) state sufficient conditions for accurate recovery of noisy signals in such linear-regression setting. A natural question to ask is whether one can accurately recover sparse signals under different noise assumptions. Herein, we extend the results of [7] to the general case of *exponential-family noise* that includes Gaussian noise as a particular case; the recovery problem is then formulated as l_1-regularized *Generalized Linear Model (GLM)* regression. We show that, under standard restricted isometry property (RIP) assumptions on the design matrix, l_1-minimization can provide stable recovery of a sparse signal in presence of exponential-family noise, and state some sufficient conditions on the noise distribution that guarantee such recovery.

3.1 Introduction

Accurate and computationally efficient recovery of sparse high-dimensional signals from low-dimensional linear measurements is the focus of compressed sensing, a rapidly developing area of signal processing [4–6, 8, 11, 13]. The ultimate goal

I. Rish (✉)
IBM T.J. Watson Research Center, Yorktown, NY, USA
e-mail: rish@us.ibm.com

G. Grabarnik
St. John's University, Queens, NY, USA
e-mail: genadyg@gmail.com

A. Y. Carmi et al. (eds.), *Compressed Sensing & Sparse Filtering*,
Signals and Communication Technology, DOI: 10.1007/978-3-642-38398-4_3,

of finding the sparsest solution satisfying a set of linear constraints is an NP-hard combinatorial problem involving cardinality, or l_0-norm, minimization. However, it is often possible to find an exact solution in a computationally efficient manner by approximating this combinatorial problem with its convex l_1-relaxation.

Herein, we focus on sparse signal recovery from *noisy* linear measurements, which is particularly important in real-life applications, such as image processing, sensor networks, biology, and medical imaging, just to name a few. This problem is typically formulated as minimization of the l_1-norm of an unobserved signal $\mathbf{x}$ subject to the sum-squared loss constraint $||\mathbf{y} - A\mathbf{x}||_{l_2} < \epsilon$ (see, for example, [7, 12]). From a probabilistic point of view, this problem is equivalent to log-likelihood maximization under the following assumptions: linear measurements disturbed by i.i.d Gaussian noise with unit variance $P(\mathbf{y}) \sim N(\mu = A\mathbf{x}, \Sigma = I)$, which results into the sum-squared (negative) log-likelihood, and the sparsity-promoting Laplace prior on the input signal, $p(\mathbf{x}) \sim e^{-\lambda||\mathbf{x}||_{l_1}}$, which produces the l_1-norm. However, in many practical applications, it might be more appropriate to use non-Gaussian models of noise: for example, Bernoulli or multinomial distributions are better suited for describing such measurements as (binary) failures or multilevel performance degradations of end-to-end test transactions ("probes") in a distributed computer systems [18, 21]; exponential distribution is better suited for describing nonnegative measurements such as end-to-end response time in such systems [3, 10]. Non-Gaussian observations, including binary, discrete, non-negative, etc., variables, are common in various other applications such as, for example, computational biology and medical imaging: e.g., predicting the presence or absence of a certain disease given DNA microarrays, or predicting various mental states (e.g., an emotional state of being angry, happy, anxious, etc.) from fMRI images [9, 14]. In these applications, sparsity constraint on the input promotes selection of most relevant variables, such as genes or brain areas, and thus improves model interpretability. Moreover, since the dimensionality of the input is often much higher than the number of samples (e.g., up to 100,000 variables corresponding to fMRI voxels, and a few hundred time points representing samples), sparsity constraint also serves as a regularizer that helps to avoid the overfitting problem.

In this paper, we will consider the exponential family of noise distributions that includes, besides Gaussian, many other commonly used distributions, such as exponential, Bernoulli, multinomial, gamma, chi-square, beta, Weibull, Dirichlet, and Poisson, just to name a few. The problem of recovering an unobserved vector $\mathbf{x}$ from the vector $\mathbf{y}$ of linear measurements $A\mathbf{x}$ contaminated by an exponential-family noise is known as *Generalized Linear Model (GLM)* regression. Solution to this problem maximizes the exponential-family loglikelhood of the observations (vector $\mathbf{y}$) with respect to the unobserved parameters (signal $\mathbf{x}$). This, in turn, is equivalent to minimizing the corresponding *Bregman divergence* $d(\mathbf{y}, \mu(A\mathbf{x}))$, where μ is the mean parameter of the exponential-family distribution, and $\theta = A\mathbf{x}$ is the corresponding natural parameter (there is a one-to-one correspondence between those two parameters). In the particular case of Gaussian likelihood we have $\mu = \theta$, and the corresponding Bregman divergence is just the squared Euclidean distance $||\mathbf{y} - A\mathbf{x}||_{l_2}^2$, assuming unit variance, i.i.d. samples. Adding l_1-norm constraint to GLM regression allows for an efficient method of sparse signal recovery, and is often used in statistical literature [17]. A natural question to ask next is whether a sparse signal can be recovered accurately when linear measurements are corrupted by an *exponential-family noise*. Herein, we answer this question positively and provide some conditions for stable sparse signal recovery from exponential-family observations. The results presented herein summarize our earlier work from [19]. Also, a more recent and closely related work by [15] is discussed at the end of this chapter.

We show that accurate recovery of sparse signals is possible when, for each measurement y_i, the log-partition function uniquely determining the corresponding exponential-family noise distribution (and thus its Legendre conjugate determining the corresponding Bregman divergence) has bounded second derivative. Also, for several specific members of the exponential family that do not always satisfy the bounded second derivative condition we provide separate proofs. Under proper conditions on the noise distribution, we show that the solution to the sparse GLM regression problem approximates the true signal $\mathbf{x}^0$ well in the l_2-sense if: (1) the true signal is sufficiently sparse, (2) the measurement noise is sufficiently small (where the noise is expressed as Bregman divergence between the measurement y and the mean μ^0 of the distribution determined by the natural parameter $\theta^0 = A\mathbf{x}^0$), and (3) the matrix A obeys the restricted isometry property (RIP) with appropriate RIP constant. Finally, we also extend to the case of exponential-family noise the results of [7] for compressible (approximately sparse) signals.

3.2 Background

3.2.1 Sparse Signal Recovery from Noisy Observations

We assume that $\mathbf{x}^0 \in R^m$ is an S-sparse signal, i.e. a signal with no more than S nonzero entries, where $S << m$. Let A be an n by m matrix that produces a vector of linear projections $\mathbf{y}^0 = A\mathbf{x}^0$, where $n << m$, and let and y be a vector of n noisy measurements that follow some noise distribution $P(\mathbf{y}|A\mathbf{x}^0)$. It is often assumed that A satisfies the so-called "restricted isometry property" (RIP) at the sparsity level S (or S-restricted isometry property), that essentially says that every subset of columns of A with cardinality less than S behaves like an almost orthonormal system. Formally, following [8]:

Definition 1 (*Restricted Isometry Property*) *Let A_T, where $T \subset \{1, ..., m\}$, denote an $n \times T$ submatrix of A that contains columns with indexes in T. The S-restricted isometry constant δ_S of A is the smallest quantity such that*

$$(1 - \delta_S)||c||_{l_2}^2 \leq ||A_T c||_{l_2}^2 \leq (1 + \delta_S)||c||_{l_2}^2 \tag{3.1}$$

for any all subsets T with $|T| \leq S$ and for any vector $(c_j)_{j\in T}$ defined over coordinates in T. The matrix A is said to satisfy the restricted isometry property if there exists such constant δ_S that the Eq. 3.1 is satisfied.

As it was shown, for example, in [8], if the following conditions is satisfied

$$\delta_S + \delta_{2S} + \delta_{3S} < 1,$$

then solving the l_1-minimization problem in Eq. 3.2 below can recover any signal $\mathbf{x}$ that is S-sparse (contains no more than S non-zero entries).

Our question is as follows: can one recover $\mathbf{x}^0$ from $\mathbf{y}$, given that noise is "sufficiently small" (to be defined precisely below)? This question has been answered in the compressed sensing literature for the particular case when the noise distribution is Gaussian. Indeed, as it was shown in [7], if (1) $||\mathbf{y} - A\mathbf{x}^0||_{l_2} \leq \epsilon$ (small noise assumption), (2) $\mathbf{x}^0$ is sufficiently sparse and the (3) matrix A obeys the restricted isometry property (RIP) with appropriate RIP constants, then the solution to the following l_1-minimization problem

$$\mathbf{x}^* = \arg\min_{\mathbf{x}} ||\mathbf{x}||_{l_1} \quad \mathit{subject\ to}\ ||\mathbf{y} - A\mathbf{x}||_{l_2} \leq \epsilon \tag{3.2}$$

approximates the true signal well. More formally, Theorem 1 in [7] states:

Theorem 1 [7] *Let S be such that $\delta_{3S} + 3\delta_{4S} < 2$, where δ_S is the S-restricted isometry constant of the matrix A, as defined above. Then for any signal $\mathbf{x}^0$ with the support $T^0 = \{t : x^0 \neq 0\}$, where $|T^0| \leq S$ and any noise vector (perturbation) e with $||e||_{l_2} \leq \epsilon$, the solution x^* to the problem in Eq. 3.2 obeys*

$$||\mathbf{x}^* - \mathbf{x}^0||_{l_2} \leq C_S \cdot \epsilon, \tag{3.3}$$

where the constant C_S may only depend on δ_{4S}. For reasonable values of δ_{4S}, C_S is well-behaved; e.g. $C_S \approx 8.82$ for $\delta_{4S} = 1/5$ and $C_S \approx 10.47$ for $\delta_{4S} = 1/4$.

Moreover, [7] show that (1) no other recovery method "can perform fundamentally better for arbitrary perturbations of size ϵ, i.e. even if an oracle would make the actual support T^0 of $\mathbf{x}^0$ available to us, making the problem well-posed, the least-squares solution $\hat{\mathbf{x}}$ (i.e., the maximum-likelihood solution which is optimal in the absence of any other information) would approximate the true signal $\mathbf{x}^0$ with the error proportional to ϵ".

Finally, [7] extend their result from sparse to approximately sparse vectors in the following

Theorem 2 [7] *Let $\mathbf{x}^0 \in R^m$ be an arbitrary vector, and let $\mathbf{x}^0_S$ be the truncated vector corresponding to the S largest values of $\mathbf{x}^0$ (in absolute value). Under the assumptions of Theorem 1, the solution $\mathbf{x}^*$ to the problem in Eq. 3.2 obeys*

$$||\mathbf{x}^* - \mathbf{x}^0||_{l_2} \leq C_{1,S} \cdot \epsilon + C_{2,S} \cdot \frac{||\mathbf{x}^0 - \mathbf{x}_S^0||_{l_1}}{\sqrt{S}}. \tag{3.4}$$

For reasonable values of δ_{4S} the constants above are well-behaved; e.g. $C_{1,S} \approx 12.04$ and $C_{2,S} \approx 8.77$ for $\delta_{4S} = 1/5$.

3.2.2 Exponential-Family Distributions and Bregman Divergences

We will now extend the standard compressed-sensing results to the case of general *exponential-family* noise distributions. Note that $||\mathbf{y} - A\mathbf{x}||_{l_2} \leq \epsilon$ constraint results from the negative log-likelihood of a Gaussian variable $y \sim N(\mu, \Sigma)$ with $\mu = A\mathbf{x}$ and $\Sigma = I$ (i.e., independent unit-variance noise):

$$-\log P(\mathbf{y}|A\mathbf{x}^0) = f(\mathbf{y}) + \frac{1}{2}||\mathbf{y} - A\mathbf{x}||_{l_2}^2. \tag{3.5}$$

Gaussian distribution is a particular member of the *exponential family* of distributions.

Definition 2 *An* **exponential family** *is a parametric family of probability distributions where the probability density has the form*

$$\log p_{\psi,\theta}(\mathbf{y}) = \mathbf{x}\theta - \psi(\theta) + \log p_0(\mathbf{y}), \tag{3.6}$$

where θ is called the **natural parameter**, *$\psi(\theta)$ is the (strictly convex and differentiable)* **cumulant function**, *or the* **log-partition function**, *that uniquely determines the member distribution of the exponential family, and $p_0(\mathbf{y})$ is a non-negative function called* **base measure** *that does not depend on the parameter θ.*

As shown by [2], there is a bijection between the exponential-family densities $p_{\psi,\theta}(\mathbf{y})$ and *Bregman divergences* $d_\phi(\mathbf{y}, \mu)$, so that each exponential-family density can be also expressed as

$$p_{\psi,\theta}(\mathbf{y}) = exp(-d_\phi(\mathbf{y}, \mu)) f_\phi(\mathbf{y}), \tag{3.7}$$

where $\mu = \mu(\theta) = E_{p_{\psi,\theta}}(Y)$ is the *expectation parameter* corresponding to θ, ϕ is the (strictly convex and differentiable) Legendre conjugate of ψ, $f_\phi(\mathbf{y})$ is a uniquely determined function, and $d_\phi(\mathbf{y}, \mu)$ is the corresponding Bregman divergence defined as follows.

Definition 3 *Given a strictly convex function $\phi : S \to \mathbb{R}$ defined on a convex set $S \subseteq \mathbb{R}$, and differentiable on the interior of S, $int(S)$ [20], the* **Bregman divergence** *$d_\phi : S \times int(S) \to [0, \infty)$ is defined as*

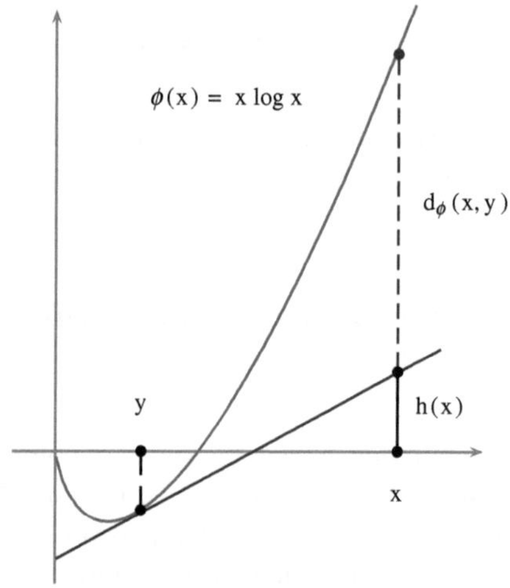

Fig. 3.1 KL-divergence

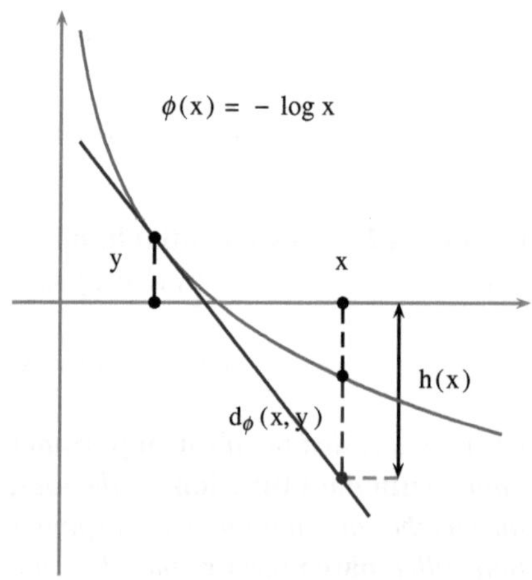

Fig. 3.2 Itakura-Saito distance

$$d_\phi(\mathbf{x}, \mathbf{y}) = \phi(\mathbf{x}) - \phi(\mathbf{y}) - < (\mathbf{x} - \mathbf{y}), \nabla\phi(\mathbf{y}) >, \tag{3.8}$$

where $\nabla\phi(\mathbf{y})$ *is the gradient of* ϕ.

In other words, the Bregman divergence can be thought of as the difference between the value of ϕ at point $\mathbf{x}$ and the value of the first-order Taylor expansion of ϕ around point $\mathbf{y}$ evaluated at point $\mathbf{x}$ (see Figs. 3.1 and 3.2, where $h(x) = \phi(y) + < (\mathbf{x} - \mathbf{y}), \nabla\phi(\mathbf{y}) >$).

Table 3.1 (derived from Tables 1 and 2 in [1]) shows particular examples of commonly used exponential-family distributions and their corresponding Bregman divergences. For example, the unit-variance Gaussian distribution leads to square loss, multivariate spherical Gaussian (diagonal covariance/independent variables)

Table 3.1 Examples of commonly-used exponential-family distributions and their corresponding Bregman divergences

Domain	Distribution	$p_\theta(y)$	μ	$\phi(\mu)$	$d_\phi(\mathbf{y}, \mu)$	Divergence
$\mathbb{R}$	$1D$ Gaussian	$\frac{1}{\sqrt{2\pi\sigma^2}} e^{-\frac{(x-a)^2}{2\sigma^2}}$	a	$\frac{1}{2\sigma^2}\mu^2$	$\frac{1}{2\sigma^2}(y-\mu)^2$	square loss
$\{0, 1\}$	Bernoulli	$q^y(1-q)^{1-y}$	q	$\mu \log \mu +$ $(1-\mu)\log(1-\mu)$	$y \log\left(\frac{y}{\mu}\right) +$ $(1-y)\log\left(\frac{1-y}{1-\mu}\right)$	Logistic loss
R_{++}	Exponential	$\lambda e^{-\lambda y}$	$1/\lambda$	$-\log \mu - 1$	$\frac{y}{\mu} - \log\left(\frac{y}{\mu}\right) - 1$	Itakura-Saito distance
n-simplex	nD Multinomial	$\frac{N!}{\prod_{j=1}^n y_j!} \prod_{j=1}^n q_j^{y_j}$	$[Nq_j]_{j=1}^{n-1}$	$\sum_{j=1}^n \mu_j \log\left(\frac{\mu_j}{N}\right)$	$\sum_{j=1}^n y_j \log\left(\frac{y_j}{\mu_j}\right)$	KL-divergence
$\mathbb{R}^n$	nD Sph. Gaussian	$\frac{1}{\sqrt{(2\pi\sigma^2)^n}} e^{-\frac{\|\mathbf{x}-\mathbf{a}\|_2^2}{2\sigma^2}}$	a	$\frac{1}{2\sigma^2}\|\mu\|_2^2$	$\frac{1}{2\sigma^2}\|\mathbf{y}-\mu\|_2^2$	Squared Euclidean distance
$\mathbb{R}^n$	nD Gaussian	$\frac{\sqrt{\det(C)}}{\sqrt{(2\pi)^n}} e^{-\frac{(\mathbf{y}-\mathbf{a})^T C (\mathbf{y}-\mathbf{a})}{2}}$	a	$\frac{\mu^T C \mu}{2}$	$\frac{(\mathbf{y}-\mu)^T C (\mathbf{y}-\mu)}{2}$	Mahalanobis distance[a]

[a]C is a symmetric positive definite matrix

gives rise to Euclidean distance, an multivariate Gaussian with the inverse-covariance (concentration) matrix C leads to Mahalanobis distance, Bernoulli distribution corresponds to logistic loss, exponential distribution leads to Itakura-Saito distance, while a multinomial distribution corresponds to the KL-divergence (relative entropy).

3.3 Main Results

We now extend the result in Theorem 1 to the case of exponential-family noise. Let us consider the following constrained l_1-minimization problem that generalizes the standard noisy compressed sensing problem of [7]:

$$\min_{\mathbf{x}} ||\mathbf{x}||_1 \; subject\ to \; \sum_i d(y_i, \mu(A_i\mathbf{x})) \leq \epsilon, \tag{3.9}$$

where $d(y_i, \mu(A_i\mathbf{x}))$ is Bregman divergence between the noisy observation y_i and the mean parameter of the corresponding exponential-family distribution with the natural parameter $\theta_i = A_i\mathbf{x}$. Note that using the Lagrangian form we can write the above problem as

$$\min_{\mathbf{x}} \lambda||\mathbf{x}||_1 + \sum_i d(y_i, \mu(A_i\mathbf{x})), \tag{3.10}$$

where the coefficient λ is the Lagrange multiplier uniquely determined by ϵ. This problem is known as an l_1-regularized *Generalized Linear Model (GLM)* regression, and includes as a particular case standard l_1-regularized linear regression, in which case $\mu(A_i\mathbf{x}) = A_i\mathbf{x}$ and the Bregman divergence simply reduces to the Euclidian distance.

We now show that, if: (1) the noise is small, (2) $\mathbf{x}^0$ is sufficiently sparse and the (3) matrix A obeys the restricted isometry property (RIP) with appropriate RIP constants, then the solution to the above problem approximates the true signal well:

Theorem 3 *Let S be such that $\delta_{3S} + 3\delta_{4S} < 2$, where δ_S is the S-restricted isometry constant of the matrix A, as defined above. Then for any signal $\mathbf{x}^0$ with the support $T^0 = \{t : \mathbf{x}^0 \neq 0\}$, where $|T^0| \leq S$, and for any vector $\mathbf{y} = (y_1, ..., y_n)$ of noisy linear measurements where*

1. *the noise follows exponential-family distributions $p_{\theta_i}(y_i)$, with the natural parameter $\theta_i = (A_{i,:}\mathbf{x}^0)$,*
2. *the noise is sufficiently small, i.e. $\forall i,\ d_{\phi_i}(y_i, \mu(A_{i,:}\mathbf{x}^0)) \leq \epsilon$, and*
3. *each function $\phi_i(\cdot)$ (i.e., the Legendre conjugate of the corresponding log-partition function, uniquely defining the Bregman divergence), satisfies the conditions imposed by at least one of the Lemmas below,*

the solution $\mathbf{x}^*$ to the problem in Eq. 3.9 obeys

$$||\mathbf{x}^* - \mathbf{x}^0||_{l_2} \leq C_S \cdot \delta(\epsilon), \tag{3.11}$$

where C_S is the constant from Theorem 1 of [7], and $\delta(\epsilon)$ is a continuous monotone increasing function of ϵ s.t. $\delta(0) = 0$ (and thus $\delta(\epsilon)$ is small when ϵ is small). A particular form of this function depends on particular members of exponential family.

Proof Following the proof of Theorem 1 in [7], we will only have to show that the "tube constraint" (condition 1) still holds (the rest of the proof remains unchanged), i.e. that

$$||A\mathbf{x}^* - A\mathbf{x}^0||_{l_2} \leq \delta(\epsilon) \tag{3.12}$$

where δ is some continuous monotone increasing function of ϵ, and $\delta(0) = 0$, so its small when ϵ is small. It was a trivial consequence of the triangle inequality in case of Euclidean distance; however, triangle inequality does not hold, in general, for Bregman divergences, and thus we must provide a different proof for the tube constraint, possibly for each type of Bregman divergence (exponential-family distribution). Since

$$||A\mathbf{x}^* - A\mathbf{x}^0||_{l_2}^2 = \sum_{i=1}^{m}(A_{i,:}\mathbf{x}^* - A_{i,:}\mathbf{x}^0)^2 = \sum_{i=1}^{m}(\theta_i^* - \theta_i^0)^2,$$

we will need to show that $|\theta_i^* - \theta_i^0| < \beta(\epsilon)$, where $\beta(\epsilon)$ is a continuous monotone increasing function of ϵ s.t. $\beta(0) = 0$ (and thus $\beta(\epsilon)$ is small when ϵ is small), then in Eq. 3.12 we get $\delta(\epsilon) = \sqrt{m \cdot \beta(\epsilon)}$. Lemma 1 provides the proof of this fact for a class of exponential-family distributions with bounded $\phi''(y)$ (where $\phi(y)$ is the Legendre conjugate of the log-partition function that uniquely determines the distribution). However, for several members of the exponential family (e.g., Bernoulli distribution) this condition is not satisfied, and those cases must be handled individually. Thus, we provide separate proofs for several different members of the exponential family in Lemmas 1, 2 and 3, and obtain particular expressions for $\beta(\epsilon)$ in each case. Note that for simplicity sake, we only consider univariate exponential-family distributions, corresponding to the case of independent noise for each measurement y_i, which was effectively assumed in standard problem formulation that used Euclidean distance corresponding to a spherical Gaussian distribution, i.e. a vector of independent Gaussian variables. However, Lemma 1 below can be extended from scalar to vector case, i.e. to multivariate exponential-family distributions that do not necessarily imply independent noise. Lemma 3 will provide a specific case of such distribution — a multivariate Gaussian with concentration matrix C.

The "cone constraint" part of the proof in [7] remains intact; it is easy to see that it does not depend on the particular constraint in the l_1-minimization problem 3.10, and only makes use of the sparsity of x^0 and l_1-optimality of $\mathbf{x}^*$. Thus, we can simply substitute $||Ah||_{l_2}$ by $\delta(\epsilon)$ in Eq. 13 on page 8 in the proof of Theorem 1 of [7], or, equivalently, replace 2ϵ (that was shown to bound $||Ah||_{l_2}$) by $\delta(\epsilon)$ in the Eq. 14.

Just like for the sparse signal case (Theorem 1 in [7]), the only change we have to make in the proof of the Theorem 2 (general case of approximable, rather than sparse, signals), when generalizing it from Euclidean distance to Bregman divergence in Eq. 3.10, is the tube constraint. Thus, once we showed it for the Theorem 3 above, the generalization to approximable signals follows automatically:

Theorem 4 *Let* $\mathbf{x}^0 \in R^m$ *be an arbitrary vector, and let* $\mathbf{x}^0_S$ *be the truncated vector corresponding to the S largest values of* $\mathbf{x}^0$ *(in absolute value). Under the assumptions of Theorem 3, the solution* $\mathbf{x}^*$ *to the problem in Eq. 3.10 obeys*

$$||\mathbf{x}^* - \mathbf{x}^0||_{l_2} \leq C_{1,S} \cdot \delta(\epsilon) + C_{2,S} \cdot \frac{||\mathbf{x}^0 - x^0_S||_{l_1}}{\sqrt{S}}. \tag{3.13}$$

where $C_{1,S}$ *and* $C_{2,S}$ *are the constants from Theorem 2 of [7], and* $\delta(\epsilon)$ *is a continuous monotone increasing function of* ϵ *s.t.* $\delta(0) = 0$ *(and thus* $\delta(\epsilon)$ *is small when* ϵ *is small). A particular form of this function depends on particular members of exponential family.*

The following lemma states the sufficient conditions for the "tube constraint" in Eq. 3.12 to hold in general case of arbitrary exponential-family noise, provided that $\phi''(y)$ exists and is bounded on the appropriate intervals.

Lemma 1 *Let y denote a random variable following an exponential-family distribution* $p_\theta(y)$*, with the natural parameter* θ*, and the corresponding mean parameters* $\mu(\theta)$*. Let* $d_\phi(y, \mu(\theta))$ *denote the Bregman divergence associated with this distribution. If*

1. $d_\phi(y, \mu^0(\theta^0)) \leq \epsilon$ *(small noise),*
2. $d_\phi(y, \mu^*(\theta^*)) \leq \epsilon$ *(constraint in GLM problem Eq. 3.10), and*
3. $\phi''(y)$ *exists and is bounded on* $[y_{min}, y_{max}]$*, where* $y_{min} = \min\{y, \mu^0, \mu^*\}$ *and* $y_{max} = \max\{y, \mu^0, \mu^*\}$*,*

then

$$|\theta^* - \theta^0| \leq \beta(\epsilon) = \sqrt{\epsilon} \cdot \frac{2\sqrt{2}\max_{\hat{\mu}\in[\mu^*;\mu^0]} |\phi''(\hat{\mu})|}{\sqrt{\min_{\hat{y}\in[y_{min};y_{max}]} \phi''(\hat{y})}} \tag{3.14}$$

Proof We prove the lemma in two steps: first, we show that $|\mu^*(\theta^*) - \mu^0(\theta^0)|$ is small if ϵ is small, and then infer $|\theta^* - \theta^0|$ is small.

1. By definition in Eq. 3.8, Bregman divergence is the non-linear tail of the Taylor expansion of $\phi(y)$ at point μ, i.e., the *Lagrange remainder* of the linear approximation:

$$d_\phi(y, \mu) = \phi''(\hat{y})(y - \mu)^2/2, \quad \hat{y} \in [y_1; y_2],$$

$$\text{where } y_1 = \min\{y, \mu\},\ \ y_2 = \max\{y, \mu\}.$$

Let $y_1^0 = \min\{y, \mu^0\}$, $y_2^0 = \max\{y, \mu^0\}$ and $y_1^* = \min\{y, \mu^*\}$, $y_2^* = \max\{y, \mu^*\}$. Using the conditions $0 \le d_\phi(y, \mu^0) \le \epsilon$ and $0 \le d_\phi(y, \mu^*) \le \epsilon$, and observing that

$$\min_{\hat{y}\in[y_{min};y_{max}]} \phi''(\hat{y}) \le \min_{\hat{y}\in[y_1^0;y_2^0]} \phi''(\hat{y})$$

$$\text{and} \quad \min_{\hat{y}\in[y_{min};y_{max}]} \phi''(\hat{y}) \le \min_{\hat{y}\in[y_1^*;y_2^*]} \phi''(\hat{y}),$$

we get

$$\phi''(\hat{y})(y-\mu^0)^2/2 \le \epsilon \;\Leftrightarrow\; (y-\mu^0)^2 \le \frac{2\epsilon}{\phi''(\hat{y})} \;\Leftrightarrow$$

$$\Leftrightarrow |y-\mu^0| \le \frac{\sqrt{2\epsilon}}{\sqrt{\min_{\hat{y}\in[y_1^0;y_2^0]} \phi''(\hat{y})}} \le$$

$$\le \frac{\sqrt{2\epsilon}}{\sqrt{\min_{\hat{y}\in[y_{min};y_{max}]} \phi''(\hat{y})}}$$

$$\text{and, similarly, } |y-\mu^*| \le \frac{\sqrt{2\epsilon}}{\sqrt{\min_{\hat{y}\in[y_1^*;y_2^*]} \phi''(\hat{y})}} \le$$

$$\le \frac{\sqrt{2\epsilon}}{\sqrt{\min_{\hat{y}\in[y_{min};y_{max}]} \phi''(\hat{y})}},$$

from which, using the triangle inequality, we conclude

$$|\mu^* - \mu^0| \le |y-\mu^*| + |y-\mu^0| \le$$

$$\le \frac{2\sqrt{2\epsilon}}{\sqrt{\min_{\hat{y}\in[y_{min};y_{max}]} \phi''(\hat{y})}} \tag{3.15}$$

Note that $\phi''(\hat{y})$ under the square root is always positive since ϕ is strictly convex.

2. The mean and the natural parameters of an exponential-family distribution relate to each other as follows: $\theta(\mu) = \phi'(\mu)$ (respectively, $\theta(\mu) = \nabla\phi(\mu)$ for vector μ), where $\phi'(\mu)$ is called the *link function*. Therefore, we can write

$$|\theta^* - \theta^0| = |\phi'(\mu^*) - \phi'(\mu^0)| = |\phi''(\hat{\mu})(\mu^* - \mu^0)|,$$
$$\text{where } \hat{\mu} \in [\mu^*; \mu^0],$$

and thus, using the above result in Eq. 3.15, we get

$$|\theta^* - \theta^0| \leq \beta(\epsilon) = \sqrt{\epsilon} \cdot \frac{2\sqrt{2} \max_{\hat{\mu} \in [\mu^*; \mu^0]} |\phi''(\hat{\mu})|}{\sqrt{\min_{\hat{y} \in [y_{min}; y_{max}]} \phi''(\hat{y})}}$$

which concludes the proof.

The condition (3) in the above lemma requires that $\phi''(y)$ exists and is bounded on the intervals between y and both μ^0 and μ^*. However, even when this condition is not satisfied, as it happens for the logistic loss, where $\phi''(y) = \frac{1}{y(1-y)}$ is unbounded at 0 and 1, and for several other Bregman divergences shown in Table 3.1, we may still be able to prove similar results using specific properties of each $\phi(y)$, as shown by the following lemmas.

Lemma 2 (Bernoulli noise/Logistic loss)
Let the conditions (1) and (2) of Lemma 1 be satisfied, and let $\phi(y) = y \log y + (1-y)\log(1-y)$, which corresponds to the logistic-loss Bregman divergence and Bernoulli distribution $p(y) = \mu^y(1-\mu)^{1-y}$, where the mean parameter $\mu = P(y=1)$. We assume that $0 < \mu^* < 1$, and $0 < \mu^0 < 1$. Then

$$|\theta^0 - \theta^*| \leq \beta(\epsilon) = 4\epsilon.$$

Proof Using the definition of the logistic-loss Bregman divergence from Table 3.1, and the conditions (1) and (2) of Lemma 1, we can write:

$$d_\phi(y, \mu^0) = y \log(\frac{y}{\mu_0}) + (1-y)\log(\frac{1-y}{1-\mu^0}) \leq \epsilon,$$
$$d_\phi(y, \mu^*) = y \log(\frac{y}{\mu_*}) + (1-y)\log(\frac{1-y}{1-\mu^*}) \leq \epsilon, \tag{3.16}$$

which implies

$$|d_\phi(y, \mu^0) - d_\phi(y, \mu^*)| \leq 2\epsilon, \tag{3.17}$$

and, after substituting the expressions 3.16 into Eq. 3.17, and simplifying, we get

$$|y \log(\frac{\mu^0}{\mu^*}) + (1-y)\log(\frac{1-\mu^0}{1-\mu^*})| \leq 2\epsilon. \tag{3.18}$$

The above must be satisfied for each $y \in \{0, 1\}$ (the domain of Bernoulli distribution). Thus, we get:

$$(1)\ |\log(\frac{1-\mu^0}{1-\mu^*})| \le 2\epsilon \ \text{ if } y=0, \text{ and}$$
$$(2)\ |\log(\frac{\mu^0}{\mu^*})| \le 2\epsilon \ \text{ if } y=1, \tag{3.19}$$

or, equivalently

$$(1)\ e^{-2\epsilon} \le \frac{1-\mu^0}{1-\mu^*} \le e^{2\epsilon} \ \text{ if } y=0, \text{and}$$
$$(2)\ e^{-2\epsilon} \le \frac{\mu^0}{\mu^*} \le e^{2\epsilon} \ \text{ if } y=1.$$

Let us first consider the case of $y=0$; subtracting 1 from the corresponding inequalities yields

$$e^{-2\epsilon}-1 \le \frac{\mu^*-\mu^0}{1-\mu^*} \le e^{2\epsilon}-1 \Leftrightarrow$$
$$\Leftrightarrow (1-\mu^*)(e^{-2\epsilon}-1) \le \mu^*-\mu^0 \le (1-\mu^*)(e^{2\epsilon}-1).$$

By the mean value theorem, $e^x-1=e^x-e^0=\frac{d(e^x)}{dx}|_{\hat{x}}\cdot(x-0)=e^{\hat{x}}x$, for some $\hat{x}\in[0,x]$ if $x>0$, or for some $\hat{x}\in[x,0]$ if $x<0$. Thus, $e^{-2\epsilon}-1=-e^{\hat{x}}\cdot 2\epsilon$, for some $\hat{x}\in[-2\epsilon,0]$, and since e^x is a continuous monotone increasing function, $e^{\hat{x}}\le 1$ and thus $e^{-2\epsilon}-1\ge -2\epsilon$. Similarly, $e^{2\epsilon}-1=e^{\hat{x}}\cdot 2\epsilon$, for some $\hat{x}\in[0,2\epsilon]$, and since $e^{\hat{x}}\le e^{2\epsilon}$, we get $e^{2\epsilon}-1\le 2\epsilon\cdot e^{2\epsilon}$. Thus,

$$-2\epsilon(1-\mu^*) \le \mu^*-\mu^0 \le 2\epsilon e^{2\epsilon}(1-\mu^*) \Rightarrow$$
$$\Rightarrow |\mu^*-\mu^0| \le 2\epsilon\cdot e^{2\epsilon}. \tag{3.20}$$

Similarly, in case of $y=1$, we get

$$e^{-2\epsilon}-1 \le \frac{\mu^0-\mu^*}{\mu^*} \le e^{2\epsilon}-1.$$

and can apply same derivation as above, and get same result for $|\mu^*-\mu^0|$ as in Eq. 3.20. Finally, since $\theta(\mu)=\phi'(\mu)=\log(\frac{\mu}{1-\mu})$, we get

$$|\theta^0-\theta^*| = |\log(\frac{\mu^0}{1-\mu^0})-\log(\frac{\mu^*}{1-\mu^*})| =$$
$$= |\log(\frac{\mu^0}{\mu^*})-\log(\frac{1-\mu^0}{1-\mu^*})|.$$

From the Eq. 3.19 we get $|\log(\frac{\mu^0}{\mu^*})|\le 2\epsilon$ and $|\log(\frac{1-\mu^0}{1-\mu^*})|\le 2\epsilon$, which implies

$$|\theta^0 - \theta^*| = |\log(\frac{\mu^0}{\mu^*}) - \log(\frac{1-\mu^0}{1-\mu^*})| \leq 4\epsilon.$$

Lemma 3 (Exponential noise/Itakura-Saito distance)
Let the conditions (1) and (2) of Lemma 1 be satisfied, and let $\phi(y) = -\log\mu - 1$, which corresponds to the Itakura-Saito distance $d_\phi(y,\mu) = \frac{y}{\mu} - \log(\frac{y}{\mu}) - 1$ and exponential distribution $p(y) = \lambda e^{\lambda y}$, where the mean parameter $\mu = 1/\lambda$. We will also assume that the mean parameter is always separated from zero, i.e. $\exists c_\mu > 0$ such that $\mu \geq c_\mu$. Then

$$|\theta^* - \theta^0| \leq \beta(\epsilon) = \frac{\sqrt{6\epsilon}}{c_\mu}.$$

Proof To establish the result of the lemma we start with inequality $|u - \log u - 1| \leq \epsilon$, where u is $\frac{y}{\mu}$. Replacing u by $z = u - 1, z > -1$ gives us $|z - \log(1+z)| \leq \epsilon$. Without loss of generality, let us assume that $\epsilon \leq \frac{1}{18}$. Then the Taylor decomposition of function $z - \log(1+z)$ at the point $z = 0$

$$z - \log(1+z) = \frac{z^2}{2} - \frac{z^3}{3} + \frac{\theta^4}{4}, \text{ for } \theta \in [0,z] \text{ or } [z,0]$$

implies that

$$\epsilon \geq z - \log(1+z) \geq \frac{z^2}{2} - \frac{z^3}{3} \text{ (since } \frac{\theta^4}{4} \geq 0 \text{).}$$

This, in turns, implies that $z \leq \frac{1}{3}$ and $\frac{z^2}{2} - \frac{z^3}{3} \geq \frac{z^2}{6}$ for $0 \leq z \leq \frac{1}{3}$. Hence

$$z - \log(1+z) \geq \frac{z^2}{2} \text{ for } -\frac{1}{3} \leq z \leq 0, \tag{3.21}$$

$$z - \log(1+z) \geq \frac{z^2}{6} \text{ for } 0 \leq z \leq \frac{1}{3}. \tag{3.22}$$

Combining together both estimates we get $|z| \leq \sqrt{6\epsilon}$, or

$$|y - \mu| \leq \sqrt{6\epsilon} \cdot \mu,$$

and

$$|\mu^0 - \mu^*| \leq \sqrt{6\epsilon} \cdot \max\{\mu^0, \mu^*\}.$$

Then

$$|\theta^* - \theta^0| = |\frac{1}{\mu^0} - \frac{1}{\mu^*}| = |\frac{\mu^* - \mu^0}{\mu^*\,\mu^0}| \leq \frac{\sqrt{6\epsilon}}{\min\{\mu^*, \mu^0\}} \leq \frac{\sqrt{6\epsilon}}{c_\mu},$$

since by the assumption of the lemma $\min\{\mu^*, \mu^0\} \geq c_\mu$.

We now consider multivariate exponential-family distributions; the next lemma handles the general case of a multivariate Gaussian distribution (not necessarily spherical one that had a diagonal covariance matrix and corresponded to the standard Euclidean distance (see Table 3.1).

Lemma 4 (Non-i.i.d. Multivariate Gaussian noise/Mahalanobis distance) *Let $\phi(\mathbf{y}) = \mathbf{y}^T C\mathbf{y}$, which corresponds to the general multivariate Gaussian with concentration matrix C, and Mahalanobis distance $d_\phi(y,\mu) = \frac{1}{2}(\mathbf{y}-\mu)^T C(\mathbf{y}-\mu)$. If $d_\phi(y,\mu^0) \le \epsilon$ and $d_\phi(y,\mu^*) \le \epsilon$, then*

$$||\theta^0 - \theta^*|| \le \sqrt{2\epsilon}||C^{-1}||^{1/2} \cdot ||C||,$$

where $||C||$ is the operator norm.

Proof Since C is (symmetric) positive definite, it can be written as $C = L^T L$ where L defines a linear operator on $\mathbf{y}$ space, and thus

$$\begin{aligned}\epsilon/2 &\ge (\mathbf{y}-\mu)^T C(\mathbf{y}-\mu) = (L(\mathbf{y}-\mu))^T(L(\mathbf{y}-\mu)) = \\ &= ||L(\mathbf{y}-\mu)||^2.\end{aligned}$$

Also, it is easy to show that $||C^{-1}||I \le C \le ||C||I$ (where $||B||$ denote the operator norm of B), and that

$$\begin{aligned}&\epsilon/2 \ge ||L(\mathbf{y}-\mu)||^2 \ge ||L^{-1}||^{-2}||\mathbf{y}-\mu||^2 \Rightarrow \\ &\Rightarrow ||\mathbf{y}-\mu|| \le \sqrt{\frac{\epsilon}{2}}||L^{-1}||.\end{aligned}$$

Then, using triangle inequality, we get

$$||\mu^* - \mu^0|| \le ||\mathbf{y}-\mu^0|| + ||\mathbf{y}-\mu^*|| \le \sqrt{2\epsilon}||L^{-1}||.$$

Finally, since $\theta(\mu) = \nabla\phi(\mu) = C\mu$, we get

$$\begin{aligned}||\theta^0 - \theta^*|| &= ||C\mu^0 - C\mu^*|| \le ||C|| \cdot ||\mu^0 - \mu^*|| = \\ &= ||C|| \cdot ||\mu^0 - \mu^*|| \le \sqrt{2\epsilon}||L^{-1}|| \cdot ||C||.\end{aligned}$$

Note that $||L^{-1}|| = ||C^{-1}||^{1/2}$, which concludes the proof.

3.4 Discussion and Conclusions

In this paper, we extend the results of [7] to the more general case of *exponential-family noise* that includes Gaussian noise as a particular case, and yields l_1-regularized *Generalized Linear Model (GLM)* regression problem. We show that, under standard restricted isometry property (RIP) assumptions on the design matrix, l_1-minimization can provide a stable recovery of a sparse signal under exponential-family noise assumptions, provided that the noise is sufficiently small and the distribution satisfies certain (sufficient) conditions, such as bounded second derivative of the Legendre conjugate $\phi(y)$ of the log-partition function that uniquely determines the distribution. We also provide distribution-specific proofs for several members of exponential family that do not satisfy the above conditions. Moreover, we show that the results of [7] for a more general case of compressible (rather than sparse) signals can be extended to the exponential-family noise in a similar way.

As we mentioned before, the results presented here are based on our earlier work in [19]. A more recent work by [15] (and its extended version [16]) is closely related to ours as it presents a unifying framework for analysis of regularized maximum-likelihood estimators (M-estimators), and states sufficient conditions that guarantee asymptotic recovery (i.e. consistency) of sparse model's parameters (i.e., sparse signals). These general conditions are: *decomposability* of the regularizer (which is satisfied for l_1-norm), and *restricted strong convexity (RSC)* of the loss function, given a regularizer. Generalized linear models are considered as a special case, and consistency results from GLMs are derived from the main result using the above two sufficient conditions. Since the l_1-regularizer is decomposable, the main challenge is establishing RSC for the exponential-family negative log-likelihood loss, and this is achieved by imposing two (sufficient) conditions, called GLM1 and GLM2, on the design matrix and on the exponential-family distribution, respectively. Bricfly, GLM1 condition requires the rows of the design matrix to be i.i.d. samples with sub-Gaussian behavior, and GLM2 condition includes as one of the alternative sufficient conditions a uniformly bounded second derivative of the cumulant function, similar to our Lemma 1 (which bounds its Legendre conjugate). Given GLM1 and GLM2 conditions, [16] derive a bound on l_2-norm of the difference between the true signal and the solution to l_1-regularized GLM regression. The result is probabilistic, with the probability approaching 1 as the number of samples increases. Our results are different in several ways. First, the bounds are deterministic and the design matrix must satisfy RIP rather than sub-Gaussianity. Second, our results are focused on the constrained l_1-norm minimization formulation rather than on its Lagrangian form; in the constrained formulation, parameter ϵ bounding the divergence between the linear projections of the signal and its noisy observations (e.g., $||\mathbf{y} - A\mathbf{x}||_{l_2} < \epsilon$) has a clear intuitive meaning, characterizing the amount of noise in measurements, while the particular values of the sparsity parameter λ in Lagrangian formulation are somewhat harder to interpret. Our results provide a very intuitive and straightforward extension of the standard compressed sensing result presented in [7]. Finally, we provide an additional treatment of some cases when the second derivative of the cumulant function (or its Legendre conjugate) are not bounded, e.g., in case of Bernoulli noise (logistic loss) or exponential noise (Itakura-Saito distance).

Another interesting topic to consider is alternative error criteria besides the l_2-norm, such as, for example, support recovery. Accurate support recovery is often a more relevant measure of success, particularly when variable selection is the main objective. However, deriving support recovery results for GLMs and other

M-estimators appears to be more challenging then the problems considered herein and in [15], and remains a direction for future work.

References

1. Banerjee A, Merugu S, Dhillon IS, Ghosh J (2005) Clustering with Bregman divergences. J Mach Learn Res 6:1705–1749
2. Banerjee A, Merugu S, Dhillon I, and Ghosh J (2004) Clustering with Bregman divergences. In: Proceedings of the fourth SIAM international conference on data mining, pp 234–245
3. Beygelzimer A, Kephart J, and Rish I (2007) Evaluation of optimization methods for network bottleneck diagnosis. In: Proceedings of ICAC-07
4. Candes E (2006) Compressive sampling. Int Cong Math 3:1433–1452
5. Candes E, Romberg J (2006) Quantitative robust uncertainty principles and optimally sparse decompositions. Found Comput Math 6(2):227–254
6. Candes E, Romberg J, Tao T (2006) Robust uncertainty principles: exact signal reconstruction from highly incomplete frequency information. IEEE Trans Inf Theory 52(2):489–509
7. Candes E, Romberg J, Tao T (2006) Stable signal recovery from incomplete and inaccurate measurements. Commun Pure Appl Math 59(8):1207–1223
8. Candes E, Tao T (2005) Decoding by linear programming. IEEE Trans Inf Theory 51(12):4203–4215
9. Carroll MK, Cecchi GA, Rish I, Garg R, Rao AR (2009) Prediction and interpretation of distributed neural activity with sparse models. Neuroimage 44(1):112–122
10. Chandalia G, Rish I (2007) Blind source separation approach to performance diagnosis and dependency discovery. In: Proceedings of IMC-2007
11. Donoho D (2006) Compressed sensing. IEEE Trans Inf Theory 52(4):1289–1306
12. Donoho D (2006) For most large underdetermined systems of linear equations, the minimal ell-1 norm near-solution approximates the sparsest near-solution. Commun Pure Appl Math 59(7):907–934
13. Donoho D (2006) For most large underdetermined systems of linear equations, the minimal ell-1 norm solution is also the sparsest solution. Commun Pure Appl Math 59(6):797–829
14. Mitchell TM, Hutchinson R, Niculescu RS, Pereira F, Wang X, Just M, Newman S (2004) Learning to decode cognitive states from brain images. Mach Learn 57:145–175
15. Negahban S, Ravikumar P, Wainwright MJ, Yu B (2009) A unified framework for the analysis of regularized M-estimators. In: Proceedings of neural information processing systems (NIPS)
16. Negahban S, Ravikumar P, Wainwright MJ, Yu B (2010) A unified framework for the analysis of regularized M-estimators. Technical Report 797, Department of Statistics, UC Berkeley
17. Park Mee-Young, Hastie Trevor (2007) An L1 regularization-path algorithm for generalized linear models. JRSSB 69(4):659–677
18. Rish I, Brodie M, Ma S, Odintsova N, Beygelzimer A, Grabarnik G, Hernandez K (2005) Adaptive diagnosis in distributed systems. IEEE Trans Neural Networks (special issue on Adaptive learning systems in communication networks) 16(5):1088–1109
19. Rish I, Grabarnik G, (2009) Sparse signal recovery with Exponential-family noise. In: Proceedings of the 47-th annual allerton conference on communication, control and, computing
20. Rockafeller RT, (1970) Convex analysis. Princeton university press. New Jersey
21. Zheng A, Rish I, Beygelzimer A (2005) Efficient test selection in active diagnosis via entropy approximation. In: Proceedings of UAI-05

Chapter 4
Nuclear Norm Optimization and Its Application to Observation Model Specification

Ning Hao, Lior Horesh and Misha Kilmer

Abstract Optimization problems involving the minimization of the rank of a matrix subject to certain constraints are pervasive in a broad range of disciples, such as control theory [6, 26, 31, 62], signal processing [25], and machine learning [3, 77, 89]. However, solving such rank minimization problems is usually very difficult as they are NP-hard in general [65, 75]. The nuclear norm of a matrix, as the tightest convex surrogate of the matrix rank, has fueled much of the recent research and has proved to be a powerful tool in many areas. In this chapter, we aim to provide a brief review of some of the state-of-the-art in nuclear norm optimization algorithms as they relate to applications. We then propose a novel application of the nuclear norm to the linear model recovery problem, as well as a viable algorithm for solution of the recovery problem. Preliminary numerical results presented here motivates further investigation of the proposed idea.

4.1 Introduction

Identification, specification and exploitation of structure plays a central part in simulation-based optimization of large-scale complex systems. To identify what structure is exploitable requires a careful observation of the various layers involved in the identification, specification and optimization processes. For example, if the simulation and/or optimization process involves the need to solve a large-scale linear system, then one interested in exploiting the matrix structure (e.g. sparsity, block structure, Toeplitz, etc.) to derive fast linear system solvers [5, 33]. In optimization, for example, exploitable structure may emerge in various forms such as

N. Hao (✉) · M. E. Kilmer
Tufts University, 503 Boston Ave, Medford, MA, 02155, USA
e-mail: ning.hao@tufts.edu

L. Horesh
T. J. Watson Research Center, Yorktown Heights, NY, 10598, USA

A. Y. Carmi et al. (eds.), *Compressed Sensing & Sparse Filtering*, 95
Signals and Communication Technology, DOI: 10.1007/978-3-642-38398-4_4,

disparity among sub-problems [54], specific structures of KKT systems, symbolic re-parametrization in automatic differentiation [37], or partial group separability [19, 36]. Other interesting examples for exploitation of structure would be model reduction techniques [7, 13, 39, 64, 90], (nearly) decomposable systems [20, 21] of stochastic processes, graph partitioning [42, 48, 74], multi-scale characteristics in computation [12, 87], and algorithmic aspects related to sequential vs. parallel processing [35].

As a general guideline, development of algorithms that account for the underlying structure of a problem is advantageous with respect to computational feasibility, stability, scalability and well-posedness. For a broad range of applications, several aspects of the structure of the problem can be specified explicitly by means of first principles [46, 72, 78, 83]. Yet, it is often the case that the inherent underlying structure may be latent, requiring (some) operations/transformations to make it more identifiable and exploitable (e.g. nodal reodering of a linear system [51, 73, 82, 94], representation and re-parameterization [34, 59, 81, 93]). Implicit forms of structure are often more complicated to specify and thereby exploit. Typically, a governing principle would serve as good candidate for that purpose. One such generic rule is Occam's razor, also known as the principle of parsimony, which states that the simplest among competing theories should be preferred [23, 30, 43, 47, 86]. This principle has proved itself highly appropriate for abroad range of natural phenomena and when describing the function of a complex system. Other naturally-inspired or physics-inspired guiding principles might include causality, conservation rules, minimal energy, the least action principle, uncertainty principle, or minimal entropy [10, 11, 22, 24, 50, 66].

As a caveat, it is important to acknowledge that postulation and imposition of inappropriate structure can equally introduce bias and deteriorate the quality of the solutions. Lastly, it is critical to acknowledge that structure only exists in a context, depending upon the intended purpose of the optimization process (inference, control, design, decision, and so forth).

This chapter is devoted to the exploitation of low rank operator structure. Following more specific background regarding low rank optimization and its tight convex relaxation, the nuclear norm optimization problem, we shall review some of the more popular applications in which these types of problems arise. We will then consider various algorithmic strategies for solving the corresponding optimization problems involving the nuclear norm. Lastly, we will describe the use of nuclear norm minimization as a generic means for resolving model inadequacies.

4.2 Background

It is well-known [33] that for every matrix, there exists a Singular Value Decomposition (SVD). One way (amongest several) of expressing the SVD of an $m \times n$ matrix X is to write it out as a sum of rank-one outer products

$$X = \sum_{i=1}^{r} \sigma_i u_i v_i^\top, \tag{4.1}$$

where the rank, r, satisfies $r \leq \min(m, n)$, the scalar values σ_i are called the singular values, and they are real and non-negative (even for complex X) and ordered such that $\sigma_1 \geq \sigma_2 \geq \cdots \geq 0$. The m-length vectors u_i form an orthonormal set, $U := \{u_1, \ldots, u_n\}$, as do the n-length vectors v_i, $V := \{v_1, \ldots, v_n\}$. The Eckart-Young theorem states that the matrix $B := \sum_{i=1}^{k} \sigma_i u_i v_i^\top$ for $k \leq r$ is the optimal (in the Frobenius and in the 2-norm sense) rank-k approximation to X. In particular, the error is given by $\|X - B\|_F^2 = \sum_{i=k+1}^{r} \sigma_i^2$. So, if k is such that the remaining singular values are relatively small compared to the largest one, B captures most of the energy in X, and the sum of the squares of the remaining singular values correspond to the energy not captured.

While the Frobenius norm of a matrix X is the square root of the sum of the squares of the singular values, the **nuclear norm** of a matrix X is just the sum of the singular values of the matrix,

$$\|X\|_* = \sum_{i=1}^{r} \sigma_i, \tag{4.2}$$

and since the nuclear ball $\{X : \|X\|_* \leq 1\}$ is the convex hull of the set of rank-one matrices with spectral norm bounded by one, this norm can be regarded as the tightest convex approximation of the rank function. For optimization purposes, it is often recast in its semidefinite programming (SDP) representation

$$\begin{aligned} \|X\|_* = \min & \frac{1}{2} \left(\text{trace}\,(W_1) + \text{trace}\,(W_2)\right) \\ \text{s.t.} & \begin{bmatrix} W_1 & X \\ X^\top & W_2 \end{bmatrix} \succeq 0, \end{aligned} \tag{4.3}$$

where the trace operation on a matrix is the sum of the diagonal entries of the matrix.

The nuclear norm can be shown to be unitarily invariant. That is, if U is an $m \times m$ unitary matrix and V is an $n \times n$ unitary matrix, then

$$\|UXV\|_* = \|X\|_*. \tag{4.4}$$

When the column dimension of the matrix collapses (i.e. $n = 1$), then X becomes a vector, x, and we observe

$$\|x\|_1 = \|\text{diag}(x)\|_*. \tag{4.5}$$

Therefore, the nuclear norm can also be regarded as the ℓ_1 norm in the spectral domain, and the tightest convex approximation for the ℓ_0 norm.

4.2.1 The Role of the Nuclear Norm in Affine Rank Minimization

The affine rank minimization problem is to find the matrix of smallest rank that satisfies a linear system,

$$\begin{aligned} &\min \operatorname{rank}(X) \\ &\text{s.t. } A(X) = b, \quad A : \mathbb{R}^{m\times n} \rightarrow \mathbb{R}^p \end{aligned} \tag{4.6}$$

for some given A and b. It was shown in [75] that this problem can be exactly solved through the minimization of the nuclear norm over the given affine space if A is a nearly isometric linear map and the number of linear constraints is within the range (depending on the dimensions of the problem). Thus, instead of solving the above problem, we may seek to solve its relaxation

$$\begin{aligned} &\min \|X\|_* = \sum_{i=1}^{min(m,n)} \sigma_i(X) \\ &\text{s.t. } A(X) = b, \quad A : \mathbb{R}^{m\times n} \rightarrow \mathbb{R}^p \end{aligned} \tag{4.7}$$

where $\sigma_i(X), i = 1, 2, ..., min(m, n)$ are the singular values of the matrix X.

4.2.2 The Nuclear Norm in Matrix Completion and Low-Rank Matrix Recovery

The matrix completion problem is to recover (implicitly or explicitly) a subset or the entire set of the entries of a matrix from a limited number of sampled entries. A well-known example of the matrix completion problem is the Netflix Prize problem [1]. The objective of the application (aka recommender system) is to provide recommendations for customers based upon knowledge regarding the current users' ratings. As such, this problem can be reduced to the problem of predicting the unknown values in a particular rating matrix based on the ratings that a given user has submitted already on a subset of the database of movies.

Mathematically, the matrix completion problem can be formulated as follows. Suppose $M \in \mathbb{R}^{n_1\times n_2}$ and let Ω be a subset of $n_1 \times n_2$ revealed entries of M. We want to find a solution of the following problem

$$\begin{aligned} &\min \operatorname{rank}(X) \\ &\text{s.t. } X_{ij} = M_{ij} \text{ for all } (i, j) \in \Omega \end{aligned} \tag{4.8}$$

Let M^{Ω} be the $m \times n$ matrix containing all the revealed entries of M, and assume it is filled with $0's$ in unrevealed entries:

$$M^{\Omega} = \begin{cases} M_{i,j} & : (i,j) \in \Omega \\ 0 & : \text{otherwise.} \end{cases} \tag{4.9}$$

Then the completion problem can be rewritten as

$$\begin{aligned} &\min \operatorname{rank}(X) \\ &\text{s.t. } X^{\Omega} = M^{\Omega}. \end{aligned} \tag{4.10}$$

The above is the ideal case, which is difficult to solve. If instead one employs the convex relaxation for the rank minimization component, we derive the alternative problem

$$\begin{aligned} &\min \|X\|_* \\ &\text{s.t. } X^{\Omega} = M^{\Omega}. \end{aligned} \tag{4.11}$$

Much of the recent literature on the matrix completion problem is devoted to efficient methods for solving this relaxed version.

4.2.3 The Nuclear Norm in Matrix Separation

The matrix separation problem also known as robust principal component analysis [16, 53], aims at separating a low-rank matrix and a sparse matrix from their sum. This recently formulated problem gains increasing research interest due to the broad range of potential applications it can tackle, such as image and model alignment as well as system identification [27, 57, 71]. For a sparse matrix S and a low-rank matrix X to which, $Y = X + S$, the problem can be defined as

$$\underset{S,X \in \mathbb{R}^{m \times n}}{\operatorname{argmin}} \ \operatorname{rank}(X) + \mu \|S\|_0 \tag{4.12}$$

or in its convex relaxation form

$$\underset{S,X \in \mathbb{R}^{m \times n}}{\operatorname{argmin}} \ \|X\|_* + \mu \|S\|_1 \tag{4.13}$$

where the term $\|S\|_1$ simply stands for the sum of absolute values of all entries in the matrix S.

4.2.4 The Nuclear Norm in Other Applications

A large volume of literature is devoted to the use of the nuclear norm for various applications in science and engineering. The scope of this chapter is to provide a brief overview and then highlight some very recent on-going research work in the context of model misspecification. However, in the interest of completeness, we shall mention few other research, to give an idea of the breadth of the usefulness of the approach. Specifically, the matrix nuclear norm has been used successfully in many applications including low rank approximation [70], system identification [38, 40, 56, 63], localization of wireless sensors [28], network traffic behavior analysis [60], low-dimensional Euclidean embedding problems [32, 75], image compression [58].

Furthermore, as the amount of data that is collected and stored in its naturally occurring high-dimensional format increases, the need to investigate the generalization of the nuclear norm to higher-order arrays (called tensors) has arisen. It turns out that such generalization is highly non-trivial, in part because even the of "rank" of a tensor is quite an subject. One must determine first which tensor rank formulation should be considered for optimization and only consequently make any attempt to define and relax to a tensor nuclear norm. For further applications and literature, the interested reader is referred to [3, 6, 26, 31, 62, 77, 89].

4.3 Methods for Nuclear Norm Optimization

In the past few years, a remarkable number of algorithms have been developed for nuclear norm minimization. The principal ideas behind a few are recent, whereas not so few leverage old, somewhat forgotten approaches. Luckily the intrinsic mechanism can be umbrellaed together under few central concepts. In this section we do not intend to provide a comprehensive review regarding all approaches, but rather to briefly introduce the most frequently used approaches and their underlying foundations.

4.3.1 Methods Based on Semi Definite Programing

Semi-Definite Programming (SDP) consists of minimization of a linear functional of a matrix subject to linear equality and inequality constraints. Most nuclear norm optimization problems can be formulated as semidefinite programs (i.e. refer to (4.3)), but the reformulation may require the utility of large number of auxiliary matrix variables. The SDP problem is computationally expensive to solve by means of general-purpose interior-point solvers, as the solution for extremely large systems of linear equations is required in order to compute search directions. Nevertheless, it is valuable to mention some background regarding SDPs, highlighting more recent work on SDPs for optimization problems involving the nuclear norm.

4.3.1.1 Interior Point Methods for SDP

Assuming a feasible initial guess,[1] interior-point methods (as the name implies) successively update the solution of the optimization problem while maintaining the solution in the interior of the feasible region at each instance. Typically, an indicator function or a logarithmic barrier associated with inequality constraints augments the objective and forces solutions towards the interior feasible domain. In terms of KKT conditions this implies that complementary slackness conditions are only satisfied approximately, where the exactness of the approximation is gradually tightened throughout the optimization process. The approach was first used as a powerful tool in linear programming. Later, it was extended to SDP in [2, 69]. For a comprehensive review of interior-point methods, the reader is refereed to [29].

In [56], Liu and Vandenberghe showed that the structure of the problem in the semidefinite programming formulation can be exploited to develop more efficient implementations of interior-point methods. The cost per iteration can be reduced to a quartic function of the problem dimensions, which makes it comparable to the cost of solving the approximation problem in the Frobenius norm.

4.3.1.2 More Recent Work Based on SDP

Considering the formal definition of the nuclear norm in (4.2), it is expected that SVD computation would play a chief computational role in the design of nuclear norm optimization schemes. Complete singular value decomposition is obviously non-tractable for a broad range of large-scale applications. An alternative approach is proposed by Jaggi and Sulovsky [44]. Their idea is to recast the optimization problem as a convex function over the set of positive semidefinite matrices with unit trace. Following the use of a scaling transformation, Hazan's [41] Sparse-SDP solver can be readily applied to the recast version of the problem. The appeal of this approach is that each iteration only involves the computation of the largest singular triple of the gradient of the function at the current iterate. This entails relatively cheap sparse-matrix operations followed by rank-one updates. The downside is that this approach produces a so-called ϵ*-accurate solution*, whose rank could be as large as $\mathcal{O}(\frac{1}{\epsilon})$. This in turn may render it infeasible for large problems in practice due to the need to hold the factorization of the solution in memory. An alternative approach is proposed in [76]. Their method directly deals with a low-rank parameterization. A pitfall of that approach is its non-convex formulation, implying that solutions are prone to fall into local minima and thereby are highly sensitive to initialization.

[1] This assumption is not mandatory for primal-dual interior point methods.

4.3.2 Projected Sub-Gradient

In optimization of differentiable objective functions, the gradient (or its estimate) whenever computable, is usually employed in some manner to determine search directions. When the objective function is non-differentiable, methods that utilize, so-called, sub-gradients can be employed. Suppose $f(X)$ denotes the (not necessarily convex) objective function. Then g is a sub-gradient of f at X if

$$f(Y) > f(X) - g^{\top}(Y - X), \ \ \forall Y.$$

When f is convex and differentiable at X, the sub-gradient coincides with the gradient. Otherwise, there may exist more than a single sub-gradient: the set of all sub-gradient vectors at X is called the sub-differential at X. For more details, see [8]. A typical sub-gradient iterative update step looks like

$$X_{k+1} = X_k + \alpha_k g_k,$$

where g_k denotes a particular sub-gradient at X_k and α_k denotes step length.

Now consider the problem

$$\min_{X \in \mathcal{C}} f(X) + \mu \|X\|_*, \tag{4.14}$$

where $\mathcal{C}$ is a convex set, and f is convex but might be non-differentiable at some points. Let $F(X)$ denote the objective function in this case. The sub-gradient of $F(X)$ can be obtained from the sub-gradient of $f(X)$ and an appropriately truncated SVD of X. A stochastic projected (i.e. each iterate is projected onto $\mathcal{C}$) sub-gradient (SSGD) approach based on these ideas for solving the above problem is presented in [4]. The key ingredient in their SSGD algorithm is derivation of an unbiased estimator for a sub-gradient. The computational difficulty associated with this approach can be somewhat relieved through the use of a matrix probing technique [17] in which the sub-gradient matrix of F is probed through its multiplication by a random low-rank matrix.

4.3.3 Singular Value Thresholding

Inspired by previous research work in the field of ℓ_1 minimization, and in particular by studies associated with linearized Bregman iterations for compressed sensing [15, 92], Cai, Candès and Shen [14] introduced a simple first-order iterative algorithm to approximate a matrix with minimum nuclear norm among all matrices obeying a set of convex constraints (i.e. can be formulated as in 4.7). The algorithm applies a soft-thresholding operation **S** upon the singular values of a sparse matrix at each step.

For a given threshold level $\tau \geq 0$ the soft-thresholding operator $\mathbf{S}_\tau$ is defined as follows:

$$\mathbf{S}_\tau(X) := U \mathrm{diag}\left(\max(0, \sigma_i - \tau)\right) V^\top \tag{4.15}$$

where U and V are usual left and right singular vector sets respectively as defined in 4.1, and the diagonal matrix of thresholded singular values essentially holds zero values for all entries smaller than τ or otherwise τ-shifted singular values. Often the soft-thresholding operation is also referred by the name shrinkage, as the thresholded values are shrunk to zero. Later, in Sect. 4.3.5 we shall see that the singular value thresholding operator is in effect the proximity operator associated with the nuclear norm.

Considering optimization problem given in Eq. (4.11)

$$\begin{aligned} &\min \|X\|_* \\ &\text{s.t. } X^\Omega = M^\Omega. \end{aligned} \tag{4.16}$$

then starting with $Y_0 = 0$, the update rule has the following form

$$X_{k+1} = \mathbf{S}_\tau\left(Y_k, \tau\right) \tag{4.17}$$

$$Y_{k+1} = Y_k + \alpha_k \left(M - X_{k+1}\right)^\Omega \tag{4.18}$$

where α_k are the step lengths, and the superscript Ω indicates orthogonal projection onto the span of matrices vanishing outside Ω (all but the sampled entries are set to zero) as defined previously in Sect. 4.2.

The method is proved to be convergent. This algorithm is relatively efficient and of low computational cost since in practice full computation of the SVD is not required. Instead, estimation of the largest singular values and vectors can be performed through Monte-Carlo sampling [57] or Lanczos bi-diagonalization [14]. Their numerical results demonstrated the utility of the algorithms for large-scale problems. Other noted algorithms involving application of a soft-thresholding operator upon the singular values of an iterate are the Soft-Impute [61] and the accelerated Proximal Gradient approach [45] that is further discussed in Sect. 4.3.5.

4.3.4 Fixed Point and Bregman Iterative Methods

The application of continuation methods for low rank approximation is a natural choice. In [57], the authors proposed the fixed point and Bregman iterative algorithms for solving the minimization problem (4.7). The basis for their algorithm is a homotopy approach together with an approximate SVD. Numerical results on real matrix completion problems are presented in [57] illustrating that their algorithm can potentially outperform SDP solvers for large problems.

Interestingly, the method has links to the Alternating Direction Method of Multipliers [9], which is based upon the classic augmented Lagrangian optimization approach. Similarly to most alternating optimization strategies, some variables are affixed at their latest values while others (such as the Lagrange multipliers) are being optimized, and then the set of updated variables is left fixed while the ones that were kept fixed previously are being updated. Yet, the main distinction is the use of seemingly redundant, dummy variables with constraints that forces the dummy variables to equate to the primary variables. By doing so, the original problem can be artificially divided into a sequence of sub-problems, each of which is simpler to handle than the original objective. This is particularly useful in situations where the optimization problem involves distinct elements for which efficient algorithms exist. For instance, consider the following objective

$$\underset{X}{\operatorname{argmin}} \frac{1}{2} \|AX - B\|_F^2 + \mu \|X\|_* \tag{4.19}$$

which can be reformulated as

$$\begin{aligned} \underset{X,Y}{\operatorname{argmin}} \quad & \frac{1}{2} \|AX - B\|_F^2 + \mu \|Y\|_* + \frac{\rho}{2} \|X - Y\|_F^2 && (4.20) \\ \text{s.t.} \quad & X = Y. && (4.21) \end{aligned}$$

The Lagrangian of this objective takes the form:

$$\mathcal{L}(X, Y, \lambda) = \frac{1}{2} \|AX - B\|_F^2 + \mu \|Y\|_* + \frac{\rho}{2} \|X - Y\|_F^2 + \operatorname{vec}(\lambda)^\top \operatorname{vec}((X - Y)). \tag{4.22}$$

The iteration would then be

$$\begin{aligned} X_{k+1} &= \underset{X}{\operatorname{argmin}} \, \mathcal{L}(X, Y_k, \lambda_k) && (4.23) \\ Y_{k+1} &= \underset{Y}{\operatorname{argmin}} \, \mathcal{L}(X_{k+1}, Y, \lambda_k) && (4.24) \\ \lambda_{k+1} &= \lambda_k + \rho(X_{k+1} - Y_{k+1}). && (4.25) \end{aligned}$$

Such framework has been implemented by Yuan and Yang [91] in a code called LRSD (Low Rank and Sparse matrix Decomposition), and by Lin, Chen, Wu and Ma [52] in a code called IALM (Inexact Augmented Lagrangian Method). In the latter study, an exact augmented Lagrangian method (EALM) is also implemented in which multiple rounds of alternating minimization are conducted before the Lagrangian multiplier is updated.

4.3.5 Proximal Gradient Algorithms

The proximity operator of a convex function is a natural extension of the notion of a projection operator onto a convex set. This numerical tool is particularly essential in the analysis and the numerical solution of convex optimization problems. Lately, its utility in the context of inverse problems and especially in the fields of signal and image processing has been considered.

The proximal mapping of a convex function h is defined as:

$$prox_h(X) = \underset{Y \in \mathbb{R}^n}{\operatorname{argmin}} \left(h(Y) + \frac{1}{2} \|Y - X\|_2^2 \right). \tag{4.26}$$

The proximal operator typically replaces a "problematic" component of the objective. For instance, let us consider an unconstrained optimization problem that can be split out into two components

$$\text{minimize} f(X) = g(X) + h(X) \tag{4.27}$$

where g is convex, differentiable and h is closed, convex, possibly differentiable function. Proximal operators are characterized by the following appealing properties:

- Inexpensiveness - $prox_h$ is firmly inexpensive. That is $prox_h$ is co-coercive

$$\|prox_h(X) - prox_h(Y)\| \leq \|X - Y\| \tag{4.28}$$

- Separability - if $h(X) = \sum_j h(X_j)$, then, $prox_h = \left[prox_{h_j}\right]_j$

More information regarding proximal methods and their properties can be found in [18].

The update step would then get the form

$$X_{k+1} = \underbrace{prox_{\alpha_k h}}_{\text{backward step}} \underbrace{(X_k - \alpha_k \nabla f(X_k))}_{\text{forward step}} \tag{4.29}$$

where α is a line search step length.

The authors of [84] considered the solution of

$$\min_X f(X) + P(X)$$

which is similar to the problem given in (4.14) for appropriate definition of P and other assumptions (e.g. f to be smooth convex). The algorithm is an accelerated proximal gradient algorithm [67, 68, 88], that converges in $\mathcal{O}(1/\sqrt{\epsilon})$ iterations to a so-called, $\epsilon - optimal$ solution. Results in that study demonstrated the potential efficiency and robustness of such accelerated proximal methods in a large-scale settings on random matrix completion problems.

Several forms of inexact proximal point algorithms were studied and implemented in [55]. In that study, the authors studied the viability of inexact proximal point algorithms in the primal, dual and primal-dual forms for solving the nuclear norm minimization with linear equality and second order cone constraints.

4.3.6 Atomic Decomposition

In [49], the authors develop an algorithm for solving a matrix form of the compressive sensing problem

$$\min_{X} \quad \|A(X) - b\|_2 \tag{4.30}$$

$$\text{s.t.} \quad rank(X) \leq r, \qquad A : \mathbb{C}^{m \times n} \to \mathbb{C}^p. \tag{4.31}$$

Their algorithm is called "Atomic Decomposition for Minimum Rank Approximation", or ADMiRA. The "atoms" in the context of this study are rank-one matrices; the set of atoms is the set of all rank one matrices that are not collinear. Clearly, a matrix must have an atomic decomposition (namely, the SVD). To summarize, this algorithm primarily makes use of least squares problems/solvers and truncated SVDs. Thus, the authors claim their algorithm can be efficient to the extent that optimized versions of subroutines for those tasks can be utilized. The algorithm is provably convergent for certain classes of operators A (which does not include the matrix completion mapping).

4.4 The Problem of Misspecified Observation Models

4.4.1 Motivation

Numerical simulators are used extensively in industry and in academia. Their main role is to imitate physical processes or systems in well-controlled and repeatable settings. Numerical simulation (i.e. defining the state of a system given input parameters and governing relations) is instrumental for description, prediction, control and design of complex systems. Fidelity of the simulation process plays a central role in attainment of meaningful predictive capabilities. Frequently, the simulation model is misspecified to a certain extent. Such misspecification may originate from incomplete or approximated physical description of the problem (i.e. governing equations, geometry, boundary conditions, input model parameters, etc.), and may also be attributed to the use of approximated numerics (i.e. floating point round-off error, truncated expansions, discretization error, numerical approximation, etc.), due to linearization of non-linear processes, or any other unknown sources of modeling error.

Whenever any of these sources of error are prominent, these errors creep into the simulation output. The ramifications of such model misspecification are vast, ranging from inaccurate state descriptions to, unstable model recovery and predictions that lead to erroneous control output or decisions.

A great body of research has been devoted to recovery/estimation of the underlying true model parameters, however, few have been underway for improving the physical model which may have further impact on estimation of the true model.

Traditionally, efforts for remediation of model misspecification have centered on explicit means, requiring the modeler to have a deep understanding regarding the most prominent model inadequacies errors, their propagation route from raw data, through estimation and end-goal objectives and lately the modeller's ability to better specify a more accurate physical model.

Moreover, in many situations one may neither have access to the simulator code, nor to sufficient documentation that elucidates the current formulation details. Furthermore, even when such formulation is known, one may often be agnostic to the relevant attributes of the simulation that exhibits the largest source of error and to the extent they deteriorate simulation and simulation-derived output.

4.4.2 Design Correction

Let us assume that for a set of simulation model input parameters, a corresponding set of high fidelity output data is attainable. For the sake of clarity and readability, we shall hereafter use the term data for the latter set. Useable set of data can be obtained in various ways, such as through experimentation with known input models, analytic derivation, or alternatively through the use of a computationally intensive high-fidelity simulation.

Given this information, along with non-intrusive[2] access to the current (low fidelity) simulator, we formulate the model correction problem as a **stochastic constrained optimization problem**, where the objective function to be optimized includes a measure of the discrepancy between the expected output of the current (low-fidelity) model along with unknown correction against the data. The correction to the operator can be represented in a number of different ways, yet, although the representation of the correction may be assumed known, the operator itself is unknown. Posing the optimization problem as one of recovering the operator's correction given only the data is often inadequate. This is because an unbounded number of potential corrections may equally fit the data. Therefore, we further incorporate some preferences regarding the structure of the correction operator through additional constraints or as a penalty on the objective function. The resulting optimization problem of recovering the correction operator given the data and these additional constraints is hereafter referred to as the **design correction optimization problem**.

[2] Non-intrusive methods relies upon black-box interface, for which output is received per input.

One such preference for an additive linear simulation model would be that the rank of the correction is small. Such an assumption can be justified in a broad range of applications in the field of ill-posed inverse problems, since the effective rank of the most comprehensive simulation models is smaller than the number of observations. The enforcement of a low-rank constraint on the objective function is computationally infeasible. However, a close computationally tractable alternative can be obtained by employing, instead of a hard rank constraint, the tightest convex relaxation to the rank, which is the nuclear norm. In other words, the problem we want to solve is to find the operator correction that satisfies a nuclear norm constraint. Other structural preferences may be more appropriate given some intuition regarding the problem under consideration. Following the solution of the design correction optimization problem, a modeling operator correction is obtained. This operator can now be used in practice to improve the fidelity of current simulation procedure.

In the next subsection, we give the details corresponding to the aforementioned description.

4.4.3 Problem Definition

Let $\mathcal{F} : \mathbb{R}^n \rightarrow \mathbb{R}^m$ be a comprehensive observation operator which transforms the model $x \in \mathbb{R}^m$ into the observable space. Let $d \in \mathbb{R}^m$ be an observation obtained through the following relation:

$$d = \mathcal{F}(x) + \epsilon \tag{4.32}$$

where ϵ stands for measurement noise. In reality, our ability to prescribe $\mathcal{F}$ fully is limited. Numerical modeling is notoriously known to be a host of various sources of error. Such inadequacies in the prescription of the observation model may arise due to various factors. Some examples include:

- Formulation - the model may cover only part of the underlying governing equations (e.g. approximated physics), or alternatively account for limited range of parameters (e.g. quasi-static approximation for problem in which the effect of of high frequency components cannot be neglected). Other problems may arise due to defective boundary conditions or any other input parameters, linearization of non-linear phenomena, truncated expansions, etc.
- Discretization - numerical errors may emerge from inadequate discrete representation of an infinite dimensional problem. Issues such as inaccurate geometrical representation of the domain, mesh quality issues, inappropriate spatio-temporal discretization, unstable numerical schemes, etc.
- Numerical solution - often solutions are attained with a prescribed numerical tolerance accuracy. Such accuracies might not be communicated consistently, and may be amplified or still be observed beyond the measurement noise level.

For the sake of simplicity, let us first assume that the observation model can be decomposed in the following additive form

$$\mathcal{F}(x) = A(x) + \eta(x) \tag{4.33}$$

where $A \in \mathbb{R}^{m \times n}$ is a discrete, incomplete observation operator that suffers from any of the aforementioned modeling inadequacies or from other sources of error. $\eta(x)$ stands for modeling error, or model misspecification (under-modeling).

If A is assumed linear (i.e independent of x), we can define a correction for the missing observation operator $\eta(x)$ in numerous ways. For the sake of simplicity we will limit this discussion to an additive model of the form

$$\mathcal{F}(x) = A(x) + B(x) \tag{4.34}$$

where A is known, B needs to be estimated. Obviously, other forms of parametrization can be considered, as long as they capture principle components of the model error in an appropriate functional form.

The key point is that given some information regarding the model space and the data, and in particular authentic paired samples from both spaces, one can design the corrective part of the observation operator.

Let us focus on the the most fundamental situation, namely

$$\begin{aligned} \hat{B} &= \underset{B}{\text{argmin}}\ \ \text{rank}(B(x)) \\ \text{s.t.}\ \ \hat{x} &= \underset{x}{\text{argmin}}\, \mathcal{D}\left((A + B)(x), d\right) + R(x) \end{aligned} \tag{4.35}$$

where $\mathcal{D}$ is a distance measure (noise model), R corresponds to a regularization operator, and B is a function of x. However, this formulation is NP-hard, and intractable even for problems of moderate size. An alternative formulation is one in which convex relaxation of the rank functional is utilized. The tightest convex relaxation for the rank operator is the nuclear norm

$$\begin{aligned} \hat{B} &= \underset{B}{\text{argmin}}\ \|B(x)\|_* \\ \text{s.t.}\ \ \hat{x} &= \underset{x}{\text{argmin}}\, \mathcal{D}\left([A + B](x), d\right) + R(x). \end{aligned} \tag{4.36}$$

This problem can be further relaxed in several ways. For instance, assuming B to be linear, and that the measurement noise level (in some measure or metric) can be bounded, we then have

$$\begin{aligned} \hat{B} &= \underset{B}{\text{argmin}}\ \|B\|_* \\ \text{s.t.}\ \ &\mathcal{D}\left((A + B)x, d\right) \le \tau. \end{aligned} \tag{4.37}$$

This can alternatively be expressed as

$$\hat{B} = \underset{B}{\operatorname{argmin}}\, \mathcal{D}\left((A+B)x, d\right)$$
$$\text{s.t.} \quad \|B\|_* \leq \frac{\delta}{2} \tag{4.38}$$

where δ and τ are linked through a Pareto curve.

4.4.4 Solution Method

We will consider a Sample Average Approximation approach (SAA) [80]. SAA solves stochastic optimization problems by Monte Carlo simulation [79]. In this framework, the expectation with respect to the model space as well as the measurement noise shall be approximated by a sample average estimate of a random sample.

We wish to solve the following:

$$\hat{B} = \underset{B}{\operatorname{argmin}}\, \frac{1}{n_x n_\epsilon} \sum_{i=1,j=1}^{n_x, n_\epsilon} \mathcal{D}\left((A+B)x_i - d_{i,j}\right)$$
$$\text{s.t.} \quad \|B\|_* \leq \frac{\delta}{2} \tag{4.39}$$

where n_x is the number of model realizations we draw and n_ϵ is the number of data realizations for each model realizations. That is for each model realization x_i, $i = 1, 2, \ldots, n_x$, we assume the availability of data realizations corresponds to n_ϵ measurement random noise samples.

One can readily observe that when the objective is a quadratic, the problem can be reformulated in matrix form as follows:

$$\hat{B} = \underset{B}{\operatorname{argmin}}\, \frac{1}{n_x n_\epsilon} \|(A+B)X - D\|_F^2$$
$$\text{s.t.} \quad \|B\|_* \leq \frac{\delta}{2} \tag{4.40}$$

where $X \in \mathbb{R}^{n \times n_x n_\epsilon}$ and $D \in \mathbb{R}^{m \times n_x n_\epsilon}$.

As this problem is a convex problem, we shall follow Jaggi and Sulovský's [44] convexification of Hazan's algorithm [41] to pursue minimization of the nuclear norm. Beyond its simplicity, the algorithm entails a convex optimization problem, and hence convergence is guaranteed. In addition, it offers means for controlling the rank.

Observe that for any nonzero matrix $B \in \mathbb{R}^{m\times n}$ and $\delta \in \mathbb{R}$: $||B||_* \leq \frac{\delta}{2}$, if and only if there exists symmetric matrices $M \in \mathbb{R}^{m\times m}$ and $N \in \mathbb{R}^{n\times n}$ such that $\begin{pmatrix} M & B \\ B^\top & N \end{pmatrix} \succeq 0$ and trace$\begin{pmatrix} M & B \\ B^\top & N \end{pmatrix} = \delta$. Let $Z = \begin{pmatrix} M & B \\ B^\top & N \end{pmatrix}$, then the problem (4.40) can be recast in the following form

$$\begin{aligned} \min_{Z} \quad & \hat{f}(Z) \\ s.t. \quad & Z \in \mathbb{S}^{(m+n)\times(m+n)} \\ & Z \succeq 0 \\ \text{trace}\,(Z) &= \delta \end{aligned} \tag{4.41}$$

where $S \in \mathbb{R}^{m+n}$ is a family of symmetric matrices and the function $\hat{f}$ applies the function $f = \mathcal{D}((A+B)x, d)$ upon the upper right $m \times n$ sub-matrix of Z (i.e. B). A description of the algorithm is given below in SAA flavor.

Algorithm 1 Low rank linear observation design correction by nuclear norm minimization.

1: Input
   (Scaled) convex function f
2: Initialization
   Set $\epsilon,\ C_f,\ \text{tol},\ v_0 \in \mathbb{R}^{(m+n)\times 1},\ \|v_0\| = 1,\ Z_1 = v_0 v_0^\top$
3: **for** $k = 1 : [\frac{4C_f}{\epsilon}]$ **do**
4: Extract $B_k = Z_k(1:m, m+1:m+n)$
5: Compute the gradient of f

$$\nabla f_k = 2\left(\left(A+B\right)X - D\right)X^\top$$

6: Assemble the gradient of $\hat{f}$

$$\nabla \hat{f}_k = \begin{pmatrix} 0 & \nabla f_k \\ \nabla f_k^\top & 0 \end{pmatrix}$$

7: Approximately (to accuracy ϵ) compute the largest (algebraic) eigen vector

$$v_k = \max_{v_k}\ eig\left(-\nabla \hat{f}_k, \text{tol}\right)$$

8: Determine step length α_k by a line search
9: Update $Z_{k+1} = Z_k + \alpha_k(v_k v_k^\top - Z_k)$
10: **end for**
11: Return $\hat{B} = Z(1:m, m+1:m+n)$;

4.5 Numerical Examples

This example is intended to mimic the so-called blind-deconvolution problem, where only an approximation to the actual blurring operator is known a-priori. To get our data D, we use 100 MR images from the Auckland MRI Research group database.[3] Of those 100 images, we randomly choose 80 to serve as the training set and the

[3] http://atlas.scmr.org/download.html

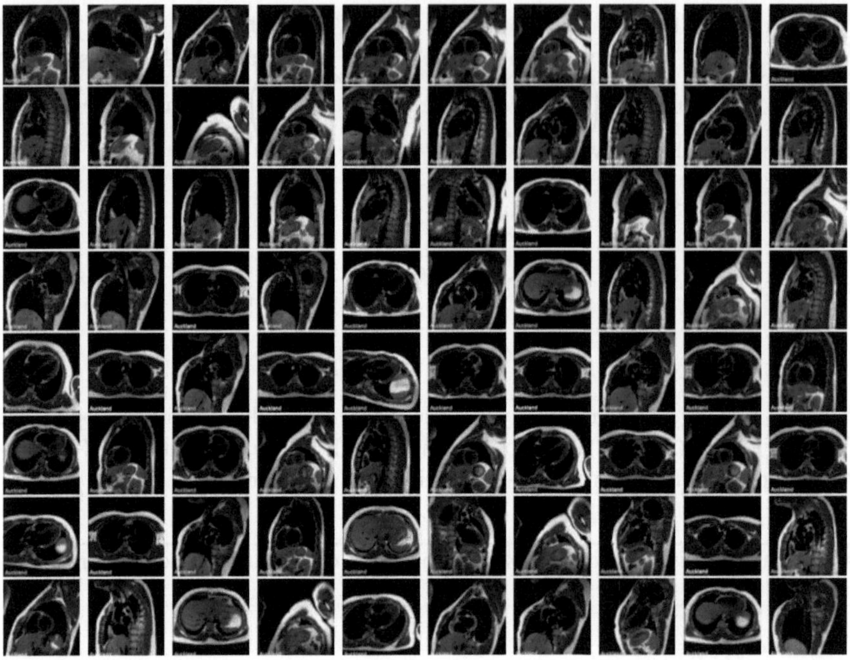

Fig. 4.1 A set of training models

remainder as the test set. In this example, we pre-process the images by cropping the watermark and resizing the cropped images to the size of 65×65. Our observation operator (blurring operator) is generated from a decomposition of a symmetric, doubly block Toeplitz matrix T that models blurring of an $N \times N$ image by a Gaussian point spread function. That is, we find the SVD of T, and use the SVD to specify the two components of our observation operator (T is not explicitly use thereafter). Our true (fully specified), rank-deficient blurring operator consists of the sum of matrices A and B, each of which is constructed as a spectral subset of the fully-specified operator by virtue of singular value decomposition of T. In this experiment, A is constructed from the first 10 singular triplets of T and is assumed to be known, whereas B is constructed from the 31st to 80th singular triples of T, and assumed to be unknown. It would then be our goal to recover B as an additive low-rank matrix correction, for the mis-specified operator A.

Mathematically, we have $T = \sum_{i=1}^{n} \sigma_i u_i v_i^\top$ and then we define

$$A = \sum_{i=1}^{10} \sigma_i u_i v_i^\top, \qquad B = \sum_{i=31}^{80} \sigma_i u_i v_i^\top,$$

so that clearly A has rank 10 and B has rank 50.

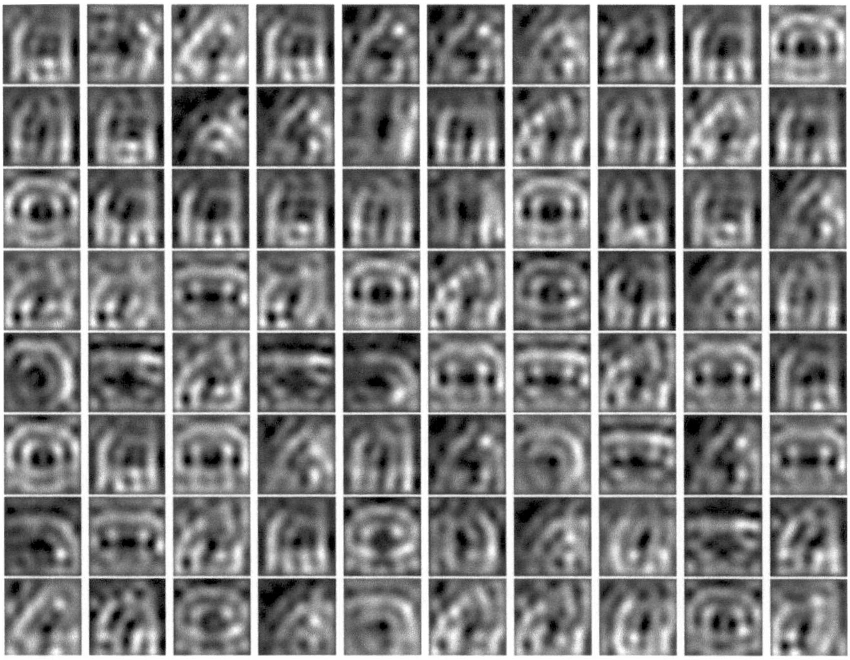

Fig. 4.2 Data (*blurred* training images) obtained using the fully specified operator $A + B$

Blurred images corresponding to a fully specified Observation operator $A + B$ and a partially specified observation operator A are shown in Figs. 4.2 and 4.3.

We use $\hat{B}$ to denote the correction for the blurring operator we obtain by the proposed algorithm. As a preparation for the low rank recovery estimate algorithm, several realizations of zero mean Gaussian noise are infused to each of data arguments. The corresponding singular value spectra of A, $A + B$ and $A + \hat{B}$ are shown in Fig. 4.4.

Figure 4.5 illustrates the convergence of the relative residual error for training set (blue) and testing set (red) respectively. The proximity of the relative residuals reassures that the recovery of the low-rank correction operator B was not over-fitted.

From Fig. 4.6, we can see that with the recovered operator correction, $\hat{B}$, the blurred test models with $A + \hat{B}$ look much closer to the true blurred test models as compared to using only operator A.

We use the (truncated) SVD algorithm to obtain the recovered model with the operators A, $A+B$ and $A+\hat{B}$. The choice of this algorithm (as opposed to such that explicitly prescribe a regularizer) was made to allow for assessment of information content arriving merely through the observation operator, as opposed to solutions reconciled with structural information that is less quantifiable. For the operators A and $A + B$, the effective condition numbers and noise levels are relatively small, so the recovered images can be obtained without truncation of any singular values, and therefore

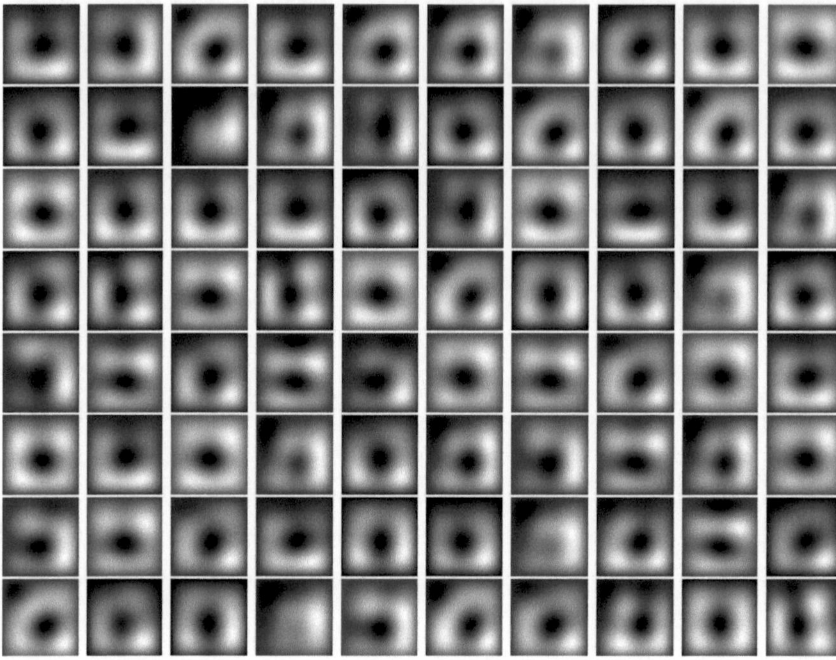

Fig. 4.3 Data (*blurred* training images) obtained using the misspecified operator A

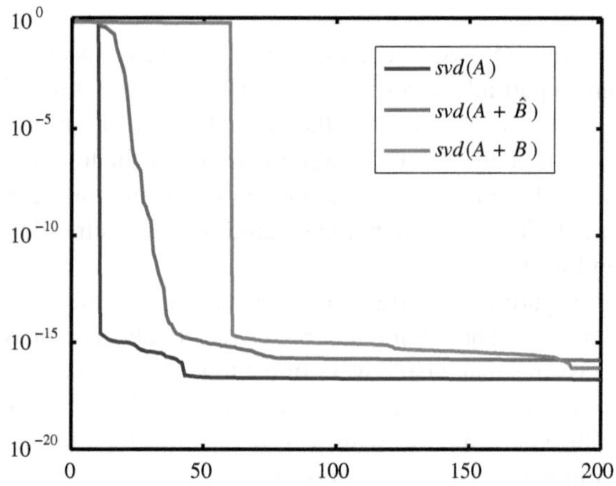

Fig. 4.4 Singular value spectrum of the misspecified operator A (*red*), the supplemented operator $A + \hat{B}$ (*blue*) and the fully specified operator $A + B$ (*green*)

correspond to the pseudo-inverses of the respective operators applied to the images. However, as is clear from Fig. 4.4, the effective condition number of $A + \hat{B}$ is large, and so we truncate the SVD of $A + \hat{B}$ and apply the pseudo-inverse of the truncated operator to D to recover the images. Some samples of the results are shown in Fig. 4.7.

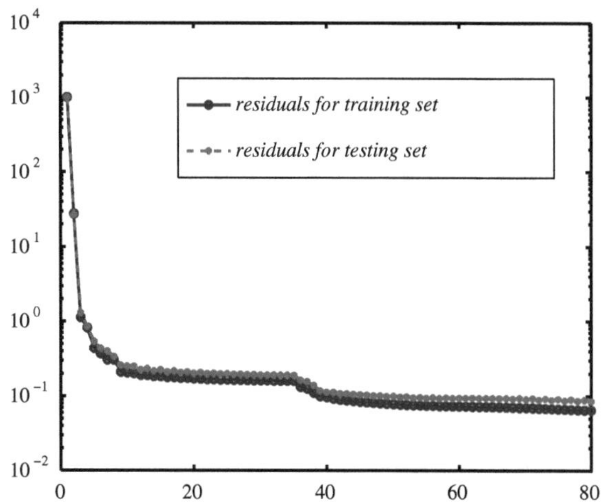

Fig. 4.5 Convergence of the relative residual error of the training set (*blue*) and the test set (*red*) of the MRI example

4.5.1 Discussion

It is evident from Fig. 4.6 that the range of the true observation operator (blurring operator in this particular case) is indeed better approximated by the corrected operator rather than with the original misspecified one.

Note that the jth column of D is constructed from the jth column of X as $D(:,j) = \sum_{i=1}^{10} \sigma_i (v_i^\top X(:,j)) u_i + \sum_{i=31}^{80} \sigma_i (v_i^\top X(:,j)) u_i$, and since the first 80 singular vectors of this operator mainly correspond to smooth modes, the blurring effect as displayed above is not surprising. Using this formulation, it is easy to show

$$
\begin{aligned}
X(:,j)_{A,recov} &= \sum_{i=1}^{10} v_i (v_i^\top X(:,j)), \\
X(:,j)_{A+B,recov} &= \sum_{i=1}^{10} v_i (v_i^\top X(:,j)) + \sum_{i=31}^{80} v_i (v_i^\top X(:,j)).
\end{aligned}
\tag{4.42}
$$

Since, in this example, $u_i = \pm v_i$, it is not surprising that recovered images also appear blurry. On the other hand, in this example, we find that the left singular vectors of $A + \hat{B}$ are *approximately* represented by the union of the subspaces span $\{v_1, \ldots, v_{10}, v_{31}, \ldots, v_{31+p}\}$ for $p \geq 0$, and the subspace spanned by additional v_ℓ vectors correspond to higher-frequency modes (although the influence of these on the quality of the correction operator estimate $\hat{B}$ is damped by the small corresponding singular values as can be observed from Fig. 4.4). The range of the corrected operator $A + \hat{B}$ more closely resembles the range of $A + B$, as seen in Fig. 4.6. However, when it comes to inversion, such a decline in the singular values of the corrected operator corresponding to the high frequency components, manifests itself through amplification of these frequency components. This behavior is evident

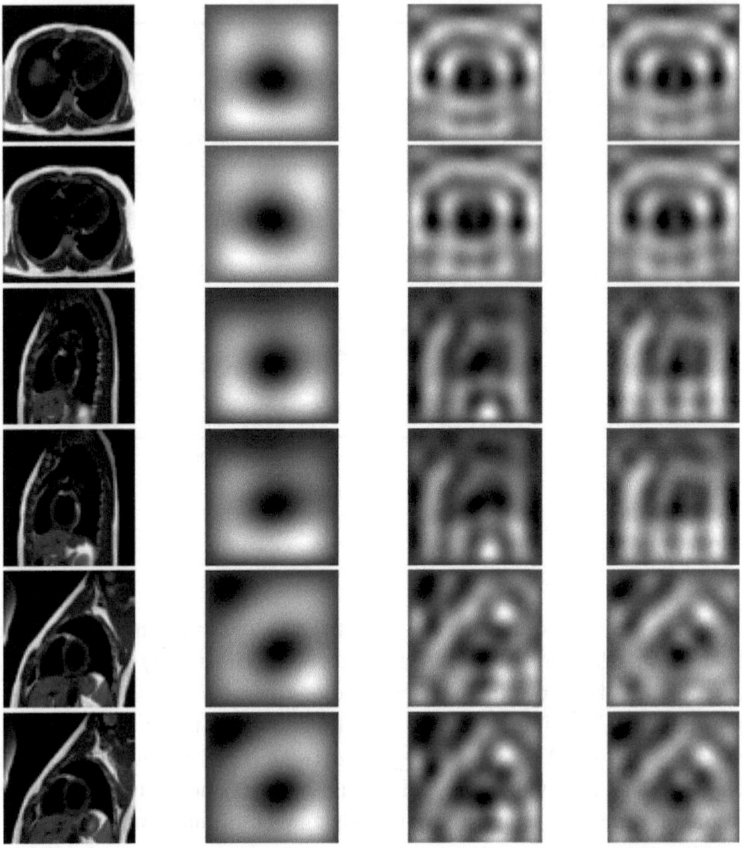

Fig. 4.6 From *left* to *right*: original test models, models observed with the mis-specified operator A, models observed with the fully specified operator $A+B$, models observed with the supplemented operator $A+\hat{B}$

from the fact that the truncated SVD filter is inversely proportional to the singular values $\tilde{\sigma_i}$. That is, the truncated SVD solution is

$$X(:,j)_{A+\hat{B},recov} = \sum_{i=1}^{s} \frac{\tilde{u_i}^\top D(:,j)}{\tilde{\sigma_i}} \tilde{v}_i. \tag{4.43}$$

where s represents the truncation level. In this example, the $\tilde{v}_i$ approximately span the union of the two subspaces mentioned above.

The significance of this in deblurring is evident: information is contained in the modes corresponding to mid-range singular values that A alone does not include. Thus, it is impossible to recover all admittable information if we cannot reconstruct the missing part of the spectrum. Note that if we had recovered B perfectly, we

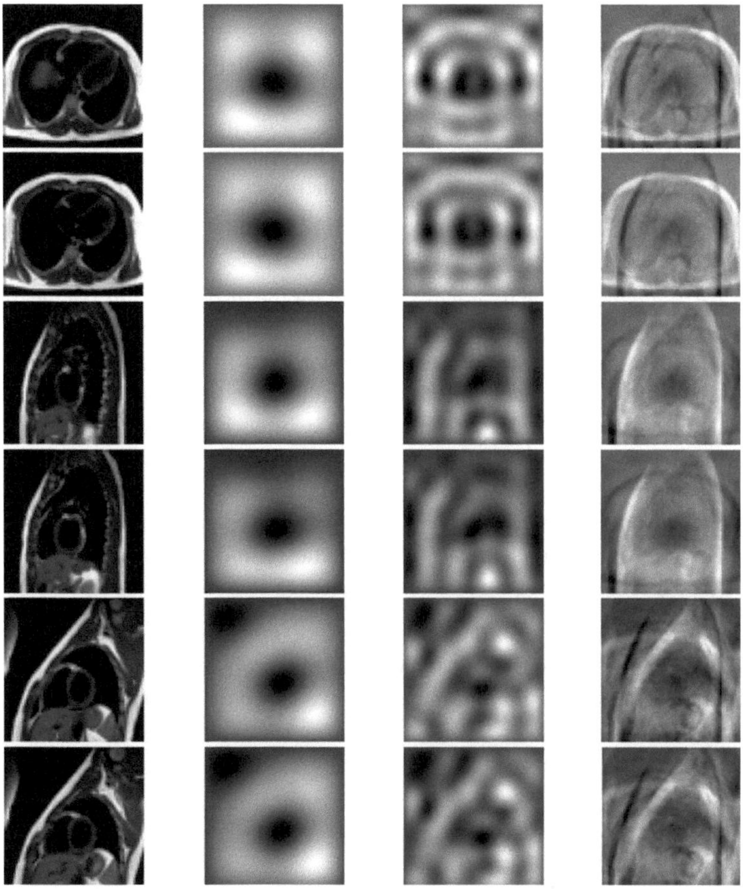

Fig. 4.7 From *left* to *right*: test set models, recovered models with the mis-specified operator A, recovered models with the fully specified operator $A + B$, recovered models with the supplemented operator $A + \hat{B}$

still could not have recovered the edge information exactly, as is obvious from the expression in (4.42). However, if the truncation index is set such that one or more higher frequency vectors $\tilde{v}_i$ gets included in the sum (4.43), edge information may be visible. Indeed, in this example, edges appear in the restoration because basis vectors with higher frequency information were able to creep into the process of recovering $\hat{B}$. This occured due to the limited number of samples and large number of degrees of freedom, which were not all truncated when the truncated SVD solution was formed.

While spectral analysis in this blurring case was possible, for other observation operators, the contribution of the correction to the simulated data output and the subsequent recovered models might require other analysis tools. Other possible assessment measures could be the expected value of the relative residual error for the training and the test set, or even more importantly, the spread of the posterior. As the correction procedure effectively minimizes the null-space of the observation operator, the spread of the posterior is expected to be far more concentrated. Frequently in the context of ill-posed inverse problems, point estimates as maximum a posteriori or maximum likelihood do not suffice, since uncertainty must be quantified. As can be observed from Fig. 4.4, model correction in the form of rank correction/augmentation, essentially increases the range space of the observation operator, at the expense of reducing the dimension of the intrusive null space. The implications of reducing the null-space dimensions are two-fold:

- The use of regularization in order to account for the ill-posed nature of the problem introduces inevitable bias [85]. The amount of structure (and consequently bias) imposed upon the inverse solutions is proportional to the null space dimension. Extracting more measurable information using the a better specified observation model minimizes our reliance upon what is often bogus and artificial a-priori information.
- In terms of the resulting posterior distribution, the concept of low rank model correction is advantageous not merely in getting improvements in recovered estimates, but also in mitigating unnecessary components of uncertainty.

We note that it is possible to improve the performance of the algorithm (with respect to exact recovery of B) by inclusion of more images in the training set. Indeed, the number of degrees of freedom in this problem are such that we would have needed at least 60 training images to have expected a more accurate recovery of B.

4.6 Summary

This chapter began by explaining the significance of the nuclear norm in certain applications where low rank (i.e. implicitly sparse) matrix approximations are desirable. This background was followed by a brief glossary of algorithmic strategies for solving the nuclear norm optimization problems. The remaining focus of the paper was on motivating and describing a problem in model misspecification and outlining the value of the nuclear norm in this context. A computationally efficient algorithm for the corresponding convex optimization problem was presented. A small set of numerical experiments showed the value in using this approach to better capture the true operator. Specifically, the examples illustrated that the range of the operator augmented by the recovered correction is superior to the representation when it is ignored. Research on the model misspecification problem is ongoing, but the preliminary results are indeed promising.

Acknowledgments The authors wish to acknowledge the valuable advice and insights of David Nahamoo and Raya Horesh. In addition, the authors wish to thank Jayant Kalagnanam, and Ulisses Mello for the infrastructural support in fostering the collaboration between Tufts University and IBM Research.

References

1. ACM Sigkdd and Netflix (2007) Proceedings of KDD Cup and Workshop
2. Alizadeh F (1995) Interior point methods in semidefinite programming with applications to combinatorial optimization. SIAM J Opt 5:13–51
3. Argyriou A, Micchelli CA, Pontil M (2008) Convex multi-task feature learning. Mach Learn http://www.springerlink.com/
4. Avron H, Kale S, Prasad S, Sindhwani V (2012) Efficient and practical stochastic subgradient descent for nuclear norm regularization. In: Proceedings of the 29th international conference on machine learning
5. Barrett R, Berry M, Chan TF, Demmel J, Donato J, Dongarra J, Eijkhout V, Pozo R, Romine C, Van der Vorst H (1994) Templates for the solution of linear systems: building blocks for iterative methods. SIAM, 2 edn
6. Beck C, D'Andrea, R (1998) Computational study and comparisons of lft reducibility methods. In: Proceedings of the American control conference
7. Belkin M, Niyogi P (2003) Laplacian eigenmaps for dimensionality reduction and data representation. Neural Comput 15:1373–1396
8. Boyd S Subgradients. Lecture notes, EE392o. http://www.stanford.edu/class/ee392o/subgrad.pdf
9. Boyd S, Parikh N, Chu E, Peleato B, Eckstein J (2011) Distributed optimization and statistical learning via the alternating direction method of multipliers. Found Trends Mach Learn 3(1):1–122
10. Boykov Y, Kolmogorov V (2001) An experimental comparison of min-cut/max-flow algorithms for energy minimization in vision. IEEE Trans Pattern Anal Mach Intell 26:359–374
11. Boykov Y, Veksler O, Zabih R (2001) Fast approximate energy minimization via graph cuts. IEEE Trans Pattern Anal Mach Intell 23:2001
12. Briggs WL, Henson VE, McCormick SF (2000) A multigrid tutorial, 2 edn Society for Industrial and Applied Mathematics, Philadelphia
13. Bui-Thanh T, Willcox K, Ghattas O (2008) Model reduction for large-scale systems with high-dimensional parametric input space. SIAM J Sci Comput 30(6):3270–3288
14. Cai J, Candès EJ, Shen Z (2010) A singular value thresholding algorithm for matrix completion. SIAM J Opt 20(4):1956–1982
15. Cai JF, Osher S, Shen Z (2009) Convergence of the linearized bregman iteration for ℓ_1-norm minimization. Math Comp 78:2127–2136
16. Candès EJ, Li XD, Ma Y, Wrightes J (2011) Robust principal component analysis. J ACM 58(11):1–37
17. Chiu JW, Demanet L (2012) Matrix probing and its conditioning. SIAM J Num Anal 50(1):171–193
18. Combettes PL, Pesquet J-C (2011) Proximal splitting methods in signal processing. In: Bauschke HH, Burachik RS, Combettes PL, Elser V, Luke DR, Wolkowicz H (eds) Fixed-point algorithms for inverse problems in science and engineering, Springer, Berlin, pp 185–212
19. Conn AR, Gould N, Toint PhL (1993) Improving the decomposition of partially separable functions in the context of large-scale optimization: a first approach
20. Courtois PJ (1975) Error analysis in nearly-completely decomposable stochastic systems. Econometrica 43(4):691–709
21. Courtois PJ (1977) Decomposability : queueing and computer system applications, volume ACM monograph series of 012193750X. Academic Press
22. Dempster AP, Laird NM, Rubin DB (1977) Maximum likelihood from incomplete data via the em algorithm. J R Stat Soc Ser B 39(1):1–38
23. Domingos P (1999) Data Min Knowl Discov 3:409–425
24. Donoho DL, Stark PhB (1989) Uncertainty principles and signal recovery. SIAM J Appl Math 49(3):906–931
25. Elman JL (1990) Finding structure in time. Cogn Sci 14(2):179–211

26. Fazel M, Hindi H, Boyd S (2001) A rank minimization heuristic with application to minimum order system approximation. In: Proceedings of the American control conference
27. Fazel M, Hindi H, Boyd S (2003) Log-Det heuristic for matrix rank minimization with applications to Hankel and Euclidean distance matrices. Proc Am Control Conf 3:2156–2162
28. Feng C (2010) Localization of wireless sensors via nuclear norm for rank minimization. In: Global telecommunications conference IEEE, pp 1–5
29. Freund RM, S Mizuno (1996) Interior point methods: current status and future directions. OPTIMA — mathematical programming society newsletter
30. Fuchs J-J (2004) On sparse representations in arbitrary redundant bases. IEEE Trans Inf Theory 50:1341–1344
31. Ghaoui LE, Gahinet P (1993) Rank minimization under lmi constraints: a framework for output feedback problems. In: Proceedings of the European control conference
32. Globerson A, Chechik G, Pereira F, Tishby N (2007) Euclidean embedding of co-occurrence data. J Mach Learn Res 8:2265–2295
33. Golub GH, Van Loan CF (1996) Matrix computations, 3 edn. Johns Hopkins University Press
34. Gonzalez-Vega L, Rúa IF (2009) Solving the implicitization, inversion and reparametrization problems for rational curves through subresultants. Comput Aided Geom Des 26(9):941–961
35. Grama A, Karypis G, Gupta A, Kumar V (2003) Introduction to parallel computing: design and analysis of algorithms. Addison-Wesley
36. Griewank A, Toint PhL (1981) On the unconstrained optimization of partially separable objective functions. In: Powell MJD (ed) Nonlinear optimization. Academic press, London, pp 301–312
37. Griewank A, Walther A (2008) Evaluating derivatives: principles and techniques of algorithmic differentiation. Soc Indus Appl Math
38. Grossmann C (2009) System identification via nuclear norm regularization for simulated moving bed processes from incomplete data sets. In: Proceedings of the 48th IEEE conference on decision and control, 2009 held jointly with the 28th Chinese control conference. CDC/CCC 2009
39. Gugercin S, Willcox K (2008) Krylov projection framework for fourier model reduction
40. Hansson A, Liu Z, Vandenberghe L (2012) Subspace system identification via weighted nuclear norm optimization. CoRR, abs/1207.0023
41. Hazan E (2008) Sparse approximation solutions to semidefinite programs. In: LATIN, pp 306–316
42. Hendrickson B, Leland R (1995) A multilevel algorithm for partitioning graphs. In: Proceedings of the 1995 ACM/IEEE conference on supercomputing, supercomputing 1995, ACM, New York
43. Horesh L, Haber E (2009) Sensitivity computation of the ℓ_1 minimization problem and its application to dictionary design of ill-posed problems. Inverse Prob 25(9):095009
44. Jaggi M, Sulovsky M (2010) A simple algorithm for nuclear norm regularized problems. In: Proceedings of the 27th international conference on machine learning
45. Ji S, Ye J (2009) An accelerated gradient method for trace norm minimization. In: Proceedings of the 26th annual international conference on machine learning, ICML 2009, ACM, New York, pp 457–464
46. Kaipio JP, Kolehmainen V, Vauhkonen M, Somersalo E (1999) Inverse problems with structural prior information. Inverse Prob 15(3):713–729
47. Kanevsky D, Carmi A, Horesh L, Gurfil P, Ramabhadran B, Sainath TN (2010) Kalman filtering for compressed sensing. In: 13th conference on information fusion (FUSION), pp 1–8
48. Karypis G, Kumar V (1998) A fast and high quality multilevel scheme for partitioning irregular graphs. SIAM J Sci Comput 20(1):359–392
49. Lee K, Bresler Y (2010) Admira: atomic decomposition for minimum rank approximation. IEEE Trans Inf Theory 56(9):4402–4416
50. Li HF (2004) Minimum entropy clustering and applications to gene expression analysis. In: Proceedings of IEEE computational systems bioinformatics conference, pp 142–151

51. Liiv I (2010) Seriation and matrix reordering methods: an historical overview. Stat Anal Data Min 3(2):70–91
52. Lin ZC, Chen MM, Ma Y (2009) The augmented Lagrange multiplier method for exact recovery of a corrupted low-rank matrices. Technical report
53. Lin ZC, Ganesh A, Wright J, Wu LQ, Chen MM, Ma Y (2009) Fast convex optimization algorithms for exact recovery of a corrupted low-rank matrix. In: Conference version published in international workshop on computational advances in multi-sensor adaptive processing
54. Liu J, Sycara KP (1995) Exploiting problem structure for distributed constraint optimization. In: Proceedings of the first international conference on multi-agent systems. MIT Press, pp 246–253
55. Liu Y-J, Sun D, Toh K-C (2012) An implementable proximal point algorithmic framework for nuclear norm minimization. Mathe Program 133:399–436
56. Liu Z, Vandenberghe L (2009) Interior-point method for nuclear norm approximation with application to system identification. SIAM J Matrix Anal Appl 31:1235–1256
57. Ma SQ, Goldfarb D, Chen LF (2011) Fixed point and bregman iterative methods for matrix rank minimization. Math Program 128:321–353
58. Majumdar A, Ward RK (2012) Nuclear norm-regularized sense reconstruction. Magn Reson Imaging 30(2):213–221
59. Mallat SG (1989) A theory for multiresolution signal decomposition: the wavelet representation. IEEE Trans Pattern Anal Mach Intell 11:674–693
60. Mardani M, Mateos G, Giannakis GB (2012) In-network sparsity-regularized rank minimization: algorithms and applications. CoRR, abs/1203.1570
61. Mazumder R, Hastie T, Tibshirani R (2010) Spectral regularization algorithms for learning large incomplete matrices. J Mach Learn Res 99:2287–2322
62. Mesbahi M, Papavassilopoulos GP (1997) On the rank minimization problem over a positive semi-definite linear matrix inequality. IEEE Trans Autom Control 42(2):239–243
63. Mohan K, Fazel M (2010) Reweighted nuclear norm minimization with application to system identification. In: American control conference (ACC), pp 2953–2959
64. Moore BC (1981) Principal component analysis in linear systems: controllability, observability, and model reduction. IEEE Trans Autom Cont AC-26:17–32
65. Natarajan BK (1995) Sparse approximate solutions to linear systems. SIAM J Comput 24:227–234
66. Neal R, Hinton GE (1998) A view of the EM algorithm that justifies incremental, sparse, and other variants. In: Learning in graphical models. Kluwer Academic Publishers, pp 355–368
67. Nemirovsky AS, Yudin DB (1983) Problem complexity and method efficiency in optimization. Wiley-Interscience series in discrete mathematics, Wiley
68. Nesterov Y (1983) A method of solving a convex programming problem with convergence rate $\mathcal{O}\left(\frac{1}{\sqrt{k}}\right)$. Sov Math Dokl 27:372–376
69. Nesterov Y, Nemirovskii A (1994) Interior-point polynomial algorithms in convex programming. In: Studies in applied and numerical mathematics. Soc for Industrial and Applied Math
70. Olsson C, Oskarsson M (2009) A convex approach to low rank matrix approximation with missing data. In: Proceedings of the 16th Scandinavian conference on image, analysis, SCIA '09, pp 301–309
71. Peng YG, Ganesh A, Wright J, Xu WL, Ma Y (2010) RASL: robust alignment by sparse and low-rank decomposition for linearly correlated images. In: IEEE conference on computer vision and pattern recognition (CVPR), pp 763–770
72. Phillips DL (1962) A technique for the numerical solution of certain integral equations of the first kind. J ACM 9(1):84–97
73. Pichel JC, Rivera FF, Fernández M, Rodríguez A (2012) Optimization of sparse matrix-vector multiplication using reordering techniques on GPUs. Microprocess Microsyst 36(2):65–77
74. Pothen A, Simon HD, Liou K-P (1990) Partitioning sparse matrices with eigenvectors of graphs. SIAM J Matrix Anal Appl 11(3):430–452
75. Recht B, Fazel M, Parillo P (2010) Guaranteed minimum rank solutions to linear matrix equations via nuclear norm minimization. SIAM Rev 52(3):471–501

76. Recht B, Ré C (2011) Parallel stochastic gradient algorithms for large-scale matrix completion. In: Optimization (Online)
77. Rennie JDM, Srebro N (2005) Fast maximum margin matrix factorization for collaborative prediction. In: Proceedings of the international conference of Machine Learning
78. Rudin LI, Osher S, Fatemi E (1992) Nonlinear total variation based noise removal algorithms. Phys D 60:259–268
79. Shapiro A, Homem de Mello T (2000) On rate of convergence of Monte Carlo approximations of stochastic programs. SIAM J Opt 11:70–86
80. Shapiro A, Dentcheva D, Ruszczyński A (eds) (2009) Lecture notes on stochastic programming: modeling and theory. SIAM, Philadelphia
81. Speer T, Kuppe M, Hoschek J (1998) Global reparametrization for curve approximation. Comput Aided Geom Des 15(9):869–877
82. Strout MM, Hovland PD (2004) Metrics and models for reordering transformations. In: Proceedings of the 2004 workshop on Memory system performance, MSP '04, ACM, New York, pp 23–34
83. Tikhonov AN (1963) Solution of incorrectly formulated problems and the regularization method. Sov Math Dokl 4:1035–1038
84. Toh K-C, Yun S (2010) An accelerated proximal gradient algorithm for nuclear norm regularized least squares problems. Pacific J Optim 6:615–640
85. Tor AJ (1997) On tikhonov regularization, bias and variance in nonlinear system identification. Automatica 33:441–446
86. Tropp JA (2004) Greed is good: algorithmic results for sparse approximation. IEEE Trans Inform Theory 50:2231–2242
87. Trottenberg U, Oosterlee CW, Schüller A (2000) Multigrid. Academic Press, London
88. Tseng P (2008) On accelerated proximal gradient methods for convex-concave optimization. SIAM J Optim (submitted)
89. Weinberger KQ, Saul LK (2006) Unsupervised learning of image manifolds by semidefinite programming. Int J Comput Vis 70(1):77–90
90. Willcox K, Peraire J (2002) Balanced model reduction via the proper orthogonal decomposition. AIAA J 40:2323–2330
91. Yang JF, Yuan XM (2013) Linearized augmented Lagrangian and alternating direction methods for nuclear norm minimization. Math Comp 82(281):301–329
92. Yin W, Osher S, Goldfarb D, Darbon J (2008) Bregman iterative algorithms for l1-minimization with applications to compressed sensing. SIAM J Imaging Sci 1:143–168
93. Yomdin Y (2008) Analytic reparametrization of semi-algebraic sets. J Complex 24(1):54–76
94. Zhang J (1999) A multilevel dual reordering strategy for robust incomplete lu factorization of indefinite matrices

Chapter 5
Nonnegative Tensor Decomposition

N. Hao, L. Horesh and M. E. Kilmer

Abstract It is more and more common to encounter applications where the collected data is most naturally stored or represented in a multi-dimensional array, known as a tensor. The goal is often to approximate this tensor as a sum of some type of combination of basic elements, where the notation of what is a basic element is specific to the type of factorization employed. If the number of terms in the combination is few, the tensor factorization gives (implicitly) a sparse (approximate) representation of the data. The terms (e.g. vectors, matrices, tensors) in the combination themselves may also be sparse. This chapter highlights recent developments in the area of non-negative tensor factorization which admit such sparse representations. Specifically, we consider the approximate factorization of third and fourth order tensors into non-negative sums of types of outer-products of objects with one dimension less using the so-called t-product. A demonstration on an application in facial recognition shows the potential promise of the overall approach. We discuss a number of algorithmic options for solving the resulting optimization problems, and modification of such algorithms for increasing the sparsity.

5.1 Introduction and Aims

The non-negative matrix factorization (NMF) problem is a well-known, well-researched problem, and it has been shown as a useful tool for describing or decomposing multi-variable data into its constitutive parts. As we highlight briefly below, it is often desirable and indeed not uncommon for the individual non-negative factors to be *sparse*. By sparse, we mean that the percentage of zero elements in the matrix is

N. Hao (✉) · M. E. Kilmer
Tufts University, 503 Boston Ave, Medford, MA 02155, USA
e-mail: ning.hao@tufts.edu

L. Horesh
T. J. Watson Research Center, Yorktown Heights, NY 10598, USA

A. Y. Carmi et al. (eds.), *Compressed Sensing & Sparse Filtering*,
Signals and Communication Technology, DOI: 10.1007/978-3-642-38398-4_5,

quite high relative to the total number of possible non-zero entries (e.g. nm possible non-zero entries in an $n \times m$ matrix). The NMF problem first appears in [35], and there has been considerable literature on the subject since that time: see [10] for example, and the references therein. In the context of this paper, a *tensor* refers to a multidimensional array. For example, a third order tensor refers to a 3-way array, a fourth order tensor is a 4-way array, and so forth. Non-negative tensor factorization (NTF), the natural generalization of NMF to higher dimensional arrays, is a field that has been less well explored, but is ever evolving. Therefore, it is worthwhile to provide a timely supplement to the existing literature (see, for instance, [9–12, 15, 16]) on the subject.

To set the stage for the discussion of the NTF problem, it behoves us to briefly highlight the NMF problem. The basic non-negative matrix factorization model involves decomposing a non-negative matrix $\mathbf{A} \in \mathbb{R}^{n \times m}$ into two non-negative matrices $\mathbf{G} \in \mathbb{R}^{n \times p}$ and $\mathbf{H} \in \mathbb{R}^{m \times p}$ such that

$$\mathbf{A} \approx \mathbf{G}\mathbf{H}^{\top}. \tag{5.1}$$

In practice, this approximate factorization is obtained by minimizing some distance function $\mathcal{D}$, between $\mathbf{A}$ and the product $\mathbf{G}\mathbf{H}^{\top}$:

$$\min_{\mathbf{G},\mathbf{H}} \mathcal{D}(\mathbf{A}; \mathbf{G}, \mathbf{H}) \;||\; \text{s. t. constraints on } \mathbf{G} \text{ and/or } \mathbf{H}. \tag{5.2}$$

A typical choice for $\mathcal{D}$ is $\mathcal{D}(\mathbf{A}; \mathbf{G}, \mathbf{H}) = \|\mathbf{A} - \mathbf{G}\mathbf{H}^{\top}\|_F$, although other metrics based on different statistical assumptions of the model may be used. The constraints are necessary to encourage additional conditions on the factors. In many applications, additional constraints such as smoothness, sparsity, symmetry, and orthogonality are applied to $\mathbf{G}$ and $\mathbf{H}$ [10]. These may be hard constraints, or the constraints may be added as penalties to the objective function directly.

To understand why non-negativity and sparsity of factors tend to go hand in hand with respect to the NMF problem, consider the following. In applications, the matrix $\mathbf{A}$ often represents measured or sampled data. For instance, the jth column of $\mathbf{A}$ might correspond to a particular pixel in an image (i.e. the non-negative image intensity values), and the kth row would correspond to a particular spectral band. In the following, we use MATLAB notation to index into matrices and arrays. In particular, $\mathbf{A}_{:,j}$ means the jth column of the matrix $\mathbf{A}$. Thus,

$$\mathbf{A} \approx \mathbf{G}\mathbf{H}^{\top} \rightarrow \mathbf{A}_{:,j} \approx \sum_{i=1}^{p} \mathbf{G}_{:,i} h_{j,i},$$

where $h_{j,i}$ are scalars corresponding to the j, i position in $\mathbf{H}$. That is, the jth column of $\mathbf{A}$ is a linear combination of the columns of $\mathbf{G}$. The columns of $\mathbf{G}$ are usually referred to as the feature vectors, and the scalars $h_{j,i}$ are the weights. Therefore, the NMF problem is equivalent to finding a non-negatively weighted representation of the p non-negative feature vectors $\mathbf{G}_{:,i}$. The feature vectors are meant to represent something distinctive, such as a chemical signature. Each pixel comes from some different element in the real world (i.e. grass) and therefore would be a mixture of some other base chemical compounds as described over the measured bands. Now, since the model does not allow for subtraction, and presumably, not every sample is comprised of every feature, some of the $h_{j,i}$ are zero. If the features themselves are distinctive, they may also well be sparse vectors. For example, as noted in [10], in the case of facial image data "the additive or part-based nature of NMF has been shown to result in a basis of facial features, such as eyes, nose, and lips."

When the data is already high dimensional by nature, however, it seems more natural to represent that information in a high dimensional space, rather than in a 2D space by flattening (i.e. collapsing) all the information to a matrix form. High dimensional representations offer consistent means for preserving inherent multilinear model structures. Multi-way analysis often provides unique insights into the relations between the entities that span the various dimensions. This is especially crucial when these dimensions can be interpreted in a meaningful way (e.g. correspond to some physical entities). Consider, for example, the problem of 2D facial recognition. A database of images itself could be considered at least as a 3rd order data cube, with each 2D image making up a slice of that cube. Indeed, there are facial recognition papers [1, 32–34] wherein the images are represented using a higher-dimensional array, where the groupings in other dimensions are made according to lighting and pose, for example. Intuitively, "flattening" the data into a matrix and seeking an NMF of that data looses something in the translation. Certainly, recent work on 2D facial recognition has shown that retaining the data as multi-way arrays and seeking PCA-like decompositions of those arrays can lead to significant compression over matrix-based PCA approaches [1, 18, 32–34]. Alternatively, one may want to decompose 2D signals with a time component that accounts for a 3rd dimension. There are many other examples that illustrate that the modeler stands to gain by keeping and factoring the multi-way model, and looking for features which are themselves somehow multidimensional (see [8, 25] for example).

Indeed, a comprehensive overview, at the time of its publication, of non-negative tensor factorizations and applications exists [10]. Rather than regurgitate all of the existing NTF literature. Because of the prevalence of such applications, the goal in the present chapter is to augment the existing literature on NTF with the latest NTF factorizations based on a different tensor framework first elucidated in [23, 24] and expanded in [21].

For the purposes of the present chapter, we focus the discussion on non-negative factorization of third and fourth order tensors with the factorization being built around the framework in [21, 23]. Yet, in moving from third to fourth order tensors, we will be able to show how the approach can be generalized recursively to apply to higher-order tensors. The beauty of building around the aforementioned framework is that the algorithms look much like the familiar NMF approaches. We will show how certain constraints tend to enforce additional sparsity in the factors. A few examples will demonstrate the promise of the new approach.

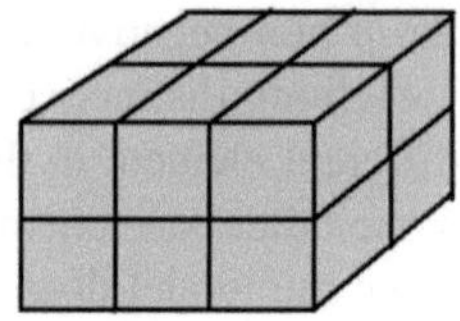

Fig. 5.1 An example of a third order tensor in $\mathbb{R}^{2\times3\times2}$

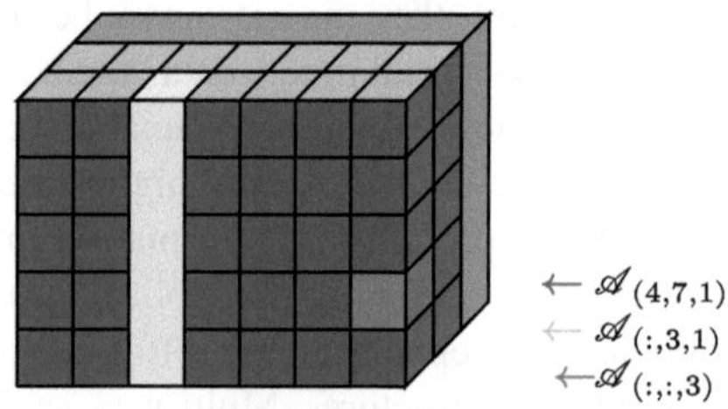

Fig. 5.2 Illustration of indexing schemes for a third order tensor

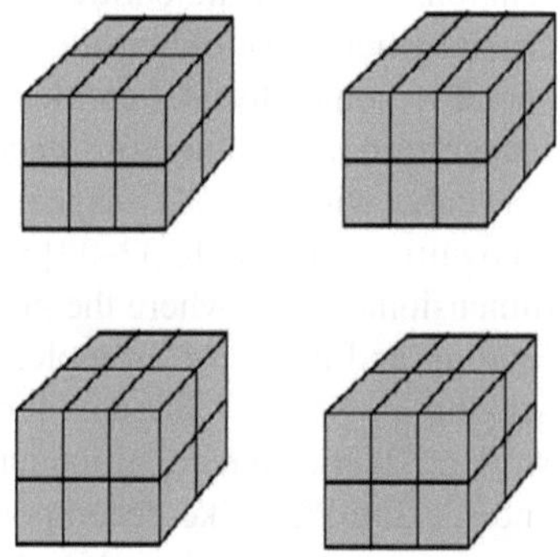

Fig. 5.3 A $2\times3\times2\times4$, fourth order tensor, represented visually as a grouping of 4, third order tensors. The $2 \times 3 \times 2$ box in the upper left would be $\mathscr{A}_{:,:,:,1}$, while the box in the lower right would be $\mathscr{A}_{:,:,:,4}$

5.2 Notation and Motivation

We begin with a presentation of notation and basic definitions. A visual interpretation of a third order tensor is given in Fig. 5.1 as a starting point.

Throughout this work, scalars, vector, matrices, and tensors are denoted as lowercase letters (a), boldface lowercase letters (**a**), boldface uppercase letters (**A**), and boldface script letters ($\mathscr{A}$), respectively. The ith entry of a vector **a** is denoted a_i, the i, j^{th} entry of a matrix **B** is B_{ij}, the i, j, k^{th} entry of a third order tensor $\mathscr{C}$ is $\mathscr{C}_{ijk}$, etc. We may also use MATLAB colon notation to index into objects. For example, $\mathscr{C}_{:,3,4}$ would denote the column vector that corresponds to all entries of the 3rd column, 4th frontal slice of a third order tensor $\mathscr{C}$, while $\mathscr{C}_{:,1,:}$ would be a matrix, oriented into the page, corresponding to the first lateral slice of that tensor. The jth column of a matrix **A** is denoted $\mathbf{a}_j$, or, when ambiguity arises, $\mathbf{A}_{:,j}$.

Suppose that $\mathbf{g} \in \mathbb{R}^m$, $\mathbf{h} \in \mathbb{R}^\ell$, $\mathbf{w} \in \mathbb{R}^n$. Then we can form a third order, $m \times \ell \times n$ tensor using the outer-product $\circ$ between the three vectors, as follows:

Definition If $\mathscr{A} = \mathbf{g} \circ \mathbf{h} \circ \mathbf{w}$, then the i, j, k^{th} scalar entry of $\mathscr{A}$ is given by $\mathscr{A}_{i,j,k} = g_i h_j w_k$ and we say that $\mathscr{A}$ is a **rank-1 tensor**. Here $1 \leq i \leq m, 1 \leq j \leq \ell, 1 \leq k \leq n$. Similarly, a fourth order rank-1 tensor, $\mathscr{C}$, would result from the 4-way outer product $\mathbf{g} \circ \mathbf{h} \circ \mathbf{w} \circ \mathbf{z}$.

5.2.1 Popular Existing Tensor Models

One of the most well-known decompositions is the CANDECOMP/PARAFAC[1], or CP, decomposition. An exact CP decomposition for a third order tensor $\mathscr{A} \in \mathbb{R}^{m \times \ell \times n}$ is given as

$$\mathscr{A} = \sum_{i=1}^{r} \mathbf{g}_i \circ \mathbf{h}_i \circ \mathbf{w}_i, \qquad \mathbf{g}_i \in \mathbb{R}^m, \mathbf{h}_i \in \mathbb{R}^\ell, \mathbf{w}_i \in \mathbb{R}^n.$$

If r is minimal in this expression for $\mathscr{A}$, then r is called the **tensor-rank**. Unfortunately, while rank-revealing factorizations of matrices exist (the best known of these being the SVD), in general there is no closed form solution to determine the rank of a tensor of order three or higher a priori, and there is no straightforward algorithm to compute the tensor rank (it is an NP-hard problem) [25]. In practice, one often guesses a value for r, and searches for the **factor matrices** $\mathbf{G} \in \mathbb{R}^{m \times r}$, $\mathbf{H} \in \mathbb{R}^{\ell \times r}$, $\mathbf{W} \in \mathbb{R}^{n \times r}$, whose columns correspond to the vectors in the approximate decomposition, that best fit $\mathscr{A}$. The best fit measure is often the Frobenius norm.

The most obvious NTF analog (for the third order case) to the NMF problem uses this CP model. The problem to solve in this case is

$$\min_{\mathbf{G},\mathbf{H},\mathbf{W}} \mathcal{D}(\mathscr{A} - \sum_{i=1}^{r} \mathbf{g}_i \circ \mathbf{h}_i \circ \mathbf{w}_i), \quad \| \text{ s. t. non-negativity constraints}$$

where $\mathcal{D}$ denotes some distance measure, often the Frobenius norm. Additional information regarding studies that focuses on algorithms for solving the above problem, with the possible addition of further constraints to encourage sparsity, can be found, for example in [10]. More recently, there has been work for sparse non-negative tensors, where the distance measure is replaced by optimization of the K-L divergence, motivated by statistical considerations relative to the distribution of non-zeros [9].

There are many other types of tensor decompositions (see [25] for a partial list), for each of these, non-negativity constraints can be imposed. Unfortunately, the notations used to express these decompositions is not completely consistent throughout the literature. We shall highlight only two additional decompositions here, and for the sake of brevity we will express these using only outer-product notation and only

[1] R. Harshman, '70; J. Carroll and J. Chang, '70.

for third order tensors. We refer the interested reader to [10, 25] for additional factorizations/expressions.

For $\mathscr{A} \in \mathbb{R}^{m \times \ell \times n}$, the Tucker-3 decomposition[2] is

$$\mathscr{A} = \sum_{i=1}^{r_1} \sum_{j=1}^{r_2} \sum_{k=1}^{r_3} c_{i,j,k} \mathbf{g}_i \circ \mathbf{h}_j \circ \mathbf{w}_k, \tag{5.3}$$

where $\mathscr{C} \in \mathbb{R}^{r_1 \times r_2 \times r_3}$ is called the **core tensor**. Note that if $r_1 = r_2 = r_3$ and the core tensor is super-diagonal (i.e. only the (i,i,i) components are potentially non-zero), this reduces to a CP decomposition. A Tucker-3 decomposition can be found in closed form in a straightforward manner, and the values r_i are bounded by the ranks of the various matrices resulting from the corresponding flattenings the tensor. Because of the additional degrees of freedom in this model and the relationship between the tensor flattenings, it is possible to find an exact decomposition where the factor matrices $\mathbf{G}$, $\mathbf{H}$, $\mathbf{W}$ have orthogonal (even orthonormal) columns. One way of obtaining such an exact orthogonal Tucker-3 is the HOSVD [26]. Note that unlike the matrix SVD, entries in the core tensor are not guaranteed to be non-negative (although they will be real), nor is the core necessarily diagonal. In practice, one looks for an approximate factorization by fixing the r_i to be relatively small and minimizing the distance between $\mathscr{A}$ and the model on the right-hand side of 5.3, with suitable constraints, such as non-negativity.

A related decomposition is the Tucker-2 factorization [3], given by the expression

$$\mathscr{A} \approx \sum_{i=1}^{r_1} \sum_{j=1}^{r_2} \mathbf{g}_i \circ \mathbf{h}_j \circ \mathscr{C}_{i,j,:}. \tag{5.4}$$

The difference is that the sum over k has disappeared, and now the third term in the expression depends both on i and j and the scaling that was attributable to the core tensor has been wrapped into the last term, which is (under the correct orientation) just a vector stored as a so-called **tube fiber** of the core tensor $\mathscr{C}$. Clearly, if $\mathscr{C}$ is "diagonal" in the sense that the tube fiber is zero if $i \neq j$, this representation also collapses to a CP representation. Another way to visualize this decomposition is that each of the n frontal slices can be written as $\mathbf{G}\mathbf{C}^{(k)}\mathbf{H}^\top$, where $\mathbf{C}^{(k)}$ is a (possibly dense) matrix that corresponds to the kth frontal slice of $\mathscr{C}$. Compression is only achieved if $\mathscr{C}$ is sparse, and/or r_1, r_2 are small relative to the tensor dimensions.

[2] If $\mathscr{A}$ is a fourth order tensor, the core tensor will be fourth order and there will be an additional summand and vector outer product.

5.2.2 The Generic Optimization Model

The point is, no matter which tensor model we use to fit the data, the problem we wish to solve can be framed generically as

$$\min_{\mathscr{B}\in\mathscr{C}} \mathcal{D}(\mathscr{A} - \mathscr{B})\|\text{s.t. constraints on } \mathscr{B}, \tag{5.5}$$

where $\mathscr{C}$ represents the specific tensor model of interest (e.g. CP, Tucker-3, Tucker-2), and $\mathcal{D}$ denotes the distance measure (e.g. Frobenius norm).

Following this definition, two immediate questions arise:

1. What class $\mathscr{C}$ should we use ?
2. What type of constraints should be considered ?

To answer the 2nd question, we turn to the vast literature on NMF. In the context of this book, the two types of constraints of interest to us are non-negativity and sparsity. Indeed, we know that [10].

> "matrix factorization methods that exploit non-negativity and sparsity constraints usually lead to estimation of the hidden components with specific structures and physical interpretations, in contrast to other blind source separation method."

On the face of it, as discussed in the introduction, the non-negativity constraints themselves often induce sparsity. However, we also know that [10]

> ... "solutions obtained by NMF algorithms may not be unique, and to this end it is often necessary to impose additional constraints (which arise naturally from the data considered) such as sparsity or smoothness. Therefore, special emphasis in this chapter is put on various regularization and penalty terms together with local learning rules in which we update sequentially one-by-one vectors of factor matrices. By incorporating the regularization and penalty terms into the weighted Frobenius norm, we show that it is possible to achieve sparse, orthogonal, or smooth representations, thus helping to obtain a desired global solution."

The question of which class $\mathscr{C}$ should be used is more difficult to answer, and may be driven by the application. In chemometrics, for example, the CP may arguably be the correct decomposition to use [8]. For other applications, however, the choice is less clear. If the end goal is simply compression, with non-negative and sparse factors, it might be worth considering the Tucker factorizations. On the other hand, if a "parts based" representation makes more sense from a physical perspective, what are those parts/components/features? Need they be represented as rank-one outer products?

In the present research, we present a new, state-of-the-art model for parts based representation in the third and fourth order tensor case, **which extends naturally to higher order tensors as well**. The beauty of this approach is that in the unconstrained case, when the distance measure is the Frobenius norm, a unique solution is guaranteed. When non-negativity constraints are added, the problem greatly resembles the NMF problem (indeed, when $n = 1$, our NTF problem will collapse to the NMF problem), and we can use our intuition in the NMF problem to inform what to do in the tensor case.

5.3 A Newer Tensor Framework

The definition of matrix-matrix product is such that if you multiply two $n \times n$ matrices together, the result is an $n \times n$ matrix. Thus, matrix multiplication is closed over the set of all $n \times n$ matrices. Furthermore, with respect to the set of $n \times n$ matrices, there is a well defined notation of identity and inverse. Despite a burgeoning literature on tensors in the past decade, until recently, there were no tensor multiplication definitions in the literature that gave rise to similar properties.

In [24], the authors introduced a new definition of third order tensor multiplication, called the **t-product**, along with a corresponding definition of tensor identity and tensor inverse, such that the set of $m \times m \times n$ tensors equipped with these definitions forms a ring. In [23], the authors build on this formalism to derive a new type of tensor SVD, called the t-SVD, that is reminiscent of matrix SVD, and offer an optimality result similar to the Eckart-Young theorem. In [18], we derive PCA-like algorithms based around the t-SVD for compression and facial recognition. Other types of third order tensor factorizations (such as QR, PQR, [18]) can be defined a similar manner. Tensor eigen-computation based around the t-product is the subject of [7, 17]. The work in [20] carefully outlines the linear-algebraic implications of the t-product framework, considering how to define range and nullspace, dimension and multi-rank, and extending familiar numerical linear algebra algorithms within this scope.

In the present work, we build on the t-product and aforementioned studies, using the t-product to formulate a new type of third order NTF problem. As the t-product was shown to generalize to higher-order tensors via a recursive definition [30], we will discuss how our third order method will generalize to higher-order methods by discussing the fourth order case in some detail.

5.3.1 Background Notation and Definitions

We begin this section with the definition of the t-product between two tensors that was introduced in [22, 23], and its extension to fourth order tensors in [30]. In order to do that, we will need to introduce a bit of notation first.

If $\mathscr{A} \in \mathbb{R}^{m \times \ell \times n}$ with $m \times \ell$ frontal slices denoted $\mathbf{A}^{(i)}$ then

$$\text{circ}(\mathscr{A}) = \begin{bmatrix} \mathbf{A}^{(1)} & \mathbf{A}^{(n)} & \mathbf{A}^{(n-1)} & \dots & \mathbf{A}^{(2)} \\ \mathbf{A}^{(2)} & \mathbf{A}^{(1)} & \mathbf{A}^{(n)} & \dots & \mathbf{A}^{(3)} \\ \vdots & \ddots & \ddots & \ddots & \vdots \\ \mathbf{A}^{(n)} & \mathbf{A}^{(n-1)} & \ddots & \mathbf{A}^{(2)} & \mathbf{A}^{(1)} \end{bmatrix},$$

is a block circulant matrix of size $mn \times \ell n$.

If $\mathscr{A} \in \mathbb{R}^{m\times\ell\times n\times k}$, then $\mathscr{A}^{(j)} := \mathscr{A}_{:,:,:,j}$ is on of the third order $m\times\ell\times n$ tensors that comprise the fourth order tensor (see Fig. 5.3). Then we recursively represent $\mathscr{A}$ using a doubly block-circulant matrix

$$\mathrm{circ}_2(\mathscr{A}) = \begin{bmatrix} \mathrm{circ}(\mathscr{A}^{(1)}) & \mathrm{circ}(\mathscr{A}^{(k)} & \mathrm{circ}(\mathscr{A}^{(k-1)}) & \dots & \mathrm{circ}(\mathscr{A}^{(2)}) \\ \mathrm{circ}(\mathscr{A}^{(2)}) & \mathrm{circ}(\mathscr{A}^{(1)}) & \mathrm{circ}(\mathscr{A}^{(k)}) & \dots & \mathrm{circ}(\mathscr{A}^{(3)} \\ \vdots & \ddots & \ddots & \ddots & \vdots \\ \mathrm{circ}(\mathscr{A}^{(k)}) & \mathrm{circ}(\mathscr{A}^{(k-1)}) & \ddots & \mathrm{circ}(\mathscr{A}^{(2)}) & \mathrm{circ}(\mathscr{A}^{(1)}) \end{bmatrix},$$

where each $\mathrm{circ}(\mathscr{A}^{(j)}$ is an $mn\times\ell n$ block circulant matrix, $\mathrm{circ}_2(\mathscr{A})$ is thus $mnk \times \ell nk$ block-circulant matrix with $mn\times\ell n$ block-circulant blocks.

Now if $\mathscr{A} \in \mathbb{R}^{m\times\ell\times n}$, then the operation $\mathtt{Vec}(\mathscr{A})$ takes an $m\times\ell\times n$ tensor and returns a block $mn\times\ell$ matrix, whereas the `Fold` operation undoes this operation:

$$\mathtt{Vec}(\mathscr{A}) = \begin{bmatrix} \mathbf{A}^{(1)} \\ \mathbf{A}^{(2)} \\ \vdots \\ \mathbf{A}^{(n)} \end{bmatrix}, \qquad \mathtt{Fold}(\mathtt{Vec}(\mathscr{A})) = \mathscr{A}.$$

Similarly, for $\mathscr{A} \in \mathbb{R}^{m\times\ell\times n\times k}$,

$$\mathtt{Vec}_2(\mathscr{A}) = \begin{bmatrix} \mathtt{Vec}(\mathscr{A}^{(1)}) \\ \mathtt{Vec}(\mathscr{A}^{(2)}) \\ \vdots \\ \mathtt{Vec}(\mathscr{A}^{(k)}) \end{bmatrix}$$

and $\mathtt{Fold}_2$ undoes this operator to return a fourth order tensor.

The definition of the $*$ multiplication in the third and fourth order cases [22, 30] is as follows:

Definition Let $\mathscr{A}$ be $m\times p\times n$ and $\mathscr{B}$ be $p\times\ell\times n$. Then the **t-product** $\mathscr{A} * \mathscr{B}$ is the $m\times\ell\times n$ tensor

$$\mathscr{A} * \mathscr{B} = \mathtt{Fold}\left(\mathrm{circ}(\mathscr{A}) \cdot \mathtt{Vec}(\mathscr{B})\right).$$

Similarly, if $\mathscr{A}$ is $m\times p\times n\times k$ and $\mathscr{B}$ is $p\times\ell\times n\times k$ then the t-product is the $m\times\ell\times n\times k$ tensor given by

$$\mathscr{A} * \mathscr{B} = \mathtt{Fold}_2\left(\mathrm{circ}_2(\mathscr{A}) \cdot \mathtt{Vec}_2(\mathscr{B})\right).$$

A few more definitions from [23, 30] are in order.

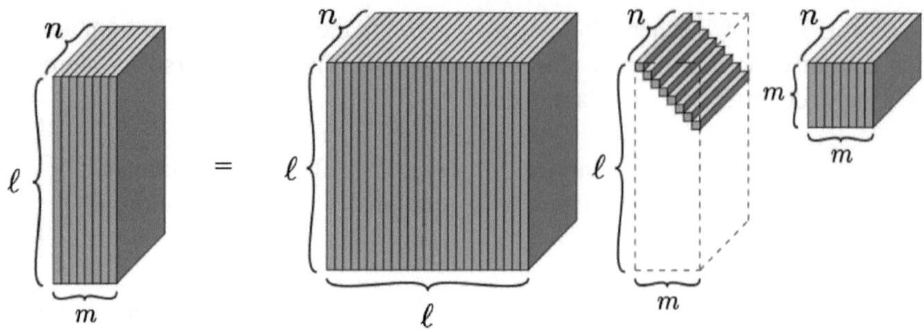

Fig. 5.4 The t-SVD of an $l \times m \times n$ tensor [20]

If $\mathscr{A}$ is $m \times \ell \times n$, then $\mathscr{A}^\top$ is the $\ell \times m \times n$ tensor obtained by transposing each of the frontal slices and then reversing the order of transposed frontal slices 2 through n. If $\mathscr{A}$ is $m \times \ell \times n \times k$, then $\mathscr{A}^\top$ is obtained by reversing the order of each third order $(\mathscr{A}^{(j)})^\top$ for $j = 2, \ldots, k$.

The $m \times m \times n$ identity tensor $\mathscr{I}_{mmn}$ is the tensor whose first frontal slice is the $m \times m$ identity matrix, and whose other frontal slices are all zeros. Similarly, the $m \times m \times n \times k$ fourth order identity tensor is $\mathscr{I}_{mmnk}$ is such that $\mathscr{I}_{:,:,1,1} = \mathbf{I}_{m\times m}$, but all the other entries in $\mathscr{I}_{mmnk}$ are zero.

An $m \times m \times n$ $(m \times m \times n \times k)$ tensor $\mathscr{Q}$ is orthogonal if $\mathscr{Q}^\top * \mathscr{Q} = \mathscr{Q} * \mathscr{Q}^\top = \mathscr{I}$.

We conclude this section with the tensor SVD, or T-SVD from [24] (See Fig. 5.4). In the case $\mathscr{A} \in \mathbb{R}^{m\times\ell\times n}$, there exists an $m \times m \times n$ orthogonal $\mathscr{U}$, an $m \times \ell \times n$ frontal face-wise diagonal $\mathscr{S}$ and $\ell \times \ell \times n$ orthogonal $\mathscr{V}$ such that

$$\mathscr{A} = \mathscr{U} * \mathscr{S} * \mathscr{V}^\top.$$

Note that this is an exact decomposition that can be computed in the time it takes to do n, $m \times \ell$ matrix SVDs. In particular, if $m = \ell = n$, the compute time is proportional to n^4. In the fourth order case, there is also a decomposition of $\mathscr{A}$ as a t-product of orthogonal $\mathscr{U}$, diagonal $\mathscr{S}$ and orthogonal $\mathscr{V}$, where the individual factors are all now fourth order of the appropriate dimension, and by diagonal $\mathscr{S}$, we mean that each 3rd order component of $\mathscr{S}$ that is obtained by holding the last index fixed is a frontal face-wise diagonal third order tensor.

We have more to say about how the T-SVD can be used to give optimal compressed factorizations in the next section, and we will use that discussion to segue into NTF algorithms.

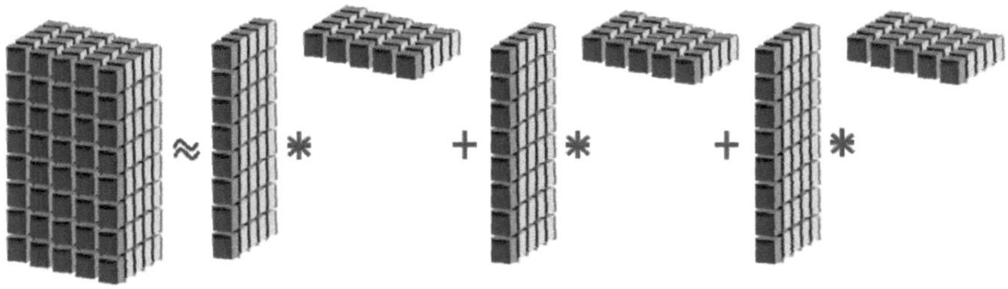

Fig. 5.5 A visual interpretation of a third order tensor approximated as a sum of t-outer-products. Note that if the third dimension, n, is one, the t-product becomes regular matrix-multiplication, and then this illustration collapses to an approximation of a matrix as a sum of outer products of vectors, as represented using the blue

5.3.2 *Defining Outer-products of Tensors*

Now that we have defined the t-product of two tensors of third or fourth order and we have defined tensor transpose, we are in a position to describe a tensor outer product under these two operations.

Definition Let $\mathscr{A} \in \mathbb{R}^{m\times 1\times n}$ and $\mathscr{B} \in \mathbb{R}^{p\times 1\times n}$. Note that these tensors are actually $m \times n$ and $p \times n$ matrices, respectively, but oriented into the page (See Fig. 5.5) then the t-outer-product, $\mathscr{A} * \mathscr{B}^\top$ is a third order tensor of dimensions $m \times p \times n$.

The implication is that the t-SVD can be written as a sum of t-outer-products of tensors (with one vanishing dimension). For example, for $\mathscr{A} \in \mathbb{R}^{m\times \ell\times n}$ with $p := \min(m, \ell)$,

$$\mathscr{A} = \mathscr{U} * \mathscr{S} * \mathscr{V}^\top = \sum_{i=1}^{p} \mathscr{U}_{:,i,:} * \mathscr{S}_{i,i,:} * \mathscr{V}^\top_{:,i,:}.$$

Notice that $\mathscr{U}_{:,i,:}$ and $\mathscr{V}_{:,i,:}$ are matrices, oriented as third order tensors. The respective collections represent, in a sense formalized in [20] the multi-dimensional analogue to basis elements for particular subspaces related to the operators $\mathscr{A}$ and $\mathscr{A}^\top$. The entries $\mathscr{S}_{i,i,:}$ are called the singular-tuples, and are themselves $1 \times 1 \times n$ tensors (or, rather, vectors in $\mathbb{R}^n$, oriented into the page).

The analogy for a fourth order tensor would be to consider the outer product of two, third order tensors of appropriate dimensions.

Definition Let $\mathscr{A} \in \mathbb{R}^{m\times 1\times n\times k}$ and $\mathscr{B} \in \mathbb{R}^{p\times 1\times n\times k}$. Then the t-outer-product $\mathscr{A} * \mathscr{B}^\top$ is a fourth order tensor of dimensions $m \times p \times n \times k$.

It follows that for the fourth order case,

$$\mathscr{A} = \sum_{i=1}^{p} \mathscr{U}_{:,i,:,:} * \mathscr{S}_{i,i,:,:} * \mathscr{V}^\top_{:,i,:,:}.$$

Here, the singular tuples are $1 \times 1 \times n \times k$ tensors.

Now, consider the following unconstrained problem for $\mathscr{A} \in \mathbb{R}^{m \times \ell \times n}$ and integer $p < \min(m, \ell)$:

$$\min_{\mathscr{G} \in \mathbb{R}^{m \times p \times n}, \mathscr{H} \in \mathbb{R}^{\ell \times p \times n}} \|\mathscr{A} - \mathscr{G} * \mathscr{H}^\top\|_F.$$

This problem has been shown in [23] to have a solution; for example, one solution is $\mathscr{G} = \mathscr{U}_{:,1:p,:} * \mathscr{S}_{1:p,1:p,:}$ and $\mathscr{H} = \mathscr{V}_{:,1:p,:}$. This is equivalent to saying there exists a solution to

$$\min_{\mathscr{B}} \|\mathscr{A} - \mathscr{B}\|_F, \quad \| \text{ s.t. } \mathscr{B} \text{ is a sum of p, t-outer-products of tensors.}$$

A similar analysis holds for the 4th order case.

5.3.3 A Parts-based Perspective

Consider the case of matrix-based PCA for the facial recognition problem. In the literature, the basis vectors are typically displayed as images, so that one can concieve how a single image is comprised of several basis images. On the other hand, the t-outer-product representation above leaves something to be desired in terms of visualizing the representation as a parts-based representation.

A similar perspective can be obtained by considering the representation of the *lateral* slices of a third order tensor. If $\mathscr{A} = \mathscr{G} * \mathscr{H}^\top$, where $\mathscr{A} \in \mathbb{R}^{m \times \ell \times n}$, then it can be shown (see [18, 20]) that

$$\texttt{squeeze}(\mathscr{A}_{:,j,:}) = \sum_{i=1}^{p} \texttt{squeeze}(\mathscr{G}_{:,i,:})\text{circ}(\texttt{squeeze}(\mathscr{H}_{j,i,:})),$$

where the `squeeze` operation on the leftmost term twists the matrix clockwise into an $m \times n$ matrix, and the squeeze act upon the tube in the rightmost term converts the $1 \times 1 \times n$ tensor into a column vector.

If $\texttt{squeeze}(\mathscr{H}_{j,i,:})$ where a multiple of $\mathbf{e}_1$, then this expression would say that the jth lateral slice of $\mathscr{A}$ is a linear combination of the columns of basis matrices given by $\texttt{squeeze}(\mathscr{G}_{:,i,:})$. When this is not the case, the basis matrices are weighted from the right by a circulant matrix. Since each $n \times n$ circulant matrix can be decomposed as a sum of at most n powers of the down-shift matrix, it follows that our method does, in fact give a type of parts-based decomposition, and it is evident why entries in $\mathscr{H}$ might indeed be sparse.

5.4 New Non-negative, Constrained, Tensor Factorizations

Piggybacking on the material from the previous section, we are in a position to establish our newest non-negative tensor factorization. Let us deal with the third order case first, where $\mathscr{A} \in \mathbb{R}^{m\times\ell\times n}$. From (5.5), if we take the class $\mathscr{C}$ now to be the set of all $m \times \ell \times n$ non-negative tensors that can be written as a sum of p, for $p \leq \min(m, \ell)$, t-outer-products of tensors of dimension one less, we get the optimization problem

$$\min_{\mathscr{G}\in\mathbb{R}_+^{m\times p\times n},\mathscr{H}\in\mathbb{R}_+^{\ell\times p\times n}} \|\mathscr{A} - \mathscr{G} * \mathscr{H}^\top\|_F \tag{5.6}$$

where

$$\mathscr{G} * \mathscr{H}^T = \sum_{i=1}^{p} \mathscr{G}_{:,i,:} * \mathscr{H}_{:,i,:}^\top.$$

As noted in [23], if the third dimension (n, in this example), is one, the t-product reduce to the matrix product. Hence, when $n = 1$, **the optimization problem would reduce to the standard non-negative matrix factorization problem**.

When $n > 1$, the Frobenius norm expression for (5.6) can be re-written in matrix form, using the definitions from previous sections, as

$$\|\texttt{Vec}(\mathscr{A}) - \text{circ}(\mathscr{G})\texttt{Vec}(\mathscr{H}^\top)\|_F, \tag{5.7}$$

Thus, (5.6) is still a NMF problem, but with a certain additional structure imposed on the first factor.

It is shown in [23] that the transpose operation satisfies $(\mathscr{G} * \mathscr{H}^\top)^\top = \mathscr{H} * \mathscr{G}^\top$. Therefore, the optimization problem above is equal to

$$\min_{\mathscr{G}\in\mathbb{R}_+^{m\times p\times n},\mathscr{H}\in\mathbb{R}_+^{\ell\times p\times n}} \|\mathscr{A}^\top - \mathscr{H} * \mathscr{G}^\top\|_F.$$

Similar to above, the term this Frobenius norm can be expressed in matrix notation as

$$\|\texttt{Vec}(\mathscr{A}^\top) - \text{circ}(\mathscr{H})\texttt{Vec}(\mathscr{G}^\top)\|_F. \tag{5.8}$$

Not surprisingly, when $\mathscr{A} \in \mathbb{R}^{m\times\ell\times n\times k}$, the fourth order version of (5.6) becomes

$$\min_{\mathscr{G}\in\mathbb{R}_+^{m\times p\times n\times k},\mathscr{H}\in\mathbb{R}_+^{\ell\times p\times n\times k}} \|\mathscr{A} - \mathscr{G} * \mathscr{H}^\top\|_F, \tag{5.9}$$

where the last term on the right can be expressed in matrix form as

$$\|\texttt{Vec}_2(\mathscr{A}) - \text{circ}_2(\mathscr{G})\texttt{Vec}_2(\mathscr{H}^\top)\|_F.$$

Presently, we discuss the relatively inexpensive (but non-ideal) method of solving (5.6) through the use of a sequential convex iteration (Sect. 5.4.1.1) and with the aid of Anderson Acceleration (Sect. 5.4.1.5).

5.4.1 The Optimization Problem

The problem (5.6) (likewise, (5.9)) is a non-convex optimization problem. An insight to the entailed non-uniqueness can be gained by observing that for any non-negative, invertible tensor of appropriate dimensions, $\mathscr{R}$, a solution of the form $\mathscr{G} * \mathscr{R}^{-1} * \mathscr{R} * \mathscr{H}^\top$ is equally valid. In more general settings, any non-negative monomial (generalized permutation) tensor, would cause such rotational ambiguity. So far we have considered the objective to be the Frobenius norm of the distance between the tensor $\mathscr{A}$ and its factors $\mathscr{G}$ and $\mathscr{H}$. Yet, other distance choices may be considered, and justified by alternative assumptions regarding the noise model $\mathcal{D}$. Some examples would be the KL-divergence, α divergence, β divergence, Pearson distance, or the Hellinger distance [10] see eqn. (5.10).

$$\min_{\mathscr{G}\in\mathbb{R}_+^{m\times p\times n\times k},\mathscr{H}\in\mathbb{R}_+^{\ell\times p\times n\times k}} \mathcal{D}\left(\mathscr{A}, \mathscr{G} * \mathscr{H}^\top\right), \tag{5.10}$$

Other than the non-negativity constraints, various choices of constraints and regularization schemes can be incorporated into the definition of the tensor factorization optimization problem. Incorporation of such preferences regarding the factors can alleviate the inherent non-uniqueness up to permutational or scaling indeterminacies. That is, we might consider

$$\min_{\mathscr{G}\in\mathbb{R}_+^{m\times p\times n\times k},\mathscr{H}\in\mathbb{R}_+^{\ell\times p\times n\times k}} \mathcal{D}\left(\mathscr{A}, \mathscr{G} * \mathscr{H}^\top\right) + \mathcal{R}(\mathscr{G}, \mathscr{H}), \tag{5.11}$$

where $\mathcal{R}$ denotes a differentiable penalty or regularization operator that is applied to one or both of the factors. Alternatively, we can consider directly adding constraints to (5.10); for example, we might add a sparsity (relaxation of an ℓ_0 type norm) or (tensor) rank constraint to one or both of the factors.

In the following sub-sections we shall briefly review some algorithmic strategies for handling the resulting optimization problem.

5.4.1.1 ALS-based Algorithms

In this subsection, let us consider the case that $\mathcal{D}$ in (5.11) is the Frobenius norm. Due to the symmetry noted in the opening discussion of Sect. 5.4, a very simple Alternating Least Squares (ALS) approach is possible (see also [21] for another representation of this problem). The ALS algorithm, provided below, can be regarded as an instance of sequential convex programming [6], in the sense that we iterate over a sequence (two, in this case) of locally convex problems. This approach is not ideal because it is not a global approach. In other words, this algorithm need not converge to a global minimum, if one exists; and if it does converge to one, it is difficult if not impossible to verify that it has done so. Furthermore, the starting guess can affect the solution. Nevertheless, this heuristic approach is quite popular due to its relative simplicity and it often produces reasonable results. As noted in the previous section, when $n = 1$, (5.6) is exactly the non-negative matrix factorization problem. Thus, when Algorithm ALS below is used for the $n = 1$ case, Algorithm ALS reduces to the alternating non-negative LS algorithm for NMF which is well known in the literature (see [6, 14]). As discussed in the introduction, NMF often admits sparse factors. In the small illustration (see Sect. 5.4.1.5 on accelerating Algorithm ALS) in Figs. 5.11 and 5.12, we see that this is also the case for Algorithm ALS; namely, the tensor factors $\mathscr{G}, \mathscr{H}$ tend to be sparse.

Algorithm 5.1 Alternating Least Squares

1: Fix $\mathscr{G}$ to have non-negative entries
2: For $i = 1$ until convergence do
3: Solve $\min_{\mathscr{H} \in \mathbb{R}_+^{\ell \times p \times n}} \|\mathscr{A} - \mathscr{G} * \mathscr{H}^\top\|_F$
4: Solve $\min_{\mathscr{G} \in \mathbb{R}_+^{m \times p \times n}} \|\mathscr{A}^\top - \mathscr{H} * \mathscr{G}^\top\|_F$

The key to solving the intermediate non-negative least squares problems is rewriting them using their matrix equivalents (5.7),(5.8). Thus, we need only solve two, non-negative least squares problems per iteration, and we can take advantage of the fact that circ($\mathscr{G}$) (circ($\mathscr{H}$)) is a structured matrix. Clearly, the algorithm can be slightly modified to incorporate either a penalty term or hard constraint on each of the subproblems. For example, it might be desirable to enforce sparsity on one of the factors, say $\mathscr{H}$, similar to what is often done in the NMF problem. Since the subproblems are equivalent to non-negative least squares matrix problems, standard techniques can be used.

Other variants of the algorithm exist. The most popular variants include weighted ALS in which covariance is used; incorporation of line search rather than maintaining a fixed point; fixed step iterations; acceleration through the incorporation of various regularization schemes. One of these acceleration strategies is further detailed in the next section.

Note that the ALS Algorithm is suitable for both the third order and fourth order tensor cases.

To compare the performance of the new nonnegative tensor decomposition, and traditional tensor decomposition as well as nonnegative matrix decomposition, we test on a part of the CBCL database which contains 200 gray-level images of faces represented by 19×19 pixels. The sample images are shown in Fig. 5.6. We took the first 100 images and look for an approximation with $p = 5$.

The reconstructed images based on different decompositions are shown in Fig. 5.7:

Fig. 5.6 CBCL database image samples

5.4.1.2 Multiplicative Algorithms

Rather than alternating minimization upon a single objective function, in this class of algorithms the alternating minimization procedures are applied for each subset, while assuming that the two problems share a joint global minimum.

$$\min_{\mathscr{H}\in\mathbb{R}_+^{\ell\times p\times n\times k}} \mathcal{D}_{\mathcal{G}}\left(\mathcal{A}_c^\top, \mathcal{H}_c * \mathcal{G}_c^\top\right) + \mathcal{R}_{\mathscr{H}} \tag{5.12}$$

$$\min_{\mathscr{G}\in\mathbb{R}_+^{m\times p\times n\times k}} \mathcal{D}_{\mathcal{H}}\left(\mathscr{A}_r, \mathscr{G}_r * \mathscr{H}^\top\right) + \mathcal{R}_{\mathscr{G}}, \tag{5.13}$$

where $\mathcal{D}_{\mathscr{G}}$, $\mathcal{D}_{\mathscr{H}}$, $\mathcal{R}_{\mathscr{G}}$, $\mathcal{R}_{\mathscr{H}}$ are prescribed distance measures and regularization operators, respectively, and the subscripts c and r imply that the minimization is performed with respect to subsets of the columns or the rows of the complete tensors.

In the following, for simplicity, we assume that there is no regularization term. In such settings the first order necessary conditions for stationarity of the above term can be phrased as follows:

$$\mathscr{G} \in \mathbb{R}_+^{m\times p\times n\times k} \tag{5.14}$$

$$\nabla_{\mathscr{G}}\mathcal{D}_{\mathscr{G}} \geq 0 \tag{5.15}$$

$$\mathscr{G} \odot \nabla_{\mathscr{G}}\mathcal{D}_{\mathscr{G}} = 0 \tag{5.16}$$

and

$$\mathscr{H} \in \mathbb{R}_+^{\ell\times p\times n\times k} \tag{5.17}$$

$$\nabla_{\mathscr{H}}\mathcal{D}_{\mathscr{H}} \geq 0 \tag{5.18}$$

$$\mathscr{H}^\top \odot \nabla_{\mathscr{H}}\mathcal{D}_{\mathscr{H}} = 0 \tag{5.19}$$

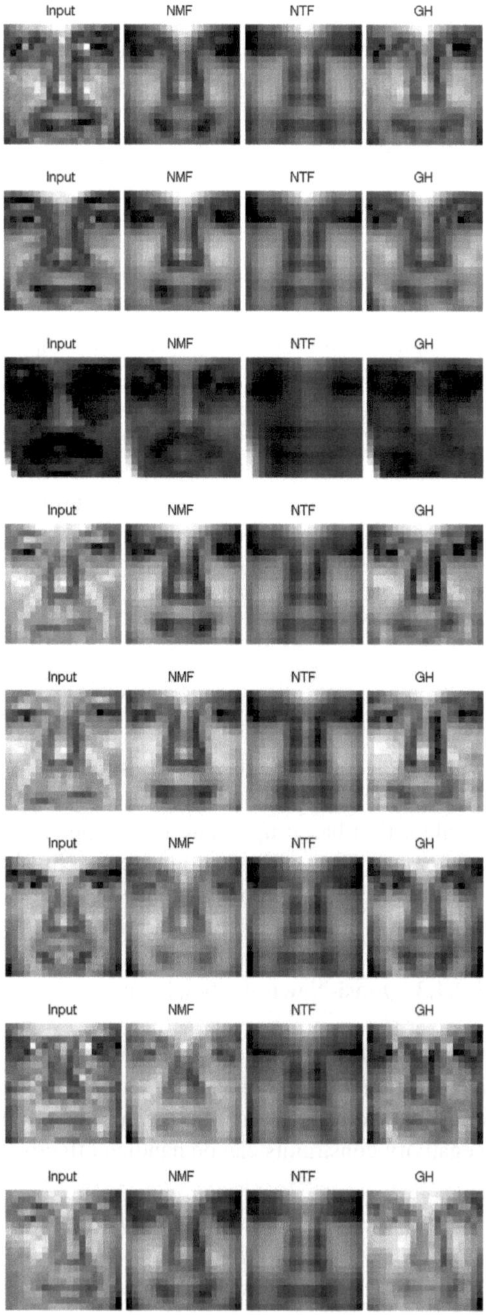

Fig. 5.7 From left to right (repeatedly): Input images, reconstructed images based upon NMF, NTF (based on CP), and upon the new NTF ($\mathscr{G} * \mathscr{H}^{\top}$)

where $\odot$ denotes the Hadamard product.

Considering the Frobenius norm distance measure, Lee and Seung [28], had proposed multiplicative update rules for minimization of the aforementioned objectives. Under our current framework, the gradients of the factors in this context are given by the expressions:

$$\nabla_{\mathscr{G}} \mathcal{D}_{\mathscr{G}} = \left(\mathcal{G} * \mathcal{H}^{\top} - \mathcal{A}\right) * \mathcal{H}^{\top} \tag{5.20}$$

$$\nabla_{\mathscr{H}} \mathcal{D}_{\mathscr{H}} = \mathcal{G}^{\top} * \left(\mathcal{G} * \mathcal{H}^{\top} - \mathcal{A}\right) \tag{5.21}$$

Through substitution of these gradient components into the complementarity conditions above we obtain

$$\mathscr{G} \odot \left(\mathcal{G} * \mathcal{H}^{\top} - \mathcal{A}\right) * \mathcal{H}^{\top} = 0 \tag{5.22}$$

$$\mathscr{H}^{\top} \odot \mathcal{G}^{\top} * \left(\mathcal{G} * \mathcal{H}^{\top} - \mathcal{A}\right) = 0 \tag{5.23}$$

which leads to he following multiplicative update formulas

$$\mathscr{G} \leftarrow \mathscr{G} \odot \mathrm{A}_{+} * \mathscr{H} \oslash \left(\mathscr{G} * \mathscr{H}^{\top} * \mathscr{H}\right) \tag{5.24}$$

$$\mathscr{H}^{\top} \leftarrow \mathscr{H}^{\top} \odot \mathscr{G}^{\top} * \mathrm{A}_{+} \oslash \left(\mathscr{G}^{\top} * \mathscr{G} * \mathscr{H}^{\top}\right) \tag{5.25}$$

where $\oslash$ denotes a point-wise division.

Despite the proven monotonicity property of these expressions in the matrix case, an algorithm based upon alternated application of these rules is not guaranteed to converge to a first-order stationary point [5], although with a slight modification [29], such convergence can be guaranteed.

5.4.1.3 Quasi-Newton Algorithms

Considerable acceleration of the convergence rate can be obtained by incorporation of curvature information in the optimization process. Due to the scale of the problem, quasi-Newton approximation for the Hessian or its inverse can be considered. Nonnegativity constraints can be handled effectively through projections.

$$\mathtt{Vec}(\mathscr{G}) \leftarrow \mathcal{P}\left(\mathtt{Vec}(\mathscr{G}) - \nabla^{-1}_{\mathscr{G}\mathscr{G}} \mathtt{Vec}(\nabla_{\mathscr{G}})\right) \tag{5.26}$$

$$\mathtt{Vec}(\mathscr{H}) \leftarrow \mathcal{P}\left(\mathtt{Vec}(\mathscr{H}) - \nabla^{-1}_{\mathscr{H}\mathscr{H}} \mathtt{Vec}(\nabla_{\mathscr{H}})\right) \tag{5.27}$$

where $\mathcal{P}$ represents projection into a non-negative feasible set, and $\nabla_{\mathscr{G}\mathscr{G}}$ and $\nabla_{\mathscr{H}\mathscr{H}}$ are approximated Hessians with respect to the tensor factors $\mathscr{G}$ and $\mathscr{H}$ respectively. Computationally, this class of algorithms are effective as long as storage of the

Hessian (or its inverse) components can be accommodated effectively. For most practical large-scale applications, limited memory construction of the Hessian (or its inverse) would be a tractable method of choice [31].

5.4.1.4 Provably Globally Convergent Algorithms

Surprisingly, many heuristic methods for matrix and tensor decomposition perform well in practice. An insight into the actual effectiveness of such procedures can be gained through the notion of separability [13]. The notion of separability was initially introduced to determine when NMF is unique. As a reminder from the NMF case 5.1, the matrix to be factored, $\mathbf{A}$, is associated with its non-negative factors $\mathbf{G}, \mathbf{H}$ through the relation $\mathbf{A} = \mathbf{G}\mathbf{H}^\top$. Imagine that $\mathbf{A}$ represents words-by-documents. Think of the columns of $\mathbf{G}$ as representing topics. The separability assumption means that for each topic (i.e. column), there is some word (i.e. row element), that appears only in that topic. For example, if the first topic has an anchor word appearing in the 2nd row, then the 2nd row of $\mathbf{G}$ is a multiple of $\mathbf{e}_1^T$, where $\mathbf{e}_1$ is the first canonical unit vector. Consequently, the 2nd row of $\mathbf{A}$ would have to be a multiple of the first row of $\mathbf{H}^\top$.

Alternatively, suppose $\mathbf{A}$ represents documents-by-words. Then separability corresponds to being able to factor $\mathbf{A} = \mathbf{A}_{\mathcal{I}}\mathbf{H}^\top$ where $\mathcal{I}$ denotes a subset of the columns of $\mathbf{A}$ containing the r anchor terms, and now $\mathbf{H}^\top$ must have a corresponding $r \times r$ diagonal matrix among its columns.

Suppose there are r anchor points. Then separability, in geometric terms, means that data is contained in a cone generated by the r anchor rows (or columns, depending on your view) of $\mathbf{A}$. Removal of a row of $\mathbf{A}$ strictly changes the convex hull if and only if it is an anchor word. Thus, anchor words can be identified via linear programming as follows: one sets up a linear program to see if it is possible to express a given row of $\mathbf{A}$ as a convex combination of the other rows. Any points for which the linear program declares infeasibility must be points that cannot be expressible as a convex combination of others, and therefore must represent anchor points. Several algorithms that try to solve the separable NMF that follow this general argument in their solution approach are found in the very recent literature (see [27] and the references therein).

If one of the factors $\mathbf{G}$ or $\mathbf{H}^\top$ satisfies the separability condition, an algorithm for exact decomposition can provably (in the context of NMF) run in polynomial-time [4]. Interestingly, although the separability condition might seem rather restrictive at first glance, in a broad range of settings these were observed to hold empirically.

In the context of NTF, using the t-product formulation, some implied questions are:

1. How does one specify a separability condition ?
2. Do examples of such occur in settings of interest ?
3. Can the state-of-the-art algorithms for solving the separable NMF be generalized to the present case ?

In order to answer this question we should rely upon the analysis in [20]. The benefit of that analysis is that it allows one to treat (in the third order case) tensors as matrices, whose elements are tubal-scalars. We expect that t-linear combinations replace linear combinations in the convex combinations, though care must be taken because tubal-scalars do not form a field. Our work in this area is on-going.

5.4.1.5 Anderson Acceleration

In a fixed point method, one aims to find the solution $\mathbf{x}$ to $\mathbf{x} = \mathbf{g}(\mathbf{x})$ for some given $g : \mathbb{R}^n \rightarrow \mathbb{R}^n$, using the fixed point iteration,

---

Algorithm 5.2 Fixed Point Iteration

1: Initialize x_0
2: For $k = 1, 2, ...$
3: Set $\mathbf{x}_{k+1} = \mathbf{g}(\mathbf{x}_k)$

Anderson Acceleration is a fixed point acceleration technique due to D.G. Anderson [2]. Recently, Walker and Ni applied Anderson Acceleration to fixed point iteration, and proved the equivalence of Anderson Acceleration without truncation and the generalized minimal residual (GMRES) method on linear problems in [19]. In terms of accelerating a fixed point iteration, the idea is simply that instead of taking $\mathbf{x}_k = \mathbf{g}(\mathbf{x}_{k-1})$, so that the kth step is a function only of the most previous iterate, one should define the kth iterate as an "optimal" linear combination of this fixed point step and some number of previous steps. This increases slightly the amount of storage associated with the method, since some number of previous steps must be stored, and there is some overhead associated with solving the optimization problem associated with computing the optimal combination, but in terms of the overall savings, the method can be quite effective. Mathematically, Anderson Acceleration can be formulated as follows.

Algorithm 5.3 Anderson Acceleration

1: Initialize x_0 and $m \geq 1$
2: For $k = 1, 2, \cdots$
3: Set $m_k = \min\{m, k\}$
4: Set $\mathbf{F}_k = (\mathbf{f}_{k-m_k}, \cdots, \mathbf{f}_k)$, where $\mathbf{f}_i = \mathbf{g}(\mathbf{x}_i) - \mathbf{x}_i$
5: Determine $\alpha^{(k)} = (\alpha_0^{(k)}, \cdots, \alpha_{m_k}^{(k)})^\top$ for the problem
$\min_{\alpha=(\alpha_0,\cdots,\alpha_{m+k})^\top} ||\mathbf{F}_k\alpha||_2 \; s.t. \sum_{i=0}^{m_k} \alpha_i = 1$
6: Set $x_{k+1} = \sum_{i=0}^{m_k} \alpha_i^{(k)} \mathbf{g}(x_{k-m_k+i})$

ALS algorithms for nonnegative matrix decomposition and nonnegative tensor decomposition can both be viewed as fixed point problems. In the NMF case, determining the kth fixed point iterate is consistent with solving the kth step of an ALS algorithm for the pair $(\mathbf{G}, \mathbf{H})$. In the third order NTF case, if we use the decomposition from Sect. 5.4.1.1, then determining the kth fixed point iterate is the same as computing the kth ALS step to find the new $(\mathscr{G}, \mathscr{H})$ in Algorithm ALS above. Therefore, the ALS schemes can be altered to include Anderson Acceleration.

In this example, we randomly generated a third order tensor $\mathscr{A} \in \mathbb{R}^{10\times 10\times 5}$ with integer entries between 1 and 10 as:

$$\mathscr{A}(:,:,1) = \begin{bmatrix} 1 & 5 & 2 & 10 & 5 & 10 & 1 & 10 & 8 & 8 \\ 1 & 5 & 2 & 9 & 1 & 4 & 2 & 8 & 9 & 4 \\ 9 & 4 & 9 & 8 & 1 & 10 & 4 & 7 & 4 & 9 \\ 4 & 9 & 5 & 10 & 2 & 10 & 8 & 6 & 5 & 1 \\ 5 & 7 & 10 & 6 & 3 & 1 & 6 & 4 & 1 & 7 \\ 8 & 10 & 9 & 9 & 4 & 5 & 3 & 2 & 8 & 5 \\ 6 & 8 & 8 & 10 & 9 & 7 & 4 & 8 & 2 & 5 \\ 5 & 10 & 1 & 7 & 8 & 9 & 6 & 5 & 8 & 4 \\ 4 & 10 & 3 & 9 & 3 & 5 & 9 & 3 & 10 & 6 \\ 8 & 5 & 8 & 8 & 5 & 2 & 6 & 8 & 1 & 9 \end{bmatrix}, \mathscr{A}(:,:,2) = \begin{bmatrix} 6 & 4 & 7 & 3 & 7 & 7 & 1 & 5 & 9 & 1 \\ 2 & 1 & 6 & 4 & 3 & 10 & 5 & 6 & 7 & 1 \\ 10 & 7 & 3 & 9 & 6 & 3 & 9 & 6 & 3 & 2 \\ 10 & 4 & 6 & 2 & 9 & 7 & 3 & 2 & 1 & 2 \\ 1 & 9 & 7 & 7 & 8 & 7 & 9 & 5 & 6 & 8 \\ 5 & 2 & 1 & 1 & 2 & 3 & 8 & 2 & 2 & 3 \\ 3 & 9 & 3 & 3 & 4 & 7 & 9 & 3 & 2 & 10 \\ 1 & 10 & 4 & 2 & 10 & 4 & 10 & 1 & 5 & 1 \\ 1 & 3 & 8 & 8 & 8 & 5 & 9 & 2 & 4 & 7 \\ 2 & 1 & 10 & 7 & 1 & 9 & 9 & 10 & 8 & 10 \end{bmatrix}$$

$$\mathscr{A}(:,:,3) = \begin{bmatrix} 6 & 3 & 2 & 6 & 10 & 8 & 9 & 5 & 6 & 4 \\ 8 & 9 & 1 & 7 & 8 & 2 & 9 & 8 & 7 & 4 \\ 3 & 4 & 2 & 1 & 8 & 1 & 1 & 2 & 3 & 2 \\ 8 & 4 & 7 & 5 & 9 & 1 & 10 & 7 & 4 & 2 \\ 8 & 1 & 8 & 1 & 2 & 1 & 6 & 4 & 2 & 6 \\ 7 & 9 & 4 & 10 & 4 & 8 & 1 & 1 & 8 & 2 \\ 6 & 10 & 8 & 6 & 4 & 6 & 1 & 10 & 3 & 7 \\ 4 & 2 & 7 & 6 & 3 & 4 & 10 & 2 & 2 & 1 \\ 9 & 8 & 7 & 9 & 6 & 8 & 5 & 7 & 5 & 3 \\ 4 & 5 & 5 & 7 & 7 & 2 & 7 & 5 & 7 & 4 \end{bmatrix}, \mathscr{A}(:,:,4) = \begin{bmatrix} 2 & 10 & 10 & 10 & 1 & 2 & 9 & 5 & 2 & 6 \\ 6 & 8 & 5 & 5 & 3 & 4 & 5 & 3 & 1 & 9 \\ 7 & 8 & 7 & 5 & 3 & 5 & 6 & 9 & 5 & 10 \\ 7 & 6 & 2 & 7 & 10 & 8 & 7 & 10 & 10 & 2 \\ 1 & 2 & 2 & 2 & 2 & 2 & 4 & 2 & 7 & 2 \\ 6 & 3 & 3 & 9 & 10 & 2 & 5 & 6 & 5 & 8 \\ 5 & 6 & 1 & 8 & 2 & 5 & 1 & 9 & 5 & 1 \\ 7 & 6 & 4 & 6 & 4 & 2 & 8 & 4 & 8 & 3 \\ 4 & 7 & 10 & 5 & 6 & 2 & 4 & 3 & 7 & 5 \\ 7 & 10 & 8 & 6 & 10 & 1 & 6 & 2 & 3 & 7 \end{bmatrix}$$

$$\mathscr{A}(:,:,5) = \begin{bmatrix} 10 & 7 & 2 & 2 & 6 & 7 & 5 & 3 & 7 & 2 \\ 2 & 2 & 9 & 4 & 5 & 2 & 2 & 9 & 8 & 9 \\ 7 & 5 & 9 & 8 & 2 & 2 & 3 & 7 & 6 & 3 \\ 9 & 8 & 3 & 4 & 10 & 5 & 5 & 9 & 9 & 5 \\ 3 & 10 & 1 & 2 & 7 & 6 & 2 & 6 & 8 & 8 \\ 3 & 5 & 1 & 3 & 8 & 4 & 10 & 3 & 8 & 4 \\ 1 & 1 & 8 & 7 & 1 & 10 & 4 & 7 & 10 & 10 \\ 7 & 2 & 10 & 8 & 4 & 2 & 4 & 10 & 3 & 4 \\ 1 & 7 & 10 & 4 & 6 & 2 & 8 & 9 & 6 & 2 \\ 5 & 2 & 5 & 9 & 1 & 7 & 2 & 10 & 8 & 2 \end{bmatrix}$$

We set $p = 3$, so $\mathscr{G}$ is of size $10 \times 3 \times 5$ and $\mathscr{H}^\top$ is of size $3 \times 10 \times 5$. In determining the Anderson Acceleration step, we used a history of up to 3 previous steps. We adopted the SVD-based initialization of the iterations. In each iteration, we implemented alternating nonnegative least-squares (ANNLS) to update $\mathscr{G}$ and $\mathscr{H}$. The Figure 5.8 shows the convergence of the ANNLS iterates with and without Anderson acceleration. Further, Figs. 5.9 and 5.10 and respectively Figs. 5.11 and 5.12 display the sparsity pattern of the tensor factors $\mathscr{G}$ and $\mathscr{H}$ with and without the Anderson acceleration respectively.

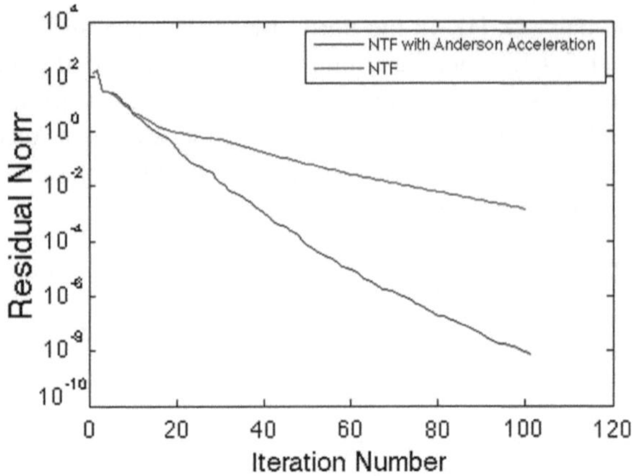

Fig. 5.8 Convergence of NTF with and without Anderson Acceleration

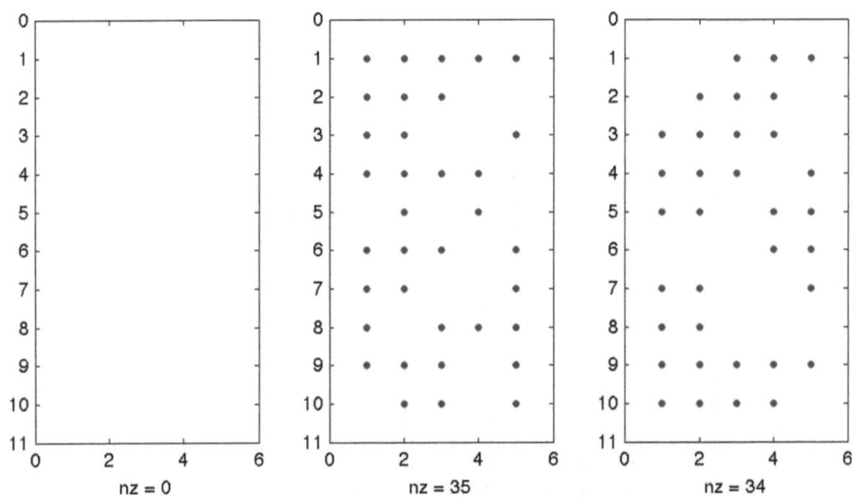

Fig. 5.9 Sparsity pattern of the tensor factor $\mathscr{G}$, shown by frontal slice, with Anderson Acceleration

5.5 Summary

Representation of multi-dimensional data can be performed natively in tensor structures. Such representations lends themselves to powerful multi-way analysis. Unlike its 2nd degree case, the matrix, the definition of tensor operations (e.g. product, decomposition) for higher degrees has been so far been an active research topic. In this chapter, various definitions of the tensor product were described, among which the relatively new notion of t-product, given in [23]. We further discussed the problem of nonnegative tensor factorization an focused in formulaiton and solution in

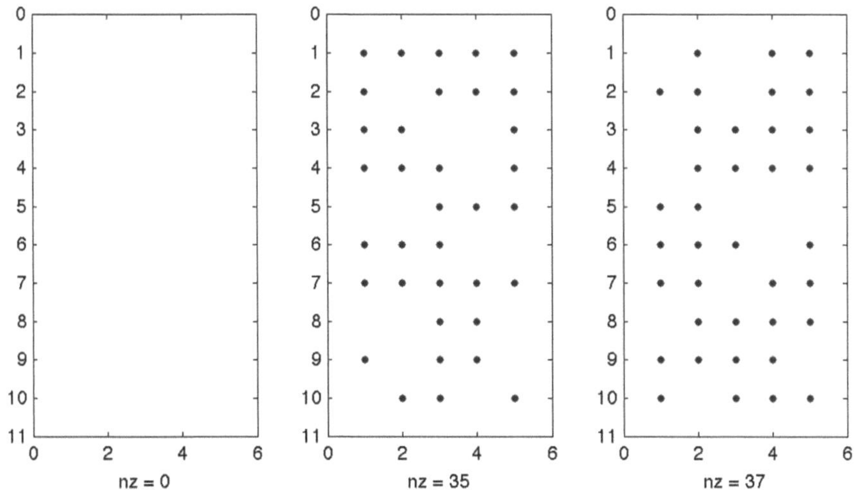

Fig. 5.10 Sparsity pattern of the tensor factor $\mathscr{H}$, shown by frontal slice, with Anderson Acceleration

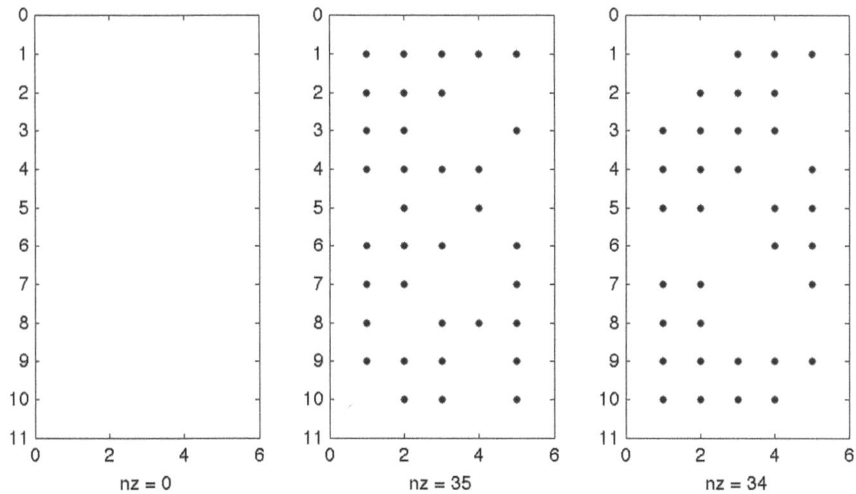

Fig. 5.11 Sparsity pattern of $\mathscr{G}$, shown by frontal slice, without Anderson Acceleration

the context of t-product tensor-tensor relations. As the test cases here illustrate, this model has the potential for producing sparse representations of the data, whether “sparse" is interpreted to mean a few non-zeros or “compact” representation, or both. The beauty of the approach is in its similarity to the NMF problem. Further, the intuition - as we have tried to develop here by moving between the third and fourth order tensor cases - readily generalizes to higher order tensors. It is possible to leverage the structure in the problems to obtain fast algorithms by exploiting the

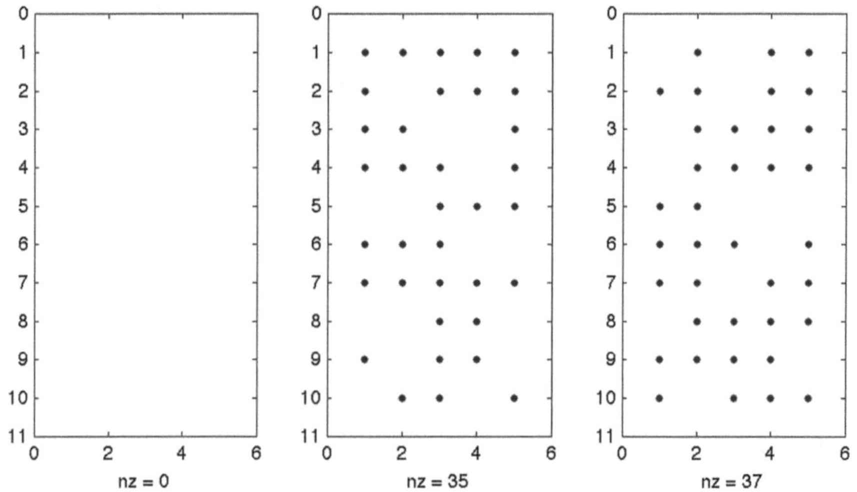

Fig. 5.12 Sparsity pattern of $\mathcal{H}$, shown by frontal slice, without Anderson Acceleration

decoupling in Fourier space. Admittedly, decompositions based upon the t-product are orientation dependent; thus, certain rotations of the tensor will result in different factorizations. On the other hand, in many applications (data mining, image compression), the orientation of the data prior to decomposition is pre-determined by the nature of the data itself.

As outlined in the various subsections, there are many avenues that have yet to be fully explored in the context of these types of NTFs, from investigations of suitable regularization terms up to and including generalizing and utilizing novel concepts such as separability.

Acknowledgments We would like to thank Homer F. Walker from the Mathematical Science Department, Worcester Polytechnic Institute for his assistant with Anderson Acceleration. We would also like to acknowledge Hao and Kilmer's support from NSF-DMS 0914957.

References

1. Alex M, Alex OM (2002) Vasilescu, and Demetri Terzopoulos. Multilinear image analysis for facial recognition
2. Anderson DG (1965) Iterative procedures for nonlinear integral equations. J ACM 12(4):547–560
3. Andersson CA, Bro R (1998) Improving the speed of multi-way algorithms: part i. tucker3. Chemometr Intell Lab Syst 42:93–103
4. Arora S, Ge R, Kannan R, Moitra A (2012) Computing a nonnegative matrix factorization - provably. In: Proceedings of the 44th symposium on theory of computing, STOC '12 ACM, New York, 145–162

5. Berry MW, Browne M,Langville AN, Paul Pauca V, Plemmons RJ (2006) Algorithms and applications for approximate nonnegative matrix factorization. In: Computational statistics and data analysis, pp 155–173
6. Boyd S. Sequential convex programming. Lecture slides
7. Braman K (2010) Third-order tensors as linear operators on a space of matrices. Linear Algebra Appl 433(7):1241–1253, Dec 2010
8. Bro R, De Jong S (1997) A fast non-negativity-constrained least squares algorithm. J Chemom 11:393–401
9. Chi EC, Kolda TG (2012) On tensors, sparsity, and nonnegative factorizations. http://arxiv.org/abs/1112.2414. August 2012 (AarXiv:1112.2414 [math.NA])
10. Cichocki A, Phan AH, Zdunek R, Amari S (2009) Nonnegative matrix and tensor factorizations: applications to exploratory multiway data analysis and blind source separation . Wiley, NJ (Preprint)
11. Cichocki A, Zdunek R, Choi S, Plemmons R, Amari S (2007) Nonnegative tensor factorization using alpha and beta divergencies. In: Proceedings of the 32nd international conference on acoustics, speech, and signal processing (ICASSP), Honolulu, April 2007
12. Cichocki A, Zdunek R, Choi S, Plemmons R, Amari S (2007) Novel multi-layer nonnegative tensor factorization with sparsity constraints. In: 8th international conference on adaptive and natural computing algorithms, Warsaw, MIT Press, Cambridge, April 2007
13. Donoho D, Stodden V (2003) When does non-negative matrix factorization give correct decomposition into parts?, p 2004. MIT Press, Cambridge
14. Elden L (2007) Matrix methods in data mining and pattern recognition. Publisher: society for industrial and applied mathematics (April 9 2007)
15. Friedlander MP, Hatz K (2006) Computing nonnegative tensor factorizations. Technical Report TR-2006-21, Computer Science Department, University of British Columbia
16. Gillis N, Plemmons RJ (2013) Sparse nonnegative matrix underapproximation and its application to hyperspectral image analysis. Linear Alg Appl 438(10):3991–4007
17. Gleich DF, Greif C, Varah JM (2012)The power and Arnoldi methods in an algebra of circulants. Numer. Linear Algebra Appl.2012. Published online in Wiley Online Library (www.wileyonlinelibrary.com). doi:10.1002/nla.1845
18. Hao N, Kilmer ME, Braman K, Hoover RC (2013) Facial recognition using tensor–tensor decompositions, SIAM J Imaging Sci 6(1):437–463
19. Homer Peng Ni, Walker F (July 2011) Anderson acceleration for fixed-point iterations. SIAM J Numer Anal 49(4):1715–1735
20. Kilmer ME, Braman K, Hao N, Hoover RC (2013) Third order tensors as operators on matrices: a theoretical and computational framework with applications in imaging. SIAM J Matrix Anal Appl 34(1):148–172
21. Kilmer ME, Kolda TG (2011) Approximations of third order tensors as sums of (non-negative) low-rank product cyclic tensors. In: Householder Symposium XVIII Plenary Talk. (there seems to be no live page with the book of abstracts), June 2011
22. Kilmer ME, MartinCarla D, Perrone L (2008) A third-order generalization of the matrix SVD as a product of third-order tensors. Technical report TR-2008-4, Department of Computer Science, Tufts University
23. Kilmer ME, Martin CD (2011) Factorization strategies for third-order tensors. Linear Algebra Appl 435(3):641–658. doi:10.1016/j.laa.2010.09.020 (Special Issue in Honor of Stewart's GW 70th birthday)
24. Kilmer ME, Martin CD, Perrone L (2008) A third-order generalization of the matrix svd as a product of third-order tensors. In: Tufts computer science technical, report, 10 2008
25. Kolda TG, Bader BW (2009) Tensor decompositions and applications. SIAM Rev 51(3):455–500
26. Kruskal JB (1989) Rank, decomposition, and uniqueness for 3-way and n-way arrays. In: Coppi R, Bolasco S (eds) Multiway data analysis. Elsevier, Amsterdam, pp 7–18
27. Kumar A, Sindhwani V, Kambadur P (2012) Fast conical hull algorithms for near-separable non-negative matrix factorization. arXiv:1210.1190 [stat.ML], Oct 2012

28. Lee D, Seung H (1999) Learning the parts of objects by non-negative matrix factorization. Nature 401:788791
29. Lin Chih-Jen (2007) On the convergence of multiplicative update algorithms for nonnegative matrix factorization. IEEE Trans Neural Netw 18(6):1589–1596
30. Martin CD, Shafer R, LaRue B (2011) A recursive idea for multiplying order-p tensors. SIAM J Sci Comput (submitted July 2011)
31. Nocedal J, Wright SJ (2006) Numerical optimization, 2nd edn. Springer, New York
32. Vasilescu MAO, Terzopoulos D(2002) Multilinear analysis of image ensembles: Tensorfaces. In: Proceedings of the 7th European conference on computer vision ECCV 2002. Lecture notes in computer science, Vol 2350, 447–460
33. Vasilescu MAO, Terzopoulos D (2002) Multilinear image analysis for face recognition. In: Proceedings of the International conference on patter recognition ICPR 2002, vol 2. Quebec City, pp 511–514
34. Vasilescu MAO, Terzopoulos D (2003) Multilinear subspace analysis of image ensembles. In: Proceedings of the 2003 IEEE Computer society conference on computer vision and pattern recognition CVPR 2003, 93–99
35. Sylvestre EA, Lawton WH (1971) Self modeling curve resolution. Technometrics 13(3):617–633

Chapter 6
Sub-Nyquist Sampling and Compressed Sensing in Cognitive Radio Networks

Hongjian Sun, Arumugam Nallanathan and Jing Jiang

Abstract Cognitive radio has become one of the most promising solutions for addressing the spectral under-utilization problem in wireless communication systems. As a key technology, spectrum sensing enables cognitive radios to find spectrum holes and improve spectral utilization efficiency. To exploit more spectral opportunities, wideband spectrum sensing approaches should be adopted to search multiple frequency bands at a time. However, wideband spectrum sensing systems are difficult to design, due to either high implementation complexity or high financial/energy costs. Sub-Nyquist sampling and compressed sensing play crucial roles in the efficient implementation of wideband spectrum sensing in cognitive radios. In this chapter, Sect. 6.1 presents the fundamentals of cognitive radios. A literature review of spectrum sensing algorithms is given in Sect. 6.2. Wideband spectrum sensing algorithms are then discussed in Sect. 6.3. Special attention is paid to the use of Sub-Nyquist sampling and compressed sensing techniques for realizing wideband spectrum sensing. Finally, Sect. 6.4 shows an adaptive compressed sensing approach for wideband spectrum sensing in cognitive radio networks.

6.1 Cognitive Radio Networks

Nowadays, radio frequency (RF) spectrum is a scarce and valuable natural resource due to its unique character in wireless communications. Under the current policy, the primary user of a frequency band has exclusive rights of using the licensed band. With

H. Sun (✉) · A. Nallanathan
Institute of Telecommunications, King's College London, London WC2R 2LS, UK
e-mail: mrhjsun@hotmail.com

Arumugam Nallanathan
e-mail: nallanathan@ieee.org

Jing Jiang
Center for Communication Systems Research, University of Surrey, Guildford GU2 7XH, UK
e-mail: jing.jiang@surrey.ac.uk

A. Y. Carmi et al. (eds.), *Compressed Sensing & Sparse Filtering*,
Signals and Communication Technology, DOI: 10.1007/978-3-642-38398-4_6,

the explosive growth of wireless communication applications, the demands for the RF spectrum are constantly increasing. It becomes evident that such spectral demands cannot be met under the exclusive spectral allocation policy. On the other hand, it has been reported that the temporal and geographic spectral utilization efficiency is very low. For example, the maximal occupancy of the frequency spectrum between 30 MHz and 3 GHz (in New York City) has been reported to be only 13.1 %, with the average occupancy of 5.2 % [1]. As depicted by Fig. 6.1, the spectral under-utilization problem can be addressed by allowing secondary users to dynamic access the licensed band when its primary user is absent. *Cognitive radio* is one of the key technologies that could improve the spectral utilization efficiency as suggested by Prof. S. Haykin [2]:

> Cognitive radio is viewed as a novel approach for improving the utilization of a precious natural resource: the radio electromagnetic spectrum.

6.1.1 Cognitive Radio Definition and Components

The term *cognitive radio*, first coined by Dr. J. Mitola [4], has the following formal definition [2]:

> Cognitive radio is an intelligent wireless communication system that is aware of its surrounding environment (i.e., outside world), and uses the methodology of understanding-by-building to learn from the environment and adapt its internal states to statistical variations in the incoming RF stimuli by making corresponding changes in certain operating parameters (e.g., transmit-power, carrier-frequency, and modulation strategy) in real-time, with two primary objectives in mind:
>
> - highly reliable communications whenever and wherever needed;
> - efficient utilisation of the radio spectrum.

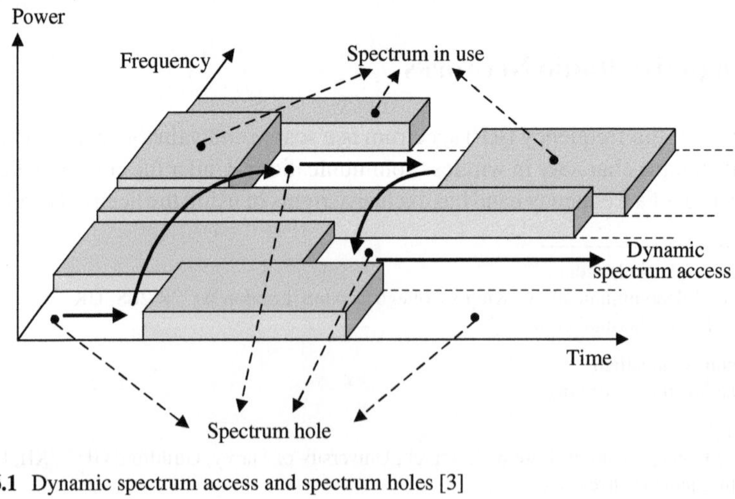

Fig. 6.1 Dynamic spectrum access and spectrum holes [3]

From the definition, the key characteristic of cognitive radio is cognitive capability. It means that cognitive radio should interact with its environment, and intelligently determine appropriate communication parameters based on quality of service (QoS) requirements. These tasks can be implemented by a basic cognitive cycle as illustrated in Fig. 6.2:

- *Spectrum sensing*: To improve the spectral utilization efficiency, cognitive radio should regularly monitor the RF spectral environment. Cognitive radio should not only find spectrum holes, which are not currently used by primary users, by scanning the whole RF spectrum, but also needs to detect the status of primary users for avoiding causing potential interference.
- *Spectrum analysis*: After spectrum sensing, the characteristics of spectrum holes should be estimated. The following parameters need to be known, e.g., channel side information, capacity, delay, and reliability, and will be delivered to the spectrum decision step.
- *Spectrum decision*: Based on the characteristics of spectrum holes, an appropriate spectral band will be chosen for a particular cognitive radio node according to its QoS requirement while considering the whole network fairness. After that, cognitive radio could determine new configuration parameters, e.g., data rate, transmission mode, and bandwidth of the transmission, and then reconfigure itself by using software defined radio techniques.

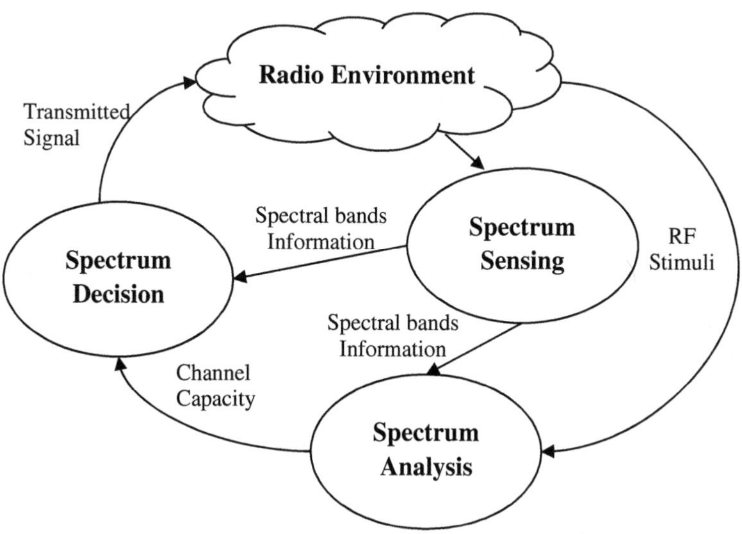

Fig. 6.2 The cognitive capability of cognitive radio enabled by a basic cognitive cycle [5]

6.1.2 Applications of Cognitive Radio Networks

Because cognitive radio is aware of the RF spectral environment and is capable of adapting its transmission parameters to the RF spectral environment, cognitive radio and the concepts of cognitive radio can be applied to a variety of wireless communication environments, especially in commercial and military applications. A few of applications are listed below:

- *Coexistence of wireless technologies*: Cognitive radio techniques were primarily considered for reusing the spectrum that is currently allocated to the TV service. Wireless regional area network (WRAN) users can take advantage of broadband data delivery by the opportunistic usage of the underutilized spectrum. Additionally, the dynamic spectrum access techniques will play an important role in full interoperability and coexistence among diverse technologies for wireless networks. For example, cognitive radio concepts can be used to optimize and manage the spectrum when the wireless local area network (WLAN) and the Bluetooth devices coexist.
- *Military networks*: In military communications, bandwidth is often at a premium. By using cognitive radio concepts, military radios can not only achieve substantial spectral efficiency on a noninterfering basis, but also reduce implementation complexity for defining the spectrum allocation for each user. Furthermore, military radios can obtain benefits from the opportunistic spectrum access function supported by the cognitive radio [6]. For example, the military radios can adapt their transmission parameters to use Global System for Mobile (GSM) bands, or other commercial bands when their original frequencies are jammed. The mechanism of spectrum management can help the military radios achieve information superiority on the battlefield. Furthermore, from the soldiers' perspective, cognitive radio can help the soldiers to reach an objective through its situational awareness.
- *Heterogeneous wireless networks*: From a user's point of view, a cognitive radio device can dynamically discover information about access networks, e.g., WiFi and GSM, and makes decisions on which access network is most suitable for its requirements and preferences. Then the cognitive radio device will reconfigure itself to connect to the best access network. When the environmental conditions change, the cognitive radio device can adapt to these changes. The information as seen by the cognitive radio user is as transparent as possible to changes in the communication environment.

6.2 Traditional Spectrum Sensing Algorithms

As a key technology in cognitive radio, spectrum sensing should sense spectrum holes and detect the presence/absence of primary users. The most efficient way to sense spectrum holes is to detect active primary transceivers in the vicinity of cognitive radios. However, as some primary receivers are passive, such as TVs, some are

Table 6.1 Summary of advantages and disadvantages of traditional spectrum sensing algorithms

Spectrum sensing algorithm	Advantages	Disadvantages
Matched filter [7]	Optimal performance Low computational cost	Require prior information of the primary user
Energy detection [8]	Do not require prior information Low computational cost	Poor performance for low SNR Cannot differentiate users
Cyclostationary [9]	Valid in slow SNR region Robust against interference	Require partial prior information High computational cost
Wavelet based detection [10]	Valid for dynamic and wideband spectrum sensing	High sampling rate High computational cost

difficult to detect in practice. Tractional spectrum sensing techniques can be used to detect the primary transmitters, i.e., matched filtering [7], energy detection [8], cyclostationary detection [9], and wavelet based detection [10]. The implementation of these algorithms requires different conditions, and their detection performance are correspondingly distinguished. The advantages and disadvantages of these algorithms are summarized in Table 6.1.

6.2.1 Matched Filter

A block diagram of a matched filter is shown in Fig. 6.3a.The matched filter method is an optimal approach for spectrum sensing in the sense that it maximizes the signal-to-noise ratio (SNR) in the presence of additive noise [11]. Another advantage of the matched filter method is that it requires less observation time since the high processing gain can be achieved by coherent detection. For example, to meet a given probability of detection, only $\mathcal{O}(1/\mathrm{SNR})$ samples are required [7]. This advantage is achieved by correlating the received signal with a template to detect the presence of a known signal in the received signal. However, it relies on prior knowledge of the primary user, such as modulation type, and packet format, and requires cognitive radio to be equipped with carrier synchronization and timing devices. With more types of primary users, the implementation complexity grows making the matched filter impractical.

6.2.2 Energy Detection

If the information about the primary user is unknown in cognitive radio, a commonly used method for detecting the primary users is energy detection [8]. Energy detection is a non-coherent detection method that avoids the need for complicated receivers required by a matched filter. An energy detector can be implemented in both the time

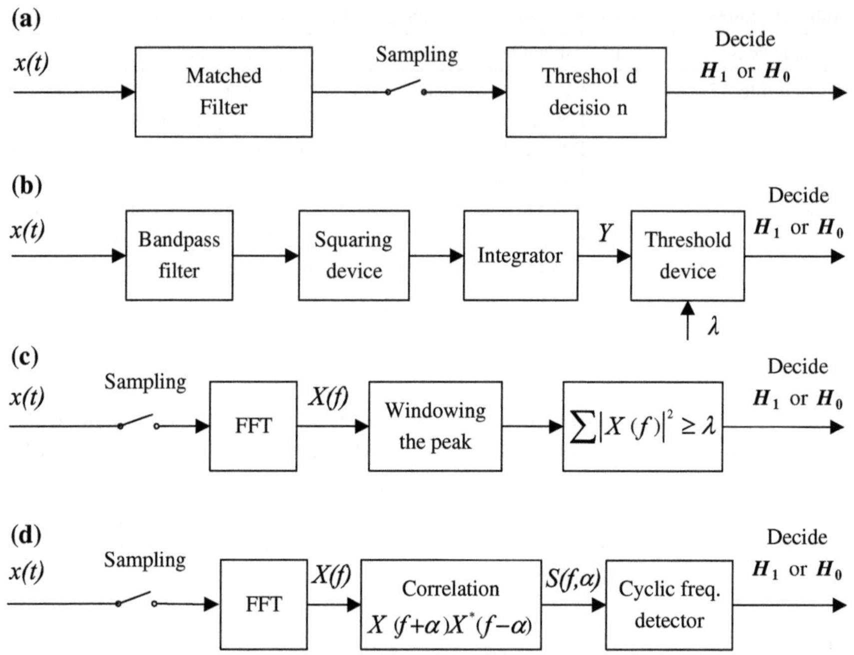

Fig. 6.3 Block diagrams for traditional spectrum sensing algorithms: **a** Matched filter. **b** Time domain energy detection. **c** Frequency domain energy detection. **d** Cyclostationary detection

and the frequency domain. For time domain energy detection as shown in Fig. 6.3b, a bandpass filter (BPF) is applied to select a center frequency and bandwidth of interest. Then the energy of the received signal is measured by a magnitude squaring device, with an integrator to control the observation time. Finally, the energy of the received signal will be compared with a predetermined threshold to decide whether the primary user is present or not. However, to sense a wide spectrum span, sweeping the BPF will result in a long measurement time. As shown in Fig. 6.3c, in the frequency domain, the energy detector can be implemented similarly to a spectrum analyzer with a fast Fourier transform (FFT). Specifically, the received signal is sampled at or above the Nyquist rate over a time window. Then the power spectral density (PSD) is computed using an FFT. The FFT is employed to analyze a wide frequency span in a short observation time, rather than sweeping the BPF in Fig. 6.3b. Finally, the PSD will be compared with a threshold, λ, to decide whether the corresponding frequency is occupied or not.

The advantages of energy detection are that prior knowledge of the primary users is not required, and both the implementation and the computational complexity are generally low. In addition, a short observation time is required, for example, $\mathcal{O}(1/\text{SNR}^2)$ samples are required to satisfy a given probability of detection [7]. Although energy detection has a low implementation complexity, it has some drawbacks. A major drawback is that it has poor detection performance under low SNR scenarios as it is a non-coherent detection scheme. Another drawback is that it cannot differentiate between the signal from a primary user and the interference from other cognitive radios, thus, it cannot take advantage of adaptive signal processing, such as interference cancelation. Furthermore, noise level uncertainty can lead to further performance loss. These disadvantages can be overcome by using two-stage spectrum sensing technique, i.e., coarse spectrum sensing and fine spectrum sensing. Coarse spectrum sensing can be implemented by energy detection or wideband spectrum analyzing techniques. The aim of coarse spectrum sensing is to quickly scan the wideband spectrum and identify some possible spectrum holes in a short observation time. By contrast, fine spectrum sensing further investigates and analysis these suspected frequencies. More sophisticated detection techniques can be used at this stage, such as cyclostationary detection described below.

6.2.3 Cyclostationary Detection

A block diagram of cyclostationary detection is shown in Fig. 6.3d. Cyclostationary detection is a method for detecting the primary users by exploiting the cyclostationary features in the modulated signals. In most cases, the received signals in cognitive radios are modulated signals, which in general exhibit built-in-periodicity within the training sequence or cyclic prefixes. This periodicity is generated by the primary transmitter so that the primary receiver can use it for parameter estimation, such as channel estimation, and pulse timing [12]. The cyclic correlation function, also called cyclic spectrum function (CSF), is used for detecting signals with a particular modulation type in the presence of noise. This is because noise is usually wide-sense stationary (WSS) without correlation, by contrast, modulated signals are cyclostationary with spectral correlation. Furthermore, since different modulated signals will exhibit different characteristics, cyclostationary detection can be used for distinguishing between different types of transmitted signals, noise, and interference in low SNR environments. One of the drawbacks of cyclostationary detection is that it still requires partial information of the primary user. Another drawback is that the computational cost is high as the CSF is a two-dimensional function dependent on frequency and cyclic frequency [9].

6.2.4 Wavelet Based Spectrum Sensing

In [10], Tian and Giannakis proposed a wavelet-based spectrum sensing approach. It provides an advantage of flexibility in adapting to a dynamic spectrum. In this approach, the PSD of the Fourier spectrum is modeled as a train of consecutive frequency subbands, where the PSD is smooth within each subband but exhibits discontinuities and irregularities on the border of two neighboring subbands as shown in

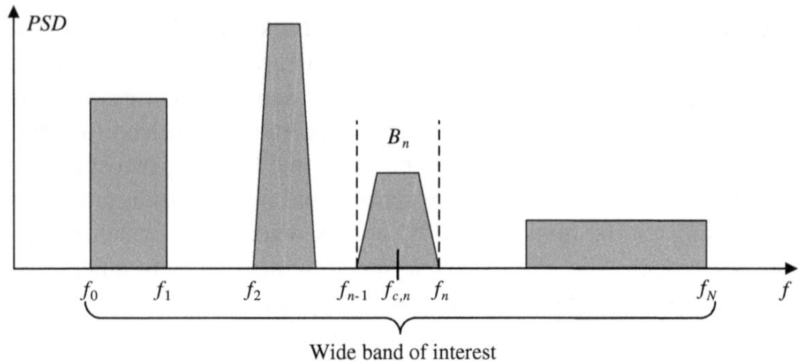

Fig. 6.4 Demonstration of the Fourier spectrum of interest. The PSD is smooth within each subband, and exhibits discontinuities and irregularities with the adjacent subbands [10, 13]

Fig. 6.4. The wavelet transform of the wideband PSD is used to locate the singularities of the PSD.

Let $\varphi(f)$ be a wavelet smoothing function, the dilation of $\varphi(f)$ is given by

$$\varphi_d(f) = \frac{1}{d}\varphi\left(\frac{f}{d}\right) \tag{6.1}$$

where d is a dyadic scale that can take values that are powers of 2, i.e., $d = 2^j$. The continuous wavelet transform (CWT) of the PSD is given by [10]

$$\mathrm{CWT}\{S(f)\} = S(f) * \varphi_d(f) \tag{6.2}$$

where "$*$" denotes the convolution and $S(f)$ is the PSD of the received signal.

Then the first and second derivative of the $\mathrm{CWT}\{S(f)\}$ are used to locate the irregularities and discontinuities in the PSD. Specifically, the boundaries of each subbands are located by using the local maxima of the first derivative of $\mathrm{CWT}\{S(f)\}$, and locations of the subbands are finally tracked by finding zero crossings in the second derivative of $\mathrm{CWT}\{S(f)\}$. By controlling the wavelet smoothing function, the wavelet-based spectrum sensing approach has flexibility in adapting to the dynamic spectrum.

6.3 Wideband Spectrum Sensing Algorithms

As the discussions in previous section, spectrum sensing is composed of data acquisition (sampling) process and decision-making process. For implementing wideband data acquisition, cognitive radio needs some essential components, i.e., wideband antenna, wideband RF front end, and high speed analog-to-digital converter (ADC).

Considering the Nyquist sampling theory, the sampling rate of ADC is required to exceed $2W$ samples per second (known as Nyquist rate), if W denotes the bandwidth of the received signal (e.g., bandwidth $W = 10$ GHz). In [14], Yoon et al. have shown that the -10 dB bandwidth of the newly designed antenna can be 14.2 GHz. Hao and Hong [15] have designed a compact highly selective wideband bandpass filter with a bandwidth of 13.2 GHz. By contrast, the development of ADC technology is relatively behind. When we require an ADC to have a high resolution and a reasonable power consumption, the achievable sampling rate of the state-of-the-art ADC is 3.6 Gsps [16]. Thus, ADC becomes a bottleneck in such a wideband data acquisition system. Even if there exists ADC with more than 20 Gsps sampling rate, the real-time digital signal processing of 20 Gb/s of data could be very expensive. This dilemma motivates researchers to look for technologies to reduce the sampling rate while retaining W by using sub-Nyquist sampling techniques.

Sub-Nyquist sampling refers to the problem of recovering signals from partial measurements that are obtained by using sampling rate lower than the Nyquist rate [17]. Three important sub-Nyquist sampling techniques are: multi-coset sub-Nyquist sampling, multi-rate sub-Nyquist sampling, and compressed sensing based sub-Nyquist sampling.

6.3.1 Multi-Coset Sub-Nyquist Sampling

Multi-coset sampling is a selection of some samples from a uniform grid, which can be obtained when uniformly sampling signal at a rate of f_N greater than the Nyquist rate. The uniform grid is then divided into blocks of L consecutive samples, and in each block $v(v < L)$ samples are retained while the rest of samples, i.e., $L - v$ samples, are skipped. A constant set C that describes the indexes of these v samples in each block is called a sampling pattern as

$$C = \{t^i\}_{i=1}^{v}, \quad 0 \leq t^1 < t^2 < \cdots < t^v \leq L - 1. \tag{6.3}$$

As shown in Fig. 6.5, the multi-coset sampling can be implemented by using v sampling channels with sampling rate of $\frac{f_N}{L}$, where the i-th sampling channel is offset by $\frac{t^i}{f_N}$ from the origin as below

$$x^i[n] = \begin{cases} x(\frac{n}{f_N}), & n = mL + t^i, \ m \in \mathbb{Z} \\ 0, & \text{otherwise} \end{cases} \tag{6.4}$$

where $x(t)$ denotes the received signal to be sampled.

The discrete-time Fourier transform (DTFT) of the samples can be linked to the unknown Fourier transform of signal $x(t)$ by

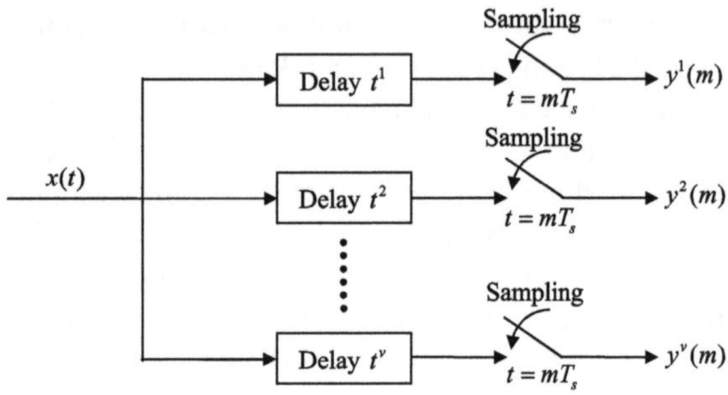

Fig. 6.5 Block diagram of multi-coset sub-Nyquist sampling

$$\mathbf{Y}(f) = \boldsymbol{\Phi}\mathbf{X}(f) \tag{6.5}$$

where $\mathbf{Y}(f)$ denotes a vector of DTFT of these measurements from v channels, $\mathbf{X}(f)$ is a vector of the Fourier transform of $x(t)$, and $\boldsymbol{\Phi}$ is the measurement matrix whose elements are determined by the sampling pattern C. The problem of wideband spectrum sensing is thus equivalent to recovering $\mathbf{X}(f)$ from $\mathbf{Y}(f)$. In order to get a unique solution from (6.5), every set of v columns of $\boldsymbol{\Phi}$ should be linearly independent. However, searching for this sampling pattern is a combinatorial problem.

In [18, 19], some sampling patterns are proved to be valid for reconstruction. The advantage of multi-coset sampling is that the sampling rate in each channel is L times lower than the Nyquist rate. Moreover, the number of measurements is $\frac{v}{L}$ lower than the Nyquist sampling case. One drawback of the multi-coset sampling is that accurate time offsets between sampling channels are required to satisfy a specific sampling pattern. Another one is that the number of sampling channels should be sufficiently high [20].

6.3.2 Multi-Rate Sub-Nyquist Sampling

An alternative model for compressing the wideband spectrum in the analog domain is a multirate sampling system as shown in Fig. 6.6. Asynchronous multirate sampling (MRS) and synchronous multirate sampling (SMRS) were used for reconstructing sparse multiband signals in [22] and [23], respectively. In addition, MRS has been successfully implemented in experiments using an electro-optical system with three sampling channels as described in [21]. Both systems employ three optical pulsed sources that operate at different rates and at different wavelengths. The received signal is modulated with optical pulses, which provided by an optical pulse generator (OPG), in each channel. In order to reconstruct a wideband signal with an 18 GHz bandwidth, the modulated pulses are amplified, and sampled by an ADC at a rate of 4 GHz in each channel.

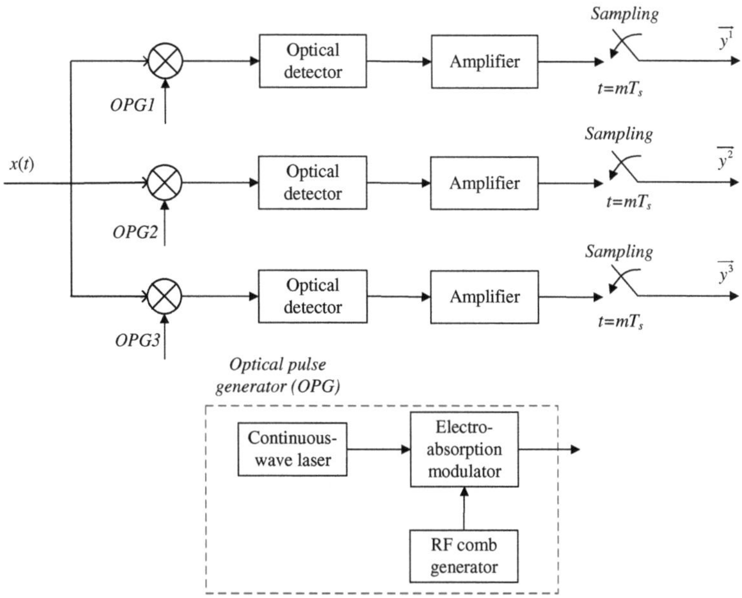

Fig. 6.6 Multirate sampling system implemented by electro-optical devices [21]. In each channel, the received signal is modulated by a train of short optical pulses. The modulated signal is then detected by an optical detector, amplified, and sampled by a low-rate ADC

In [22], the sampling channels of the MRS can be implemented separately without synchronisation. However, reconstruction of the spectrum requires that each frequency of the signal must be non-aliased in at least one of the sampling channels. In [23] SMRS reconstructs the spectrum from linear equations, which relate the Fourier transform of the signal to the Fourier transform of its samples. Using compressed sensing theory, sufficient conditions for perfectly reconstructing the spectrum are obtained; $v \geq 2k$ (the Fourier transform of the signal is k-sparse) sampling channels are required. In order to reconstruct the spectrum using MRS with fewer sampling channels, the spectrum to be recovered should possess certain properties, e.g., minimal bands, and uniqueness. Nonetheless, the spectral components from primary users may not possess these properties. Obviously, even though the multirate sampling system has broad application, there is a long way to go to implement it in a cognitive radio network because of its stringent requirements on both optical devices and the number of sampling channels.

6.3.3 Compressed Sensing Based Sub-Nyquist Sampling

In the classic work [13], Tian and Giannakis introduced compressed sensing theory to realize wideband spectrum sensing by exploiting the sparsity of radio signals. The

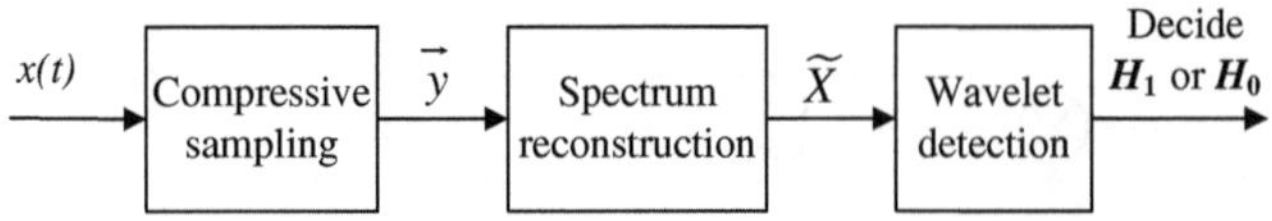

Fig. 6.7 Block diagram of the compressed sensing based wideband spectrum sensing algorithm

technique takes advantage of using fewer samples closer to the information rate, rather than the inverse of the bandwidth, to perform wideband spectrum sensing. After reconstruction of the wideband spectrum, wavelet-based edge detection was used to detect the wideband spectrum as shown in Fig. 6.7.

Let $x(t)$ represent a wideband signal received at cognitive radio. If $x(t)$ is sampled at the Nyquist sampling rate, the sequence vector, i.e., $\mathbf{x}$ ($\mathbf{x} \in \mathbb{C}^N$), will be obtained. The Fourier transform of the sequence, $\mathbf{X} = \mathbf{F}\mathbf{x}$, will therefore be alias-free, where $\mathbf{F}$ denotes the Fourier matrix. When the spectrum, $\mathbf{X}$, is k-sparse ($k \ll N$), which means k out of N values in $\mathbf{X}$ are not neglectable, $x(t)$ can be sampled at a sub-Nyquist rate while its spectrum can be reconstructed with a high probability. The sub-sampled/compressed signal, $\mathbf{y} \in \mathbb{C}^M$ ($k < M \ll N$), is linked to the Nyquist sequence $\mathbf{x}$ by [13],

$$\mathbf{y} = \boldsymbol{\Phi}\mathbf{x} \tag{6.6}$$

where $\boldsymbol{\Phi} \in \mathbb{C}^{M \times N}$ is the measurement matrix, which is a selection matrix that randomly chooses M columns of the size-N identity matrix. Namely, $N - M$ samples out of N samples are skipped. The relationship between the spectrum $\mathbf{X}$ and the compressed sequence $\mathbf{y}$ is given by [13]

$$\mathbf{y} = \boldsymbol{\Phi}\mathbf{F}^{-1}\mathbf{X} \tag{6.7}$$

where $\mathbf{F}^{-1}$ denotes the inverse Fourier matrix.

Approximating $\mathbf{X}$ from $\mathbf{y}$ in (6.7) is a linear inverse problem and is NP-hard. The basis pursuit (BP) [24] algorithm can be used to solve $\mathbf{X}$ by linear programming [13]:

$$\widetilde{X} = \arg\ \min \|\mathbf{X}\|_1, \quad \text{s. t.} \quad \mathbf{y} = \boldsymbol{\Phi}\mathbf{F}^{-1}\mathbf{X}. \tag{6.8}$$

After reconstructing the full spectrum $\mathbf{X}$, the PSD is calculated using $\widetilde{X}$. Then the wavelet detection approach can be used to analyze the edges in the PSD. Although less measurements are used for characterizing the wideband spectrum, the requirement of high sampling rate on ADC is not relaxed. By contrast, in [25], Polo et al. suggested using an analog-to-information converter (AIC) model (also known as random demodulator, [26]) for compressing the wideband signal in the analog domain. The block diagram of AIC is given in Fig. 6.8.

A pseudorandom number generator is used to produce a discrete-time sequence $\varepsilon_0, \varepsilon_1, \ldots$, called a chipping sequence, the number of which takes values of ± 1 with equal probability. The waveform should randomly alternate at or above the Nyquist rate, i.e., $\varpi \geq 2W$, where W is the bandwidth of signal. The output of the pseu-

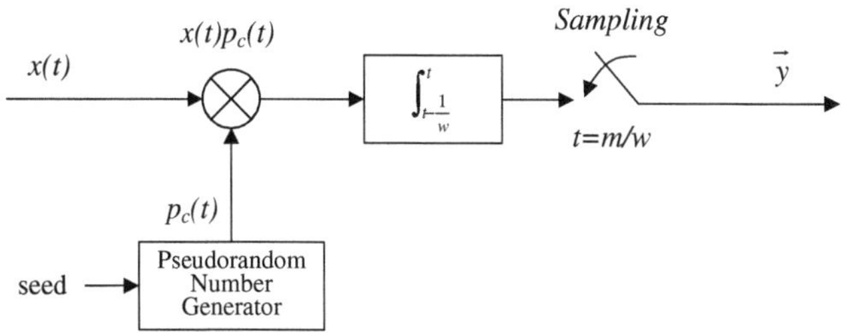

Fig. 6.8 Block diagram for the analog-to-information converter [26]. The received signal, $x(t)$, is randomly demodulated by a pseudorandom chipping sequence, integrated by an accumulator, and sampled at a sub-Nyquist rate

dorandom number generator, i.e., $p_c(t)$, is employed to demodulate a continuous-time input $x(t)$ by a mixer. Then an accumulator sums the demodulated signal for $1/w$ seconds, and the filtered signal is sampled at a sub-Nyquist rate of w. This sampling approach is called integrate-and-dump sampling since the accumulator is reset after each sample is taken. The samples acquired by the AIC, $\mathbf{y} \in \mathbb{C}^w$, can be related to the received signal, $\mathbf{x} \in \mathbb{C}^{\varpi}$, by

$$\mathbf{y} = \boldsymbol{\Phi}\mathbf{x} \tag{6.9}$$

where $\boldsymbol{\Phi} \in \mathbb{C}^{w \times \varpi}$ is the measurement matrix describing the overall action of the AIC system on the input signal $\mathbf{x}$. The signal $\mathbf{x}$ can be identified by solving the convex optimization problem,

$$\widetilde{x} = \arg\ \min \|\mathbf{x}\|_1, \quad \text{s. t.} \quad \mathbf{y} = \boldsymbol{\Phi}\mathbf{x}, \tag{6.10}$$

by BP or other greedy pursuit algorithms. The PSD of the wideband spectrum can be estimated using the recovered signal $\widetilde{x}$, followed by a hypothesis test on the PSD. Alternatively, the PSD can be directly recovered from the measurements using compressed sensing algorithms [25]. Although the AIC bypasses the requirement for a high sampling rate ADC, it leads to a high computational complexity as the huge-scale of the measurement matrix. Furthermore, it has been identified that the AIC model can easily be influenced by design imperfections or model mismatches [27].

In [27], Mishali and Eldar proposed a parallel implementation of the AIC model, called modulated wideband converter (MWC), as shown in Fig. 6.9. The key difference is that in each channel the accumulator for integrate-and-dump sampling is replaced by a general low-pass filter. One of the benefits of introducing parallel structure is that the dimension of the measurement matrix is reduced making the reconstruction easier. Another benefit is that it provides robustness to noise and model mismatch. On the other hand, the implementation complexity increases as multiple sampling channels are involved. An implementation issue of using MWC is that the storage and transmission of the measurement matrix must be considered when

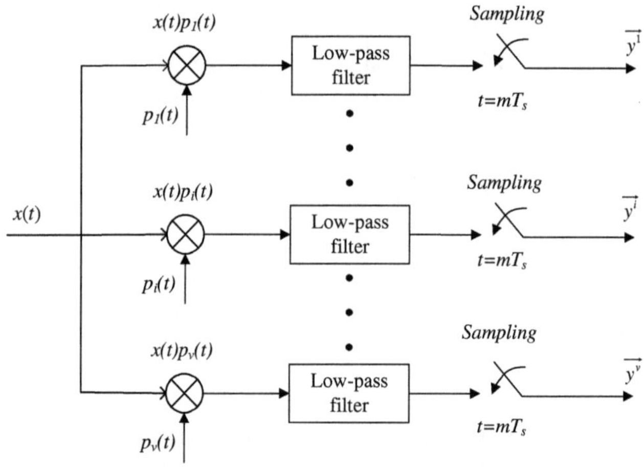

Fig. 6.9 Block diagram for the modulated wideband converter [27]. In each channel, the received signal is demodulated by a pseudorandom sequence, filtered by a low-pass filter, and sampled at a sub-Nyquist rate $\frac{1}{T_s}$

it is used in a distributed cognitive radio network under a data fusion collaborative scheme.

6.4 Adaptive Compressed Sensing Framework for Wideband Spectrum Sensing

The compressed sensing technologies require that the signal to be sampled should be sparse in a suitable basis. If it is sparse, the signal can be reconstructed from partial measurements by using some recovery algorithms, e.g., orthogonal matching pursuit (OMP) or compressive sampling matching pursuit (CoSaMP) [28]. Given the low spectral occupancy, the wideband signal that is received by cognitive radios can be assumed to be sparse in the frequency domain [13]. If this sparsity level (denoted by k) is known, we can choose an appropriate number of measurements M to secure the quality of spectral recovery, e.g., $M = C_0 k \log(N/k)$, where C_0 denotes a constant and N denotes the number of measurements when using the Nyquist rate [13]. However, in order to avoid incorrect spectral recovery in the cognitive radio system, traditional compressed sensing approaches must pessimistically choose the parameter C_0, which results in excessive number of measurements. As shown in Fig. 6.10, considering $k = 10$, traditional compressed sensing approaches tend to choose $M = 37\,\%N$ measurements for achieving a high successful recovery rate. We note that, with $20\,\%N$ measurements, we can still achieve $50\,\%$ successful recovery rate. If these $50\,\%$ successful recovery cases can be identified, we could save the number of measurements. In addition, in a practical cognitive radio system, the

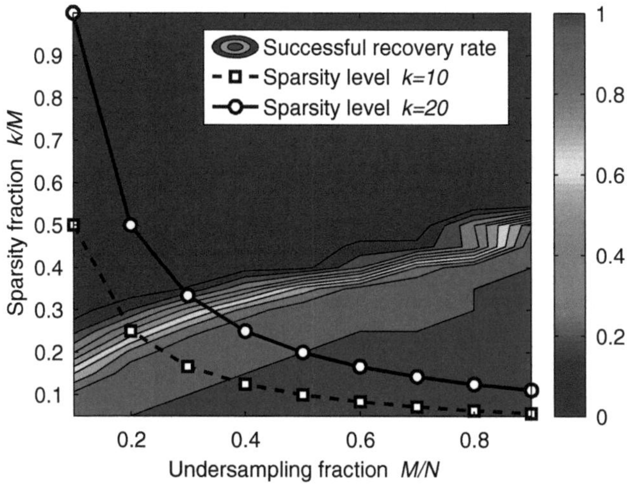

Fig. 6.10 An example of a traditional compressed sensing system, where the successful recovery rate varies when the number of measurements and the sparsity level vary. In simulations, considering $N = 200$, we varied the number of measurements M from 20 to 180 in eight equal-length steps. The sparsity level k was set to between 1 and M. The measurement matrix was assumed to be Gaussian. The figure was obtained with 5,000 trials of each parameter setting

sparsity level of the instantaneous spectrum is often unknown or difficult to estimate because of either the dynamic activities of primary users or the time-varying fading channels between the primary users and cognitive radios. Due to this sparsity level uncertainty, traditional compressed sensing approaches should further increase the number of measurements. For example, in Fig. 6.10, if k is known to be $10 \leq k \leq 20$, traditional compressed sensing approaches would select $M = 50\,\%N$, which does not fully exploit the advantages of using compressed sensing technologies for wideband spectrum sensing. Further, the sparsity level uncertainty could also result in early or late termination of greedy recovery algorithms. Due to the effects of underfitting or over-fitting caused by the early or late iteration termination, traditional compressed sensing recovery algorithms will lead to unfavorable spectral recovery quality.

To address these challenges, adaptive compressed sensing approach should be adopted for reconstructing the wideband spectrum by using an appropriate number of compressive measurements without prior knowledge of the instantaneous spectral sparsity level. Specifically, the adaptive framework divides the spectrum sensing interval into several equal-length time slots, and performs compressive measurements in each time slot. The measurements are then partitioned into two complementary subsets, performing the spectral recovery on the training subset, and validating the recovery result on the testing subset. Both the signal acquisition and the spectral estimation will be terminated if the designed ℓ_1 norm validation parameter meets certain requirements. In the next section, we will introduce the adaptive compressed sensing approach in detail for addressing wideband spectrum sensing issues in cognitive radios.

6.4.1 Problem Statement

Suppose that an analog primary signal $x(t)$ is received at a cognitive radio, and the frequency range of $x(t)$ is $0 \sim W$ (Hz). If the signal $x(t)$ were sampled at the sampling rate f (Hz) in the observation time τ (seconds), a signal vector $\mathbf{x} \in \mathbb{C}^{N\times 1}$ would be obtained, where N denotes the number of samples and can be written as $N = f\tau$. Without loss of generality, we assume that N is an integer number. However, here we consider that the signal is sampled at sub-Nyquist rate as enhanced by compressed sensing.

The compressed sensing theory relies on the fact that we can represent many signals using only a few non-zero coefficients in a suitable basis or dictionary. Such signals may therefore be acquired by sub-Nyquist sampling, which leads to fewer samples than predicted on the basis of Nyquist sampling theory. The sub-Nyquist sampler, e.g., the random demodulator [26, 29, 30], will generate a vector of compressive measurements $\mathbf{y} \in \mathbb{C}^{M\times 1}$ $(M \ll N)$ via random projections of the signal vector $\mathbf{x}$. Mathematically, the compressive measurement vector $\mathbf{y}$ can be written as

$$\mathbf{y} = \boldsymbol{\Phi}\mathbf{x} \tag{6.11}$$

where $\mathbf{x}$ denotes the signal vector obtained by using sampling rate higher than or equal to the Nyquist rate (i.e., $f \geq 2W$), and $\boldsymbol{\Phi}$ denotes an $M \times N$ measurement matrix. Of course, there is no hope to reconstruct an arbitrary N-dimensional signal $\mathbf{x}$ from partial measurements $\mathbf{y}$. However, if the signal $\mathbf{x}$ is k-sparse $(k < M \ll N)$ in some basis, there do exist measurement matrices that allow us to recover $\mathbf{x}$ from $\mathbf{y}$ using some recovery algorithms.

Based on the fact of spectral sparseness in a cognitive radio system [13], the compressed sensing technologies can be applied for signal acquisition at cognitive radios. A block diagram of a typical compressed sensing based spectrum sensing infrastructure is shown in Fig. 6.11. The goal is to reconstruct the Fourier spectrum

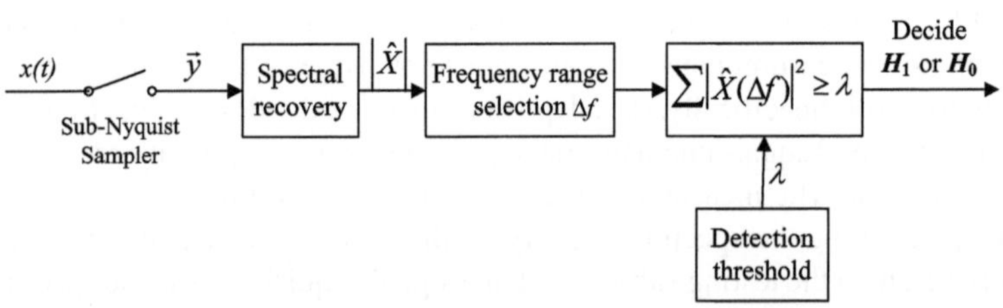

Fig. 6.11 Diagram of compressed sensing based spectrum sensing approach when using the spectral domain energy detection approach

$\mathbf{X} = \mathbf{Fx}$ from partial measurements $\mathbf{y}$, and to perform spectrum sensing based on the reconstructed spectrum $\hat{X}$. Due to the advantages of short running time and good sampling efficiency, greedy recovery algorithms are often used in some practical scenarios where the signal processing should be performed on a near real-time basis in addition to computational capability constraints.

After the spectral recovery, spectrum sensing approaches can be performed by using the reconstructed spectrum $\hat{X}$. A typical spectrum sensing approach is spectral domain energy detection as the discussions in Sect. 6.2. As depicted in Fig. 6.11, this approach extracts the reconstructed spectrum in the frequency range of interest, e.g., Δf, and then calculates the signal energy in the spectral domain. The output energy will be compared with a detection threshold (denoted by λ) to decide whether the corresponding frequency band is occupied or not, i.e., choosing between hypotheses $\mathcal{H}_1$ (presence of primary users) and $\mathcal{H}_0$ (absence of primary users).

It can be easily understood that the performance of such an infrastructure will highly depend on the recovery quality of the Fourier spectrum $\mathbf{X}$. From the compressed sensing theory, we know that the recovery quality is determined by: the sparsity level, the choice of measurement matrix, the recovery algorithm, and the number of measurements. The spectral sparsity level in a cognitive radio system is mainly determined by the activities of primary users within a specific frequency range and the medium access control (MAC) of the cognitive radios. One elegant metric for evaluating the suitability of a chosen measurement matrix is the restricted isometry property (RIP) [31]. For a comprehensive understanding of RIP and measurement matrix design, we refer the reader to [32] and references therein. In the following, we will concentrate on addressing: the choice of the number of measurements and the design of the recovery algorithm. We will discuss an adaptive sensing framework enabling us to gradually acquire spectral measurements. Both the signal acquisition and the spectral estimation will be terminated when certain halting criterions are met, thereby avoiding the problems of excessive or insufficient numbers of compressive measurements.

6.4.2 System Description

Consider a cognitive radio system using a periodic spectrum sensing infrastructure in which each frame is comprised of a spectrum sensing time slot and a data transmission time slot, as shown in Fig. 6.12. The length of each frame is A (seconds), and the duration of spectrum sensing is T ($0 < T < A$). The remaining time $A - T$ is used for data transmission. Further, we assume that the spectrum sensing duration T is carefully chosen so that the symbols from primary users, and the channels between the primary users and cognitive radios are quasi-stationary. We propose to divide the spectrum sensing duration T into P equal-length mini-time slots, each of which has length $\tau = \frac{T}{P}$, as depicted in Fig. 6.12. As enforced by protocols, e.g., at the MAC layer [33], all cognitive radios can keep quiet during the spectrum sensing interval. Therefore, the spectral components of the Fourier spectrum $\mathbf{X} = \mathbf{Fx}$ arise only from

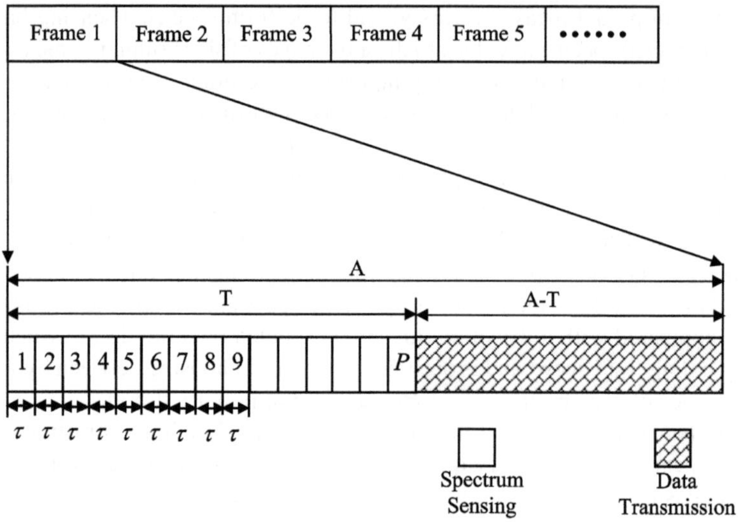

Fig. 6.12 Frame of periodic spectrum sensing in cognitive radio networks

primary users and background noise. Due to the low spectral occupancy [13], the Fourier spectrum **X** can be assumed to be k-sparse, which means it consists only of k largest values that are not negligible. The spectral sparsity level k is unknown except that $k \leq k_{\max}$, where $k_{\max}$ is a known parameter. This assumption is reasonable because the maximal occupancy of the spectrum can be estimated by long-term spectral usage measurements.

For simplicity, we name the adaptive compressed sensing-based wideband spectrum sensing approach as: compressed adaptive sensing (CASe). The aim of CASe is to gradually acquire compressive measurements, reconstruct the wideband spectrum **X**, and terminate the signal acquisition if and only if the current spectral recovery performance is satisfactory. The work procedure of CASe is shown in Table 6.2. We assume that cognitive radio performs compressive measurements using the same sub-Nyquist sampling rate f_s $(f_s < 2W)$ in all P mini time slots. In each time slot, an m-length measurement vector would be obtained, where $m = f_s\tau = \frac{f_sT}{P}$ is assumed to be an integer. Without loss of generality, the measurement matrices of P time slots are assumed to follow the same distribution, e.g., the standard normal distribution, or the Bernoulli distribution with equal probability of ± 1. We partition the measurement set of the first time slot into two complementary subsets, i.e., validating the spectral recovery result using the testing subset **V** ($\mathbf{V} \in \mathbb{C}^{r\times 1}, 0 < r < m$) which is given by

$$\mathbf{V} = \boldsymbol{\Psi}\mathbf{F}^{-1}\mathbf{X} \tag{6.12}$$

and performing the spectral recovery using the training subset $\mathbf{y}_1$ ($\mathbf{y}_1 \in \mathbb{C}^{(m-r)\times 1}$), where $\boldsymbol{\Psi} \in \mathbb{C}^{r\times N}$ denotes the testing matrix. The measurements of other time slots, i.e., $\mathbf{y}_i, \forall i \in [2, P]$, are used only as the training subsets for spectral recovery. We

Table 6.2 Compressed adaptive sensing (CASe) framework

Input: Sensing duration T, N, noise variance δ^2, threshold ϖ
accuracy ε in the noiseless case, accuracy ϵ in the noisy case.

1. *Initialize:*
   Divide T into P time slots, each has length $\tau = \frac{T}{P}$, index $p = 0$.
2. *While* the halting criterion is false and $p < P$, *do*
   (a) Increment p by 1.
   (b) Perform compressive sampling in the time slot p using rate f_s.
   (c) If $p = 1$, partition the measurement vector into:
   the training set $\mathbf{y}_1$ and testing set $\mathbf{V}$ as in (6.12–6.13).
   (d) Concatenate the training sets from the time slots $1, \ldots, p$
   to form $\mathbf{Y}_p$ as in (6.13).
   (e) Estimate the spectrum from $\mathbf{Y}_p$ using spectral recovery algorithm
   resulting in the spectral estimate $\hat{X}_p$.
   (f) Calculate the validation parameter using $\mathbf{V}$ and $\boldsymbol{\Psi}$:
   $\rho_p = \frac{\|\mathbf{V}-\boldsymbol{\Psi}\mathbf{F}^{-1}\hat{X}_p\|_1}{r}$.
3. *Check and make decision:*

If the halting criterion is true
   (a) Terminate the signal acquisition.
   (b) Perform spectrum sensing using the reconstructed spectrum $\hat{X}_p$.
   (c) Choose un-occupied bands, and start the data transmission.

Else if $p = P$
   (a) Terminate the signal acquisition.
   (b) Report its reconstruction is not trustworthy.
   (c) Increase f_s and wait for next spectrum sensing frame.

end

Halting Criterion: $\frac{\sqrt{\frac{\pi N}{2}}\rho_p}{1-\varepsilon} \leq \varpi$, in the noiseless measurement case.
$|\rho_p - \sqrt{\frac{\pi}{2}}\delta| \leq \epsilon$, in the noisy measurement case.

concatenate the training subsets of all p time slots as

$$\mathbf{Y}_p \triangleq \begin{pmatrix} \mathbf{y}_1 \\ \mathbf{y}_2 \\ \vdots \\ \mathbf{y}_p \end{pmatrix} = \boldsymbol{\Phi}_p \mathbf{F}^{-1} \mathbf{X}_p \tag{6.13}$$

where $\mathbf{Y}_p \in \mathbb{C}^{(pm-r)\times 1}$ denotes the concatenated measurement vector, $\boldsymbol{\Phi}_p$ denotes the measurement matrix after p time slots, and $\mathbf{X}_p$ denotes the signal spectrum. It should be noted that $\boldsymbol{\Phi}_p$ and the testing matrix $\boldsymbol{\Psi}$ are chosen to be different but have the same distribution, and the signal spectrum $\mathbf{X}_p$ is always noisy, e.g., due to the receiver noise. We then gradually estimate the spectrum from $\mathbf{Y}_1, \mathbf{Y}_2, \ldots, \mathbf{Y}_p$ using a certain compressed sensing recovery algorithm, leading to a sequence of spectral estimates $\hat{X}_1, \hat{X}_2, \ldots, \hat{X}_p$.

6.4.3 Acquisition Termination Metric

We hope that the signal acquisition procedure can be terminated if we find a good spectral approximation $\hat{X}_p$ that makes the spectral recovery error $\|\mathbf{X} - \hat{X}_p\|_2$ sufficiently small. The remaining spectrum sensing time slots, i.e., $p+1, \ldots, P$, can be used for data transmission. If this target can be achieved, we could not only improve the cognitive radio system throughput (due to the longer data transmission time), but could also obtain measurement savings, leading to both energy and computational savings. However, the spectral recovery error $\|\mathbf{X} - \hat{X}_p\|_2$ is typically not known as $\mathbf{X}$ is unknown under the sub-Nyquist sampling rate. Hence, when using traditional compressed sensing approaches, we do not know when we should terminate the signal acquisition procedure. In this chapter, we propose to use the following validation parameter as a proxy for $\|\mathbf{X} - \hat{X}_p\|_2$:

$$\rho_p \triangleq \frac{\|\mathbf{V} - \boldsymbol{\Psi}\mathbf{F}^{-1}\hat{X}_p\|_1}{r} \tag{6.14}$$

and terminate the signal acquisition if the validation parameter ρ_p is smaller than a predetermined threshold. This is based on the following observation:

Theorem 1 *Assume that $\boldsymbol{\Phi}_1, \ldots, \boldsymbol{\Phi}_P$ and $\boldsymbol{\Psi}$ follow the same distribution, i.e., either the standard normal distribution or the Bernoulli distribution with equal probability of ± 1. Let $\varepsilon \in (0, \frac{1}{2})$, $\xi \in (0, 1)$, and $r = C\varepsilon^{-2}\log\frac{4}{\xi}$ (C is a constant). Then using $\mathbf{V}$ for testing the spectral estimate $\hat{X}_p$, the validation parameter ρ_p satisfies:*

$$\Pr\left[(1-\varepsilon)\|\mathbf{X} - \hat{X}_p\|_2 \le \sqrt{\frac{\pi N}{2}}\rho_p \le (1+\varepsilon)\|\mathbf{X} - \hat{X}_p\|_2\right] \ge 1 - \xi \tag{6.15}$$

where ξ can also be written as $\xi = 4\exp(-\frac{r\varepsilon^2}{C})$.

The proof of Theorem 1 is given in Appendix A.

Remark 1 In Theorem 1, we can see that, with either higher ε or greater r, we have higher confidence for estimating the actual spectral recovery error $\|\mathbf{X} - \hat{X}_p\|_2$. Figure 6.13a shows the influence of using different number of measurements for testing the spectral estimate when the number of time slots increases. The spectral occupancy is assumed to be 6 %, which means the spectral sparsity level $k = 6\,\%N = 120$ where $N = 2000$. It can be seen that with more testing data, the validation result is more trustworthy. Furthermore, we can find that even with $r = 5$ measurements for testing, the validation result is still very close to the actual recovery error. The

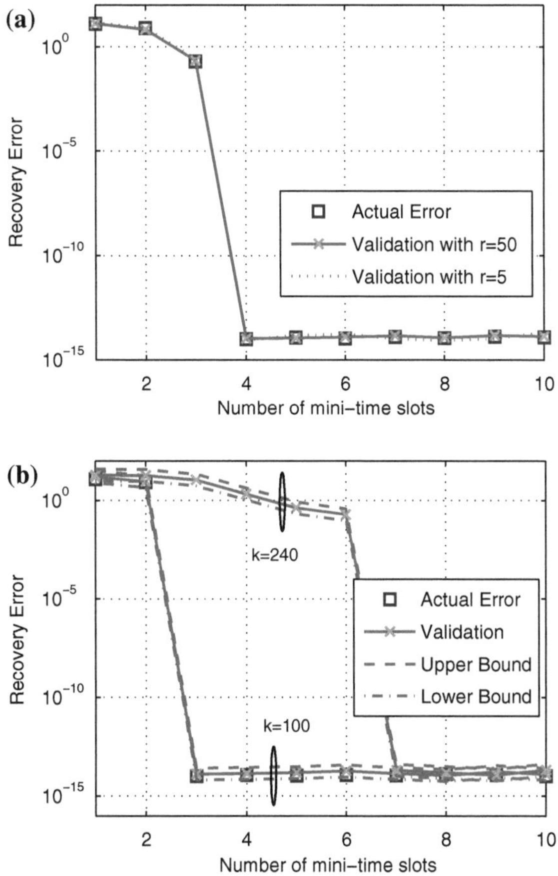

Fig. 6.13 Comparison of the actual recovery error and the proposed validation parameter when the number of mini time slots increases. **a** Different number of measurements for validation when the spectral sparsity level $k = 120$. **b** Different spectral sparsity levels when $r = 50$. It was assumed that there is no measurement noise in the compressive measurements. The upper and lower bounds on the actual recovery error are given in (6.16)

choice of parameter C in Theorem 1 depends on the concentration property of random variables in the measurement matrix $\boldsymbol{\Psi}$. For a good $\boldsymbol{\Psi}$, e.g., the measurement matrix with random variables following either the Gaussian or Bernoulli distribution, C could be a small number.

Remark 2 Theorem 1 can be used to provide tight upper and lower bounds on the unknown recovery error $\|\mathbf{X} - \hat{X}_p\|_2$ by using (6.15) such that

$$\frac{\sqrt{\frac{\pi N}{2}}\rho_p}{1+\varepsilon} \leq \|\mathbf{X} - \hat{X}_p\|_2 \leq \frac{\sqrt{\frac{\pi N}{2}}\rho_p}{1-\varepsilon}. \tag{6.16}$$

Figure 6.13b compares the actual recovery error $\|\mathbf{X} - \hat{X}_p\|_2$ and the validation parameter $\sqrt{\frac{\pi N}{2}}\rho_p$ when the spectral sparsity level varies. It is evident that the validation parameter can closely fit the unknown actual recovery error. The upper and lower bounds on the actual recovery error that we obtained in (6.16) can correctly predict the trend of the actual recovery error even if either p or k vary. Figure 6.13b also illustrates that the lower the sparsity level, the fewer time slots (thereby the fewer compressive measurements) are required for reconstructing the spectrum. When the spectral occupancy is 12 % (i.e., $k = 12\,\%N = 240$), the CASe framework requires $p = 7$ mini-time slots, i.e., $M = pm = 1400$ measurements in total. On the other hand, when $k = 100$, only $p = 3$ time slots and $M = pm = 600$ measurements are required. The remaining time slots can be used for data transmission, which can therefore lead to higher throughput than the cognitive radio system using traditional compressed sensing approaches. If we require $\|\mathbf{X} - \hat{X}_p\|_2$ (unknown) to be less than a tolerable recovery error threshold ϖ, we can let the upper bound on (6.16) to be a proxy for $\|\mathbf{X} - \hat{X}_p\|_2$. As shown in Table 6.2, we choose the upper bound on (6.16) as the signal acquisition termination metric in the noiseless case. If it is less than or equal to the threshold ϖ, i.e., $\|\mathbf{X} - \hat{X}_p\|_2 \leq \frac{\sqrt{\frac{\pi N}{2}}\rho_p}{1-\varepsilon} \leq \varpi$, the signal acquisition can be terminated. This approach, to some extent, decreases the probabilities of excessive or insufficient numbers of measurements.

6.4.4 Noisy Compressed Adaptive Sensing

Due to either the quantization error of ADC or the imperfect design of sub-Nyquist sampler, the measurement noise may exist when performing compressive measurements. In this section, the ℓ_1 norm validation approach is further studied to fit the CASe framework in the noisy case. After that, we present a sparsity-aware recovery algorithm that can correctly terminate greedy iterations when the spectral sparsity level is unknown and the effects of measurement noise are not negligible.

In the noisy signal measurement case, the concatenated training set $\mathbf{Y}_p$ and the testing subset $\mathbf{V}$ can be written as

$$\mathbf{Y}_p = \boldsymbol{\Phi}_p \mathbf{F}^{-1} \mathbf{X}_p + \mathbf{n} \tag{6.17}$$

and

$$\mathbf{V} = \boldsymbol{\Psi} \mathbf{F}^{-1} \mathbf{X} + \mathbf{n} \tag{6.18}$$

respectively, where the measurement noise $\mathbf{n}$ is additive noise (added to the real compressed signal after the random projection) generated by the signal measurement procedure, i.e., signal quantization. The measurement noise can be modeled by circular complex additive white Gaussian noise (AWGN). Without loss of generality, we assume that $\mathbf{n}$ has an upper bound $\bar{n}$, and has zero mean and known variance δ^2, i.e., $\mathbf{n} \sim \mathcal{CN}(0, \delta^2)$. For example, if the measurement noise $\mathbf{n}$ is generated by the quantization noise of a uniform quantizer, the noise variance δ^2 can be estimated by $\Delta^2/12$ and $\mathbf{n} \leq \bar{n} = \Delta$, where Δ denotes the cell width.

If ρ_p is close enough to $\sqrt{\frac{\pi}{2}}\delta$, the signal acquisition procedure can be safely terminated. This observation is due to the following theorem:

Theorem 2 *Let* $\epsilon > 0$, $\delta > 0$, $\varrho \in (0,1)$, $\nu \geq \frac{\sqrt{2/\pi}}{\delta}\bar{n} - 1$, *and* $r = \ln\left(\frac{2}{\varrho}\right)\frac{3(4-\pi)\delta^2+\sqrt{2\pi}\epsilon\delta\nu}{3\epsilon^2}$. *If the best spectral approximation exists within the sequence of spectral estimates* $\hat{X}_1, \cdots, \hat{X}_P$, *then there exists a validation parameter* ρ_p *that satisfies*

$$\Pr\left[\sqrt{\frac{\pi}{2}}\delta - \epsilon \leq \rho_p \leq \sqrt{\frac{\pi}{2}}\delta + \epsilon\right] > 1 - \varrho, \tag{6.19}$$

where ϱ *is given by* $\varrho = 2\exp\left(-\frac{3r\epsilon^2}{3(4-\pi)\delta^2+\sqrt{2\pi}\epsilon\delta\nu}\right)$.

The proof of Theorem 2 is given in Appendix B.

Remark 3 It is worthwhile to note that Theorem 2 addresses the problem of finding the best spectral approximation, i.e., $\hat{X}_p = \mathbf{X}^\star$, that minimizes $\|\mathbf{X} - \hat{X}_p\|_2$ among all possible spectral estimates in the noisy case. This is different from Theorem 1, which focuses on finding a satisfactory spectral estimate $\hat{X}_p$ that makes $\|\mathbf{X} - \hat{X}_p\|_2 \leq \varpi$ in the noiseless case. Using Theorem 1, we should carefully choose the tolerable recovery error threshold ϖ in order to avoid excessive or insufficient numbers of measurements. In addition, in Theorem 1, the relation between the tolerable recovery error threshold ϖ and the probability of finding the best spectral approximation is unknown. By contrast, Theorem 2 shows that if there exists a best spectral approximation, the corresponding validation parameter should be within a certain small range with a probability greater than $1 - \varrho$. Thus, if the result of Theorem 2 is used as the signal acquisition termination metric, the issues of excessive or insufficient numbers of measurements can be solved.

Remark 4 If the best spectral approximation exists, the probability of finding it exponentially increases as the size of testing set (i.e., r) increases. It means that if we monitor ρ_p, we have a higher probability of finding the best spectral approximation when using more measurements for validation. However, we should note that there is a trade off between the size of the training set and the size of the testing set for a fixed sub-Nyquist sampling rate. On the one hand, a smaller r (i.e., larger training set for a fixed m) could result in better spectral recovery, while on the other hand, the probability of finding the best spectral approximation decreases as r becomes small. In addition, for a fixed degree of confidence $1 - \varrho$, we face a trade off between the accuracy ϵ and the size of the testing set r, as shown in Theorem 2. At the expense of the accuracy ϵ (i.e., larger ϵ), r can be small. We should also emphasize that, as we can see in (6.32), linear increase of the standard deviation δ will lead to quadratic growth in the size of the testing set. This is the reason why we should carefully consider the effects of measurements noise in the validation approach.

6.4.5 Sparsity-Aware Recovery Algorithm

As the above discussions indicated, Theorem 2 can be used for identifying the best spectral approximation to $\mathbf{X}$ from the spectral estimate sequence $\hat{X}_1, \hat{X}_2, \ldots, \hat{X}_p$, which is calculated by increasing the number of measurements in the proposed CASe framework. We note that Theorem 2 can also be used for preventing over-fitting or under-fitting in greedy recovery algorithms. Greedy recovery algorithms iteratively generate a sequence of estimates $\hat{X}_p^1, \hat{X}_p^2, \ldots, \hat{X}_p^t$, where the best spectral estimate may exist under certain system parameter choices. For example, the OMP algorithm chooses one column from the measurement matrix at a time for reconstructing $\mathbf{X}$ from $\mathbf{y}$. After $t = k$ iterations, the k-sparse vector $\hat{X}^k$ will be returned as an approximation to $\mathbf{X}$. Note that OMP requires the sparsity level k as an input, and such an input is commonly needed by most greedy recovery algorithms. However, the sparsity level k of the spectrum in the cognitive radio system is often unknown, and therefore traditional greedy compressed sensing algorithms will result in either early or late termination of greedy algorithms. Then the problems of under-fitting and over-fitting arise, leading to inferior spectral recovery performance. In order to reconstruct the full spectrum in the case of unknown k, we propose to use the testing set for validating the spectral estimate sequence $\hat{X}_p^1, \hat{X}_p^2, \ldots, \hat{X}_p^t$, and terminate the iterations if the current validation parameter satisfies the conditions given in Theorem 2.

As shown in Table 6.3, we present a sparsity-aware OMP algorithm. One important advantage of the proposed algorithm is that it does not require the instantaneous spectral sparsity level k, but requires instead its upper bound $k_{\max}$ which can be easily known. In each iteration, the column index $\lambda^t \in [1, N]$ that has the maximum correlation between the residual and the measurement matrix will be found, and be merged with the previously computed spectral support to form a new spectral support Λ^t. After that, the full spectrum is recovered by solving a least squares problem as shown in the step 2-*d*) of Table 6.3. Note that $\boldsymbol{\Theta}_p^t \triangleq \boldsymbol{\Phi}_p(\Lambda^t)$ is the sub-matrix obtained by only selecting the columns whose indices are within Λ^t in the matrix $\boldsymbol{\Phi}_p$, while other columns are set to all zeros. For a spectral estimate $\hat{X}_p^t$, we validate it by using the validation parameter ρ_p^t, which can be calculated by using the testing set $\mathbf{V}$ and the spectral estimate $\hat{X}_p^t$ as shown in the step 2-*e*) of Table 6.3. The residual is then updated. We emphasize that the proposed algorithm monitors the validation parameter ρ_p^t, instead of the residual $\|R_p^t\|_2 \leq \varpi$ as used in the traditional greedy recovery algorithms. Based on Theorem 2, if the best spectral estimate is included in the spectral estimate sequence $\hat{X}_p^1, \hat{X}_p^2, \ldots, \hat{X}_p^t$, the probability of finding it will be greater than $1 - 2\exp\left(-\frac{3r\epsilon^2}{3(4-\pi)\delta^2+\sqrt{2\pi}\epsilon\delta\upsilon}\right)$. In other words, the probability of under-/over-fitting is less than or equal to $2\exp\left(-\frac{3r\epsilon^2}{3(4-\pi)\delta^2+\sqrt{2\pi}\epsilon\delta\upsilon}\right)$, and becomes smaller as r increases.

For the proposed spectral recovery algorithm, there is a key parameter we need to know, i.e., ϵ. The following quadratic equation regarding ϵ holds by using (6.31):

Table 6.3 Sparsity-Aware OMP Algorithm

Input: training set $\mathbf{Y}_p$, testing set $\mathbf{V}$, measurement matrix $\boldsymbol{\Phi}_p$,
testing matrix $\boldsymbol{\Psi}$, noise variance δ^2, accuracy ϵ, $k_{\max}$.

1. *Initialize:*
   Index set $\Lambda^0 = \emptyset$, residual $\mathbf{R}_p^0 = \mathbf{Y}_p$, and iteration index $t = 0$.
   Let $\rho_p^t = C_1$ ($\forall t \in [0, k_{\max}]$), where C_1 is a large constant.
2. *While* $|\rho_p^t - \sqrt{\frac{\pi}{2}}\delta| > \epsilon$ and $t < k_{\max}$, *do*
   (a) Increment t by 1.
   (b) Find the index λ^t that solves the optimization problem:
   $\lambda^t = \arg\ \max_{j=1,\cdots,N} | < \mathbf{R}_p^{t-1}, \boldsymbol{\Phi}_p^j > |$.
   (c) Augment the index set $\Lambda^t = \Lambda^{t-1} \cup \{\lambda^t\}$, and revise
   the matrix $\boldsymbol{\Theta}_p^t = \boldsymbol{\Phi}_p(\Lambda^t)$ by only selecting the column
   index belongs to Λ^t, other columns are all zeros.
   (d) Solve a least squares problem:
   $\hat{X}_p^t = \arg\ \min_{\mathbf{X}} \|\mathbf{Y}_p - \boldsymbol{\Theta}_p^t \mathbf{F}^{-1}\mathbf{X}\|_2$.
   (e) Calculate the validation parameter using $\mathbf{V}$ and $\boldsymbol{\Psi}$:
   $\rho_p^t = \frac{\|\mathbf{V} - \boldsymbol{\Psi}\mathbf{F}^{-1}\hat{X}_p^t\|_1}{r}$.
   (f) Update residual:
   $\mathbf{R}_p^t = \mathbf{Y}_p - \boldsymbol{\Phi}_p \mathbf{F}^{-1}\hat{X}_p^t$.

Output: $\hat{X}_p = \arg\ \min_{\hat{X}_p^t} |\rho_p^t - \sqrt{\frac{\pi}{2}}\delta|, \quad \forall t \in [1, k_{\max}]$

$$r \cdot \epsilon^2 - \frac{\sqrt{2\pi}}{3} \ln\left(\frac{2}{\varrho}\right) \delta\nu \cdot \epsilon - (4 - \pi) \ln\left(\frac{2}{\varrho}\right) \delta^2 = 0. \tag{6.20}$$

It can be easily determined that the discriminant of the above quadratic equation is positive, so there are two distinct real roots. The following positive root can be used to determine ϵ:

$$\epsilon = \left[\frac{\sqrt{2\pi} \ln\left(\frac{2}{\varrho}\right) \delta\nu \pm \delta \sqrt{2\pi \ln^2\left(\frac{2}{\varrho}\right) \nu^2 + 36(4 - \pi) \ln\left(\frac{2}{\varrho}\right) r}}{6r} \right]^+ \tag{6.21}$$

where $[x]^+$ denotes $\max(x, 0)$.

6.4.6 Numerical Results

In our simulations, we adopt the wideband analog signal model in [27] and let the received signal $x(t)$ at a cognitive radio to be of the form

$$x(t) = \sum_{l=1}^{N_b} \sqrt{E_l B_l} \cdot \operatorname{sinc}\left(B_l(t-\alpha)\right) \cdot \cos\left(2\pi f_l(t-\alpha)\right) + z(t) \tag{6.22}$$

where $\operatorname{sinc}(x) = \frac{\sin(\pi x)}{\pi x}$, α denotes a random time offset smaller than $T/2$, $z(t)$ is AWGN (i.e., $z(t) \sim \mathcal{N}(0, 1)$), and E_l is the received power for the subband l at cognitive radio. The received signal $x(t)$ consists of $N_b = 8$ non-overlapping subbands. The l-th subband is in the frequency range of $[f_l - \frac{B_l}{2}, f_l + \frac{B_l}{2}]$, where the bandwidth $B_l = 10 \sim 30$ MHz and f_l denotes the center frequency. The center frequency of the subband l is randomly located within $[\frac{B_l}{2}, W - \frac{B_l}{2}]$ (i.e., $f_l \in [\frac{B_l}{2}, W - \frac{B_l}{2}]$), where the overall signal bandwidth $W = 2$ GHz. Therefore, the Nyquist rate is $f = 2W = 4$ GHz, and the spectral occupancy (i.e., $\frac{\sum_{l=1}^{8} B_l}{W}$) is a random number between 4 % and 12 %. We emphasize that the spectral occupancy of 4 % $\sim$ 12 % in our simulations is very close to the spectral measurements in New York City as noted above. The received signal-to-noise ratios (SNRs) of these 8 active subbands are random natural numbers between 5 dB and 25 dB. The spectrum sensing duration is chosen to be $T = 5\ \mu$s, during which the symbols from primary users and the channels between the primary users and cognitive radios are assumed to be quasi-stationary. We then divide T into $P = 10$ mini time slots, each of which has $\tau = \frac{T}{P} = 0.5\ \mu$s. If the received signal $x(t)$ were sampled at the Nyquist rate, the number of Nyquist samples in each time slot would be $N = 2W\tau = 2,000$. It can be calculated that the spectral sparsity level k is in the range of $4\,\% \times N = 80 \le k \le 12\,\% \times N = 240$. In the proposed framework, rather than using the Nyquist sampling rate, we adopt the sub-Nyquist sampling rate $f_s = 400$ MHz; thus, the number of measurements in each time slot is $m = f_s\tau = 200$. In other words, the undersampling fraction in each time slot is $m/N = 10\,\%$. For the purpose of testing/validation, $r = 50$ measurements in the first time slot are reserved, while the remaining measurements are used for reconstructing the spectrum. The measurement matrices, i.e., $\boldsymbol{\Phi}_p$ and $\boldsymbol{\Psi}$, follow the standard normal distribution with zero mean and unit variance. Due to the imperfect design of signal measurement devices, the measurement noise may exist. In the noisy case, the measurement noise is assumed to be circular complex AWGN, i.e., $\mathbf{n} \sim \mathcal{CN}(0, \delta^2)$. As the measurement noise in this chapter is mainly due to the signal quantization in the ADCs, we set the signal-to-measurement-noise ratios (SMNR) to be 50 dB and 100 dB. This is because the SMNR of the uniform quantization increases 6 dB for each one-bit; thus, the SMNR of 8-bit quantization is 48 dB and the SMNR of 16-bit quantization is 96 dB, which are approximately 50 dB and 100 dB.

Firstly, we consider the effects of measurement noise to both the spectral recovery quality and the validation parameter. In Fig. 6.14, the spectral sparsity level is set to $k = 120$. We can see that, in either the noiseless measurement case or the noisy measurement case, the proposed CASe framework can reconstruct the spectrum using 6 time slots. The spectral recovery quality becomes worse when the measurement noise level increases. In the noiseless case, the proposed validation parameter can closely fit the actual recovery error. By contrast, there is a gap between the actual

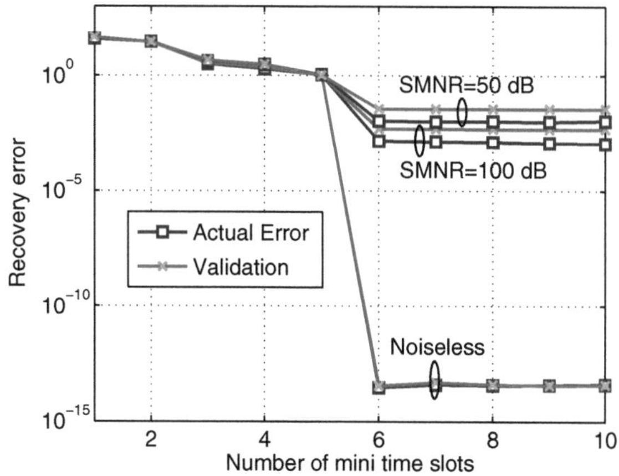

Fig. 6.14 The effects of measurement noise on both the actual recovery error and the proposed validation parameter when the SMNR varies. The spectral sparsity level was set to $k = 120$

recovery error and the validation result when the measurement noise exists. This is because, on the one hand, the actual recovery error $\|\mathbf{X} - \hat{X}_p\|_2$ can be very small, e.g., 10^{-14} in the case of best spectral approximation, on the other hand, the validation parameter is mainly determined by the noise level as shown in Theorem 2. This implies that the effects of measurement noise should be carefully considered even if $\hat{X}_p$ is the best spectral approximation. In Fig. 6.15, it is seen that when the best spectral approximation occurs (i.e., the actual recovery error is small enough), the validation parameter is very close to the scaled noise standard deviation, i.e., $\sqrt{\frac{\pi}{2}}\delta$. This observation validates the results of Theorem 2. If the validation method is used for designing the termination metric of the signal acquisition, such as in the algorithm given in Table 6.2, the problems of insufficient or excessive numbers of measurements can be solved.

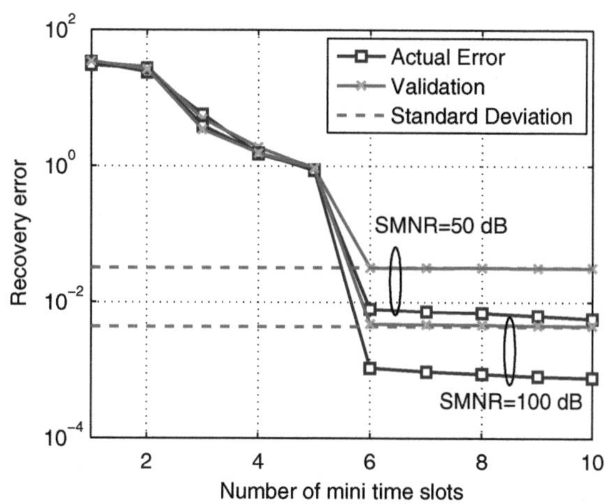

Fig. 6.15 Comparison of the validation parameter and the actual recovery error when the best spectral approximation occurs. The *dash line*denotes the predicted validation value, i.e., $\sqrt{\frac{\pi}{2}}\delta$ (scaled standard deviation), as used in Theorem 2

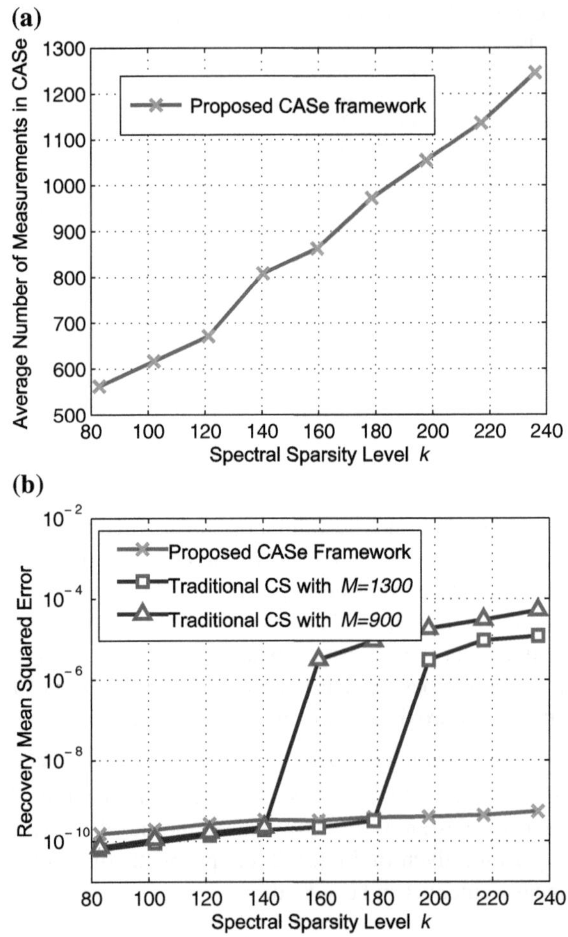

Fig. 6.16 Performance analysis of spectral recovery when using different compressed sensing approaches. **a** The average number of measurements required by CASe. **b** The spectral recovery mean square error. The SMNR was set to 100 dB

Secondly, Fig. 6.16 analyzes the spectral recovery performance when using different compressed sensing approaches. In these simulations, in order to find the best spectral approximation with high confidence, the accuracy parameter ϵ in (6.19) is set to $\delta/2$ and the number of testing measurements is $r = 50$. As depicted in Fig. 6.16a, the proposed CASe framework can adaptively adapt its number of measurements to the unknown spectral sparsity level k. The corresponding spectral recovery performance is shown in Fig. 6.16b, where the spectral recovery mean square error (MSE) of different compressed sensing approaches is given. We can see that, even with the total number of measurements $M = 1300$, the performance of the traditional compressed sensing system is inferior to that of the proposed CASe framework as the traditional compressed sensing system cannot deal with the case of $k \geq 200$. Note that, if we assume that the spectral sparsity level k has a uniform distribution between 80 and 240, the average number of measurements required by CASe is 900.

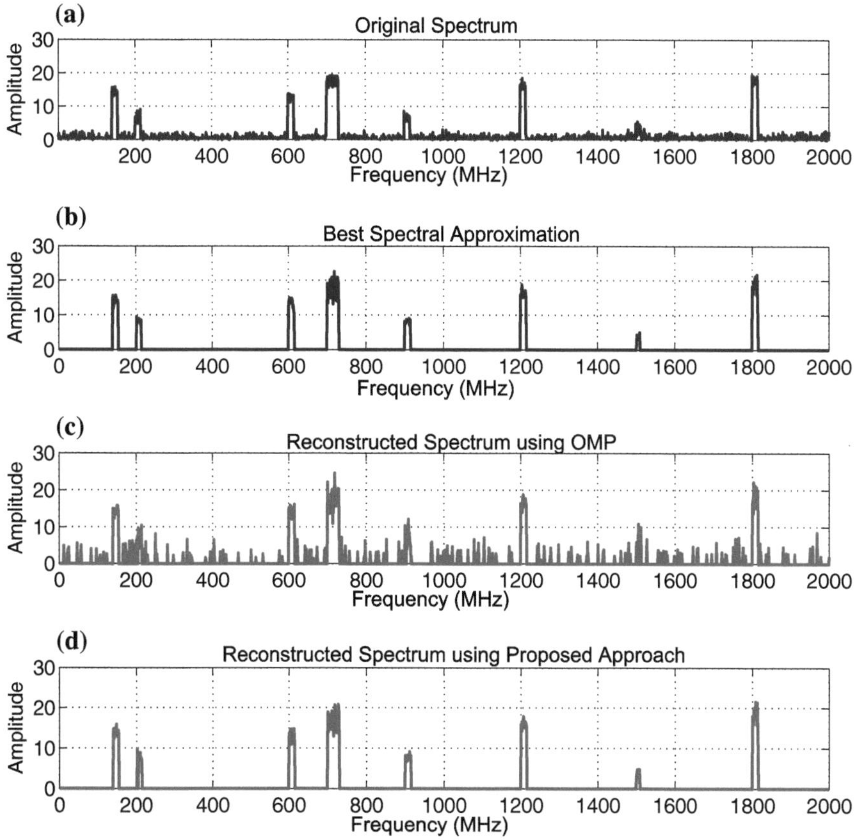

Fig. 6.17 Examples of the reconstructed spectrum when using different recovery algorithms. The spectral sparsity level was assumed to be $k = 150$, with the total number of measurements $M = 800$. The received SNRs of these 8 active subbands were set to random natural numbers between 5 dB and 25 dB. The SMNR was set to 50 dB

Compared to the traditional compressed sensing system with $M = 900$, it is obvious that the CASe framework has much lower MSE for most of $k \in [80, 240]$.

Thirdly, Fig. 6.17 shows examples of the original spectrum and the reconstructed spectrum when using different spectral recovery algorithms, i.e., OMP and the proposed algorithm. We can see that the recovery performance of the proposed algorithm is superior to that of the traditional OMP algorithm. As the sparsity level is unknown and has the range of $80 \leq k \leq 240$, if the OMP algorithm is used, the problems of either under-fitting (i.e., iteration is terminated earlier as k is under-estimated) or over-fitting exist. As the problem of under-fitting could lead to the missed detection of primary users which may cause harmful interference to primary users, the traditional OMP algorithm should prevent the under-fitting from occurring, and tends to choose more number of iterations. In the case of over-fitting, the traditional OMP algorithm will result in a "noisy" reconstructed spectrum as depicted in Fig. 6.17c. With the aid of the testing set, the proposed approach has an improved recovery performance as shown in Fig. 6.17d. Compared with the OMP algorithm, the proposed algorithm provides better spectral estimate, and is much more similar to the best spectral approximation in Fig. 6.17b. It is worthwhile to emphasize that the proposed algorithm will have more noticeable improvement over the OMP algorithm when there is larger uncertainty in the spectral sparsity level k.

Finally, Fig. 6.18 further explores the performance of different recovery algorithms. In order to illustrate the performance of CASe when using different recovery algorithms, the MSE of the reconstructed spectrum is given in Fig. 6.18a. It can be seen that the gain of using the proposed algorithm over OMP is approximately one order of magnitude in MSE. This is because the proposed algorithm can terminate the iteration at the right iteration index; by contrast, when using OMP, the

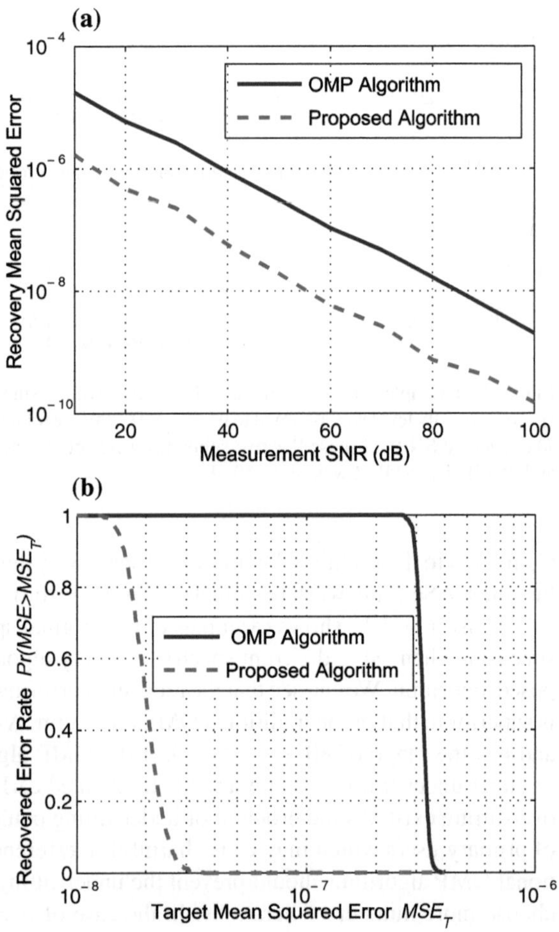

Fig. 6.18 Performance comparison of different recovery algorithms. **a** The spectral recovery mean square error when the SMNR increases. **b** The recovered error rate $\Pr(MSE > MSE_T)$ when SMNR = 50 dB. The spectral sparsity level was assumed to be $k = 120$, with the average number of measurements $M = 800$

problems of either under-fitting or over-fitting exist, leading to either incomplete spectral recovery or noisy spectral recovery. As a consequence, we can see from Fig. 6.18b that, for a fixed SMNR=50 dB, the proposed algorithm has much lower recovered error rate than the OMP algorithm. We note that the recovered error rate is defined as the probability of simulated mean MSE larger than the target MSE.

6.4.7 Discussions and Conclusions

6.4.7.1 Discussions

The CASe framework shares its goals with some recent efforts that have looked at testing the actual error directly from compressed data. The ℓ_2 norm cross validation approach for compressed sensing has been studied by Ward [34], and Boufounos et al. [35]. These results are very remarkable as they allow us to verify the actual decoding error with almost no effort (i.e., a very few measurements are reserved for testing). We note that the results here are different from those in these papers. In particular, we have studied a different validation approach, i.e., the ℓ_1 norm is used for validating the recovery result, rather than the ℓ_2 norm. In addition, the effects of measurement noise were carefully considered in our analysis. By contrast, Ward's validation approach did not model the effects of measurement noise. When the proposed ℓ_1 norm validation approach is used in compressed sensing technologies, it could be a useful complement to the work in [34, 35]. It should also be emphasized that, compared to the ℓ_2 norm validation approach, the proposed ℓ_1 norm validation approach is less sensitive to outliers. As shown in Fig. 6.19a, when outliers exist in the testing set, the validation parameter of using the ℓ_1 norm is one order in magnitude lower than that of using the ℓ_2 norm. Moreover, we note that using compressed sensing technologies for wideband spectrum sensing in a cognitive radio system, we cannot avoid outliers. This is because the ADC is not a noise-free device, and the non-linearity of ADC could be a source of generating outliers. Furthermore, in a real-time compressed sensing device such as the random demodulator in [26, 29, 30], imperfect synchronization of the pseudo-random sequence generator and the low-rate ADC could result in outliers.

A natural technique for choosing the stopping time of the measurement would be sequential detection [36], in which we collect one sample at a time until we have enough observations to generate a final decision. However, we note that, in the compressed sensing-based spectrum sensing system, the sequential measurements cannot be directly used for performing sequential test. This is because, due to the sub-Nyquist sampling, there exists spectral aliasing phenomenon, which makes frequencies become indistinguishable. Thus, in order to apply sequential detection, the wideband spectrum should be reconstructed before each sequential test for avoiding spectral aliasing. In such a scenario, sequential detection could lead to high computational costs. Malioutov et al. [37] have studied a typical compressed sensing-based sequential measurement system, where the decoder can receive compressed samples

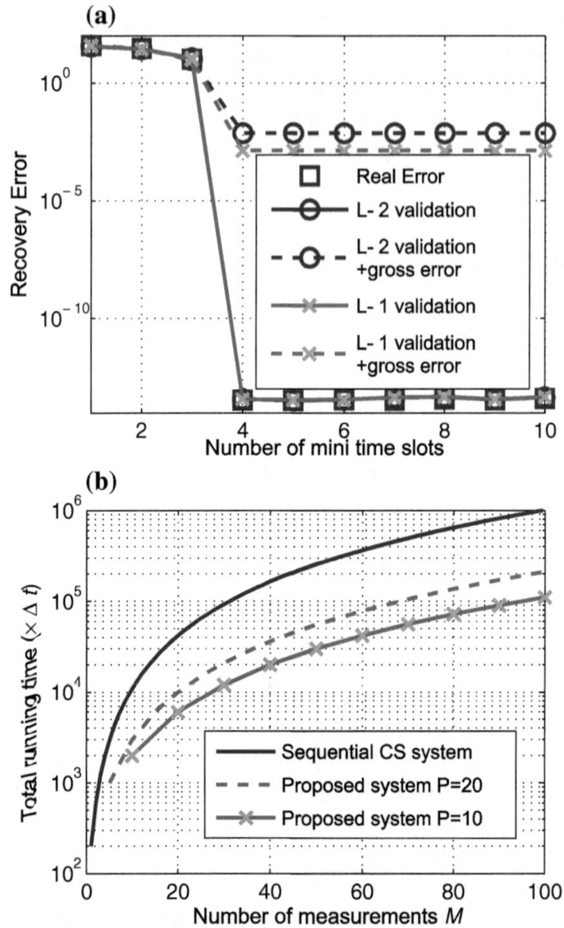

Fig. 6.19 Comparison between the proposed system and the existing systems. **a** Sensitivity test of both the ℓ_1 norm validation and the ℓ_2 norm validation approaches against outliers. In simulations, the measurement error was added to a single sample of the testing set, and the magnitude of the measurement error was set to 100 dB lower than that of the sample. **b** Total running time of reconstructing the spectrum for both the sequential compressed sensing measurement setup and the proposed system when using the CoSaMP algorithm. In simulations, $N = 200$, and $M = Pm = 100$ where m denotes the number of measurements in each mini-time slot and P is the number of mini-time slots

sequentially. It has been shown that such a system can successfully estimate the current decoding error by using some additional samples. Nevertheless, it is not proper to apply the compressed sensing-based sequential measurement setup in cognitive radio systems. Because, in this scheme, the wideband spectrum should be repeatedly reconstructed for each additional measurement that could lead to high computational costs and large spectrum sensing overhead in cognitive radios. For example, using the CoSaMP algorithm [28], the running time in each reconstruction is $\mathcal{O}(\beta N)$, where β denotes the current number of measurements. Thus, the total running time for the sequential measurement setup is $\mathcal{O}(\frac{M(M+1)N}{2})$, where M denotes the number of measurements till the termination of measurement. By contrast, in our proposed system, the spectrum sensing time slot is divided into P equal-length mini-time slots, and the wideband spectrum is reconstructed after each mini-time slot. The total running time of the proposed system is therefore $\mathcal{O}(\frac{M(P+1)N}{2})$, where

$P \ll M$. Figure 6.19b shows that the spectrum sensing overhead (due to the spectral reconstruction) of the sequential compressed sensing system is several times higher than that of the proposed system. Furthermore, another advantage of the proposed system is that, by changing the length of mini-time slot (thus the value of P because $P = \frac{M}{m}$), we can control the trade-off between the cost of computation and the cost of acquiring additional measurements.

6.4.7.2 Conclusions

We have presented a novel framework, i.e., CASe, for wideband spectrum sensing in cognitive radio systems. It has been shown that CASe can considerably improve the spectral recovery performance when the sparsity level of the spectrum is unknown, thanks to the ℓ_1 norm validation approach. We have shown that the proposed validation parameter can be a very good proxy for the actual spectral recovery error in the noiseless measurement case even if the testing set is small. The proper use of the validation approach could solve the problems of excessive or insufficient numbers of measurements, thereby improving not only the energy-efficiency of cognitive radio, but also the throughput of cognitive radio networks. In addition, we have shown that, in the case of noisy compressive measurements, if the best spectral approximation exists, then the corresponding validation parameter has a very large probability of being within a certain small range. Based on this property, we have proposed a sparsity-aware recovery algorithm for reconstructing the wideband spectrum without the knowledge of the spectral sparsity level. In the proposed algorithm, if the best spectral approximation exists, then the correct iteration termination index can be found with high probability; therefore, the issues of under-/over-fitting are addressed.

Simulation results have shown that the proposed framework can correctly terminate the signal acquisition that saves both spectrum sensing time slots and signal acquisition energy, while providing better spectral recovery performance than traditional compressed sensing approaches. Compared with the existing greedy recovery algorithm, the proposed sparsity-aware algorithm can achieve lower MSE for reconstructing the spectrum and better spectrum sensing performance. As the RF spectrum is the lifeblood of wireless communication systems and the wideband techniques could potentially offer greater capacity, we expect that the proposed framework has a broad range of applications, e.g., broadband spectral analyzers, signals-intelligence receivers, and ultra wideband radars. Moreover, the proposed ℓ_1 norm validation approach can be used in other compressed sensing applications, e.g., a compressed sensing based communication system where we need to terminate the decoding algorithm with high confidence and small predictable decoding error.

Acknowledgments H. Sun and A. Nallanathan acknowledge the support of the UK Engineering and Physical Sciences Research Council (EPSRC) with Grant No. EP/I000054/1.

Appendix

Proof of Theorem 1

Using a variant of the Johnson-Lindenstrauss lemma as shown in Theorem 5.1 of [38], we have

$$\Pr\left[(1-\varepsilon)\|\mathbf{x}\|_2 \le \frac{\|\boldsymbol{\Psi}\mathbf{x}\|_1}{\sqrt{2/\pi}\, r} \le (1+\varepsilon)\|\mathbf{x}\|_2\right] \ge 1-\xi. \tag{6.23}$$

Defining $\mathbf{x} \triangleq \mathbf{F}^{-1}(\mathbf{X}-\hat{X}_p)$ in (6.23), we obtain

$$\Pr\left[(1-\varepsilon)\|\mathbf{F}^{-1}(\mathbf{X}-\hat{X}_p)\|_2 \le \frac{\|\boldsymbol{\Psi}\mathbf{F}^{-1}(\mathbf{X}-\hat{X}_p)\|_1}{\sqrt{2/\pi}\, r} \le (1+\varepsilon)\|\mathbf{F}^{-1}(\mathbf{X}-\hat{X}_p)\|_2\right] \ge 1-\xi. \tag{6.24}$$

The above inequality can be rewritten by using (6.12) and (6.14)

$$\Pr\left[(1-\varepsilon)\|\mathbf{F}^{-1}(\mathbf{X}-\hat{X}_p)\|_2 \le \sqrt{\frac{\pi}{2}}\rho_p \le (1+\varepsilon)\|\mathbf{F}^{-1}(\mathbf{X}-\hat{X}_p)\|_2\right] \ge 1-\xi. \tag{6.25}$$

Applying Parseval's relation to (6.25), we have

$$\Pr\left[(1-\varepsilon)\|\mathbf{X}-\hat{X}_p\|_2 \le \sqrt{\frac{\pi N}{2}}\rho_p \le (1+\varepsilon)\|\mathbf{X}-\hat{X}_p\|_2\right] \ge 1-\xi. \tag{6.26}$$

Thus, Theorem 1 follows.

Proof of Theorem 2

The best spectral approximation $\mathbf{X}^\star$ means that $\|\mathbf{X}^\star - \mathbf{X}\|_2$ is sufficiently small. Without loss of generality, we approximate $\mathbf{X}^\star$ by $\mathbf{X}$. Thus, if $\hat{X}_p$ is the best spectral approximation, the validation parameter can be rewritten by using (6.18)

$$\rho_p = \frac{\|\mathbf{V}-\boldsymbol{\Psi}\mathbf{F}^{-1}\hat{X}_p\|_1}{r} = \frac{\|\mathbf{n}\|_1}{r} = \frac{\sum_{i=1}^{r}|n^i|}{r}. \tag{6.27}$$

As the measurement noise $n^i \sim \mathcal{CN}(0,\delta^2)$, its absolute value $|n^i|$ follows the Rayleigh distribution with mean $\sqrt{\frac{\pi}{2}}\delta$ and variance $\frac{4-\pi}{2}\delta^2$. Using the cumulative distribution function of the Rayleigh distribution, we have $\Pr(|n^i| \le x) = 1 -$

$\exp(-\frac{x^2}{2\delta^2})$. Further, as the measurement noise level has an upper-bound $\bar{n}$ in practice, there exists a sufficiently large parameter ν that makes $|n^i| \leq \bar{n} \leq (\nu+1)\sqrt{\frac{\pi}{2}}\delta$ almost surely. If we define a new variable $D_i = |n^i| - \sqrt{\frac{\pi}{2}}\delta$, we obtain $\mathbb{E}[D_i] = 0$, $\mathbb{E}[D_i^2] = \frac{4-\pi}{2}\delta^2$, and $|D_i| \leq \sqrt{\frac{\pi}{2}}\delta\nu$. Based on the Bernstein's inequality [39], the following inequality holds

$$\begin{aligned} \Pr\left[\left|\sum_{i=1}^{r} D_i\right| > \varepsilon\right] &= \Pr\left[\left|\sum_{i=1}^{r} |n^i| - r\sqrt{\frac{\pi}{2}}\delta\right| > \varepsilon\right] \\ &\leq 2\exp\left(-\frac{\varepsilon^2/2}{\sum_{i=1}^{r}\mathbb{E}[D_i^2] + \overline{D}\varepsilon/3}\right) \\ &\leq 2\exp\left(-\frac{3\varepsilon^2}{3(4-\pi)r\delta^2 + \sqrt{2\pi}\,\varepsilon\delta\nu}\right) \end{aligned} \tag{6.28}$$

where $\overline{D} = \sqrt{\frac{\pi}{2}}\delta\nu$ denotes the upper-bound on $|D_i|$.

Simply replacing ε by $r\epsilon$ in (6.28) while using (6.27), we can rewrite (6.28) as

$$\Pr\left[\left|\rho_p - \sqrt{\frac{\pi}{2}}\delta\right| > \epsilon\right] \leq 2\exp\left(-\frac{3r\epsilon^2}{3(4-\pi)\delta^2 + \sqrt{2\pi}\epsilon\delta\nu}\right). \tag{6.29}$$

Using (6.29), we end up with

$$\Pr\left[\left|\rho_p - \sqrt{\frac{\pi}{2}}\delta\right| \leq \epsilon\right] > 1 - 2\exp\left(-\frac{3r\epsilon^2}{3(4-\pi)\delta^2 + \sqrt{2\pi}\epsilon\delta\nu}\right). \tag{6.30}$$

To derive the required r, we set the lower probability bound in (6.30) as

$$1 - 2\exp\left(-\frac{3r\epsilon^2}{3(4-\pi)\delta^2 + \sqrt{2\pi}\epsilon\delta\nu}\right) = 1 - \varrho. \tag{6.31}$$

Solving the above equation, we obtain

$$r = \ln\left(\frac{2}{\varrho}\right)\frac{3(4-\pi)\delta^2 + \sqrt{2\pi}\epsilon\delta\nu}{3\epsilon^2}. \tag{6.32}$$

This completes the proof of Theorem 2.

References

1. McHenry MA (2005) NSF spectrum occupancy measurements project summary. Technical Report, Shared Spectrum Company
2. Haykin S (2005) Cognitive radio: brain-empowered wireless communications. IEEE J Sel Areas Commun 23(2):201–220
3. Akyildiz I, Lee W-Y, Vuran M, Mohanty S (2008) A survey on spectrum management in cognitive radio networks. IEEE Commun Mag 46(4):40–48
4. Mitola J (2000) Cognitive radio: an integrated agent architecture for software defined radio. Ph.D. dissertation, Department of Teleinformatics, Royal Institute of Technology Stockholm, Sweden, 8 May 2000
5. Akyildiz IF, Lee W-Y, Vuran MC, Mohanty S (2006) NeXt generation/dynamic spectrum access/cognitive radio wireless networks: a survey. Comput Netw 50(13):2127–2159
6. Ekram H, Bhargava VK (2007) Cognitive wireless communications networks. In: Bhargava VK (ed) Springer Publication, New York
7. Cabric D, Mishra SM, Brodersen RW (2004) Implementation issues in spectrum sensing for cognitive radios. Proc Asilomar Conf Signal Syst Comput 1:772–776
8. Sun H, Laurenson D, Wang C-X (2010) Computationally tractable model of energy detection performance over slow fading channels. IEEE Commun Lett 14(10):924–926
9. Hossain E, Niyato D, Han Z (2009) Dynamic spectrum access and management in cognitive radio networks. Cambridge University Press, Cambridge
10. Tian Z, Giannakis GB (2006) A wavelet approach to wideband spectrum sensing for cognitive radios. In: Proceedings of IEEE cognitive radio oriented wireless networks and communications, Mykonos Island, pp 1–5
11. Proakis JG(2001) Digital communications, 4th edn. McGraw-Hill, New York
12. Yucek T, Arslan H (2009) A survey of spectrum sensing algorithms for cognitive radio applications. IEEE Commun Surv Tutor 11(1):116–130
13. Tian Z, Giannakis GB (2007) Compressed sensing for wideband cognitive radios. In: Proceedings of IEEE international conference on acoustics, speech, and signal processing, Hawaii, April 2007, pp 1357–1360
14. Yoon M-H, Shin Y, Ryu H-K, Woo J-M (2010) Ultra-wideband loop antenna. Electron Lett 46(18): 1249–1251
15. Hao Z-C, Hong J-S (2011) Highly selective ultra wideband bandpass filters with quasi-elliptic function response. IET Microwaves Antennas Propag 5(9):1103–1108
16. [Online]. Available: http://www.national.com/pf/DC/ADC12D1800.html
17. Sun H, Chiu W-Y, Jiang J, Nallanathan A, Poor HV (2012) Wideband spectrum sensing with sub-Nyquist sampling in cognitive radios. IEEE Trans Sig Process 60(11):6068–6073
18. Venkataramani R, Bresler Y (2000) Perfect reconstruction formulas and bounds on aliasing error in sub-Nyquist nonuniform sampling of multiband signals. IEEE Trans Inf Theory 46(6):2173–2183
19. Tao T (2005) An uncertainty principle for cyclic groups of prime order. Math Res Lett 12:121–127
20. Mishali M, Eldar YC (2009) Blind multiband signal reconstruction: compressed sensing for analog signals. IEEE Trans Signal Process 57(3):993–1009
21. Feldster A, Shapira Y, Horowitz M, Rosenthal A, Zach S, Singer L (2009) Optical undersampling and reconstruction of several bandwidth-limited signals. J Lightwave Technol 27(8):1027–1033
22. Rosenthal A, Linden A, Horowitz M (2008) Multi-rate asynchronous sampling of sparse multiband signals, arXiv.org:0807.1222
23. Fleyer M, Rosenthal A, Linden A, Horowitz M (2008) Multirate synchronous sampling of sparse multiband signals, arXiv.org:0806.0579
24. Chen SS, Donoho DL, Saunders MA (2001) Atomic decomposition by basis pursuit. SIAM Review, 43(1):129–159. [Online]. Available: http://www.jstor.org/stable/3649687

25. Polo YL, Wang Y, Pandharipande A, Leus G (2009) Compressive wide-band spectrum sensing. In: Proceedings of IEEE international conference on acoustics, speech, and signal processing, Taipei, pp 2337–2340
26. Tropp JA, Laska JN, Duarte MF, Romberg JK, Baraniuk R (2010) Beyond Nyquist: efficient sampling of sparse bandlimited signals. IEEE Trans Inf Theory 56(1):520–544
27. Mishali M, Eldar Y (2010) From theory to practice: sub-Nyquist sampling of sparse wideband analog signals. IEEE J Sel Top Signal Process 4(2):375–391
28. Needell D, Tropp J (2009) Cosamp: iterative signal recovery from incomplete and inaccurate samples. Appl Comput Harmon Anal 26(3):301–321 . [Online]. Available: http://www.sciencedirect.com/science/article/B6WB3-4T1Y404-1/2/a3a764ae1efc1bd0569dcde301f0c6f1
29. Laska J, Kirolos S, Massoud Y, Baraniuk R, Gilbert A, Iwen M, Strauss M (2006) Random sampling for analog-to-information conversion of wideband signals. In: Proceedings of IEEE DCAS, pp 119–122
30. Laska JN, Kirolos S, Duarte MF, Ragheb TS, Baraniuk RG, Massoud Y (2007) Theory and implementation of an analog-to-information converter using random demodulation. Proc IEEE Int Symp Circ Syst ISCAS 2007(27–30):1959–1962
31. Candes EJ, Tao T (2005) Decoding by linear programming. IEEE Trans Inf Theory 51(12):4203–4215
32. Haupt J, Bajwa WU, Rabbat M, Nowak R (2008) Compressed sensing for networked data. IEEE Signal Process Mag 25(2):92–101
33. Quan Z, Cui S, Sayed AH, Poor HV (2009) Optimal multiband joint detection for spectrum sensing in cognitive radio networks. IEEE Trans Signal Process 57(3):1128–1140
34. Ward R (2009) Compressed sensing with cross validation. IEEE Trans Inf Theory 55(12):5773–5782
35. Boufounos P, Duarte M, Baraniuk R (2007) Sparse signal reconstruction from noisy compressive measurements using cross validation. In: Proceedings of IEEE/SP 14th workshop on statistical signal processing, Madison, pp 299–303
36. Chaudhari S, Koivunen V, Poor HV (2009) Autocorrelation-based decentralized sequential detection of OFDM signals in cognitive radios. IEEE Trans Signal Process 57(7):2690–2700
37. Malioutov D, Sanghavi S, Willsky A (2010) Sequential compressed sensing. IEEE J Sel Top Signal Process 4(2):435–444
38. Matousek J (2008) On variants of the Johnson-Lindenstrauss lemma. Random Struct Algor 33:142–156
39. Hazewinkel M (ed) (1987) Encyclopaedia of mathematics vol 1. Springer, New York

Chapter 7
Sparse Nonlinear MIMO Filtering and Identification

G. Mileounis and N. Kalouptsidis

Abstract In this chapter system identification algorithms for sparse nonlinear multi input multi output (MIMO) systems are developed. These algorithms are potentially useful in a variety of application areas including digital transmission systems incorporating power amplifier(s) along with multiple antennas, cognitive processing, adaptive control of nonlinear multivariable systems, and multivariable biological systems. Sparsity is a key constraint imposed on the model. The presence of sparsity is often dictated by physical considerations as in wireless fading channel–estimation. In other cases it appears as a pragmatic modelling approach that seeks to cope with the curse of dimensionality, particularly acute in nonlinear systems like Volterra type series. Three identification approaches are discussed: conventional identification based on both input and output samples, Semi-Blind identification placing emphasis on minimal input resources and blind identification whereby only output samples are available plus a–priori information on input characteristics. Based on this taxonomy a variety of algorithms, existing and new, are studied and evaluated by simulations.

7.1 Introduction

System nonlinearities are present in many practical situations and remedies based on linear approximations often degrade system performance. A popular model that captures system nonlinearities is Volterra series [69, 71, 77]. This model is employed in communications, digital magnetic recording, physiological systems, control of multivariable systems, etc. Volterra series constitute a class of polynomial models

G. Mileounis (✉) · N. Kalouptsidis
Department of Informatics and Telecommunications, Division of Communications and Signal Processing, University of Athens, Panepistimiopolis 15784, Ilisia, Greece
e-mail: gmil@di.uoa.gr

N. Kalouptsidis
e-mail: kalou@di.uoa.gr

A. Y. Carmi et al. (eds.), *Compressed Sensing & Sparse Filtering*,
Signals and Communication Technology, DOI: 10.1007/978-3-642-38398-4_7,

that can be regarded as a Taylor series with memory. An attractive feature of this model is that the unknown parameters enter linearly at the output. On the other hand, the number of terms increases exponentially with the order and memory of the model.

Most of the work reported in the literature focuses on modelling and identification of single input single output (SISO) Volterra systems. When the underlying nonlinear system is a MIMO system, the resulting model is more complicated and has received little attention. MIMO models are addressed in this chapter. Nonlinear MIMO systems involve a large number of parameters to be estimated, which increases exponentially with the order, the memory and the number of inputs. Therefore, there is a strong need to reduce complexity by considering those terms that strongly contribute to the outputs. This leads naturally to a sparse approximation of the underlying nonlinear MIMO system. Identification of sparse nonlinear MIMO systems is approached under three different settings: conventional, Semi-Blind and blind. Blind methods identify the unknown system parameters merely based on the output signals. On the other hand, conventional and Semi-Blind methods, require a training or a pilot sequence.

The objective of this chapter is twofold. First, it extends existing algorithms for adaptive filtering of SISO models to the MIMO case and demonstrates their applicability to nonlinear MIMO systems. Secondly, it presents new algorithms for blind and Semi-Blind identification of nonlinear MIMO systems excited by finite alphabet inputs. The chapter is divided into four sections. The sparse nonlinear MIMO models under consideration are presented in Sect. 7.2. Adaptive filters for sparse MIMO systems are discussed in Sect. 7.3. Then, algorithms for blind and Semi-Blind identification are addressed in Sect. 7.4. Finally, summary and future work are discussed in Sect. 7.5.

7.2 System Model

MIMO polynomial systems form the basic class of models we shall be working with. These finitely parametrizable recursive structures are defined next. First the basic notation from SISO Volterra series is reviewed. Then MIMO extensions are considered and some special cases of interest are introduced. Finally, various applications which employ MIMO Volterra models are briefly reviewed.

Volterra series constitute a popular model for the description of nonlinear behaviour [69, 71]. A SISO discrete–time Volterra model has the following form

$$y(n) = \sum_{p=1}^{\infty} \sum_{\tau_1=-\infty}^{\infty} \cdots \sum_{\tau_p=-\infty}^{\infty} h_p(\tau_1, \ldots, \tau_p) \left[\prod_{i=1}^{p} x(n-\tau_i) \right]. \tag{7.1}$$

Each output is formed by weighting the input shifted samples $x(n-\tau_i)$ and their products. The weights $h_p(\tau_1, \ldots, \tau_p)$ constitute the *Volterra kernels* of order p. Well possessedness conditions ensuring that inputs give rise to well defined outputs are given in [13, 51]. If only a finite number of nonlinearities enters Eq. (7.1), the resulting expression defines a finite Volterra system. Suppose the kernels of a finite Volterra system are causal and absolutely summable. Then Eq. (7.1) defines a bounded input bounded output (BIBO) stable system and can be approximated by the polynomial system

$$y(n) = \sum_{p=1}^{P} \sum_{\tau_1=0}^{M} \cdots \sum_{\tau_p=0}^{M} h_p(\tau_1, \ldots, \tau_p) \left[\prod_{i=1}^{p} x(n-\tau_i) \right]. \tag{7.2}$$

Equation (7.2) is parametrized by the finite Volterra kernels and has finite memory M. A more general result established by Boyd and Chua [13, 14] states that any shift invariant causal BIBO stable system with *fading memory* can be approximated by Eq. (7.2). The fading memory is a continuity property with respect to a weighted norm which penalizes the remote past in the formation of the current output. The reader may consult [13, 14, 51] for more details.

A key feature of Eq. (7.2) is that it is *linear in the parameters*. For estimation purposes it is useful to write Eq. (7.2) in matrix form using Kronecker products [15]. Indeed, let $\mathbf{x}(n) = [x(n), x(n-1), \cdots, x(n-M)]^T$ (the superscript T denotes the transpose operation) and the pth–order Kronecker power

$$\mathbf{x}_p(n) = \underbrace{\mathbf{x} \otimes \cdots \otimes \mathbf{x}}_{p \text{ times}}, \quad p = 2, \ldots, P.$$

The Kronecker power contains all pth–order products of the input. Likewise $\mathbf{h} = \left[\mathbf{h}_1(\cdot), \cdots, \mathbf{h}_p(\cdot)\right]^T$ is obtained by treating the p–dimensional kernel as a M^p column vector. We rewrite Eq. (7.2) as follows

$$y(n) = \left[\mathbf{x}^T(n)\mathbf{x}_2^T(n) \cdots \mathbf{x}_p^T(n)\right] \begin{bmatrix} \mathbf{h}_1 \\ \mathbf{h}_2 \\ \vdots \\ \mathbf{h}_p \end{bmatrix} = \mathbf{x}^T(n)\mathbf{h}. \tag{7.3}$$

Collecting n successive output samples from the above equation into the vector $\mathbf{y}(n) = [y(1), \ldots, y(n)]$ results in the following system of linear equations:

$$\mathbf{y}(n) = \mathbf{X}(n)\mathbf{h}$$

when

$$\mathbf{X}(n) = \left[\mathbf{x}^T(1), \ldots, \mathbf{x}^T(n)\right]^T.$$

From a practical viewpoint, Volterra models of order higher than three are rarely considered. This is due to the fact that the number of parameters ($\sum_{p=1}^{P} M^p$) involved in the model of Eq. (7.2) grows exponentially as a function of the memory size and the order of nonlinearity. To cope with this complexity several sub–families of Eq. (7.2) have been considered, most notable Wiener, Hammerstein and Wiener–Hammerstein models. In all cases the universal approximation capability is lost. A Wiener system is the cascade of a linear filter followed by a static nonlinearity. If we approximate the static nonlinearity with its Taylor expansion up to a certain order, we obtain the following expression for the output of the Wiener system

$$y(n) = \sum_{p=1}^{P} \left[\sum_{\tau=0}^{M} h_p(\tau) x(n-\tau) \right]^p. \tag{7.4}$$

The Hammerstein system (or memory polynomial) is composed of a memoryless nonlinearity (a Taylor approximation of the static nonlinearity is employed) followed by a linear filter, and has the following form

$$y(n) = \sum_{p=1}^{P} \sum_{\tau=0}^{M} h_p(\tau) x^p(n-\tau). \tag{7.5}$$

A Wiener–Hammerstein or sandwich model is composed of a memoryless nonlinearity sandwiched between two linear filters with impulse responses $h(\cdot)$ and $g(\cdot)$ and is defined as

$$y(n) = \sum_{p=1}^{P} \sum_{\tau_1=0}^{M} \cdots \sum_{\tau_p=0}^{M} \sum_{k=0}^{M_{h_p}+M_{g_p}} g_p(k) \prod_{l=1}^{p} h_p(\tau_l - k) x(n-\tau_l). \tag{7.6}$$

The above models have been employed in a wide range of applications including: satellite, telephone channels, mobile cellular communications, wireless LAN devices, radio and TV stations, digital magnetic systems and others [8, 32, 71, 77, 80].

7.2.1 Nonlinear MIMO Systems With Universal Approximation Capability

The discussion of the previous subsection is next extended to MIMO nonlinear systems. Attention is limited to MIMO polynomial systems. These are finitely parametrizable structures that naturally extend Eq. (7.2) and preserve a universal approximation capability over a broad class of multivariables systems. We start our discussion by considering cases where either the MIMO system has a single input or a single output. In the end, sparsity is imposed in order to reduce the number of unknown parameters.

The input–output relationship of nonlinear single input multiple output (SIMO) system is

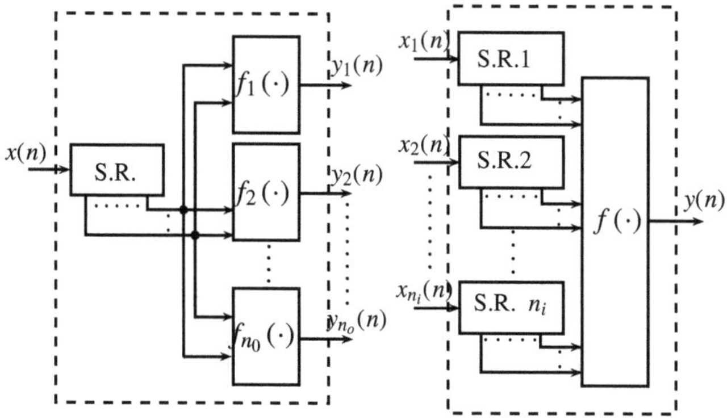

Fig. 7.1 SIMO and MISO polynomial systems (SR denotes a shift register)

$$y_r(n) = \sum_{p=1}^{p} \sum_{\tau_1=0}^{M} \cdots \sum_{\tau_p=0}^{M} h_p^{(r)}(\tau_1, \ldots, \tau_p) \prod_{i=1}^{p} x(n - \tau_i) \tag{7.7}$$

where $y_r(n)$ is the output associated with the rth output signal and $h_p^{(r)}(\tau_1, \ldots, \tau_p)$ is the *pth–order Volterra kernel* of the rth output. The difference between Eq. (7.2) and Eq. (7.7) is that a distinct kernel $h_p^{(r)}(\tau_1, \ldots, \tau_p)$ is associated with each output signal $y_r(n)$. This is illustrated in Fig. 7.1. SIMO systems can be obtained by oversampling the output signal of a SISO system at a sufficiently high rate and demultiplexing the samples [44].

A multiple input single output (MISO) system comprises n_i input signals and a single output. The input–output of a MISO system has the form

$$y(n) = \sum_{p=1}^{P} \sum_{t=1}^{n_i} \sum_{\tau_1=0}^{M} \cdots \sum_{\tau_p=0}^{M} h_p(\tau_1, \ldots, \tau_p) \prod_{i=1}^{p} x_t(n - \tau_i) \tag{7.8}$$

where $x_t(n)$ is the tth input signal ($1 \leq t \leq n_i$). A shift register (SR) is associated with each input. The contents of all registers are then converted into the output by means of a feed forward polynomial as shown in Fig. 7.1.

The general MIMO case is readily construed from the above special cases. A MIMO finite support Volterra system with n_i inputs and n_o outputs has the following form:

$$\begin{aligned} y_r(n) &= \\ f_r(x_1(n), x_1(n-1), \ldots, x_1(n-M), &\cdots, x_{n_i}(n), x_{n_i}(n-1), \ldots, x_{n_i}(n-M)), \\ r = 1, \ldots, n_o. \end{aligned} \tag{7.9}$$

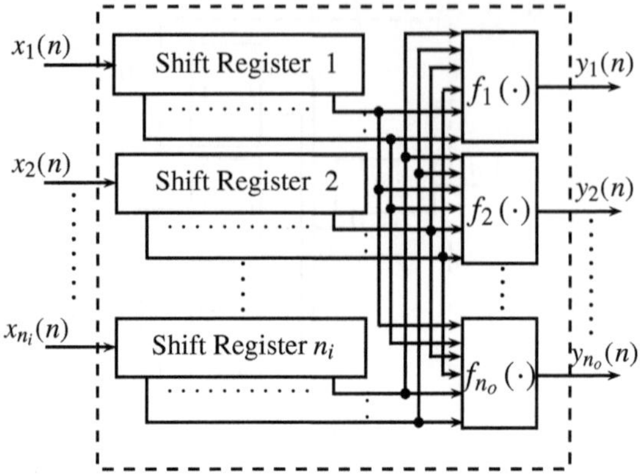

Fig. 7.2 A nonlinear MIMO Volterra

Each output $y_r(n)$ is obtained by a polynomial combination of the n_i inputs and their shifts. The parameter M specifies the memory of the n_i registers associated with each input. The MIMO finite support Volterra architecture is depicted in Fig. 7.2. This model is capable of capturing nonlinear effects resulting from any product combinations of the n_i inputs and their shifts. Expanding $f_r(\cdot)$ as a polynomial of degree P gives rise to the nonlinear MIMO Volterra model with n_i inputs and n_o outputs defined as

$$y_r(n) = \sum_{p=1}^{P}\sum_{t_1=1}^{n_i}\cdots\sum_{t_p=1}^{n_i}\sum_{\tau_1=0}^{M}\cdots\sum_{\tau_p=0}^{M} h_p^{(r,t_1\cdots t_p)}(\tau_1,\ldots,\tau_p)\prod_{i=1}^{p} x_{t_i}(n-\tau_i) \quad (7.10)$$

where $h_p^{(r,t_1\cdots t_p)}(\tau_1,\ldots,\tau_p)$ is the pth order Volterra kernel associated with the rth output and the $(t_1\cdots t_p)$ inputs. In this case, the Volterra kernels have multidimensional indices $(r, t_1\cdots t_p)$.

The above expressions are made complicated by the presence of multiple summations. Kronecker products alleviate this problem. Let

$$\bar{\mathbf{x}}(n) = [x_1(n), x_1(n-1),\ldots,x_1(n-M),\cdots,x_{n_i}(n), x_{n_i}(n-1),\ldots,x_{n_i}(n-M)]^T$$

and hence the nonlinear input vector is given by

$$\mathbf{x}(n) = [\bar{\mathbf{x}}(n), \bar{\mathbf{x}}_2(n),\cdots,\bar{\mathbf{x}}_p(n)]^T. \quad (7.11)$$

Then Eq. (7.10) takes the form:

$$\mathbf{y}(n) = \mathbf{H}\mathbf{x}(n) \quad (7.12)$$

where $\mathbf{y}(n) = [y_1(n),\ldots,y_{n_o}(n)]^T$ is the output vector, and the system matrix is $\mathbf{H} = [\mathbf{h}_{1:},\ldots,\mathbf{h}_{n_o:}]^T$, with $\mathbf{h}_{n_o:}$ containing all the Volterra kernels associated with the rth output. In this case the parameter matrix contains

$$\sum_{i=1}^{p}(n_i \times M)^p$$

parameters. The MIMO polynomial family of Eq. (7.9) has a universal approximation capability in the following sense: every nonlinear system with more than one inputs and outputs that is causal, shift invariant, bounded input bounded output stable and has fading memory can be approximated by a MIMO polynomial system of the form given in Eq. (7.10). This assertion is established if the same statement is proved for MISO systems. The latter follows with straightforward modifications of the proof for the SISO case.

7.2.1.1 Sparsity Aware Volterra Kernels

A major obstacle in using Volterra series in practical applications is the exponential growth of the model parameters (as a function of the order, the memory length of the systems and the number of inputs). Thus models of order $p > 3$ and memory length $M > 5$, translate into increased computational complexity cost and data requirements for identification purposes. For this reason, parsimonious, reduced order alternatives become relevant.

Sparse representations provide a viable alternative. The parameter matrix $\mathbf{H}$ in Eq. (7.12) is s–sparse if the number of non–zero elements is less than s, *i.e.*

$$\|\text{vec}[\mathbf{H}]\|_{\ell_0} = \{\#(i, j) : H_{ij} \neq 0\} \leq s.$$

7.2.2 Special Classes of MIMO Nonlinear Systems

In this section, some special classes of MIMO Volterra systems are studied. We start with a simplified version of the MIMO Volterra model. Then structured nonlinear models like Wiener, Hammerstein and Wiener–Hammerstein are extended to the MIMO case. These models are formed by the cascade connection of linear MIMO filters and MIMO static nonlinearities.

7.2.2.1 Parallel Cascade MIMO Volterra

In MIMO systems the signals from the n_i inputs interact with each other and the resulting mixture is received at each output. A special case of Eq. (7.10) results when the MIMO system obtains from the parallel connection of SISO systems, where each SISO system is often referred to as a path or parallel system. If the path between each input and each output is modelled as a Volterra system, then the rth output is expressed as follows

$$y_r(n) = \sum_{p=1}^{P}\sum_{t=1}^{n_i}\sum_{\tau_1=0}^{M}\cdots\sum_{\tau_p=0}^{M} h_p^{(r,t)}(\tau_1,\ldots,\tau_p)\prod_{i=1}^{p} x_t(n-\tau_i) \tag{7.13}$$

where $h_p^{(r,t)}(\tau_1,\ldots,\tau_p)$ is the *pth–order Volterra kernel* between the tth input and the rth output for all $t = 1,\ldots,n_i$ and $r = 1,\ldots,n_o$. The above model does not allow product combinations along different inputs. Instead each input is nonlinearly transformed and then all different inputs are linearly mixed. Such a model can be considered as a parallel cascade of n_i SIMO Volterra models.

Equation (7.13) can be written in a form identical to that of Eq. (7.12). Define the tth input regressor vector as

$$\mathbf{x}^{(t)}(n) = [x^{(t)}(n), x^{(t)}(n-1),\ldots,x^{(t)}(n-M)]^T.$$

Then the linearly mixed input vector takes the form:

$$\mathbf{x}(n) = [\mathbf{x}_1^{(1)}(n), \mathbf{x}_2^{(1)}(n),\ldots,\mathbf{x}_p^{(1)}(n),\cdots,\mathbf{x}_1^{(n_i)}(n), \mathbf{x}_2^{(n_i)}(n),\ldots,\mathbf{x}_p^{(n_i)}(n)]^T.$$

The total number of parameters of the above linearly mixed model is

$$n_i\sum_{i=1}^{p} M^p$$

and is considerably reduced when compared to the general case.

The linearly mixed model finds application in nonlinear communications. Communication nonlinearities can be categorized into the following three types: transmitter nonlinearity (due to nonlinearity in amplifiers), inherent physical channel nonlinearity, and receiver nonlinearity (e.g., due to nonlinear filtering). The power amplifier (PA) (which is located at the transmitter) constitutes the main source of nonlinearity. In a system equipped with multiple transmit antennas, each transmitter amplifies the signal. Amplifiers often operate near saturation to achieve power efficiency. In those cases they introduce nolinearities which cause interference and reduce spectral efficiency. At the receiver end, each antenna receives a linear superposition of all transmitted signals, as illustrated in Fig. 7.3. It should be pointed out that the nonlinear effects are applied to each input signal individually prior to mixing the transmitted signals.

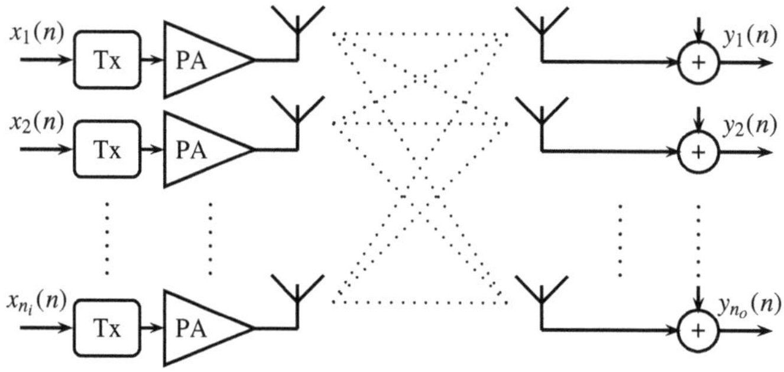

Fig. 7.3 An example of a parallel cascade MIMO Volterra channel

7.2.2.2 Block–Structured Classes of Nonlinear MIMO Systems

The *MIMO Wiener model* is shown in Fig. 7.4. It consists of a linear MIMO system in cascade with a polynomial nonlinearity for each output. The output is given by

$$y_r(n) = \sum_{p=1}^{P}\sum_{t_1=1}^{n_i}\cdots\sum_{t_p=1}^{n_i}\sum_{\tau_1=0}^{M}\cdots\sum_{\tau_p=0}^{M}\prod_{i=1}^{p} h_p^{(r,t_i)}(\tau_i)x_{t_i}(n-\tau_i). \tag{7.14}$$

This model is a special subclass of the MIMO Volterra series model. The relationship between the pth–order Volterra kernel and pth–order Wiener kernel is

$$h_p^{(rt_1\cdots t_p)}(\tau_1,\ldots,\tau_p) = \prod_{i=1}^{p} h_p^{(r,t_i)}(\tau_i).$$

Thus a MIMO Wiener model is equivalent to a MIMO Volterra system with separable kernels. The MIMO Hammerstein model is one of the simplest and most popular subclasses of MIMO Volterra models. As the diagram of Fig. 7.5 shows, the MIMO Hammerstein is a cascade connection of a static polynomial nonlinearity for each input connected in series by a linear MIMO system. It consists of the same building blocks as the Wiener model, but connected in reverse order. It has the following form:

Fig. 7.4 A MIMO Wiener system

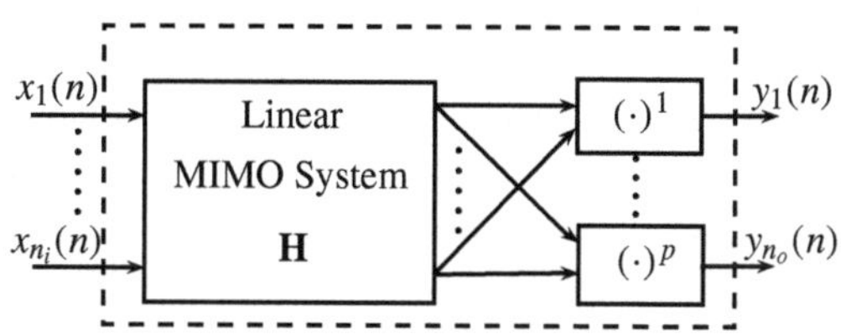

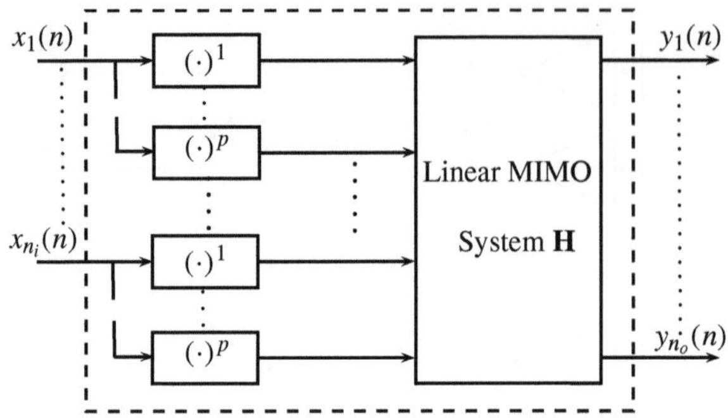

Fig. 7.5 A MIMO Hammerstein system

$$y_r(n) = \sum_{p=1}^{P}\sum_{t=1}^{n_i}\sum_{\tau=0}^{M} h_p^{(r,t)}(\tau)x^p(n-\tau_i). \tag{7.15}$$

The pth–order Volterra kernel of a Hammerstein model is given by

$$h_p^{(rt_1\cdots t_p)}(\tau_1,\ldots,\tau_p) = h_p(\tau_1)\delta(\tau_2-\tau_1)\cdots\delta(\tau_p-\tau_1)\delta(t_2-t_1)\cdots\delta(t_p-t_1) \tag{7.16}$$

A Hammerstein system prohibits product interactions between different inputs and hence corresponds to a diagonal MIMO Volterra model.

We finally consider the case where the MIMO Volterra kernels have factorable form:

$$h_p^{(rt_1\cdots t_p)}(\tau_1,\ldots,\tau_p) = \sum_{k=0}^{M_h+M_g} g_p^r(k)\prod_{i=1}^{p} h_p^{t_i}(\tau_i-k)$$

Substituting the above form into Eq. (7.10), we obtain:

$$y_r(n) = \sum_{p=1}^{P}\sum_{t_1=1}^{n_i}\cdots\sum_{t_p=1}^{n_i}\sum_{\tau_1=0}^{M}\cdots\sum_{\tau_p=0}^{M}\sum_{k=0}^{M_h+M_g} g_p^r(k)\prod_{i=1}^{p} h_p^{t_i}(\tau_i-k)x_{t_i}(n-\tau_i). \tag{7.17}$$

The pth–order kernel corresponds to a cascade connection of a linear MIMO system followed by a memoryless nonlinearity followed by another linear MIMO system and is known as *MIMO Wiener–Hammerstein* or sandwich model. In its simplest form a MIMO Wiener–Hammerstein system has a sandwiched structure with a single input single output static nonlinearity placed between a MISO and a SIMO linear systems. In the general case, illustrated in Fig. 7.6, the two linear filters can have arbitrary input and output dimensions. Compatibility is secured by proper dimensioning of the MIMO static nonlinearity. The Wiener–Hammerstein has been widely employed in satellite transmission, where both the earth station and the satellite repeater employ (nonlinear) power amplifiers. In such cases the signal bandwidth is very carefully

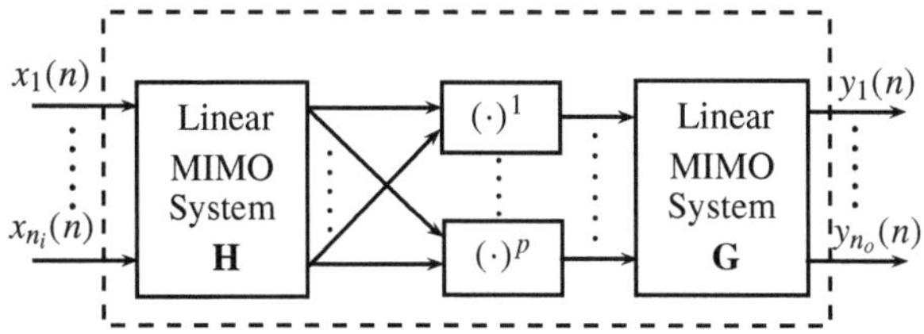

Fig. 7.6 A MIMO Wiener–Hammerstein system

defined depending on the application so that the output signal contains only spectral components near the carrier frequency ω_c. This leads to the *MIMO baseband Wiener–Hammerstein* system [8, Chap. 14], given by

$$y_r(n) = \sum_{p=1}^{\lfloor \frac{P-1}{2} \rfloor} \sum_{t_1=1}^{n_i} \cdots \sum_{t_{2p+1}=1}^{n_i} \sum_{\tau_1=0}^{M} \cdots \sum_{\tau_{2p+1}=0}^{M} \sum_{k=0}^{M_h+M_g} g_p^r(k) \prod_{i=1}^{p} \\ \times \prod_{i=1}^{p+1} h_{2p+1}^{t_i}(\tau_i - k) x_{t_i}(n-\tau_i) \prod_{j=p+2}^{2p+1} h_{2p+1}^{t_j}(\tau_j - k) x_{t_j}^*(n-\tau_j) \tag{7.18}$$

where $\lfloor \cdot \rfloor$ denote the floor operation. The above representation only considers odd–order powers with one more unconjugated input than conjugated input. This way the output does not create spectral components outside the frequency band of interest.

7.2.3 Practical Applications of MIMO Volterra Systems

Nonlinear MIMO systems are found in a range of communication and control applications. These are shortly reviewed next.

7.2.3.1 Nonlinear Communication Systems

Communication systems equipped with multiple transmit and/or receive antennas are MIMO systems that help provide spatial diversity. Exploitation of spatial diversity results in higher capacity and performance improvements in interference reduction, fading mitigation and spectral efficiency. Most of existing MIMO schemes are limited to linear systems. However, in many cases, system nonlinearities are present and possible remedies based on linear MIMO approximations degrade performance significantly.

In a communication system, there are often limited resources (power, frequency, and time slots) which have to be efficiently shared by many users. Quite often in practice we encounter a situation whereby the number of users exceeds the number of available frequency or time slots. In infrastructure–based networks, a base station or an access point is responsible for allocating resources among the users, thereby reducing the access delays/transmission latency and improving quality–of–service (QoS). This is established through a variety of *multiple access* schemes. Two key multiple access technologies suitable for higher data rates are: orthogonal frequency–division multiple access (OFDMA) and code–division multiple access (CDMA).

OFDMA dynamically allocates resources both in frequency (by dividing the available bandwidth into a number of subbands, called subcarriers) and in time (via OFDM symbols). The transmission system assigns different users to groups of orthogonal subcarriers and thus allows them to be spaced very close together with no overhead as in frequency division multiple access. Furthermore it prevents interference between adjacent subcarriers. OFDMA has been implemented in several wireless communication standards (IEEE 802.11a/g/n wireless local area networks (WLANs), IEEE 802.16e/m worldwide interoperability for microwave access (WiMAX), Hiperlan II), high–bit–rate digital subscriber lines (HDSL), asymmetric digital subscriber lines (ADSL), very high-speed digital subscriber lines (VHDSL), digital audio broadcasting (DAB), digital television and high-definition television (HDTV).

OFDMA is capable of mitigating intersymbol interference (ISI), (due to multipath propagation) using low–complexity/simple equalization structures. This is achieved by transforming the available bandwidth into multiple orthogonal narrowband subcarriers, where each subcarrier is sufficiently narrow to experience relatively flat fading. Nevertheless, OFDM is sensitive to synchronization issues and is characterized by high peak–to–average–power–ratio (PAPR), caused by the sum of several symbols with large power fluctuations. Such variations are problematic because practical communication systems are peak powered limited. In addition, OFDM transceivers are intrinsically sensitive to power amplifier (PA) nonlinear distortion [38], which dissipates the highest amount of power. One way to avoid nonlinear distortion is to operate the PA at the so–called "back–off" regime which results in low power efficiency. The trade–off between power efficiency and linearity motivated the development of signal processing tools that cope with MIMO–OFDM nonlinear distortion [38, 40, 45].

CDMA is based upon spread spectrum techniques. It plays an important role in third generation mobile systems (3G) and has found application in IEEE 802.11b/g (WLAN), Bluetooth, and cordless telephony. In CDMA multiple users share the same bandwidth at the same time through the use of (nearly) orthogonal spreading codes. The whole process effectively spreads the bandwidth over a wide frequency range (using pseudo–random code spreading or frequency hopping) several magnitudes higher than the original data rate. Two critical factors that limit the performance of CDMA systems are interchip and intersymbol interference (ICI/ISI), due to multipath propagation, mainly because they tend to destroy orthogonality between user codes and thus prevent interference elimination. Suppression of the detrimental effects of interference (ICI and ISI) get further complicated when nonlinear distortion is introduced due to power amplifiers. The combined effects of ICI, ISI and nonlinearities are comprehensively examined in [40, 67]. However, as recently illustrated in [22], the CDMA system model is sparse due to user inactivity/uncertainty, timing offsets and multipath propagation. CDMA system performance can be expected to improve further if nonlinearities along with sparse ICI/ISI are revisited.

7.2.3.2 MIMO Nonlinear Physiological Systems

In several physiological applications it is mandatory to gain as much insight information is possible about the functioning of the system. It is well documented in the biomedical literature that nonlinear systems can significantly enhance the quality of modelling [57, 80]. Very often linear approximations discard significant information about the nonlinearities. For this reason, several physiological systems like sensory systems (cockroach tactile spine, auditory system, retina), reflex loops (in the control of limb and eye position), organ systems (heart rate variability, renal auto–regulation) and tissue mechanics (lung tissue, skeletal muscle) have been approached via nonlinear system analysis using Volterra series [57, 80]. Many of the above physiological systems receive excitation from more than one input, and hence leads naturally to MIMO Volterra models.

7.2.3.3 Control Applications

Quite often control applications exhibit multivariable interactions and nonlinear behaviour, which make the modelling task and design more challenging. Examples of such control systems include: multivariable polymerization reactor [32], fluid catalytic cracking units (FCCU) [83, 84], and rapid thermal chemical vapor decomposition systems (RTCVD) [72].

Multivariable polymerization reactor aims to control the reactor temperature at the unstable steady state by manipulating the cooling water and monomer flow rates. MIMO Volterra models have been employed to capture/track the nonlinear plant output [32]. The FCCU unit constitutes the workhorse of modern refinery and its purpose is to convert gas oil into a range of hydrocarbon products. The major challenges related to FCCU are its internal feedback loops (interactions) and its highly nonlinear behaviour [84]. RTCVD is a process used to deposit thin films on a semiconductor wafer via thermally activated chemical mechanisms. Process and equipment models for RTCVD consist mainly of balance equations for conservation of energy, momentum and mass, along with equations that describe the relevant chemical mechanisms. An important characteristic of RTCVD systems is their wide region of operation, which requires excitation of the system with as many modes as possible and hence a nonlinear MIMO system becomes relevant. A major challenge in all the above control applications is the large number of parameters required by the nonlinear MIMO models.

7.3 Algorithms for Sparse Multivariable Filtering

Adaptive filters with a large number of coefficients are often encountered in multimedia signal processing, MIMO communications, biomedical applications, robotics, acoustic echo cancellation, and industrial control systems. Often, these applications are subject to nonlinear effects which can be captured using the models of Sect 7.2. The steady–state and tracking performance of conventional adaptive algorithms can be improved by exploiting the sparsity of the unknown system. This is achieved via two different strategies [82]. The first is based on *proportionate adaptive filters*, which update each parameter of the filter independently of the others by adjusting the step size in proportion to the magnitude of the estimated filter parameter. In this manner, the adaptation gain is "proportionately" redistributed among all parameters, emphasizing the large coefficients in order to speed up convergence and increase the overall convergence rate. The second strategy is motivated by the *compressed sensing* framework [16, 36, 76]. Compressed sensing approaches follow two main paths: (a) the ℓ_1 minimization (also referred to as basis pursuit) and (b) greedy algorithms (matching pursuit). Basis pursuit penalizes the cost function by the ℓ_1–norm of the unknown parameter vector (or a weighted ℓ_1–norm), as the ℓ_1–norm (unlike the ℓ_2–norm) favours sparse solutions. These methods combine conventional adaptive filtering algorithms such as LMS, RLS, etc with a sparsity promoting operation. Additional operations include the soft–thresholding (originally proposed for denoising by D. L. Donoho in [30]) and the metric projection onto the ℓ_1–ball [25, 33]. Greedy algorithms, on the other hand, iteratively compute the support set of the signal and construct an approximation of the parameters until convergence is reached. Proportionate adaptive filtering was developed by Duttweiler in 2000 [34]. Thereafter, a variety of improved versions has been proposed [64]. A connection between proportionate adaptive filtering and compressed sensing is discussed in [64].

7.3.1 Sparse Multivariable Wiener Filter

The block diagram of Fig. 7.7 shows a discrete–time MIMO filter with n_i inputs and n_o outputs [7, 47]. The output $\mathbf{y}(n)$, the impulse response matrix $\mathbf{H}$ and the input $\mathbf{x}(n)$ are related by:

$$\mathbf{y}(n) = \mathbf{H}\mathbf{x}(n) + \mathbf{v}(n) \tag{7.19}$$

where $\mathbf{x}(n)$ is defined in Eq. (7.11) and $\mathbf{v}(n) = [v_1(n), v_2(n), \cdots, v_{n_o}]^T$ is a Gaussian white noise vector. The following equation shows the rth output signal

$$\begin{aligned} y_r(n) &= \sum_{\tau=1}^{n_i} \mathbf{h}_{r\tau}^T \mathbf{x}_\tau(n) + v_r(n) &(7.20)\\ &= \mathbf{h}_{r:}^T \mathbf{x}(n) + v_r(n), \quad r = 1, \ldots, n_o. &(7.21) \end{aligned}$$

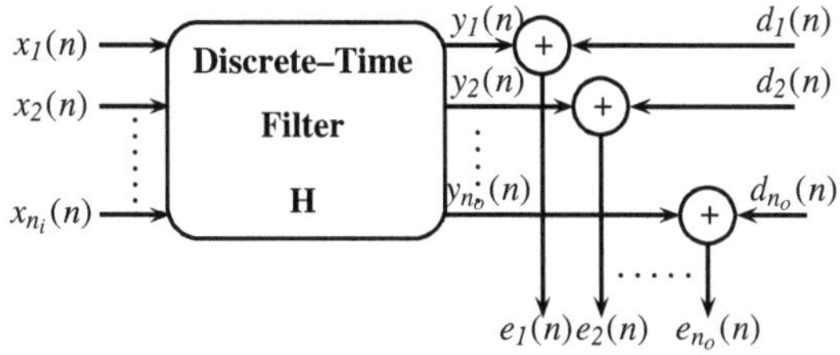

Fig. 7.7 MIMO filtering

and

$$\mathbf{H} = \begin{bmatrix} \mathbf{h}_{11}^T & \cdots & \mathbf{h}_{1n_i}^T \\ \vdots & \ddots & \vdots \\ \mathbf{h}_{n_o1}^T & \cdots & \mathbf{h}_{n_on_i}^T \end{bmatrix} = \begin{bmatrix} \mathbf{h}_{1:}^T \\ \vdots \\ \mathbf{h}_{n_o:}^T \end{bmatrix}. \tag{7.22}$$

An adaptive process is employed to cause the rth output to agree as closely as possible with the desired response signal $d_r(n)$. This is accomplished by comparing the outputs with the corresponding desired responses and by adjusting the parameters to minimize the resulting estimation error. More specifically, given an estimate $\widehat{\mathbf{H}}(n)$ of $\mathbf{H}$ the estimation error is:

$$e_r(n) = y_r(n) - d_r(n) = y_r(n) - \widehat{\mathbf{h}}_{r:}^T(n)\mathbf{x}(n), \quad r = 1, \ldots, n_o \tag{7.23}$$

and in vector form:

$$\mathbf{e}(n) = \mathbf{y}(n) - \widehat{\mathbf{H}}(n)\mathbf{x}(n). \tag{7.24}$$

The performance of a filter is assessed by a functional of the estimation error. LS filters, minimize the total squared error:

$$J_{LS}(n) = \sum_{i=1}^{n} \mathbf{e}H^1(i)\mathbf{e}(i) = \sum_{i=1}^{n} \|\mathbf{e}(i)\|_{\ell_2}^2 \tag{7.25}$$

$$= \sum_{r=1}^{n_o} J_{\mathbf{h}_{r:}}(n). \tag{7.26}$$

The optimum MIMO filter is given by the system of linear equations

$$\mathbf{H}_o(n)\mathbf{R}_{\mathbf{xx}}(n) = \mathbf{P}_{\mathbf{yx}}(n) \tag{7.27}$$

where $\mathbf{R}_{\mathbf{xx}}(n)$ is the input sample covariance matrix (which has block Toeplitz structure) with

$$\mathbf{R}_{x_ix_j}(n) = \sum_{t=1}^{n} \mathbf{x}_i(t)\mathbf{x}_j^H(t),$$

and

$$\mathbf{P}_{\mathbf{yx}}(n) = \sum_{i=1}^{n} \mathbf{y}(i)\mathbf{x}^H(i) = \left[\mathbf{p}_{yx_1}(n)\ \mathbf{p}_{yx_2}(n) \cdots \mathbf{p}_{yx_{n_i}}(n)\right]. \tag{7.28}$$

Under broad conditions the solutions of Eq. (7.27) tends to be the optimum mean squared filter (occasionally referred to as Wiener filter) that minimizes the mean squared error $\mathbb{E}\{\|\mathbf{e}(i)\|_{\ell_2}^2\}$ and satisfies the system of linear equations given by Eq. (7.27) ($\mathbf{P}_{\mathbf{yx}}(n) = \mathbb{E}\{\mathbf{y}(i)\mathbf{x}^H(i)\}$ and $\mathbf{R}_{\mathbf{xx}}(n) = \mathbb{E}\{\mathbf{x}(i)\mathbf{x}^H(i)\}$). Equation (7.27) can be decomposed in n_o independent MISO equations each corresponding to an output signal [9, 47], as follows:

$$\mathbf{h}_{r:,o}(n)\mathbf{R}_{\mathbf{xx}}(n) = \mathbf{p}_{y_r\mathbf{x}}(n), \quad r = 1, \ldots, n_o. \tag{7.29}$$

Consequently, minimizing $J_{LS}(n)$ or minimizing each $J_{\mathbf{h}_{r:}}(n)$ independently gives exactly the same results.

Two popular algorithms for adaptive filtering are the Least Mean Squares (LMS) algorithm and the Recursive Least Squares (RLS) algorithm. The LMS follows a stochastic gradient method and has a computationally simpler implementation. On the other hand, the more complex RLS has better convergence rate.

The LMS seeks to minimize the instantaneous error

$$J_{LMS}(n) = \mathbf{e}^H(n)\mathbf{e}(n). \tag{7.30}$$

The LMS estimate for the impulse response matrix $\mathbf{H}$ is based on the following update equation:

$$\mathbf{H}(n) = \mathbf{H}(n-1) + \mu\mathbf{e}(n)\mathbf{x}^H(n) \tag{7.31}$$

where the step size μ determines the convergence rate of the algorithm. To achieve convergence in the mean to the optimal Wiener solution, μ should be chosen so that:

$$0 < \mu < \frac{2}{M\sum_{\tau}^{n_i}\sigma_{x_\tau}^2}. \tag{7.32}$$

The RLS algorithm attempts to minimize the exponentially weighted cost function:

$$J_{RLS}(n) = \sum_{t=1}^{n} \lambda^{n-t}\mathbf{e}^H(t)\mathbf{e}(t) \tag{7.33}$$

where λ denotes the forgetting factor. The RLS estimates are updated as follows:

$$\mathbf{H}(n) = \mathbf{H}(n-1) + \mathbf{e}(n)\mathbf{k}^T(n) \tag{7.34}$$

where

$$\mathbf{k}(n) = \frac{\mathbf{R}_{\mathbf{xx}}^{-1}(n)\mathbf{x}^*(n)}{\lambda + \mathbf{x}^T(n)\mathbf{R}_{\mathbf{xx}}^{-1}(n)\mathbf{x}^*(n)}$$

is known as the *Kalman gain* [46, 70]. The matrix inversion lemma [46, 70], leads to:

$$\mathbf{R}_{\mathbf{xx}}^{-1}(n) = \lambda^{-1}\mathbf{R}_{\mathbf{xx}}^{-1}(n-1) - \mathbf{k}(n)\mathbf{x}^T(n)\mathbf{R}_{\mathbf{xx}}^{-1}(n-1). \tag{7.35}$$

7.3.2 $\mathcal{L}_1$ Constrained Adaptive Filters

These algorithms are based on the minimization of cost functions penalized by the ℓ_1–norm (or a weighted ℓ_1–norm or an approximate ℓ_0–norm) and are inspired by the fact that the ℓ_1–norm promotes sparse solutions and is the best convex relaxation to the ℓ_0 quasi–norm.

7.3.2.1 LMS–Type Filters

The sparse cost function combines the instantaneous error with a sparseness inducing penalty term

$$J_{ZA-LMS}(n) = \frac{1}{2}\left(\mathbf{e}^H(n)\mathbf{e}(n)\right) + \tau \text{pen}\left(\mathbf{H}(n)\right) \tag{7.36}$$

τ is a positive scalar regularization parameter which provides a trade–off between penalization and signal reconstruction error. The most well–known sparsity inducing penalty term is the ℓ_1–norm (pen $(\mathbf{H}(n)) = \|\text{vec}\,[\mathbf{H}(n)]\,\|_{\ell_1}$). Although a large portion of the literature focuses on the ℓ_1–norm there are other functions which promote sparsity [42, 48]. In fact, any penalization term, with pen $(\mathbf{H}(n))$ being symmetric, monotonically non–decreasing, and with decreasing derivative will serve the same purpose [36].

Sparse LMS–type variants obey the following updating scheme

$$\left\{\begin{matrix} new \\ parameter \\ estimate \end{matrix}\right\} = \left\{\begin{matrix} old \\ parameter \\ estimate \end{matrix}\right\} + \left\{stepsize\right\}\left\{\begin{matrix} new \\ information \end{matrix}\right\} + \left\{\begin{matrix} zero \\ attraction \\ term \end{matrix}\right\}$$

where the new information term is the error vector between the outputs of the filter and the desired signal vector. The *Zero–Attraction* (ZA) term is a norm related regularization function which exerts an attraction to zero on small parameters. Convergence of the recursion may be slow because the two parts are hard to balance. This issue is addressed in some detail later in the subsection.

The first of this type of algorithms (originally developed in [18, 19] for SISO systems) minimizes Eq. (7.36). The filter parameter matrix is updated by

Table 7.1 ZA–LMS algorithm

Algorithm description
$\mathbf{H}(0)=\mathbf{0}$
For n:=1,2,…**do**
1: $\mathbf{e}(n) = \mathbf{d}(n) - \mathbf{H}(n-1)\mathbf{x}(n)$
2: $\mathbf{H}(n) = \mathbf{H}(n-1) + \mu\mathbf{e}(n)\mathbf{x}^H(n) - \gamma \operatorname{sgn}(\mathbf{H}(n-1))$
End For

$$\begin{aligned}\mathbf{H}(n) &= \mathbf{H}(n-1) - \mu\nabla J_{ZA-LMS}(n) \\ &= \mathbf{H}(n-1) + \mu\mathbf{e}(n)\mathbf{x}^H(n) - \gamma\nabla^s \text{pen}\,(\mathbf{H}(n-1)) \end{aligned} \tag{7.37}$$

where $\nabla^s \text{pen}\,(\mathbf{H}(n-1))$ is the sub–gradient of the convex function pen $(\mathbf{H}(n-1))$, $\gamma = \mu\tau$ is the regularization parameter. In the adaptive filtering context γ is also referred to as *regularization step size*. Usually the regularization step size is fine tuned offline (via exhaustive simulations) or in an ad–hoc manner. A systematic approach to choosing γ is developed in [19].

Under the standard compressive sensing setting, the penalty is given by the ℓ_1–norm and the resulting algorithm is shown in Table 7.1. Note that sgn$(\cdot)$ is a *component–wise sign function* defined as

$$\operatorname{sgn}\left(H_{ij}\right) = \begin{cases} H_{ij}/|H_{ij}| & \text{if } H_{ij} \neq 0, \\ 0 & \text{if } H_{ij} = 0. \end{cases} \tag{7.38}$$

It is well known that the LMS, in a stationary environment, achieves unbiased convergence in the mean to the Wiener solution (using the independence assumption) [46]. However, unlike the conventional LMS, ZA–LMS leads to a biased behaviour [48], that is

$$\mathbb{E}\left[\mathbf{H}(n)\right] = \mathbf{H}_o - \frac{\gamma}{\mu}\mathbb{E}\left[\mathbf{H}(n)\right]\mathbf{R}_{\mathbf{xx}}^{-1}(n), \quad \text{as } n \to \infty \tag{7.39}$$

Recall that a key difference between the ℓ_0 norm and the ℓ_1 norm penalty, is that the ℓ_1 norm depends on the magnitudes of the non–zero components, whereas the ℓ_0–norm penalty does not. As a result, the larger a component is, the heavier it is penalized by the ℓ_1 penalty. To overcome this often unfair penalization two different penalty terms are introduced in the conventional LMS cost function. Both form better approximations to the ℓ_0 norm. The first is based on an approximation of the step function [79]

$$\text{pen}\,(\mathbf{H}(n)) = \sum_i \left(1 - \exp^{-a'|\text{vec}_i[\mathbf{H}(n)]|}\right) \tag{7.40}$$

where $a > 0$ is a parameter that must be chosen. The authors in [49] reduce the computational complexity of the resulting zero attraction term by considering the first order Taylor series expansion of exponential functions. The resulting filter update iteration (named ℓ_0–LMS) becomes

$$\mathbf{H}(n) = \mathbf{H}(n-1) + \mu\mathbf{e}(n)\mathbf{x}^H(n) - \gamma a\left(1 - a|\mathbf{H}(n-1)|\right)_+ \operatorname{sgn}\left(\mathbf{H}(n-1)\right) \quad (7.41)$$

where $(x)_+ = \max\{x, 0\}$. Motivated by the re–weighted ℓ_1 cost function in [17], the authors in [18] follow this approach in order to reinforce the ZA–LMS by re–weighting the sparse penalty term. The proposed penalty term is given by

$$\operatorname{pen}\left(\mathbf{H}(n)\right) = \sum_i \log\left(1 + \epsilon'^{-1}|\operatorname{vec}_i\left[\mathbf{H}(n)\right]|\right). \quad (7.42)$$

According to the stochastic gradient approach, the resulting filter update iteration is

$$\mathbf{H}(n) = \mathbf{H}(n-1) + \mu\mathbf{e}(n)\mathbf{x}^H(n) - \gamma\frac{\operatorname{sgn}\left(\mathbf{H}(n-1)\right)}{1 + \epsilon|\mathbf{H}(n-1)|} \quad (7.43)$$

and the algorithm is named RZA–LMS. Small coordinates of the estimated matrix are more heavily weighted (by $1/(1+\epsilon|\mathbf{H}(n-1)|)$) towards zero, and small weights encourage larger coordinates. As a result, the bias of the mean value of the converged matrix for RZA–LMS is reduced.

So far we have examined how to solve the penalized LMS cost function of Eq. (7.36) by embedding additional terms to the update formula. A different viewpoint arises by considering *proximity splitting* methods [21]. The proximity operator of a (possibly non–differentiable) convex function $\Omega\left(\mathbf{H}(n)\right)$ is defined as

$$\operatorname{prox}_{\tau,\Omega}\left(\mathbf{H}(n)\right) := \underset{\mathbf{H}(n)}{\operatorname{argmin}} \ \frac{1}{2\tau}\|\mathbf{Y}(n) - \mathbf{H}(n)\|_{\ell_2}^2 + \Omega\left(\mathbf{H}(n)\right).$$

Proximity operators are the main ingredient of proximal methods [21] which arise in many well–known algorithms (*e.g.*, iterative thresholding, projected Landweber, projected gradient, alternating projections, alternating-direction method of multipliers, alternating split Bregman). In these algorithms, proximal methods can be understood as generalizations of quasi–Newton methods to non–differentiable convex problems. An important example is the Iterative Thresholding procedure [24] which solves problems of the form:

$$\min_{\mathbf{H}} \ J\left(\mathbf{H}\right) + \Omega\left(\mathbf{H}\right), \quad (7.44)$$

$J\left(\mathbf{H}\right)$ is differentiable with Lipschitz gradient. By iterating the fixed point equation

$$\mathbf{H}(n) := \underbrace{\operatorname{prox}_{\mu,\Omega}}_{\text{backward step}} \ \underbrace{\left[\mathbf{H}(n-1) - \mu\nabla J\left(\mathbf{H}(n-1)\right)\right]}_{\text{forward step}} \quad (7.45)$$

for values of the step–size parameter μ in a suitable bounded interval. This scheme is known as a forward–backward splitting algorithm. In some cases, the proximity operator $\text{prox}_{\mu,\Omega}$ can be evaluated in closed form.

If we consider the minimization of the cost function J_{ZA-LMS} (defined in Eq. (7.36)) with pen $(\mathbf{H}(n)) = \| \operatorname{vec}[\mathbf{H}(n)] \|_{\ell_1}$ we obtain

$$\min_{\mathbf{H}} \frac{1}{2}|\mathbf{e}(n)|^2 + \tau \| \operatorname{vec}[\mathbf{H}(n)] \|_{\ell_1}.$$

We observe that the above problem is a special case of Eq. (7.44) with

$$\begin{cases} J: & \mathbf{H} \mapsto \frac{1}{2}|\mathbf{e}(n)|, \\ \Omega: & \mathbf{H} \mapsto \tau \| \operatorname{vec}[\mathbf{H}(n)] \|_{\ell_1}. \end{cases}$$

Then it follows from [21, 61] that the proximity operator $\text{prox}_{\mu,\Omega}$ leads to a non-linear component–wise shrinkage operation known as soft–thresholding [30]. The component-wise *soft–thresholding* operation is defined by

$$\mathbb{S}_\tau \left[H_{ij}\right] = \begin{cases} H_{ij} - \tau & \text{if } H_{ij} \geq \tau, \\ 0 & \text{if } |H_{ij}| \leq 0, \\ H_{ij} + \tau & \text{if } H_{ij} \leq \tau \end{cases} \tag{7.46}$$

or in compact notation $\mathbb{S}_\tau \left[H_{ij}\right] = \operatorname{sgn}\left(H_{ij}\right)\left(|H_{ij}| - \tau\right)_+$ [30]. This operation shrinks coefficients above the threshold in magnitude by an amount equal to τ. An instantaneous proximity operation leads to the soft–thresholded LMS filter

$$\mathbf{H}(n) = \mathbb{S}_\tau \left[\mathbf{H}(n-1) + \mu\mathbf{e}(n)\mathbf{x}^H(n)\right]. \tag{7.47}$$

Detailed analysis of the dynamics of Eq. (7.47) in its batch format, has shown that the algorithm converges initially relatively fast, then it overshoots the ℓ_1 penalty, and it takes very long to re–correct back. To avoid such a behaviour in the adaptive case, we force the successive iterates to remain within a particular ℓ_1 ball B_R [25]. To achieve this the thresholding operation is replaced by a projection $\mathbb{P}_{B_R}$, where, for any closed convex set C and any $\mathbf{H}$, the projection $\mathbb{P}_C(\mathbf{H})$ is defined as the unique point in C for which the ℓ_2 distance to $\mathbf{H}$ is minimal. We thus obtain the projected LMS on ℓ_1 ball

$$\mathbf{H}(n) = \mathbb{P}_{B_R} \left[\mathbf{H}(n-1) + \mu\mathbf{e}(n)\mathbf{x}^H(n)\right]. \tag{7.48}$$

The projection operator $\mathbb{P}_{B_R}\left[H_{ij}(n)\right]$ is obtained by a suitable thresholding of $H_{ij}(n)$, given

$$\mathbb{P}_{B_R}\left[H_{ij}\right] = \begin{cases} \mathbb{P}_{B_R}\left[H_{ij}\right] = \mathbb{S}_\mu\left[H_{ij}\right] & \text{if } \|\operatorname{vec}[\mathbf{H}(n)]\|_{\ell_1} > R, \text{ and choose } \mu \\ & \text{such that } \|\mathbb{S}_\mu[\operatorname{vec}[\mathbf{H}(n)]]\|_{\ell_1} = R \\ \mathbb{P}_{B_R}\left[H_{ij}\right] = \mathbb{S}_0\left[H_{ij}\right] & \text{if } \|vec[\mathbf{H}(n)]\|_{\ell_1} \le R. \end{cases} \tag{7.49}$$

Using proximal splitting methods other types of adaptive filters, such as NLMS/APA and Adaptive Projection algorithms, can be modified to promote sparsity [54, 61].

7.3.2.2 RLS–Type Filters

Sparse RLS–type filters modify the RLS cost function (7.33) by the addition of a sparsifying term:

$$J_{ZA-RLS}(n) = \frac{1}{2}J_{RLS}(n) + \tau \operatorname{pen}(\mathbf{H}(n)). \tag{7.50}$$

The regularization parameter τ controls sparsity and weighted squared error. The sparse RLS filter can be seen as an adaptive version of Gauss–Newton or Newton–Raphson search with sparse updates [55]. Alternatively, the RLS algorithm is a special case of a Kalman filter [46, 70]. The main recursion takes the following form:

$$\left\{\begin{matrix} new \\ parameter \\ estimate \end{matrix}\right\} = \left\{\begin{matrix} old \\ parameter \\ estimate \end{matrix}\right\} + \left\{\begin{matrix} Kalman \\ gain \end{matrix}\right\}\left\{\begin{matrix} innovation \\ vector \end{matrix}\right\} + \left\{\begin{matrix} zero \\ attraction \\ term \end{matrix}\right\}$$

The correction term is proportional to the innovation error vector between the predicted observations and the actual observations. The coefficients of this correction are provided by the Kalman gain. For the regularized Recursive Least Square problem of Eq. (7.50) the solution of the Wiener equation takes the form [35]:

$$\mathbf{H}(n) = \mathbf{P_{yx}}(n)\mathbf{C}(n) - \gamma(1-\lambda)\nabla^s \operatorname{pen}(\mathbf{H}(n-1))\,\mathbf{C}(n) \tag{7.51}$$

where $\mathbf{C}(n) = \mathbf{R}_{\mathbf{xx}}^{-1}(n)$, $\lambda \in (0,1)$ is the forgetting factor and $\nabla^s \operatorname{pen}(\mathbf{H}(n-1))$ is a subgradient since $J_{ZA-RLS}(n)$ is non–differentiable at any point where $H_{ij}(n) = 0$ [10, p. 227]. The exponentially weighted autocorrelation and cross–correlation matrices are recursively updated as:

$$\mathbf{R_{xx}}(n) = \sum_{t=1}^{n} \lambda^{n-t}\mathbf{x}(t)\mathbf{x}^H(t) = \lambda \mathbf{R_{xx}}(n-1) + \mathbf{x}(n)\mathbf{x}^H(n) \tag{7.52}$$

$$\mathbf{P_{yx}}(n) = \sum_{t=1}^{n} \lambda^{n-t}\mathbf{y}(t)\mathbf{x}^H(t) = \lambda \mathbf{P_{yx}}(n-1) + \mathbf{y}(n)\mathbf{x}^H(n). \tag{7.53}$$

The regularized RLS filter relies on the following recursion [35]:

$$\mathbf{H}(n) = \mathbf{H}(n-1) + \mathbf{e}(n)\mathbf{k}^T(n) - \gamma(1-\lambda)\nabla^s \operatorname{pen}(\mathbf{H}(n-1))\,\mathbf{C}(n) \tag{7.54}$$

where the regularization parameter γ is usually fine tuned offline or using the selection rule proposed in [35] (for white inputs). In this case the corresponding subgradient is $\nabla^s \| H_{ij}(n-1)\|_{\ell_1} = \operatorname{sgn}(H_{ij}(n-1))$. Instead we may utilize the penalty functions, suggested in the LMS context, given by Eqs. (7.40) and (7.42).

RLS algorithms are developed based on the batch LASSO estimator in [3, 4]. This method modifies the LASSO cost function to include a forgetting factor:

$$\underset{\mathbf{H}(n)}{\operatorname{argmin}} \ \frac{1}{\sigma^2}\sum_{i=1}^{n} \lambda^{n-i} \|\mathbf{y}(i) - \mathbf{H}(i)\mathbf{x}(i)\|_{\ell_2}^2 + \gamma \operatorname{pen}(\mathbf{H}(n)). \tag{7.55}$$

The first order subgradient based optimality conditions for the exponentially weighted LASSO cost imply:

$$\begin{cases} \nabla_{ij} J_{RLS}(n) + \tau \operatorname{sgn}\left(H_{ij}(n)\right), & \text{if } H_{ij}(n) \neq 0 \\ |\nabla_{ij} J_{RLS}(n)| \leq \tau & \text{if } H_{ij}(n) = 0. \end{cases}$$

These conditions and the value of $\nabla_{ij} J_{RLS}(n)$ are used to define a pseudo–gradient for each component of $\mathbf{H}$ [2]. The pseudo–gradient of $J_{R-LASSO}(n)$ is the element of the sub–differential of $J_{R-LASSO}(n)$ at $\mathbf{H}(n)$ with minimum norm and is given by:

$$\nabla_{ij} J_{R-LASSO}(n) = \begin{cases} \nabla_{ij} J_{RLS}(n) + \tau \operatorname{sgn}\left(H_{ij}(n)\right), & \text{if } H_{ij}(n) \neq 0 \\ \nabla_{ij} J_{RLS}(n) + \tau & \text{if } H_{ij}(n) = 0,\ \nabla_{ij} J_{RLS}(n) < -\tau \\ \nabla_{ij} J_{RLS}(n) - \tau & \text{if } H_{ij}(n) = 0,\ \nabla_{ij} J_{RLS}(n) > \tau \\ 0 & \text{if } H_{ij}(n) = 0,\ -\tau \leq \nabla_{ij} J_{RLS}(n) \leq \tau. \end{cases}$$

In the first case the function is differentiable, so the pseudo–gradient is simply the gradient with respect to ij (the only element of the sub–gradient). In the remaining three cases we obtain the minimum–norm solution by the soft–thresholding operation to $\nabla_{ij} J_{RLS}(n)$. The global solution to the smooth part of the LASSO cost function is the Wiener equation, where the autocorrelation and the cross–correlation matrices are recursively updated from Eqs. (7.52) and (7.53).

Using the subgradient, an instantaneous subgradient descent strategy is employed for online updating as follows

$$\mathbf{H}(n) = \mathbf{H}(n-1) + \mu \nabla J_{R-LASSO}(n). \tag{7.56}$$

Table 7.2 R-LASSO algorithm

Algorithm description
$\mathbf{R_{xx}}(0) = \mathbf{0}, \mathbf{P_{yx}}(0) = \mathbf{0}, \mathbf{H}(0) = \mathbf{0}$
For $n := 1, 2, \ldots$ **do**
1: $\mathbf{R_{xx}}(n) = \lambda \mathbf{R_{yx}}(n-1) + \mathbf{x}(n)\mathbf{x}^H(n)$
2: $\mathbf{P_{yx}}(n) = \lambda \mathbf{P_{yx}}(n-1) + \mathbf{y}(n)\mathbf{x}^H(n)$
3: $\nabla J_{R-LASSO}(n) = \begin{cases} \mathbf{H}(n-1)\mathbf{R_{xx}}(n) - \mathbf{P_{yx}}(n) + \tau \mathrm{sgn}(\mathbf{H}(n-1)) & \text{if } H_{ij} \neq 0, \\ \mathbb{S}_\tau \left[\mathbf{H}(n-1)\mathbf{R_{xx}}(n) - \mathbf{P_{yx}}(n)\right] & \text{if } H_{ij} = 0. \end{cases}$
4: $\mathbf{H}(n) = \mathbf{H}(n-1) + \mu_n \nabla J_{R-LASSO}(n)$
End For

The Recursive LASSO (R–LASSO) filter outlined here is summarized in Table 7.2. As with the batch LASSO estimator, the R–LASSO does not necessarily converge to the true parameter $\mathbf{H}$ since it fails to recover the correct support and at the same time estimate the non–zero entries of $\mathbf{H}$ consistently [4].

In order to improve the performance of the R–LASSO filter, one could use a different penalty term which is signal dependent and weights differently the entries in the ℓ_1 norm, that is pen $(\mathbf{H}(n)) = \sum_i w_\tau(|\,\mathrm{vec}_i \left[\widehat{\mathbf{H}}^{RLS}(n)\right]|)\|\,\mathrm{vec}_i\,[\mathbf{H}(n)]\,\|_{\ell_1}$. By generalizing the *Smoothly Clipped Absolute Deviation* (SCAD) regularizer introduced for the batch weighted LASSO estimator to its adaptive case, the following weight function is obtained

$$w_\tau(|\,\mathrm{vec}_i\,[\mathbf{H}(n)]\,|) = \frac{\left[\alpha\tau - |\,\mathrm{vec}_i\,[\mathbf{H}(n)]\,|\right]_+}{\tau(\alpha-1)} u(|\,\mathrm{vec}_i\,[\mathbf{H}(n)]\,| - \tau) + u(\tau - |\,\mathrm{vec}_i\,[\mathbf{H}(n)]\,|)$$

$u(\cdot)$ stands for the step function and α is usually set to 3.7. The reweighted LASSO estimator (RW–LASSO) places higher weight to small entries, and lower weight to entries with large amplitudes. In fact, the estimates of size less than τ are penalized as in R–LASSO, while estimates between τ and $\alpha\tau$ are penalized in a linearly decreasing manner. Estimates larger than $\alpha\tau$ are not penalized at all. The implementation of RW–LASSO is established using an instantaneous pseudo–gradient descent strategy, similar to R–LASSO. The downside of this estimator is its high complexity because it requires running in parallel an RLS algorithm to supply the needed weights.

A different viewpoint to sparse RLS algorithms is provided in [5] (and its MIMO extension in [53]). This approach makes use of the Expectation Maximization (EM) method [27] to derive an adaptive filter that solve a penalized Maximum Likelihood problem. The penalized Recursive Least Squares problem may be posed as a penalized Maximum Likelihood problem [41]. This penalized ML problem can be efficiently solved by an EM algorithm following the noise decomposition idea (proposed in [41]) in order to divide the optimization problem into a denoising and a filtering problem. Consider the following decomposition for $\mathbf{V}(n)$

$$\mathbf{V}(n) = \alpha \mathbf{V}_1(n)\mathbf{X}(n) + \mathbf{V}_2(n). \tag{7.57}$$

The noise matrices are ensembles of Gaussian–distributed random matrices

$$\mathbf{V}_1(n) = \left(\mathbf{0}, \mathbf{I}_{n_i} \otimes \mathbf{I}_{n_o}\right)$$
$$\mathbf{V}_2(n) = \left(\mathbf{0}, \left(\sigma^2 \Lambda^{-1} - \alpha^2 \mathbf{X}(n)\mathbf{X}^H(n)\right)^T \otimes \mathbf{I}_{n_o}\right)$$

where $\Lambda := \text{diag}\left[\lambda^{n-1} \cdots \lambda^0\right]$ and α is a constant which must fulfil $\alpha \le \sigma^2/\lambda_{max}$ $[\mathbf{X}(n)\mathbf{X}^H(n)]$ with $\lambda_{max}[\cdot]$ being the maximum eigenvalue. Since $\lambda_{max}[\mathbf{X}(n)\mathbf{X}^H(n)]$ $\approx n_i$ for large n and for independent input, $\alpha^2 = \sigma^2/5n_i$ satisfies this condition with high probability. Therefore the model is rewritten as follows:

$$\begin{cases} \mathbf{Y}(n) = \mathbf{G}(n)\mathbf{X}(n) + \mathbf{V}_2(n) \\ \mathbf{G}(n) = \mathbf{H}(n) + \alpha \mathbf{V}_1 \end{cases}. \tag{7.58}$$

The EM algorithm is used to solve the following penalized ML problem

$$\mathbf{H}(n) = \underset{\mathbf{H}(n)}{\text{argmax}} \quad \log P(\mathbf{Y}(n), \mathbf{V}(n), |\mathbf{H}(n)) - \gamma \text{pen}(\mathbf{H}(n))) \tag{7.59}$$

which is easier to solve, by employing $\mathbf{V}(n)$ as the auxiliary variable. The λth iteration of the EM algorithm is defined as [5]:

$$\begin{cases} \text{E–Step} \quad Q(\mathbf{H}, \mathbf{H}(n)) = -\frac{1}{2\alpha^2}\|\mathbf{G}^{(\lambda)}(n) - \mathbf{H}\|_{\ell_2}^2 - \gamma \| \text{vec}[\mathbf{H}] \|_{\ell_1} \\ \text{M–Step} \quad \mathbf{H}^{(\lambda+1)}(n) = \underset{\mathbf{H}(n)}{\text{argmax}} \quad Q(\mathbf{H}, \mathbf{H}(n)) = \mathbb{S}_{\gamma\alpha^2}\left(\mathbf{G}^{(\lambda)}(n)\right) \end{cases} \tag{7.60}$$

where

$$\mathbf{G}^{(\lambda)}(n) = \mathbf{H}^{(\lambda)}(n)\left(\mathbf{I} - \frac{\alpha^2}{\sigma^2}\mathbf{X}(n)\Lambda\mathbf{X}^H(n)\right) + \frac{\alpha^2}{\sigma^2}\mathbf{Y}(n)\Lambda\mathbf{X}^H(n)$$

The above algorithm is an iterated shrinkage method. The soft thresholding function tends to decrease the support of $\mathbf{H}(n)$, since it shrinks the support to those elements whose absolute value is greater than $\gamma\alpha^2$. The algorithm described above can be further simplified by considering only the corresponding positions of the non–zero entries within the thresholding step [5]. The autocorrelation and cross–corellation matrices, which appear in the E–step of the algorithm, can be obtained recursively and the resulting algorithm (known as spaRLS) is summarized in Table 7.3.

Another algorithm related to the EM approach is presented in [52]. Unlike the noise decomposition idea which is followed in [5], their approach uses normal priors on the unknown parameter matrix. In the EM approach the individual parameters are treated as missing variables, and the E–step computes the conditional expectation of the missing variables given past observations. Subsequently, the M–Step maximizes this expectation minus a sparsity inducing penalty (like the ℓ_1 norm). To apply the

Table 7.3 spaRLS algorithm

Algorithm description
$\mathbf{R}_{xx}(0) = \mathbf{0}, \mathbf{P}_{\mathbf{yx}}(0) = \mathbf{0}, \mathbf{H}(0) = \mathbf{0}$
For $n := 1, 2, \ldots$ **do**
1: $\mathbf{R}_{xx}(n) = \lambda \mathbf{R}_{xx}(n-1) + \frac{a^2}{\sigma^2}\mathbf{x}(n)\mathbf{x}^H(n)$
2: $\mathbf{P}_{yx}(n) = \lambda \mathbf{P}_{yx}(n-1) + \frac{a^2}{\sigma^2}\mathbf{y}(n)\mathbf{x}^H(n)$
3: **Repeat**
4: $\widehat{\mathbf{G}}^{(\lambda)}(n) = \widehat{\mathbf{H}}^{(\lambda)}(n)\,(\mathbf{I} - \mathbf{R}_{xx}(n)) + \mathbf{P}_{yx}(n)$
5: $\widehat{\mathbf{H}}^{(\lambda)}(n) = \mathbb{S}_{\gamma a^2}\left[\widehat{\mathbf{G}}^{(\lambda)}(n)\right]$
6: **Until** $\lambda = k$
End For

EM approach the complete and incomplete data must be specified. The matrix $\mathbf{H}(n)$ at time n is taken to represent the complete data vector, whereas $\mathbf{Y}(n-1)$ accounts for the incomplete data [39, pp. 31–33]. The resulting EM approach is summarized by the following equation:

$$\mathbf{G}(n) = \arg\max_{\mathbf{G}} \left\{\mathbb{E}_{p(\mathbf{H}(n)|\mathbf{Y}(n-1);\mathbf{G}(n-1))}\big[\log p(\mathbf{H}(n);\mathbf{G})\big] - \gamma \|\operatorname{vec}[\mathbf{G}]\|_{\ell_1}\right\}. \tag{7.61}$$

The EM algorithm aims to maximize the log–likelihood of the complete data, $\log p(\mathbf{H}(n); \mathbf{G})$. However, because $\mathbf{H}(n)$ is an unknown parameter, it maximizes instead its expectation given the incomplete data $\mathbf{Y}(n-1)$ and a current estimate of the parameters $\mathbf{G}(n-1)$. The E–step, computes the conditional expectation of the log–likelihood, given observations $\mathbf{Y}(n-1)$ and parameter estimate $\mathbf{G}(n-1)$ from the previous iteration

$$\begin{aligned}\textbf{E-step}: \quad Q\big(\mathbf{G}, \mathbf{G}(n-1)\big) &= \mathbb{E}_{p(\mathbf{H}(n)|\mathbf{Y}(n-1);\mathbf{G}(n-1))}\big[\log p(\mathbf{H}(n);\mathbf{G})\big] \qquad (7.62)\\ &= \text{constant} + \mathbf{G}^H\mathbf{S}^{-1}(n)\mathbb{E}[\mathbf{H}(n)|\mathbf{Y}(n-1);\mathbf{G}(n-1)] - \frac{1}{2}\mathbf{G}^H\mathbf{S}^{-1}(n)\mathbf{G}\end{aligned}$$

where $\mathbf{S}(n)$ is a diagonal covariance matrix, and the constant incorporates all terms that do not involve $\mathbf{G}$ and hence do not affect maximization. The M–step, described below, calculates the maximum of the penalized Q–function

$$\begin{aligned}\textbf{M-step}: \quad \mathbf{G}(n) &= \arg\max_{\mathbf{G}} \left\{Q\big(\mathbf{G}, \mathbf{G}(n-1)\big) - \gamma\|\operatorname{vec}[\mathbf{G}]\|_{\ell_1}\right\} \qquad (7.63)\\ &= \mathbb{S}_{\gamma \mathbf{S}_{ii}(n)}\left(\mathbb{E}\left[\mathbf{H}(n)|\mathbf{Y}(n-1);\mathbf{G}(n-1)\right]\right)\end{aligned}$$

which in turn leads to the *soft thresholding* function. In order to carry out the conditional expectation of Eq. (7.62) (essentially the E–step), one needs to assume a prior on $\mathbf{H}(n)$ given the past observations $\mathbf{Y}(n-1)$ and $\mathbf{G}(n-1)$. Consider the Gaussian prior of the form

Table 7.4 EM–RLS algorithm

Algorithm description
$\mathbf{H}(0) = \mathbf{0}$, $\mathbf{C}_0 = \delta^{-1}\boldsymbol{I}$ with δ =const.
For n:=1,2,...**do**
1: $\mathbf{k}(n) = \dfrac{\mathbf{C}(n-1)\mathbf{x}^*(n)}{\lambda + \mathbf{x}^T(n)\mathbf{C}(n-1)\mathbf{x}^H(n)}$
2: $\mathbf{G}(n) = \mathbf{H}(n-1) + (\mathbf{y}(n) - \mathbf{H}(n-1)\mathbf{x}(n))\mathbf{k}^T(n)$
3: $\mathbf{C}(n) = \lambda^{-1}\mathbf{C}(n-1) - \lambda^{-1}\mathbf{k}(n)\mathbf{x}^T(n)\mathbf{C}(n-1)$
4: $\mathbf{H}(n) = \mathbb{S}_{\gamma\lambda^{-1}\mathbf{C}(n-1)}[\mathbf{G}(n)]$
End For

$$\text{Prior} = p(\mathbf{H}(n)|\mathbf{Y}(n-1); \mathbf{G}(n-1)) \simeq \mathcal{N}(\mathbf{G}(n-1), \mathbf{S}(n))\,.$$

It is well known that this conditional expectation may be obtained recursively using the Kalman filter, if a Gaussian prior is assumed on $\mathbf{H}(n)$ given the past observation. The Kalman filter then determines the posterior probability density function for $\mathbf{H}(n)$ recursively over time. In a Bayesian context if $\mathbf{H}(n)$ is assumed to be Gaussian, the RLS filter can be regarded as a Kalman filter [55]. Therefore, the main recursion takes the form [55, 70]

$$\begin{aligned}\mathbf{H}(n) &= \mathbf{H}(n-1) + \mathbf{e}(n)\mathbf{k}^T(n)\\ \mathbf{C}(n) &= \lambda^{-1}\mathbf{C}(n-1) - \lambda^{-1}\mathbf{k}(n)\mathbf{x}^T(n)\mathbf{C}(n-1)\end{aligned}$$

where $\mathbf{k}(n)$ is the Kalman gain and $\mathbf{e}(n)$ denotes the prediction error given by $\mathbf{e}(n) = \mathbf{y}(n) - \mathbf{H}(n-1)\mathbf{x}(n)$. Hence $\mathbf{H}(n)$ depends linearly on $\mathbf{G}$. The Riccati equation that updates $\mathbf{C}(n) = \mathbf{R_{xx}}^{-1}(n)$ indicates that $\mathbf{C}(n)$ does not depend on $\mathbf{G}$. Moreover, $\mathbb{E}[\mathbf{e}(n)\mathbf{Y}(n-1)] = 0$ because the prediction error $\mathbf{e}(n)$ is uncorrelated to measurements. The ith diagonal component of the prior covariance $\mathbf{S}_i(n)$ can be computed as follows

$$\mathbf{S}_i(n) = \lambda^{-1}\mathbf{C}_i(n-1).$$

The method outlined above is named EM–RLS filter and is summarized in Table 7.4.

7.3.3 Greedy Adaptive Filters

Greedy algorithms provide an alternative approach to ℓ_1 penalization methods. For the recovery of a sparse parameter matrix in the presence of noise, greedy algorithms iteratively improve the current estimate by modifying one or more elements until a halting condition is met. The basic principle behind greedy algorithms is to iteratively find the support set of the sparse matrix and reconstruct it using the restricted support Least Squares (LS) estimate. The computational complexity depends on the number of iterations required to find the correct support set. One of the earliest algorithms proposed for sparse signal recovery is the Orthogonal Matching Pursuit (OMP) [26, 65, 75]. At each iteration, OMP finds the entry of the proxy matrix $\mathbf{P}(n) = (\mathbf{Y}(n) - \mathbf{H}\mathbf{X}(n))\mathbf{X}^H(n)$ with the largest magnitude, and adds it to the support set. Then, it solves the following least squares problem:

$$\hat{\mathbf{H}} = \arg\min_{\mathbf{H}} \|\mathbf{Y}(n) - \mathbf{H}\mathbf{X}(n)\|_{\ell_2}^2$$

and updates the residual. By repeating these steps a total of s times, the support of $\mathbf{H}$ is recovered.

Several improvements have been proposed for greedy reconstruction. The Stagewise OMP (StOMP), proposed in [31], selects all proxy components whose values are above a certain threshold. Due to the multiple selection step, StOMP achieves better runtime than OMP. On the other hand, parameter tuning in StOMP might be difficult and there are rigorous asymptotic results available. A more sophisticated algorithm was developed by Needell and Vershynin, and is known as Regularized OMP (ROMP) [63]. ROMP chooses the s largest components of the proxy, and applies a regularization step to ensure that not too many incorrect components are selected. The recovery bounds obtained in [63] are optimal up to a logarithmic factor. Tighter recovery bounds which avoid the presence of the logarithmic factor are obtained by Needell and Tropp via the Compressed Sampling Matching Pursuit algorithm (CoSaMP) [62]. CoSaMP provides tighter recovery bounds than ROMP that are optimal up to a constant factor. An algorithm similar to the CoSaMP, was presented by Dai and Milenkovic and is known as Subspace Pursuit (SP) [23].

As with most greedy algorithms, CoSaMP takes advantage of the measurement matrix $\mathbf{X}(n)$ which is assumed to be approximately orthonormal ($\mathbf{X}(n)\mathbf{X}^H(n)$ is close to the identity matrix). Hence, the largest components of the signal proxy $\mathbf{P}(n) = \mathbf{H}\mathbf{X}(n)\mathbf{X}^H(n)$ most likely correspond to the non–zero rows of $\mathbf{H}$. Next, the algorithm adds the largest components of the signal proxy to the running support set and performs least squares to get an estimate for the signal. Finally, it prunes the least square estimation and updates the error residual. The main ingredients of the CoSaMP algorithm are outlined below:

Identification of the largest $2s$ components of the proxy signal
Support Merger: forms the union of the set of newly identified components with the set of indices corresponding to the s largest components of the least squares estimate obtained in the previous iteration
Estimation via least squares on the merged set of components
Pruning: restricts the LS estimate to its s largest components
Sample update: updates the error residual.

The above steps are repeated until a halting criterion is met. The main difference between CoSaMP and SP is in the identification step where the SP algorithm chooses the s largest components.

It was established in [58] that greedy algorithms can be converted into an adaptive mode, while maintaining their superior performance gains. We demonstrate below that this conversion is applicable in the multichannel set up. We focus our analysis on CoSaMP/SP due to their superior performance, but similar ideas are applicable to other greedy algorithms as well. Multichannel greedy algorithms can be approached via two strategies. The first approach assumes that the subsystems share the same sparsity pattern. Hence the greedy algorithm simultaneously recovers the support set (also known as joint sparsity or group sparsity) [12, 75] by choosing an element which reaches the maximum value of the multichannel energy. Under the second strategy adopted here, the subsystems exhibit different sparsity patterns [56]. Next greedy versions of the main adaptive multichannel algorithms are presented based on the CoSaMP/SP platform.

7.3.3.1 Greedy LMS Filter

The multichannel adaptive greedy LMS algorithm modifies the proxy identification, estimation and error residual update. The error residual is evaluated by

$$\mathbf{v}(n) = \mathbf{y}(n) - \mathbf{H}(n)\mathbf{x}(n). \tag{7.64}$$

The above formula involves the current sample only, in contrast to the CoSaMP/SP scheme which requires all previous samples. A new proxy signal that is more suitable for the adaptive mode, is defined as:

$$\mathbf{P}(n) = \sum_{i=1}^{n-1} \lambda^{n-1-i}\mathbf{v}(i)\mathbf{x}^H(i)$$

and is updated by

$$\mathbf{P}(n) = \lambda\mathbf{P}(n-1) + \mathbf{v}(n-1)\mathbf{x}^H(n)$$

This way the algorithm is capable of capturing variations on the support of $\mathbf{H}$. The estimate $\mathbf{H}(n)$ is updated by the LMS recursion [46, 70]. At each iteration the current regressor $\mathbf{x}(n)$ and the previous estimate $\mathbf{H}(n-1)$ are restricted to the instantaneous support originated from the support merging step. However, because the row support corresponding to each output is different, some extra care is required. Recall that any MIMO filter with n_o outputs is simplified to n_o MISO adaptive filters (all of which have different row support). Let Λ denote the estimated set of indices and $\Lambda^{(r)}$ $(r = 1, 2, \ldots, n_o)$ the set of indices associated with the rth row of $\mathbf{H}(n)$. The update equation for the rth output is given by

$$\mathbf{h}_{r:|\Lambda^{(r)}}(n) = \mathbf{h}_{r:|\Lambda^{(r)}}(n-1) + \mu e_r(n)\mathbf{x}^H_{|\Lambda^{(r)}}(n), \quad \forall r = 1, \ldots, n_o \tag{7.65}$$

where $\mathbf{x}_{|\Lambda^{(r)}}(n)$ denotes the sub–vector corresponding to the index set $\Lambda^{(r)}$. If all rows of $\mathbf{H}$ share the same row support then the update step can be performed jointly for all outputs and the selection of the largest proxy signal components is simplified [75].

Table 7.5 SpAdOMP algorithm

Algorithm description	
$\mathbf{H}(0) = \mathbf{0}, \mathbf{W}(0) = \mathbf{0}, \mathbf{P}(0) = \mathbf{0}$	{Initiliazation}
$\mathbf{v}(0) = \mathbf{y}(0)$	{Initial residual}
$0 < \lambda \leq 1$	{Forgetting factor}
$0 < \mu < 2\lambda_{\max}^{-1}$	{Step size}
For $n := 1, 2, \ldots$ **do**	
1: $\mathbf{P}(n) = \lambda\mathbf{P}(n-1) + \mathbf{v}(n-1)\mathbf{x}^H(n-1)$	{Form signal proxy}
2: $\Omega = \text{supp}(\mathbf{P}_{2s}(n))$	{Identify large components}
3: $\Lambda = \Omega \cup \text{supp}(\mathbf{H}(n-1))$	{Merge supports}
4: $e_r(n) = y_r(n) - \mathbf{w}_{r:\|\Lambda^{(r)}}(n-1)\mathbf{x}_{\|\Lambda^{(r)}}(n)$	{Prediction error}
5: $\mathbf{w}_{r:\|\Lambda^{(r)}}(n) = \mathbf{w}_{r:\|\Lambda^{(r)}}(n-1) + \mu e_r(n)\mathbf{x}^H_{\|\Lambda^{(r)}}(n)$	{LMS iteration}
6: $\Lambda_s = \max(\|\mathbf{H}_{\|\Lambda}(n)\|, s)$	{Obtain the pruned support}
7: $\mathbf{H}_{\|\Lambda_s}(n) = \mathbf{W}_{\|\Lambda_s}(n),\ \ \mathbf{H}_{\|\Lambda_s^c}(n) = \mathbf{0}$	{Prune the LMS estimates}
8: $\mathbf{v}(n) = \mathbf{y}(n) - \mathbf{H}(n)\mathbf{x}(n)$	{Update error residual}
end For	

The multichannel Sparse Adaptive Orthogonal Matching Pursuit (SpAdOMP) algorithm, is presented in Table 7.5. The operator $\max(|a|, s)$ returns s indices of the largest elements of a and Λ^c represents the complement of Λ. An important point to note about step 5 of Table 7.5 is that the choice of a proper step–size μ that ensures convergence is difficult. The Normalized LMS (NLMS) addresses this issue by scaling with the input power

$$\mathbf{h}_{r:|\Lambda^{(r)}}(n) = \mathbf{h}_{r:|\Lambda^{(r)}}(n-1) + \frac{\mu}{\epsilon + \|\mathbf{x}_{|\Lambda^{(r)}}(n)\|^2} e_r(n)\mathbf{x}^H_{|\Lambda^{(r)}}(n), \quad \forall r = 1, \ldots, n_o$$

where $0 < \mu < 2$ and ϵ is a small positive constant (inserted to avoid division by small numbers). NLMS may be viewed as an LMS with time–varying step–size. This partially explains the superior tracking performance as compared to LMS in non–stationary environments.

7.3.3.2 Greedy RLS Filter

In this subsection we develop greedy adaptive schemes whose estimation part is based on rank one updates for the autocorrelation and cross–correlation matrices. A straightforward forward attempt towards this direction, would be to re–use the framework adapted by the SpAdOMP algorithm [58] (Table 7.5) and replace the estimation step with the RLS algorithm. However in doing so, we will have to update the entries of the inverse covariance matrix as well as the Kalman gain entries which are required to perform an RLS update for the currently estimated support set. A more efficient technique avoids the last action of the CoSaMP/SP framework (Sample Update) and is described next.

Consider the normal equations

$$\mathbf{H}\mathbf{X}(n)\mathbf{X}^H(n) = \mathbf{Y}(n)\mathbf{X}^H(n). \tag{7.66}$$

An iterative method known as Landweber–Fridman or Van Cittert iteration [29, 78] is incorporated in order to express Eq. (7.66) into an equivalent fixed point equation of the form

$$\mathbf{H} = \mathbf{H} + (\mathbf{Y}(n) - \mathbf{H}\mathbf{X}(n))\,\mathbf{X}^H(n).$$

The Landweber iteration starts from an initial guess $\mathbf{H}^0$ and solves $\mathbf{y}(n) = \mathbf{H}\mathbf{x}(n)$ iteratively by

$$\mathbf{H}^{(t)} = \mathbf{H}^{(t-1)} + \left(\mathbf{Y}(n) - \mathbf{H}^{(t-1)}\mathbf{X}(n)\right)\mathbf{X}^H(n) \quad t = 1, 2, \ldots$$

The above iteration requires the norm of $\mathbf{X}(n)$ to be less than or equal to one, otherwise it diverges or converges too slowly. To avoid divergence and accelerate the speed of convergence a step size term μ is introduced

$$\mathbf{H}^{(t)} = \mathbf{H}^{(t-1)} + \mu\left(\mathbf{Y}(n) - \mathbf{H}^{(t-1)}\mathbf{X}(n)\right)\mathbf{X}^H(n) \quad t = 1, 2, \ldots \tag{7.67}$$

where $\mu \in (0, 2/\|\mathbf{X}(n)\mathbf{X}^H(n)\|)$. The above iterations is similar to Steepest Descent except that the step size term is fixed. To derive an adaptive Landweber filter we rewrite Eq. (7.67) as

$$\mathbf{H}^{(t)} = \mathbf{H}^{(t-1)}\left(\mathbf{I} - \mu\mathbf{X}(n)\mathbf{X}^H(n)\right) + \mu\mathbf{Y}(n)\mathbf{X}^H(n) \quad t = 1, 2, \ldots \tag{7.68}$$

The above iteration requires the autocorrelation matrix $\mathbf{R}_{\mathbf{xx}}(n) = \mathbf{X}(n)\mathbf{X}^H(n)$ and the cross–correlation matrix $\mathbf{P}_{\mathbf{yx}}(n) = \mathbf{Y}(n)\mathbf{X}(n)^H$. In practice, the data arrive sequentially and might vary with time. For this reason we approximate $\mathbf{R}_{\mathbf{xx}}(n)$ and $\mathbf{P}_{\mathbf{yx}}(n)$ via exponentially weighted sample averages [46, 70]. Therefore the Landweber iteration takes the form

$$\mathbf{H}(n) = \mathbf{H}(n-1)\left(\mathbf{I} - \mu\mathbf{R}_{\mathbf{xx}}(n)\right) + \mu\mathbf{P}_{\mathbf{yx}}(n) \quad n = 1, 2, \ldots \tag{7.69}$$

The resulting expression is identical to the one derived in [5] (Step 4 in Table 7.3) via the EM formulation and the decomposition of the noise vector.

Finally let us take a second look at the proxy signal and the sample update, which are described in Sect. 7.3.3. The authors in [58] proposed an adaptive mechanism to estimate the signal proxy and the sample update. Examination of

$$\mathbf{P}(n) = (\mathbf{Y}(n) - \mathbf{H}\mathbf{X}(n))\,\mathbf{X}^H(n) \tag{7.70}$$

Table 7.6 SpAdOMP (RLS) algorithm

Algorithm description	
$\mathbf{H}(0) = \mathbf{0}, \mathbf{W}(0) = \mathbf{0}, \mathbf{P}(0) = \mathbf{0}, \mathbf{R}_{\mathbf{xx}}(0) = \mathbf{0}, \mathbf{P}_{\mathbf{yx}}(0) = \mathbf{0}$	{Initiliazation}
For $n := 1, 2, \ldots$ **do**	
1: $\mathbf{R}_{\mathbf{xx}}(n) = \lambda \mathbf{R}_{\mathbf{xx}}(n-1) + \mathbf{x}(n)\mathbf{x}^H(n)$	{Update autocorrelation}
2: $\mathbf{P}_{\mathbf{yx}}(n) = \lambda \mathbf{P}_{\mathbf{yx}}(n-1) + \mathbf{y}(n)\mathbf{x}^H(n)$	{Update cross–correlation}
3: $\mathbf{P}(n) = \mathbf{P}_{\mathbf{yx}}(n) - \mathbf{H}(n)\mathbf{R}_{\mathbf{xx}}(n)$	{Form signal proxy}
4: $\Omega = \text{supp}(\mathbf{P}_{2s}(n))$	{Identify large components}
5: $\Lambda = \Omega \cup \text{supp}(\mathbf{H}(n-1))$	{Merge supports}
6: $\mathbf{W}(n) = \mathbf{W}(n-1)(\mathbf{I} - \mu \mathbf{R}_{\mathbf{xx}}(n)) + \mu \mathbf{P}_{\mathbf{yx}}(n)$	{Recursive Landweber iteration}
7: $\Lambda_s = \max(\lvert \mathbf{W}_{\lvert\Lambda}(n)\rvert, s)$	{Obtain the pruned support}
8: $\mathbf{H}_{\lvert\Lambda_s}(n) = \mathbf{W}_{\lvert\Lambda_s}(n), \quad \mathbf{H}_{\lvert\Lambda_s^c}(n) = \mathbf{0}$	{Prune the Landweber estimates}
end For	

shows that the sample update constitutes an ingredient of the signal proxy. Additionally, the above equation can be re–expressed as follows

$$\mathbf{P}(n) = \mathbf{Y}(n)\mathbf{X}^H(n) - \mathbf{H}\mathbf{X}(n)\mathbf{X}^H(n) \simeq \mathbf{P}_{\mathbf{yx}}(n) - \mathbf{H}\mathbf{R}_{\mathbf{xx}}(n) \tag{7.71}$$

and hence there is no need for the sample update, since all the required information is obtained from the correlation and cross–correlation matrices. The algorithm is summarized in Table 7.6. The key difference between spaRLS and this version of SpAdOMP algorithm is that the latter has two mechanisms for support estimation (the proxy signal followed by pruning which is a special form of hard thresholding) and hence can achieve better support estimation.

7.3.4 Computer Simulations of Sparse Adaptive MIMO Filters

In this subsection we demonstrate and compare the performance of the algorithms outlined in this section. Computer simulations are conducted under different scenarios in order to evaluate performance over a wide range of conditions. The Normalized Mean Square Error (NMSE, in dB scale)

$$\text{NMSE}_{ij} := \text{MC}^{-1} \sum_{t=1}^{\text{MC}} \frac{\sum_{n=1}^{N} \lvert \hat{H}_{ij}^{(t)}(n) - H_{ij}(n) \rvert^2}{\sum_{n=1}^{N} \lvert H_{ij}(n) \rvert^2}$$

is used as performance measure, where $\hat{H}_{ij}^{(t)}(n)$ denotes the estimate of the ij subsystem for the tth Monte Carlo (MC) run. The overall NMSE is obtained by averaging over all subsystems

$$\text{NMSE} := \frac{1}{n_o \times n_i \times M} \sum_{i=1}^{n_o} \sum_{j=1}^{n_i \times M} \text{NMSE}_{ij}. \tag{7.72}$$

All NMSE results were obtained for 50 different system realizations (every non–zero parameter at each realization is assigned to random locations and their values are generated randomly from a complex normal distribution). The experiments are conducted in a moderate noise environment with Signal to Noise Ratio (SNR := $10 \log \|\mathbf{H}\|_{\ell_2}^2 / \|\mathbf{v}\|_{\ell_2}^2$) of 15 dB.

To compare the performance of different adaptive filters we use their corresponding *learning curves* which are plots of the NMSE versus the number of iterations. Learning curves help us visualize the convergence and tracking behaviour of adaptive filters. Note that although the LMS and RLS type filters are examined under the same scenarios, we have chosen to plot them separately due to different convergence speeds and computational complexity requirements.

Adaptive Identification of Linear MIMO Systems

First we consider a linear (3, 3)–MIMO system with a memory length $M = 5$ and 5 non–zero elements. The system is excited by a complex Gaussian input signal with zero mean and variance $1/5$. For a fair comparison between all competing LMS–type filters the step size is common and equal to

$$\mu_n = \frac{1}{\|\mathbf{x}(n)\|_{\ell_2}^2}.$$

The regularization step size γ for the ZA–LMS (or ℓ_1–LMS) and the RZA–LMS (or *log*–LMS) are adjusted adaptively following the systematic approach introduced in [19]. For the ℓ_0–LMS filter the regularization parameters required offline fine tuning and the best performance is obtained when $\alpha = 5$ and $\gamma = 0.01$. The SpAdOMP filter required a–priori knowledge of the sparsity level in order to perform the adaptive greedy selection procedure. Figure 7.8a shows that SpAdOMP obtains the faster convergence and better steady state accuracy, followed by the ℓ_0–LMS and log–LMS whose performance is nearly identical.

The RLS–type filters share a common forgetting factor $\lambda = 0.98$. The R–LASSO, spaRLS and SpAdOMP follow an instantaneous steepest descent pattern (that involves the autocorrelation and cross–correlation matrices) and employ a step size to accelerate convergence. The step size is set to

$$\mu_n = \frac{0.3}{\|\mathbf{x}(n)\|_{\ell_2}^2} \tag{7.73}$$

for all schemes. The EM–RLS, R–LASSO and SpaRLS required offline processing to find the optimum regularization parameter for each filter ($\gamma_{EM-RLS} = 6 \times 10^{-4}$,

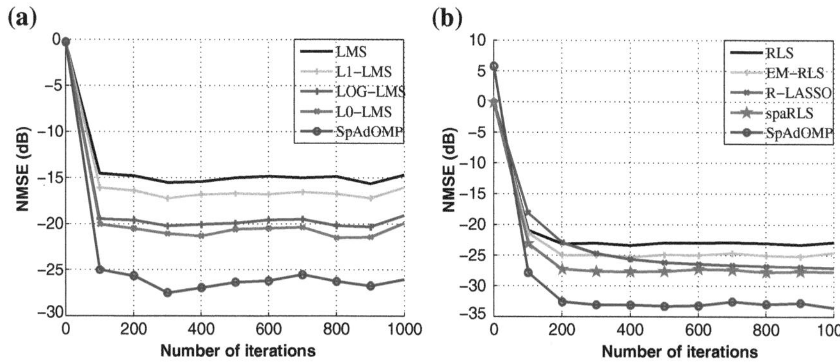

Fig. 7.8 Learning curves of adaptive MIMO filters. **a** LMS-type filters. **b** RLS-type filters

$\tau_{R-LASSO} = 0.3$ and $\alpha^2 \gamma_{spaRLS} = 0.03$). The adaptive greedy filter (SpAdOMP) is fine tuned using a–priori knowledge of the sparsity level ($s = 5$). Figure 7.8b presents the learning curves of RLS–type of filters. We observe that the adaptive greedy filter gives the best performance. It is followed by spaRLS, R–LASSO and EM–RLS. The convergence rate of spaRLS, R–LASSO and EM–RLS can be significantly improved if a more sparsity aware regularization function is employed (like those discussed in Sect. 7.3.2.1).

Adaptive Identification of Nolinear MIMO Systems

Next, we evaluate the filtering performance of sparse nonlinearly mixed MIMO systems. The MIMO system consists of 3 inputs, 3 outputs, has memory length $M = 2$ and poses a quadratic nonlinearity where all different product combinations of the inputs are allowable. The combination of sparsity with nonlinearity significantly increases the parameter space of the unknown system matrix and may give rise to *degeneracy* in the parameters. Note that degeneracy causes all important parameters to be close to zero and as a result some outputs may also be zero. To avoid this situation we consider 9 non–zero parameters, 6 of which belong to the linear part of the system (spread among different inputs) and 3 correspond to the nonlinear part. The input sequence is drawn from a complex Gaussian distribution of zero mean and variance 1/9.

Initially we compare the learning curves of LMS–type filters. The step size is common to all filters and given by Eq. (7.73). Unlike the linear case, it was experimentally found that the systematic approach (developed in [19]) for choosing the best regularization parameter for ZA–LMS and RZA–LMS (or ℓ_1–LMS and log–LMS as we will refer to respectively) does not perform that well in the case of nonlinearly mixed MIMO. Therefore, ℓ_1–LMS, log–LMS and ℓ_0–LMS are required to optimize their parameters via exhaustive simulations and the corresponding values are summarized in the following table.

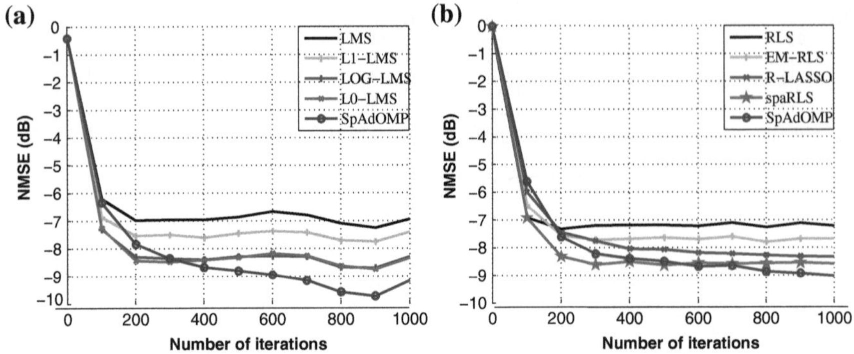

Fig. 7.9 Learning curves of adaptive nonlinear MIMO filters. **a** LMS-type filters. **b** RLS-type filters

ℓ_1–LMS (γ)	log–LMS (γ, ϵ)	ℓ_0–LMS (γ, α)	EM–RLS (γ)	R–LASSO (γ)	spaRLS ($\alpha^2\gamma$)
5×10^{-3}	1×10^{-2},10	1×10^{-3},5	2×10^{-3}	9×10^{-2}	2×10^{-3}

The conclusions drawn from inspection of Fig. 7.9a, are almost identical to those in the linear case. However, this time the convergence speed of the greedy filter is slightly worse than the one obtained by ℓ_0–LMS and log–LMS filter.

Next, we study the performance of RLS–type of filters in nonlinearly mixed MIMO systems. As in the linear case, some filters require offline processing to fine tune their regularization parameters and the optimum values are summarized in the above table. Figure 7.9b shows that almost all RLS–type of filters achieve relatively similar steady–state accuracy, and spaRLS has the fastest convergence speed.

One common conclusion for LMS and RLS type of filters, operating in nonlinear MIMO systems, is that the extraordinary good performance of adaptive greedy filters is slightly degraded. This is because greedy filters require strongly incoherent dictionaries (this has been studied for the linear case in [58]). The problem of designing input sequences with incoherent dictionary for nonlinear MIMO systems requires further work.

Tracking Performance of Sparse Adaptive Filters

The time–varying nonlinear MIMO system is initialized using the same parameters as used to generate Fig. 7.9. At the 400th iteration the system experiences a sudden change, where all active parameters of the nonlinear part randomly change locations. We note from Fig. 7.10 that spaRLS, ℓ_0–LMS and log–LMS have the fastest support tracking behaviour and that adaptive greedy filters achieve better steady state accuracy.

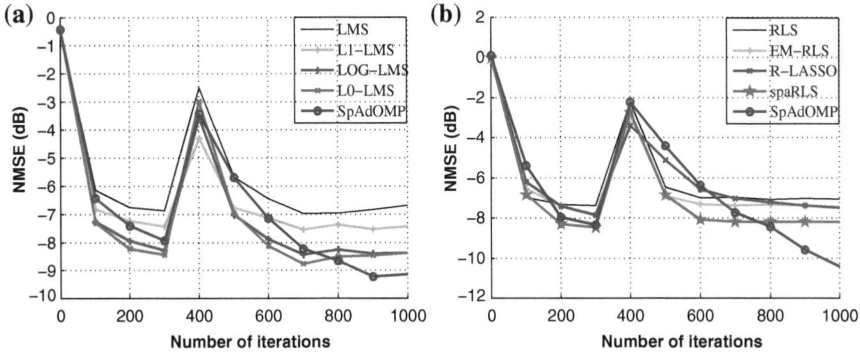

Fig. 7.10 Comparison of tracking performances on nonlinear MIMO systems. **a** LMS-type filters. **b** RLS-type filters

7.4 Blind and Semi-Blind Identification of Sparse MIMO Systems Excited by Finite-Alphabet Inputs

This section is concerned with the sparse MIMO parameter estimation problem encountered in blind system identification whereby the unknown sparse MIMO system is estimated using output information as well as some a priori knowledge of the system. This problem arises in digital communications, seismic data, image deblurring and speech coding.

The sparse blind MIMO identification problem has been approached by two different methodologies, namely: (1) dictionary learning [36, Chap. 12] and (2) maximum penalized likelihood estimator (via the Expectation Maximization algorithm) [60]. The first approach solves an optimization problem by iteratively applying two convex steps: the parameter update step on a fixed measurement matrix and the measurement matrix update step on a fixed parameter. The second approach, employs Expectation–Maximization for finding maximum penalized likelihood estimates. Both algorithms do not converge to global minima, whereas for the case of dictionary learning even a local minimum can not be guaranteed.

In this section we discuss joint state estimation and sparse parameter estimation techniques under the Finite-Alphabet property. Two different techniques are described. The first algorithm maximizes the likelihood of the received sequence over all possible input sequences and system parameters. It does so by converting the joint maximization into a two stage maximization problem. For a given parameter value at the end of the ℓ–th iteration, the most likely state (equivalently input sequence) is estimated by carrying out the inner maximization. This maximization can be performed by the Viterbi algorithm since the polynomial MIMO system is represented by a Hidden Markov Model (HMM). Once the inner maximization is completed and the most likely state sequence is determined, the outer maximization takes over. Given the state sequence at step ℓ, maximization of the penalized likelihood with respect to system parameters is effected by sparsity aware schemes. The two main stages (state estimation, parameter estimation) iterate until a stopping criterion is satisfied.

The second blind estimation method considered in this section is based on Expectation Maximization (EM). Instead of working with likelihood, EM employs the augmented likelihood formed by the so called complete data which consist of the state sequence and the output sequence. It turns out that maximization of the augmented likelihood is easier to perform. Then the EM procedure alternates between the E–step during which the log–likelihood function of the complete data is estimated, and the M–step which maximizes the augmented likelihood to generate an updated parameter matrix. Parameter sparsity is naturally embedded in the M–step by the insertion of a penalty term (typically the ℓ_1 norm of the parameters).

7.4.1 An Alternating Maximum Likelihood Procedure for State Estimation and Sparse System Estimation

Let us consider the basic set up defined in Sect. 7.2. The input–output relationship is given by

$$\mathbf{y}(n) = f(x_1(n), x_1(n-1), \ldots, x_1(n-M), \cdots, x_{n_i}(n), x_{n_i}(n-1), \ldots, x_{n_i}(n-M)) + \mathbf{v}(n). \tag{7.74}$$

The noise vector $\mathbf{v}(n)$ is a multivariate Gaussian i.i.d. with mean and covariance matrix $\mathcal{N}(0, \mathbf{Q})$. Following the analysis of Sect. 7.2.1 let

$$\bar{\mathbf{x}}(n) = [x_1(n), x_1(n-1), \ldots, x_1(n-M), \cdots, x_{n_i}(n), x_{n_i}(n-1), \ldots, x_{n_i}(n-M)]^T.$$

Hence the nonlinear input vector is given by

$$\mathbf{x}(n) = [\bar{\mathbf{x}}(n), \bar{\mathbf{x}}_2(n), \cdots, \bar{\mathbf{x}}_p(n)]^T.$$

We shall refer to $\mathbf{x}(n)$ as the augmented state or simply the state. Equation (7.74) is compactly written as

$$\mathbf{y}(n) = \mathbf{H}\mathbf{x}(n) + \mathbf{v}(n).$$

In a blind (or Semi-Blind) environment, information on the input sequence that generated a given output is not available. Suppose $\mathbf{Y}(n) = [\mathbf{y}(1), \mathbf{y}(2), \cdots, \mathbf{y}(n)]$ denotes the known $n_i \times n$ observation sequence. The task of joint state estimation and system parameter estimation is based solely on a small number of measurements n. The probability density function (PDF) of the observation matrix $\mathbf{Y}(n)$ conditioned on $(\mathbf{X}(n), \mathbf{H})$ is given by

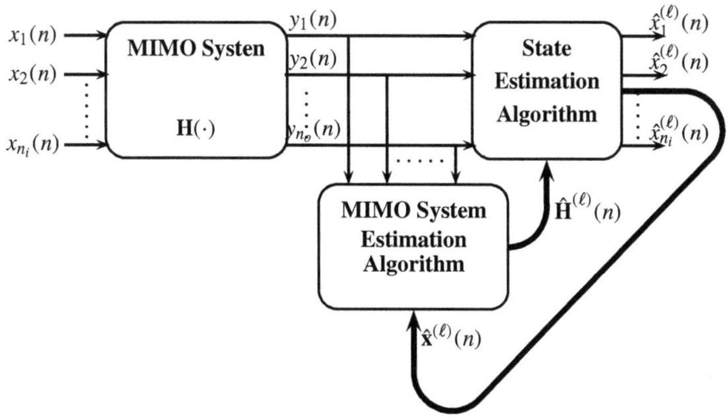

Fig. 7.11 Alternating MIMO detector–estimator

$$p(\mathbf{Y}(n)|\mathbf{H}, \mathbf{X}(n)) = \frac{1}{(2\pi\sigma^2)^{n_o \times n}} \exp\left(-\frac{1}{2\sigma^2} \sum_{t=1}^{n} \|\mathbf{y}(t) - \mathbf{H}\mathbf{x}(t)\|_{\ell_2}^2 \right) \tag{7.75}$$

The joint Maximum Likelihood (ML) estimator of $\mathbf{X}(n)$ and $\mathbf{H}$ is obtained by jointly maximizing $p(\mathbf{Y}(n)|\mathbf{H}, \mathbf{X}(n))$ over $\mathbf{X}(n)$ and $\mathbf{H}$, as follows:

$$(\hat{\mathbf{X}}(n), \hat{\mathbf{H}}) = \arg \max_{\mathbf{X}(n),\mathbf{H}} \log p(\mathbf{Y}(n)|\mathbf{H}, \mathbf{X}(n)).$$

The above optimization problem is intractable. We thus convert it into a two stage maximization problem that is iteratively performed over $\mathbf{X}(n)$ and $\mathbf{H}$ (see Fig. 7.11) as

$$(\hat{\mathbf{X}}(n), \hat{\mathbf{H}}) = \arg \max_{\mathbf{H}} \max_{\mathbf{X}(n)} \log p(\mathbf{Y}(n)|\mathbf{H}, \mathbf{X}(n)). \tag{7.76}$$

The iterative procedure alternates information between a state estimation scheme and a system parameter estimation scheme. In several applications, including communications, the input signals take values in a Finite-Alphabet. Then the state vector evolves as a Markov chain and the input–output relationship becomes a Hidden Markov Process (HMP) [37]. Therefore the inner maximization at step (ℓ) can be accomplished by dynamic programming and the Viterbi Algorithm (VA). Given $\mathbf{X}^{(\ell)}$ the iterative process updates the system parameters. The outer level maximization is equivalent to a quadratic minimization problem. Hence, the optimum solution leads to a set of normal equations:

$$\mathbf{H}^{(\ell)}\mathbf{X}^{(\ell)}(n)\,\mathbf{X}^{(\ell)H}(n) = \mathbf{Y}(n)\,\mathbf{X}^{(\ell)H}(n).$$

The algorithm is repeated until a fixed point is reached or until a stopping criterion is met. Local convergence of the algorithm can be established [37]. The above procedure is known in the literature under several different names: Baum–Viterbi [37], ML Alternating Least Squares [1, 68] and bootstrap equalization [73].

Remark: Although the above procedure can operate in a pure blind fashion, it converges very slowly and suffers from an inherent permutation and scaling ambiguity problem [74]. This ambiguity is resolved if very few training input samples are used to provide an initial parameter matrix $\mathbf{H}^{(0)}$ estimate. The initial estimate does not need to be accurate enough since it is improved through successive iterations. The minimum number of training data, namely, $T = n_i M$, is equal to the rank of the MIMO system. The training symbol matrix $\mathbf{X}^{(0)}$ can be designed to yield the optimal estimation performance.

In the MIMO models discussed in Sect. 7.2 the parameter space increases exponentially and often the number of parameters exceeds the number of available measurements and the resulting system becomes underdetermined. Additionally, the unknown system may exhibit slow time–variations, so that during the time period of n data the entries of $\mathbf{H}$ may be considered constant. Therefore, even if $\mathbf{X}(n)$ is known, estimating $\hat{\mathbf{H}}$ remains an underdetermined problem. The key observation here is to consider the parameters of $\mathbf{H}$ that actually contribute to the output (see Sect. 7.2.1.1). This motivates the addition of a regularization term into the cost function for joint state estimation and sparse parameter estimation. Following the Compressed Sensing paradigm, the ℓ_1 penalty term is added and the cost function takes the form:

$$(\hat{\mathbf{X}}(n), \hat{\mathbf{H}}) = \arg\left\{\max_{\mathbf{H}}\left[\max_{\mathbf{X}(n)\in S} \log p(\mathbf{Y}(n)|\mathbf{H}, \mathbf{X}(n)) - \tau\|\text{vec}[\mathbf{H}]\|_{\ell_1}\right]\right\}. \tag{7.77}$$

The maximization of the likelihood with respect to $\mathbf{H}$ resembles the basis pursuit or LASSO criterion and hence any compressed sensing algorithm can be used to perform this maximization [36, 76].

A two stage maximization algorithm of this type is summarized in Table 7.7. The first step involves an approximation to $\hat{\mathbf{X}}(n)$ which is obtained using a Maximum A Posteriori (MAP) criterion:

$$\arg\max_{\mathbf{X}(n)} \quad p(\mathbf{X}(n)|\mathbf{Y}(n); \mathbf{H}^{(\ell-1)}). \tag{7.78}$$

Table 7.7 Baum–Viterbi algorithm

Algorithm description	
$\ell = 0: \quad \mathbf{H}^{(0)} = \arg\max_{\mathbf{H}} \log p(\mathbf{Y}(T)|\mathbf{H}, \mathbf{X}(T)) - \tau\|\text{vec}[\mathbf{H}]\|_{\ell_1}$ {Initiliazation}	
Repeat	
$\ell = \ell + 1$	
1: $\mathbf{X}^{(\ell)}(n) = \arg\max_{\mathbf{X}(n)\in S} \log p(\mathbf{Y}(n)|\mathbf{H}^{(\ell-1)})$	{Viterbi Algorithm}
2: $\mathbf{H}^{(\ell)} = \arg\max_{\mathbf{H}} \log p(\mathbf{Y}(n)|\mathbf{X}^{(\ell)}(n)) - \tau\|\text{vec}[\mathbf{H}]\|_{\ell_1}$	{System Parameter Re–estimation}
Until $(\mathbf{X}^{(\ell)}(n), \mathbf{H}^{(\ell)}(n)) \approx (\mathbf{X}^{(\ell-1)}(n), \mathbf{H}^{(\ell-1)}(n))$	

Table 7.8 Viterbi algorithm

Algorithm description	
$\delta_1(i) = \log p(\mathbf{y}(1)\vert\mathbf{x}^{(i)}(1); \mathbf{H}^{(\ell-1)}(1)), \quad i = 1, \ldots, A^{n_i M}$	{Initiliazation}
For $t := 2, \ldots, n$ **do**	
For $j := 1, 2, \ldots, A^{n_i M}$ **do**	
1: $\delta_t(j) = \log p(\mathbf{y}(t)\vert\mathbf{x}^{(j)}(t); \mathbf{H}^{(\ell-1)}(t)) + \max_i \left[\delta_{t-1}(i)\right]$	{Recursion}
2: $\psi_t(j) = \arg\max_i \left[\delta_{t-1}(i)\right]$	
End	
End	
3: $i_n = \arg\max_i \delta_n(i), \quad \hat{\mathbf{x}}(n) = \mathbf{x}^{(i_n)}(n)$	{Termination}
4: $i_t = \psi_{t+1}(i_{t+1}), \quad \hat{\mathbf{x}}(t) = \mathbf{x}^{(i_t)}, \quad t = n-1, \ldots, 1$	{Backtracking}

The above is solved using the *Viterbi* algorithm. The Viterbi algorithm searches among all possible paths through the state trellis in order to efficiently find the most probable path. A pseudo–code is provided in Table 7.8.

Maximization of the penalized likelihood over $\mathbf{H}$ is equivalent to maximizing the auxiliary function [37, 50]:

$$\sum_{t=1}^{n} \sum_{i}^{A^{n_i M}} \delta(\hat{\mathbf{x}}^{(\ell)}(n) - \mathbf{x}_i) \log p\left(\mathbf{y}(t)\vert\mathbf{H}\right) - \tau \Vert \mathrm{vec}[\mathbf{H}] \Vert_{\ell_1} \tag{7.79}$$

where $\delta(\cdot)$ is the delta function that is equal to one when $\mathbf{x}^{(\ell)}(n) = \mathbf{x}_i$ and zero otherwise. Since the noise is Gaussian, expression (7.79) is equivalent to penalized least squares estimation. The linearity in the parameters leads to the following closed form expression

$$\mathbf{H}^{(\ell)} = \mathbb{S}_\tau \left[\left(\mathbf{X}^{(\ell)}(n)\, \mathbf{X}^{(\ell)^H}(n) \right)^{-1} \mathbf{Y}(n)\, \mathbf{X}^{(\ell)^H}(n) \right]. \tag{7.80}$$

ML Detection via sphere decoding. The state estimator based on the Viterbi algorithm requires searching over $A^{M \times n_i}$ possible trellis state. This is affordable when A and $M \times n_i$ are small but it is not realistic when M is large. An alternative decoder structure, employs a sphere decoder [11, 20]. The underlying principle of sphere decoding is to search the closest lattice point (or vector) to the output signal within a sphere of radius r centered at the output signal. Sphere decoding techniques increase the radius when there exists no vector within a sphere, and decrease the radius when there exist multiple vectors within the sphere. The main idea is to limit the search among the possible states to those located within a sphere having radius r. We write with some abuse of notation the following:

$$\hat{\mathbf{X}}(n) = \arg\min_{\mathbf{X}(n)} \ \Vert \mathbf{Y}(n) - \mathbf{H}\mathbf{X}(n) \Vert_{\ell_2}^2 \le r^2 \tag{7.81}$$

$$= \arg\min_{\mathbf{X}(n)} \quad \|\mathbf{H}(\mathbf{X}(n) - \bar{\mathbf{X}})\|_{\ell_2}^2 \leq r^2 \tag{7.82}$$

where (the MMSE estimate) $\bar{\mathbf{X}} = (\mathbf{H}^H\mathbf{H})^{-1}\mathbf{H}^H\mathbf{Y}(n)$ is the center of the sphere of radius r. The ML solution is contained in this sphere and can be found via low–complexity tree based search algorithm [43]. This way an exhaustive search procedure is avoided and the complexity is independent of alphabet size.

7.4.2 An Expectation Maximization and Smoothing Approach to MIMO Parameter Recovery

The alternating ML detector and parameter estimation procedure outlined in Sect. 7.4.1 can also be performed by employing the Expectation Maximization (EM) framework [27]. Instead of maximizing the likelihood $p(\mathbf{Y}(n)|\mathbf{H}) = \sum_{\mathbf{X}(n)} p(\mathbf{Y}(n), \mathbf{X}(n)|\mathbf{H})$ EM works with the complete likelihood $p(\mathbf{Y}(n), \mathbf{X}(n)|\mathbf{H})$. Of course, the complete likelihood can not be evaluated, since the data $\mathbf{X}(n)$ are unknown. Instead the expected value is used. The conditional log likelihood

$$\log p(\mathbf{Y}(n)|\mathbf{H}) = \log p(\mathbf{Y}(n), \mathbf{X}(n)|\mathbf{H}) - \log p(\mathbf{X}(n)|\mathbf{Y}(n), \mathbf{H})$$

is employed to evaluate the estimated complete log–likelihood

$$\begin{aligned} Q\left(\mathbf{H}, \mathbf{H}^{(\ell-1)}\right) &= \mathbb{E}_{p(\mathbf{X}(n)|\mathbf{Y}(n),\mathbf{H}^{(\ell-1)})}\left[\log p\left(\mathbf{Y}(n), \mathbf{X}(n)|\mathbf{H}\right)\right] \\ &= \sum_{\mathbf{X}(n)} p(\mathbf{X}(n)|\mathbf{Y}(n), \mathbf{H}^{(\ell-1)}) \log p(\mathbf{Y}(n), \mathbf{X}(n)|\mathbf{H}). \end{aligned} \tag{7.83}$$

The EM algorithm iterates between the E–step and the M–step until convergence (see Table 7.9). The expectation step (E–Step), where the conditional density of the unknown data, given the actual observations is estimated based on the current values of the unknown parameters is used to evaluate the expected value of the complete log–likelihood function. The maximization step (M–Step) finds the maximum of the estimated complete log–likelihood function with respect to the unknown system parameters.

Quite often in practice the number of available observations, n, is significantly smaller than $n_i M$, and the resulting system of equations is severely underdetermined. Furthermore, the effective rank of $\mathbf{H}$ is often significantly less than n. Such problems are often reduced by use of a Bayesian prior to favour some solutions over others. The prior is incorporated as a penalty term in the maximization step which is maximized to estimate the unknown system parameter matrix. Therefore the M–step seeks to solve the following problem:

Table 7.9 EM algorithm

Algorithm description	
$\ell = 0 : \mathbf{H}^{(0)}$	{Initiliazation}
Repeat	
$\ell = \ell + 1$	
1: $Q\left(\mathbf{H}, \mathbf{H}^{(\ell-1)}\right) = \mathbb{E}_{p(\mathbf{X}(n)\vert\mathbf{Y}(n), \mathbf{H}^{(\ell-1)})}\left[\log p\left(\mathbf{Y}(n), \mathbf{X}(n)\vert\mathbf{H}\right)\right]$	{E–Step}
2: $\mathbf{H}^{(\ell)} = \arg\max_{\mathbf{H}} Q\left(\mathbf{H}, \mathbf{H}^{(\ell-1)}\right)$	{M–Step}
Until $\Vert\text{vec}[\mathbf{H}^{(\ell)}] - \text{vec}[\mathbf{H}^{(\ell-1)}]\Vert_{\ell_2}^2 < \epsilon$	{Termination Condition}

$$\mathbf{H}^{(\ell)} = \arg\max_{\mathbf{H}} \quad Q\left(\mathbf{H}, \mathbf{H}^{(\ell-1)}\right) + \log p\left(\mathbf{H}\right).$$

A widely used prior which promotes sparsity and avoids underdetermined problems is the Laplacian prior

$$p\left(\mathbf{H}\right) \propto \exp\left(-\tau \Vert\text{vec}[\mathbf{H}]\Vert_{\ell_1}\right).$$

The introduction of such prior, allows the algorithm to choose only the non–zero components of $\mathbf{H}$.

The log–likelihood function increases monotonically at successive iterates $\mathbf{H}^{(\ell)}$ of the parameter vector [27], *i.e.*

$$p(\mathbf{Y}(n)|\mathbf{H}^{(\ell)}) \geq p(\mathbf{Y}(n)|\mathbf{H}^{(\ell-1)}).$$

Consequently the sequence $\{p(\mathbf{Y}(n)|\mathbf{H}^{(\ell)}), \quad \ell > 0\}$ converges as $\ell \to \infty$. For practical purposes, we truncate the number of iterations to a finite number L. Although the convergence of the likelihood values does not by itself ensure the convergence of the iterates $\mathbf{H}^{(\ell)}$, under relatively mild smoothness conditions for the log–likelihood function $p(\mathbf{Y}(n)|\mathbf{H})$ the sequence converges to a local maximum of $p(\mathbf{Y}(n)|\mathbf{H}^{(\ell)})$ [81]. However, its monotonic convergence behaviour is dependent on initialization [81]. To avoid it from being trapped to a stationary point which is not a local (global) maximum we may have to use several different initializations and also incorporate prior information about the distribution of $\mathbf{H}^{(0)}$. For Gaussian noise the corresponding log–likelihood function is log–concave. The log–concavity of the likelihood ensures convergence of the EM iteration to a stationary point, regardless of initialization.

Since $\mathbf{X}(n)$ is independent of $\mathbf{H}$ in Eq. (7.83), we can keep only the terms that depend on $\mathbf{H}$. Thus

$$Q\left(\mathbf{H}, \mathbf{H}^{(\ell-1)}\right) = \mathbb{E}_{p(\mathbf{X}(n)|\mathbf{Y}(n), \mathbf{H}^{(\ell-1)})}\left[\log p\left(\mathbf{Y}(n)|\mathbf{H}\right)\right]$$

with

$$p(\mathbf{Y}(n)|\mathbf{H}) = \frac{1}{(2\pi\sigma^2)^{n_o \times n}} \exp\left(-\frac{1}{2\sigma^2}\sum_{t=1}^{n} \Vert\mathbf{y}(t) - \mathbf{H}\mathbf{x}(t)\Vert_{\ell_2}^2\right).$$

Let us next take a closer look at the E–step. The resulting function, still denoted by Q takes the form

$$Q\left(\mathbf{H}, \mathbf{H}^{(\ell-1)}\right) = -\frac{1}{2\sigma^2}\sum_{t=1}^{n}\mathbb{E}\left\{\|\mathbf{y}(t) - \mathbf{H}\mathbf{x}(t)\|_{\ell_2}^2 | \mathbf{y}(t), \mathbf{H}^{(\ell-1)}\right\}.$$

The E–step depends on first and second order statistics of the hidden variable $\mathbf{X}(n)$, which are not available since it is unknown. Therefore the complete log likelihood is given by

$$Q\left(\mathbf{H}, \mathbf{H}^{(\ell-1)}\right) = -\frac{1}{2\sigma^2}\sum_{t=1}^{n}\sum_{i}^{A^{n_i M}} \|\mathbf{y}(t) - \mathbf{H}\mathbf{x}(t)\|_{\ell_2}^2 \gamma_{ti}^{(\ell)}$$

where

$$\gamma_{ti}^{(\ell)} = p\left(\mathbf{x}(t) = s_i | \mathbf{Y}(n); \mathbf{H}^{(\ell-1)}\right)$$

and thus a primary goal of the E–step is to compute the *a posteriori probabilities* (APPs), $\gamma_{ti}^{(\ell)}$. These in turn are computed by the forward–backward recursions presented next.

Maximization of the regularized Q–function with respect to $\mathbf{H}$ at the M–step, has a closed form expression and is given by the soft–thresholding function

$$\mathbf{H}^{(\ell)} = \mathbb{S}_\tau\left[\left(\sum_{i=1}^{A^{n_i M}} \mathbf{X}_i(n)\mathbf{X}_i^H(n)\gamma_{ni}^{(\ell)}\right)^{-1}\left(\mathbf{Y}(n)\left[\sum_{i=1}^{A^{n_i M}} \mathbf{X}_i^H(n)\gamma_{ni}^{(\ell)}\right]\right)\right] \tag{7.84}$$

The above is a convex problem and can be solved using linear programming methods, interior-point methods, and iterative thresholding [36, 76]. Note that the M–step can also be executed by Greedy algorithms.

Computation of smoothing probabilities. To implement the EM iteration, the APP's $\gamma_{ti}^{(\ell)} = p\left(\mathbf{x}(t) = s_i | \mathbf{Y}(n); \mathbf{H}^{(\ell-1)}\right)$ are needed. They correspond to the E–step of the EM algorithm, and they can be computed by soft decoders if the underlying structure of the MIMO system enables us to follow a Hidden Markov Model (HMM) formulation, where the APP's are expressed as functions of the transition probabilities. In such case the a posteriori distribution of the hidden variables is obtained using a two–stage message passing algorithm. In the context of HMM models, it is known as *forward–backward* algorithm [66], or *Baum–Welch*, or the *BCJR* algorithm [6].

A complete description by a HMM model requires a trellis diagram with state set $S = \{s_1, s_2, \ldots, s_{A^{M\times n_i}}\}$, where A is the alphabet size. The algorithm is split up into two stages. The first stage calculates the filtering probabilities $p(\mathbf{x}(t) = \cdots | \mathbf{Y}(t); \mathbf{H}^{(\ell-1)})$, while the second stage calculates the future probabilities $p(\mathbf{x}(t) = \cdots | \mathbf{Y}(t+1:n); \mathbf{H}^{(\ell-1)})$ (where $\mathbf{Y}(t+1:n) = [\mathbf{y}(t+1), \ldots, \mathbf{y}(n)]$.

Assume that the two stages are already computed for all $t \in \{1, \dots, n\}$. Then using the Markov chain rule we obtain the smoothing probabilities $p(\mathbf{x}(t) = s_i | \mathbf{Y}(n); \mathbf{H}^{(\ell-1)})$ for each $s_i \in S$

$$p(\mathbf{x}(n) = s_i | \mathbf{Y}(n); \mathbf{H}^{(\ell-1)}) = \tag{7.85}$$
$$\underbrace{p(\mathbf{x}(t) = s_i | \mathbf{Y}(t-1); \mathbf{H}^{(\ell-1)})}_{\alpha_t(\mathbf{x}(t))} \underbrace{p(\mathbf{x}(t) = s_i)}_{b_t(\mathbf{x}(t),\mathbf{x}(t+1))} \underbrace{p(\mathbf{x}(t) = s_i | \mathbf{Y}(t+1:n); \mathbf{H}^{(\ell-1)})}_{\beta_{t+1}(\mathbf{x}(t))}$$

Next forward/backward recursions are derived that allow the probabilities of Eq. (7.85) efficiently. The filtering or forward probability $\alpha_t(\mathbf{x}(t))$ is obtained by summing all the lookahead probabilities as

$$\alpha_t(\mathbf{x}(t)) = \sum_{\forall \mathbf{x}(t-1) \in S} \alpha_{t-1}(\mathbf{x}(t-1)) b_{t-1}(\mathbf{x}(t-1), \mathbf{x}(t)). \tag{7.86}$$

The derivation of the backward filtering is similar to the filtering probability

$$\beta_t(\mathbf{x}(t)) = \sum_{\forall \mathbf{x}(t+1) \in S} \beta_{t+1}(\mathbf{x}(t+1)) b_t(\mathbf{x}(t), \mathbf{x}(t+1)) \tag{7.87}$$

b is determined from the received signal and a–priori information

$$b_t(\mathbf{x}(t), \mathbf{x}(t+1)) = \exp\left\{ -\frac{1}{2\sigma^2} \|\mathbf{y}(t) - \mathbf{H}\mathbf{x}(t)\|_{\ell_2}^2 \right\} Pr(\mathbf{x}(t+1) = s | \mathbf{x}(t) = s'). \tag{7.88}$$

7.4.3 Computer Simulations of Blind Identification Algorithms

In this subsection we compare the performance of the methods outlined here under two different operating modes: Semi-Blind and blind. Performance is measured in terms of Normalized Mean Square Error (NMSE, defined in Sect. 7.3.4) and Vector Symbol Error Rate (SER, that is the probability of at least one of the transmitted symbols is in error) for a frame of 100 vector symbols from BPSK constellations averaged over 100 different system realizations. The non–zero coefficients are i.i.d. (independent, identically distributed) complex Gaussian random variables with zero mean and variance 1. The positions of the non–zero parameters are randomly selected in each realization, ensuring each output in non–zero. We consider a 2×2 linear MIMO system of memory length 4 and sparsity level 4.

We start by considering a Semi-Blind operation in which a short training sequence (consisting of five symbols) is available at the receiver side; the short training sequence is sent over the unknown system by the transmitter prior to the actual data transmission session. This training sequence, is used to initialize the algorithms.

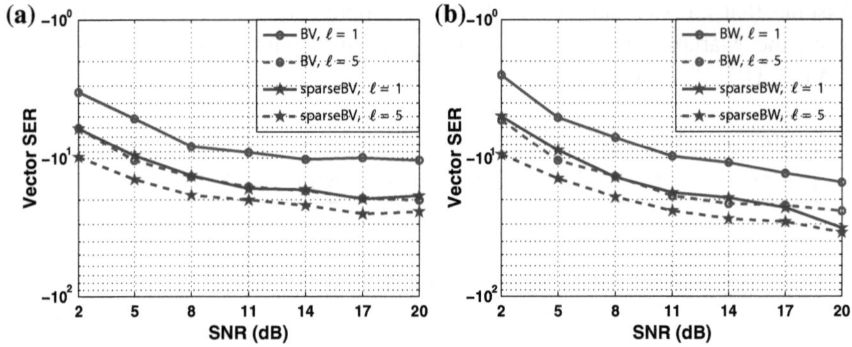

Fig. 7.12 Symbol error rate (SER) for sparse MIMO systems

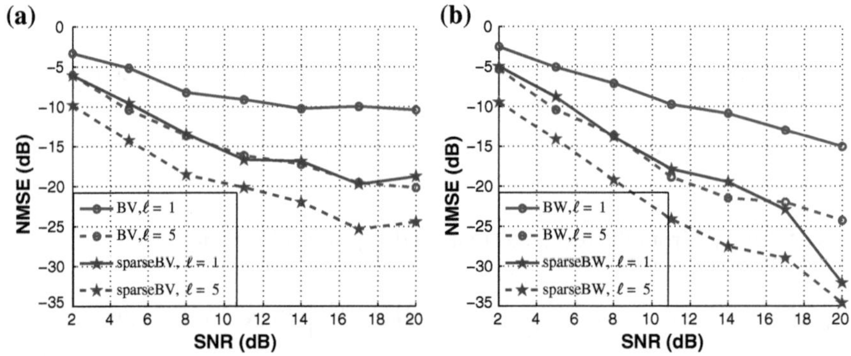

Fig. 7.13 NMSE performance comparison under different noise conditions

We note from inspection of Figs. 7.12, 7.13 that the performance of Baum–Viterbi's (sparse and non–sparse) is identical to Baum–Welch's (sparse and non–sparse) for an SNR range of $2 - 10$ dB; whereas in less noisy conditions Baum–Welch performs better. The conventional algorithms (Baum–Viterbi and Baum–Welch) la behind their sparse counterpart by approximately 5dB. We then inspect the vector SER for a maximum sequence detector Fig. 7.12a and a maximum a posteriori detector Fig. 7.12b where the sparse algorithms achieve better SER performance since they provide more accurate system estimates.

Next, we consider a blind operational mode where a key issue is how to acquire a reliable initial estimate for the parameter matrix. To avoid using different initial conditions we employ the single–spike strategy [28] in which all the parameters are set to zero except the dominant parameter which is set to ± 1, depending on its sign. By using this initialization, both algorithms converge in approximately 5 iterations. The algorithms are tested under a fixed noise condition (SNR 10dB). As it can be seen from Fig. 7.14, Baum–Viterbi fails to converge whereas Baum–Welch is more robust to initial conditions. For both algorithms (Baum–Viterbi and Baum–Welch)

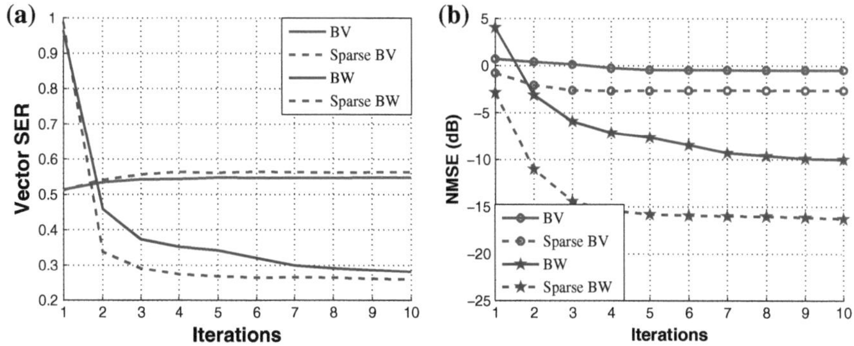

Fig. 7.14 Comparison of the two methods for fixed SNR of 10dB

the sparse versions are better than the conventional counterparts and achieve faster convergence, see Fig. 7.14b.

7.5 Summary and Future Directions

In this chapter adaptive filtering and identification for multi input multi output nonlinear polynomial systems was considered. The exponential growth of complexity was addressed by sparsity aware schemes. Sparse LMS, RLS and greedy adaptive algorithms were described and their performance was demonstrated by simulations under a wide range of operating conditions. The above methods were combined with state estimation techniques such as the Viterbi family, in a Semi-Blind context. Alternative algorithms based on expectation maximization and smoothing methods were also discussed.

A number of other methods have been developed for linear systems. These include subspace methods, second order statistics and higher order statistics [59]. Adaptation of these methods to the nonlinear case and assessment of their performance is worth to pursue.

Acknowledgments This research has been co-financed by the European Union (European Social Fund – ESF) and Greek national funds through the Operational Program "Education and Lifelong Learning" of the National Strategic Reference Framework (NSRF) – Research Funding Program: THALIS–UOA– SECURE WIRELESS NONLINEAR COMMUNICATIONS AT THE PHYSICAL LAYER.

References

1. Abuthinien M, Chen S, Hanzo L (2008) Semi-blind joint maximum likelihood channel estimation and data detection for MIMO systems. IEEE Signal Process Lett 15:202–205
2. Andrew G, Gao J (2007) Scalable training of L1 regularized log linear models. International conference on machine learning, In
3. Angelosante D, Bazerque JA, Giannakis GB (2010) Online adaptive estimation of sparse signals: Where RLS meets the ℓ_1-norm. IEEE Trans Sig Process 58(7):3436–3447
4. Angelosante D, Giannakis GB (2009) RLS-weighted lasso for adaptive estimation of sparse signals. IEEE Int Conf Acoust Speech Sig Process 2009:3245–3248
5. Babadi B, Kalouptsidis N, Tarokh V (2010) Sparls: The sparse RLS algorithm. IEEE Trans Sig Process 58(8):4013–4025
6. Bahl L, Cocke J, Jelinek F, Raviv J (1974) Optimal decoding of linear codes for minimizing symbol error rate (corresp.). IEEE Trans Inf Theory 20(2):284–287.
7. Barry JR, Lee EA, Messerschmitt DG (eds) (2003) Digital communication. Springer, New York
8. Benedetto S, Biglieri S (1998) Principles of Digital Transmission: with wireless applications. Kluwer Academic, New York
9. Benesty J, Gnsler T, Huang Y, Rupp M (2004) Adaptive algorithms for MIMO acoustic echo cancellation. In: Huang Y, Benesty J (eds) Audio Signal Processing for Next-Generation Multimedia Communication Systems. Springer, Berlin
10. Bertsekas D, Nedic A, Ozdaglar A (2003) Convex Analysis and Optimization. Athena Scientific, Cambridge
11. Biglieri E, Calderbank R, Constantinides A, Goldsmith A, Paulraj A, Poor VH (2007) MIMO Wireless Communications. Cambridge University Press, England
12. Boufounos PT, Raj B, Smaragdis P (2011) Joint sparsity models for broadband array processing. In Proc SPIE Wavelets and Sparsity 14:18–21
13. Boyd S (1985) Volterra Series: Engineering Fundamentals. PhD thesis, UC Berkeley.
14. Boyd S, Chua L (1985) Fading memory and the problem of approximating nonlinear operators with volterra series. IEEE Trans Circ Syst 32(11):1150–1161
15. Brewer J (1978) Kronecker products and matrix calculus in system theory. IEEE Trans Circ Syst 25(9):772–781
16. Bruckstein AM, Donoho DL, Elad M (2009) From sparse solutions of systems of equations to sparse modeling of signals and images. SIAM Review 51(1):34–81
17. Cands E, Wakin M, Boyd S (2008) Enhancing sparsity by reweighted l1 minimization. J Fourier Anal Appl 14(8):77–905
18. Chen Y, Gu Y, Hero AO (2009) Sparse LMS for system identification. IEEE Int Conf Acoust Speech Sig Process 2:3125–3128
19. Chen Y, Gu Y, Hero AO (2010) Regularized Least-Mean-Square algorithms. Arxiv, preprint stat. ME/1012.5066v2.
20. Fu-Hsuan Chiu. (2006) Transceiver design and performance analysis of bit-interleaved coded MIMO-OFDM systems. PhD thesis, University of Southern California.
21. . Combettes PL, Pesquet JC,(2011) Proximal splitting methods in signal processing. In: Bauschke HH, Burachik RS, Combettes PL, Elser V, Luke DR, Wolkowicz H (eds) Fixed-point algorithms for inverse problems in science and engineering. Springer, New York, pp 185–212
22. Giannakis GB, Angelosante D, Grossi E, Lops M (2010) Sparsity-aware estimation of CDMA system parameters. EURASIP J Adv Sig Process 59(7):3262–3271
23. Dai W, Milenkovic O (2009) Subspace pursuit for compressive sensing signal reconstruction. IEEE Trans Inf Theory 55(5):2230–2249
24. Daubechies I, Defrise M, De Mol C (2004) An iterative thresholding algorithm for linear inverse problems with a sparsity constraint. commun Pure Appl Math 57(11):1413–1457.
25. Daubechies I, Fornasier M, Loris I (2008) Accelerated projected gradient method for linear inverse problems with sparsity constraints. J Fourier Anal Appl 14:764–792

26. Davis S, Mallat GM, Zhang Z (1994) Adaptive time-frequency decompositions. SPIE J Opt Engin 33(7):2183–2191
27. Dempster AP, Laird NM, Rubin DB (1977) Maximum likelihood from incomplete data via the EM algorithm. J R Stat Soc B 39:1–38
28. Ding Z, Li Y (2001) Blind Equalization and Identification. Marcel Dekker, New York
29. Doicu A, Trautmann T, Schreier F (2010) Numerical Regularization for Atmospheric Inverse Problems. Springer, Heidelberg
30. Donoho DL, Johnstone IM (1994) Ideal spatial adaptation by wavelet shrinkage. Biometrika Trust 81(3):425–455
31. Donoho DL, Tsaig Y, Drori I (2013) Starck JL Sparse solution of underdetermined linear equations by stagewise orthogonal matching pursuit. Submitted for publication.
32. Doyle FJIII, Pearson RK, Ogunnaike BA (2002) Identification and control using Volterra series. Springer, New York
33. Duchi J, Shwartz S.S., Singer Y, Chandra T (2008) Efficient projections onto the l1 ball for learning in high dimensions. In: Proceedings of International Conference on Machine Learning (ICML 08) pp 272–279
34. Duttweiler DL (2000) Proportionate normalized Least-Mean-Squares adaptation in echo cancellers. IEEE Trans Speech Audio Process 8(5):508–518
35. Eksioglu EM, Tanc AK (2011) RLS algorithm with convex regularization. IEEE Signal Process Lett 18(8):470–473
36. Elad M (2010) Sparse and Redundant Representations: From Theory to Applications in Signal and Image Processing. Springer, New York
37. Ephraim Y, Merhav N (2002) Hidden markov processes. IEEE Trans Inf Theory 48(6):1518–1569
38. Werner S, Riihonen T, Gregorio F (2011) Power amplifier linearization technique with iq imbalance and crosstalk compensation for broadband MIMO-OFDM transmitters. EURASIP J Adv Sig Process 19:1–15
39. Feder M (1987) Statistical signal processing using a class of iterative estimation algorithms. PhD thesis, M.I.T Cambridge MA.
40. Fernandes, CAR (2009) Nonlinear MIMO communication systems: Channel estimation and information recovery using volterra models. PhD thesis, Universite de Nice-Sophia Antipolis.
41. Figueiredo MAT, Nowak RD (2003) An EM algorithm for wavelet-based image restoration. IEEE Trans Image Process 12(8):906–916
42. Gholami A (2011) A general framework for sparsity-based denoising and inversion. IEEE Trans Sigl Process 59(11):5202–5211
43. Giannakis G, Liu Z, Ma X (2003) Space Time Coding for Broadband Wireless Communications. Wiley-Interscience, Hoboken
44. Giannakis GB, Serpedin E (1997) Linear multichannel blind equalizers of nonlinear FIR volterra channels. IEEE Trans Sig Process 45(1):67–81
45. Gregorio F, Werner S, Laakso TI, Cousseau J (2007) Receiver cancellation technique for nonlinear power amplifier distortion in SDMA-OFDM systems. IEEE Trans Veh Technol 56(5):2499–2516
46. Haykin SO (2001) Adaptive filter theory, 4th edn. Springer, New York
47. Huang Y, Benesty J, Chen J (2006) Acoustic MIMO signal processing. Springer, Berlin
48. Hurley N, Rickard S (2009) Comparing measures of sparsity. IEEE Trans Inf Theory 55(10):4723–4741
49. Jin J, Gu Y, Mei S (2010) A stochastic gradient approach on compressive sensing signal reconstruction based on adaptive filtering framework. IEEE J Sel Topics Sig Process 4(2):409–420
50. Kaleh GK, Vallet R (1994) Joint parameter estimation and symbol detection for linear or nonlinear unknown channels. IEEE Trans Comm 42(7):2406–2413
51. Kalouptsidis N (1997) Signal Processing Systems Theory and Design. John Wiley and Sons, New York

52. Nicholas Kalouptsidis, Gerasimos Mileounis, Behtash Babadi, Vahid Tarokh (2011) Adaptive algorithms for sparse system identification. Sig Process 91(8):1910–1919
53. Koike-Akino T, Molisch AF, Pun MO, Annavajjala R, Orlik P (2011) Order-extended sparse RLS algorithm for doubly-selective MIMO channel estimation. In: IEEE international conference on communications (ICC), 1–6 June 2011.
54. Kopsinis Y, Slavakis K, Theodoridis S (2011) Online sparse system identification and signal reconstruction using projections onto weighted ℓ_1 balls. IEEE Trans Sig Process 59(3):936–952
55. Ljung L (1993) General structure of adaptive algorithms: Adaptation and tracking. Kalouptsidis N, Theodoridis S (eds) In adaptive system identification and signal processing algorithms
56. Maleh R, Gilbert AC (2007) Multichannel image estimation via simultaneous orthogonal matching pursuit. Proceedings workshop statistics signal process, In
57. Marmarelis PZ, Marmarelis VZ (1978) Analysis of physiological systems. Plenum Press, New York
58. Mileounis G, Babadi B, Kalouptsidis N, Tarokh V (2010) An adaptive greedy algorithm with application to nonlinear communications. IEEE Trans Sig Process 58(6):2998–3007
59. Mileounis G, Kalouptsidis N (2012) A sparsity driven approach to cumulant based identification. In IEEE International Workshop on Signal Processing Advances in Wireless Communications.
60. Mileounis G Kalouptsidis N, Babadi B, Tarokh V (2011) Blind identification of sparse channels and symbol detection via the EM algorithm. In: International conference on digital signal processing (DSP), 1–5 july 2011.
61. Murakami Y, Yamagishi M, Yukawa M, Yamada I (2010) A sparse adaptive filtering using time-varying soft-thresholding techniques. In: 2010 IEEE international conference on acoustics speech and signal processing (ICASSP), 3734–3737.
62. Needell D, Tropp JA (2009) CoSaMP: Iterative signal recovery from incomplete and inaccurate samples. Appl Comput Harmon Anal 26:301–321
63. Needell D, Vershynin R (2009) Uniform uncertainty principle and signal recovery via regularized orthogonal matching pursuit. Found Comput Math 9(3):317–334
64. Paleologu C, Benesty J, Ciochina S (2010) Sparse adaptive filters for echo cancellation. Morgan and Claypool Publishers, San Rafael
65. Pati YC, Rezaiifar R, Krishnaprasad PS (1993) Orthogonal matching pursuit: recursive function approximation with applications to wavelet decomposition. In: 27th asilomar Conference on signals, systems and Computation 40–44.
66. Rabiner LR (1989) A tutorial on hidden markov models and selected applications in speech recognition. IEEE Proc 77(2):257–286
67. Redfern AJ, Zhou GT (2001) Blind zero forcing equalization of multichannel nonlinear CDMA systems. IEEE Trans Sig Process 49(10):2363–2371
68. Rizogiannis C, Kofidis E, Papadias CB, Theodoridis S (2010) Semi-blind maximum-likelihood joint channel/data estimation for correlated channels in multiuser MIMO networks. Sig Process 90(4):1209–1224
69. Rugh WJ, (1981) Nonlinear System Theory. The Johns Hopkins University Press, Baltimore (in press).
70. Adaptive Filters. Wiley-Blackwell, Hoboken.
71. Schetzen M (1980) The volterra and wiener theories of nonlinear systems. Willey and Sons, New York
72. Seretis C (1997) Control-relevant identification for constrained and nonlinear systems. PhD thesis, University of Maryland.
73. Tidestav C, Lindskog E (1998) Bootstrap equalization. In: IEEE International Conference on Universal Personal Communications (ICUPC), vol 2. 1221–1225 oct 1998.
74. Tong L, Rw Liu, Soon VC, Huang YF (1991) Indeterminacy and identifiability of blind identification. IEEE Trans Circ Syst 38(5):499–509
75. Tropp JA, Gilbert AC, Strauss MJ (2006) Algorithms for simultaneous sparse approximation. part i: Greedy pursuit. Sig Process 86(3):572–588.

76. Tropp JA, Wright SJ (2010) Computational methods for sparse solution of linear inverse problems. Proceedings of the IEEE 98(6):948–958
77. Sicuranza GL, Mathews VJ (2000) Polynomial Signal Processing. Wiley-Blackwell, New York
78. Wang Y, Yagola AG, Yang C (2010) Optimization and regularization for computational inverse problems and applications. Springer, Heidelberg
79. Weston J, Elisseeff A, Schölkopf B, Tipping M (2003) Use of the zero norm with linear models and kernel methods. J Mach Learn Res 3:1439–1461
80. Westwick DT, Kearney RE (2003) Identification of nonlinear physiological systems. IEEE Press, New York
81. Wu C (1983) On the convergence properties of the EM algorithm. Ann Statist 11:95–103
82. Yukawa M (2010) Adaptive filtering based on projection method. Block Seminar in Elite Master Study Course SIM.
83. Zheng Q (1995) A volterra series approach to nonlinear process control and control relevant identification. PhD thesis, University of Maryland.
84. Zheng Q, Zafiriou E (2004) Volterra Laguerre models for nonlinear process identification with application to a fluid catalytic cracking unit. Ind Eng Chem Res 43(2):340–348

76. Tropp JA, Wright SJ (2010) Computational methods for sparse solution of linear inverse problems. Proceedings of the IEEE 98(6):948–958
[illegible]

Chapter 8
Optimization Viewpoint on Kalman Smoothing with Applications to Robust and Sparse Estimation

Aleksandr Y. Aravkin, James V. Burke and Gianluigi Pillonetto

Abstract In this chapter, we present the optimization formulation of the Kalman filtering and smoothing problems, and use this perspective to develop a variety of extensions and applications. We first formulate classic Kalman smoothing as a least squares problem, highlight special structure, and show that the classic filtering and smoothing algorithms are equivalent to a particular algorithm for solving this problem. Once this equivalence is established, we present extensions of Kalman smoothing to systems with nonlinear process and measurement models, systems with linear and nonlinear inequality constraints, systems with outliers in the measurements or sudden changes in the state, and systems where the sparsity of the state sequence must be accounted for. All extensions preserve the computational efficiency of the classic algorithms, and most of the extensions are illustrated with numerical examples, which are part of an open source Kalman smoothing Matlab/Octave package.

8.1 Introduction

Kalman filtering and smoothing methods form a broad category of computational algorithms used for inference on noisy dynamical systems. Over the last 50 years, these algorithms have become a gold standard in a range of applications, including

A. Y. Aravkin (✉)
Numerical Analysis and Optimization,
IBM T.J. Watson Research Center, Vancouver, BC, Canada
e-mail: saravkin@eos.ubc.ca

J. V. Burke
Department of Mathematics, University of Washington, Seattle, WA, USA
e-mail: jvburke@uw.edu

G. Pillonetto
Control and Dynamic Systems Department of Information Engineering,
University of Padova, Padova, Italy
e-mail: giapi@dei.unipd.it

A. Y. Carmi et al. (eds.), *Compressed Sensing & Sparse Filtering*,
Signals and Communication Technology, DOI: 10.1007/978-3-642-38398-4_8,

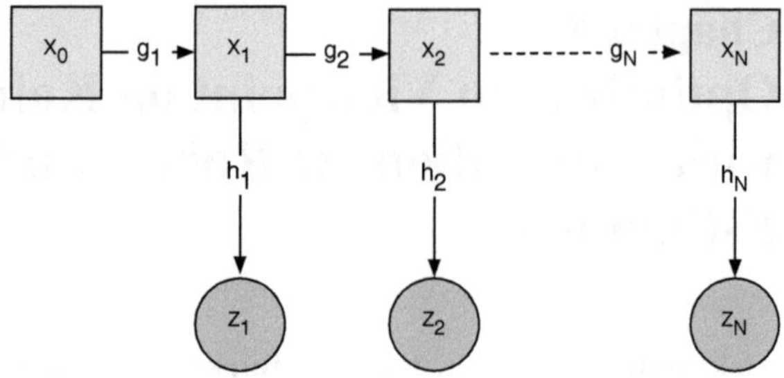

Fig. 8.1 Dynamic systems amenable to Kalman smoothing methods

space exploration, missile guidance systems, general tracking and navigation, and weather prediction. In 2009, Rudolf Kalman received the National Medal of Science from President Obama for the invention of the Kalman filter. Numerous books and papers have been written on these methods and their extensions, addressing modifications for use in nonlinear systems, smoothing data over time intervals, improving algorithm robustness to bad measurements, and many other topics.

The classic Kalman filter [29] is almost always presented as a set of recursive equations, and the classic Rauch-Tung-Striebel (RTS) fixed-interval smoother [42] is typically formulated as two coupled Kalman filters. An elegant derivation based on projections onto spaces spanned by random variables can be found in [2]. In this chapter, we use the terms 'Kalman filter' and 'Kalman smoother' much more broadly, **including any method of inference on any dynamical system fitting the graphical representation of Fig.** 8.1. Specific mathematical extensions we consider include

- Nonlinear process and measurement models.
- Inequality state space constraints.
- Different statistical models for process and measurement errors.
- Sparsity constraints.

We also show numerous applications of these extensions.

The key to designing tractable inference methods for the above applications is an optimization viewpoint, which we develop in the classic Kalman smoothing case and then use to formulate and solve *all* of the above extensions. Though it has been known for many years that the Kalman filter provides the maximum *a posteriori* estimate for linear systems subject to Gaussian noise, the optimization perspective underlying this idea has not been fully deployed across engineering applications. Notably, several groups (starting in 1977) have discovered and used variants of this perspective to implement extensions to Kalman filtering and smoothing, including singular filtering [33, 39, 40], robust smoothing [7, 22], nonlinear smoothing with inequality state space constraints [9, 11], and sparse Kalman smoothing [1].

We focus exclusively on smoothing here, leaving online applications of these ideas to future work (see [41] for an example of using a smoother for an online application). We start by presenting the classic RTS smoothing algorithm in Sect. 8.2, and show that the well-known recursive equations are really an algorithm to solve a least squares system with special structure. Once this is clear, it becomes much easier to discuss novel extensions, since as long as special structure is preserved, their computational cost is on par with the classic smoother (or, put another way, the classic smoothing equations are viewed as a particular way to solve key subproblems in the extended approaches).

In the subsequent sections, we build novel extensions, briefly review theory, discuss the special structure, and present numerical examples for a variety of applications. In Sect. 8.3, we formulate the problem for smoothing with nonlinear process and measurement models, and show how to solve it. In Sect. 8.4, we show how state space constraints can be incorporated, and the resulting problem solved using interior point techniques. In Sect. 8.5, we review two recent Kalman smoothing formulations that are highly robust to measurement errors. Finally, in Sect. 8.6, we review recent work in sparse Kalman smoothing, and show how sparsity can be incorporated into the other extensions. We end the chapter with discussion in Sect. 8.7.

8.2 Optimization Formulation and RTS Smoother

8.2.1 Probabilistic Model

The model corresponding to Fig. 8.1 is specified as follows:

$$\begin{aligned} \mathbf{x_1} &= g_1(x_0) + \mathbf{w_1}, \\ \mathbf{x_k} &= g_k(\mathbf{x_{k-1}}) + \mathbf{w_k} \quad k = 2, \ldots, N, \\ \mathbf{z_k} &= h_k(\mathbf{x_k}) + \mathbf{v_k} \quad k = 1, \ldots, N, \end{aligned} \tag{8.1}$$

where $\mathbf{w_k}$, $\mathbf{v_k}$ are mutually independent random variables with known positive definite covariance matrices Q_k and R_k, respectively. We have $\mathbf{x_k}, \mathbf{w_k} \in \mathbb{R}^n$, and $\mathbf{z_k}, \mathbf{v_k} \in \mathbb{R}^{m(k)}$, so measurement dimensions can vary between time points. The classic case is obtained by making the following assumptions:

1. x_0 is known, and g_k, h_k are known *linear* functions, which we denote by
$$g_k(x_{k-1}) = G_k x_{k-1} \quad h_k(x_k) \quad = H_k x_k \tag{8.2}$$
where $G_k \in \mathbb{R}^{n \times n}$ and $H_k \in \mathbb{R}^{m(k) \times n}$,
2. $\mathbf{w_k}$, $\mathbf{v_k}$ are mutually independent *Gaussian* random variables.

In later sections, we will show how to relax these classic assumptions, and what gains can be achieved once they are relaxed. In this section, we will formulate estimation of the *entire* state sequence, $x_1, x_2, \ldots, x_N$, as an optimization problem, and show how the RTS smoother solves it.

8.2.2 Maximum a Posteriori Formulation

To begin, we formulate the maximum *a posteriori* (MAP) problem under linear and Gaussian assumptions. Using Bayes' theorem, we have

$$
\begin{aligned}
P\left(\{x_k\}\middle|\{z_k\}\right) &\propto P\left(\{z_k\}\middle|\{x_k\}\right) P\left(\{x_k\}\right) \\
&= \prod_{k=1}^{N} P(\{v_k\})P(\{w_k\}) \\
&\propto \prod_{k=1}^{N} \exp\Big(-\frac{1}{2}(z_k - H_k x_k)^\top R_k^{-1}(z_k - H_k x_k) \\
&\quad - \frac{1}{2}(x_k - G_k x_{k-1})^\top Q_k^{-1}(x_k - G_k x_{k-1})\Big).
\end{aligned} \tag{8.3}
$$

A better (equivalent) formulation to (8.3) is minimizing its negative log posterior:

$$
\min_{\{x_k\}} f(\{x_k\}) := \sum_{k=1}^{N} \frac{1}{2}(z_k - H_k x_k)^\top R_k^{-1}(z_k - H_k x_k) + \frac{1}{2}(x_k - G_k x_{k-1})^\top Q_k^{-1}(x_k - G_k x_{k-1}). \tag{8.4}
$$

To simplify the problem, we now introduce data structures that capture the entire state sequence, measurement sequence, covariance matrices, and initial conditions.

Given a sequence of column vectors $\{u_k\}$ and matrices $\{T_k\}$ we use the notation

$$
\operatorname{vec}(\{u_k\}) = \begin{bmatrix} u_1 \\ u_2 \\ \vdots \\ u_N \end{bmatrix}, \quad \operatorname{diag}(\{T_k\}) = \begin{bmatrix} T_1 & 0 & \cdots & 0 \\ 0 & T_2 & \ddots & \vdots \\ \vdots & \ddots & \ddots & 0 \\ 0 & \cdots & 0 & T_N \end{bmatrix}.
$$

We now make the following definitions:

$$
\begin{aligned}
R &= \operatorname{diag}(\{R_k\}) & x &= \operatorname{vec}(\{x_k\}) \\
Q &= \operatorname{diag}(\{Q_k\}) & w &= \operatorname{vec}(\{g_0, 0, \ldots, 0\}) \\
H &= \operatorname{diag}(\{H_k\}) & z &= \operatorname{vec}(\{z_1, z_2, \ldots, z_N\})
\end{aligned}
\qquad
G = \begin{bmatrix} \mathrm{I} & 0 & & \\ -G_2 & \mathrm{I} & \ddots & \\ & \ddots & \ddots & 0 \\ & & -G_N & \mathrm{I} \end{bmatrix}, \tag{8.5}
$$

where $g_0 := g_1(x_0) = G_1 x_0$.

With definitions in (8.5), problem (8.4) can be written

$$\min_x f(x) = \frac{1}{2}\|Hx - z\|_{R^{-1}}^2 + \frac{1}{2}\|Gx - w\|_{Q^{-1}}^2, \tag{8.6}$$

where $\|a\|_M^2 = a^\top Ma$. We knew the MAP was a least squares problem already, but now the structure is fully transparent. In fact, we can write down the closed form solution by taking the gradient of (8.6) and setting it equal to 0:

$$\begin{aligned} 0 &= H^\top R^{-1}(Hx - z) + G^\top Q^{-1}(Gx - w) \\ &= (H^\top R^{-1}H + G^\top Q^{-1}G)x - H^\top R^{-1}z - G^\top Q^{-1}w. \end{aligned}$$

The smoothing estimate is therefore given by solving the linear system

$$(H^\top R^{-1}H + G^\top Q^{-1}G)x = H^\top R^{-1}z + G^\top Q^{-1}w. \tag{8.7}$$

8.2.3 Special Subproblem Structure

The linear system in (8.7) has a very special structure: it is a symmetric positive definite block tridiagonal matrix. This can be immediately observed from the fact that both G and Q are positive definite. To be specific, it is given by

$$C = (H^\top R^{-1}H + G^\top Q^{-1}G) = \begin{bmatrix} C_1 & A_2^{\mathrm{T}} & 0 & \\ A_2 & C_2 & A_3^{\mathrm{T}} & 0 \\ 0 & \ddots & \ddots & \ddots \\ & 0 & A_N & C_N \end{bmatrix}, \tag{8.8}$$

with $A_k \in \mathbb{R}^{n\times n}$ and $C_k \in \mathbb{R}^{n\times n}$ defined as follows:

$$\begin{aligned} A_k &= -Q_k^{-1}G_k, \\ C_k &= Q_k^{-1} + G_{k+1}^{\mathrm{T}}Q_{k+1}^{-1}G_{k+1} + H_k^\top R_k^{-1}H_k. \end{aligned} \tag{8.9}$$

The special structure of the matrix C in (8.8) can be exploited to solve the linear system equivalent to the Kalman smoother. While a structure-agnostic matrix inversion scheme has complexity $O(n^3N^3)$, exploiting the block tridiagonal structure reduces this complexity to $O(n^3N)$.

A straightforward algorithm for solving any symmetric positive definite block tridiagonal linear system is given in [10]. We review it here, since it is essential to build the connection to the standard viewpoint of the RTS smoother.

8.2.4 Block Tridiagonal (BT) Algorithm

Suppose for $k = 1, \ldots, N$, $c_k \in \mathbf{R}^{n\times n}$, $e_k \in \mathbf{R}^{n\times \ell}$, $r_k \in \mathbf{B}^{n\times \ell}$, and for $k = 2, \ldots, N$, $a_k \in \mathbf{R}^{n\times n}$. We define the corresponding block tridiagonal system of equations

$$\begin{pmatrix} c_1 & a_2^{\mathrm{T}} & 0 & \cdots & 0 \\ a_2 & c_2 & & & \vdots \\ \vdots & & \ddots & & 0 \\ 0 & & a_{N-1} & c_{N-1} & a_N^{\mathrm{T}} \\ 0 & \cdots & 0 & a_N & c_N \end{pmatrix} \begin{pmatrix} e_1 \\ e_2 \\ \vdots \\ e_{N-1} \\ e_N \end{pmatrix} = \begin{pmatrix} r_1 \\ r_2 \\ \vdots \\ r_{N-1} \\ r_N \end{pmatrix} \tag{8.10}$$

The following algorithm for (8.10) is given in [10, Algorithm 4].

Algorithm 1 *The inputs to this algorithm are* $\{a_k\}$, $\{c_k\}$, *and* $\{r_k\}$. *The output is a sequence* $\{e_k\}$ *that solves Eq.* (8.10).

1. Set $d_1 = c_1$ and $s_1 = r_1$.
2. For $k = 2, \ldots, N$, set $d_k = c_k - a_k^{\mathrm{T}} d_{k-1}^{-1} a_k$, $\;s_k = r_k - a_k^{\mathrm{T}} d_{k-1}^{-1} s_{k-1}$.
3. Set $e_N = d_N^{-1} s_N$.
4. For $k = N-1, \ldots, 1$, set $e_k = d_k^{-1}(s_k - a_{k+1} e_{k+1})$.

Note that after the first two steps of Algorithm 1, we have arrived at a linear system equivalent to (8.10) but upper triangular:

$$\begin{pmatrix} d_1 & a_2^{\mathrm{T}} & 0 & \cdots & 0 \\ 0 & d_2 & & & \vdots \\ \vdots & & \ddots & & 0 \\ 0 & & 0 & d_{N-1} & a_N^{\mathrm{T}} \\ 0 & \cdots & 0 & 0 & d_N \end{pmatrix} \begin{pmatrix} e_1 \\ e_2 \\ \vdots \\ e_{N-1} \\ e_N \end{pmatrix} = \begin{pmatrix} s_1 \\ s_2 \\ \vdots \\ s_{N-1} \\ s_N \end{pmatrix} \tag{8.11}$$

The last two steps of the algorithm then simply back-solve for the e_k.

8.2.5 Equivalence of Algorithm (1) to Kalman Filter and RTS Smoother

Looking at the very first block, we now substitute in the Kalman data structures (8.9) into step 2 of Algorithm 1:

$$\begin{aligned} d_2 &= c_2 - a_2^{\mathrm{T}} d_1^{-1} a_2 \\ &= \underbrace{\underbrace{Q_2^{-1} - \left(Q_2^{-1}G_2\right)^{\mathrm{T}} \Bigg(\underbrace{Q_1^{-1} + H_1^{\mathrm{T}} R_1^{-1} H_1}_{P_{1|1}^{-1}} + G_2^{\mathrm{T}} Q_2^{-1} G_2 \Bigg)^{-1} \left(Q_2^{-1}G_2\right)}_{P_{2|1}^{-1}} + H_2^{\mathrm{T}} R_2^{-1} H_2}_{P_{2|2}^{-1}} + G_3^{\mathrm{T}} Q_3^{-1} G_3 \end{aligned} \tag{8.12}$$

These relationships can be seen quickly from [5, Theorem 2.2.7]. The matrices $P_{k|k}$, $P_{k|k-1}$ are common to the Kalman filter framework: they represent covariances of the state at time k given the the measurements $\{z_1, \ldots, z_k\}$, and the covariance of the a priori state estimate at time k given measurements $\{z_1, \ldots, z_{k-1}\}$, respectively.

From the above computation, we see that

$$d_2 = P_{2|2}^{-1} + G_3^\top Q_3^{-1} G_3.$$

By induction, it is easy to see that in fact

$$d_k = P_{k|k}^{-1} + G_{k+1}^\top Q_{k+1}^{-1} G_{k+1}.$$

We can play the same game with s_k. Keeping in mind that $r = H^\top R^{-1} z + G^\top Q^{-1} w$, we have

$$\begin{aligned} s_2 &= r_2 - a_2^\top d_1^{-1} r_1 \\ &= \underbrace{H_2^\top R_2^{-1} z_2 + \underbrace{\left(Q_2^{-1} G_2\right)^\top \left(\underbrace{Q_1^{-1} + H_1^\top R_1^{-1} H_1}_{P_{1|1}^{-1}} + G_2^\top Q_2^{-1} G_2 \right)^{-1} \left(H_1^\top R_1^{-1} z_1 + G_1^\top P_{0|0}^{-1} x_0\right)}_{a_{2|1}}}_{a_{2|2}} \end{aligned} \tag{8.13}$$

These relationships also follow from [5, Theorem 2.2.7]. The quantities $a_{2|1}$ and $a_{2|2}$ are from the information filtering literature, and are less commonly known: they are preconditioned estimates

$$a_{k|k} = P_{k|k}^{-1} x_k, \quad a_{k|k-1} = P_{k|k-1}^{-1} x_{k|k-1}. \tag{8.14}$$

Again, by induction we have precisely that $s_k = a_{k|k}$.

When you put all of this together, you see that step 3 of Algorithm 1 is given by

$$e_N = d_N^{-1} s_N = \left(P_{N|N}^{-1} + 0\right)^{-1} P_{N|N}^{-1} x_{k|k} = x_{k|k}, \tag{8.15}$$

so in fact e_N is the Kalman filter estimate (and the RTS smoother estimate) for time point N.

Step 4 of Algorithm 1 then implements the backward Kalman filter, computing the smoothed estimates $x_{k|N}$ by back-substitution. **Therefore the RTS smoother is Algorithm 1 applied to** (8.7).

The consequences are profound—instead of working with the kinds expressions seen in (8.13) and (8.12), we can think at a high level, focusing on (8.6), and simply using Algorithm 1 (or variants) as a subroutine. As will become apparent, the key to all extensions is preserving the block tridiagonal structure in the subproblems, so that Algorithm 1 can be used.

8.2.6 Numerical Example: Tracking a Smooth Signal

In this example, we focus on a very useful and simple model: the process model for a *smooth* signal. Smooth signals arise in a range of applications: physics-based models, biological data, and financial data all have some inherent smoothness.

A surprisingly versatile technique for modeling *any* such process is to treat it as integrated Brownian motion. We illustrate on a scalar time series x. We introduce a new derivative state $\dot{x}$, with process model $\dot{x}_{k+1} = \dot{x}_k + \dot{w}_k$, and then model the signal x or interest as $x_{k+1} = x_k + \dot{x}_k \Delta t + w_k$. Thus we obtain an augmented (2D) state with process model

$$\begin{bmatrix} \dot{x}_{k+1} \\ x_{k+1} \end{bmatrix} = \begin{bmatrix} I & 0 \\ \Delta t & I \end{bmatrix} \begin{bmatrix} \dot{x}_k \\ x_k \end{bmatrix} + \begin{bmatrix} \dot{w}_k \\ w_k \end{bmatrix}. \tag{8.16}$$

Using a well-known connection to stochastic differential equations (see [11, 26, 38]) we use covariance matrix

$$Q_k = \sigma^2 \begin{bmatrix} \Delta t & \Delta t^2/2 \\ \Delta t^2/2 & \Delta t^3/3 \end{bmatrix}. \tag{8.17}$$

Model equations (8.16) and (8.17) can be applied as a process model for any smooth process. For our numerical example, we take direct measurements of the sin function, which is very smooth. Our measurement model therefore is

$$z_k = H_k x_k + v_k, \quad H_k = \begin{bmatrix} 0 & 1 \end{bmatrix}. \tag{8.18}$$

The resulting fit is shown in Fig. 8.2. The measurements guide the estimate to the true smooth time series, giving very nice results. The figure was generated using the `ckbs` package [6], specifically using the example file `affine_ok.m`. Measurement errors were generated using $R_k = 0.35^2$, and this value was given to the smoother. The σ^2 in (8.17) was taken to be 1. The program and example are available for download from COIN-OR.

8.3 Nonlinear Process and Measurement Models

In the previous section, we have shown that when g_k and h_k in model (8.1) are linear, and $\mathbf{v_k}$, $\mathbf{w_k}$ are Gaussian, then the smoother is equivalent to solving a least squares problem (8.6). We have also shown that the filter estimates appear as intermediate results when one uses Algorithm 1 to solve the problem.

In this section, we turn to the case where g_k and h_k are nonlinear. We first formulate the smoothing problem as a maximum *a posteriori* (MAP) problem, and show that it is a nonlinear least squares (NLLS) problem. To set up for later sections, we also introduce the broader class of *convex composite* problems.

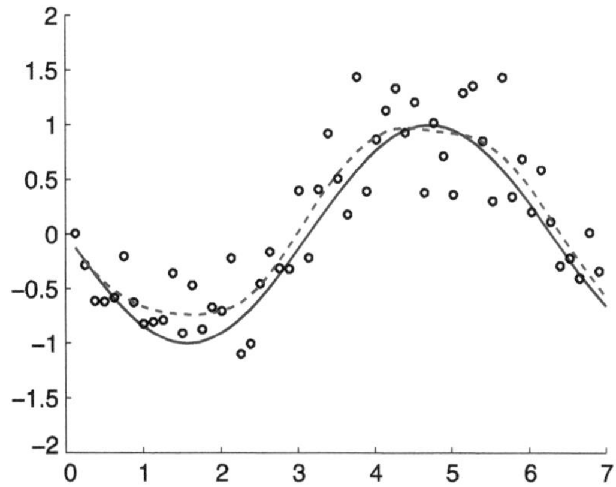

Fig. 8.2 Tracking a smooth signal (sine wave) using a generic linear process model (8.16) and direct (noisy) measurements (8.18). *Red solid line* is true signal, *blue dashed line* is Kalman (RTS) smoother estimate. Measurements are displayed as *circles*

We then review the standard Gauss-Newton method in the broader context of convex composite models, and show that when applied to the NLLS problem, each iteration is equivalent to solving (8.6), and therefore to a full execution of the RTS smoother. We also show how to use a simple line search to guarantee convergence of the method to a local optimum of the MAP problem.

This powerful approach, known for at least 20 years [9, 12, 21], is rarely used in practice; instead practitioners favor the EKF or the UKF [18, 28], neither of which converge to a (local) MAP solution. MAP approaches work very well for a broad range of applications, and it is not clear why one would throw away an efficient MAP solver in favor of another scheme. To our knowledge, the optimization (MAP) approach has never been included in a performance comparison of 'cutting edge' methods, such as [34]. While such a comparison is not in the scope of this work, we lay the foundation by providing a straightforward exposition of the optimization approach and a reproducible numerical illustration (with publicly available code) for smoothing the Van Der Pol oscillator, a well known problem where the process model is a nonlinear ODE.

8.3.1 Nonlinear Smoother Formulation and Structure

In order to develop a notation analogous to (8.6), we define functions $g : \mathbb{R}^{nN} \to \mathbb{R}^{n(N+1)}$ and $h : \mathbb{R}^{nN} \to \mathbb{R}^{M}$, with $M = \sum_k m_k$, from components g_k and h_k as follows.

$$g(x) = \begin{bmatrix} x_1 \\ x_2 - g_2(x_1) \\ \vdots \\ x_N - g_N(x_{N-1}) \end{bmatrix}, \quad h(x) = \begin{bmatrix} h_1(x_1) \\ h_2(x_2) \\ \vdots \\ h_N(x_N) \end{bmatrix}. \tag{8.19}$$

With this notation, the MAP problem, obtained exactly as in Sect. 8.2.2, is given by

$$\min_x f(x) = \frac{1}{2}\|g(x) - w\|^2_{Q^{-1}} + \frac{1}{2}\|h(x) - z\|^2_{R^{-1}}, \tag{8.20}$$

where z and w are exactly as in (8.5), so that z is the *entire* vector of measurements, and w contains the initial estimate $g_1(x_0)$ in the first n entries, and zeros in the remaining $n(N-1)$entries.

We have formulated the nonlinear smoothing problem as a nonlinear least-squares (NLLS) problem—compare (8.20) with (8.6). We take this opportunity to note that NLLS problems are a special example of a more general structure. Objective (8.20) may be written as a composition of a convex function ρ with a smooth function F:

$$f(x) = \rho(F(x)), \tag{8.21}$$

where

$$\rho\begin{pmatrix} y_1 \\ y_2 \end{pmatrix} = \frac{1}{2}\|y_1\|^2_{Q^{-1}} + \frac{1}{2}\|y_2\|^2_{R^{-1}}, \quad F(x) = \begin{bmatrix} g(x) - w \\ h(x) - z \end{bmatrix}. \tag{8.22}$$

As we show in the next sub-section, problems of general form (8.21) can be solved using the Gauss-Newton method, which is typically associated specifically with NLLS problems. Presenting the Gauss-Newton right away in the more general setting will make it easier to understand extensions in the following sections of the chapter.

8.3.2 Gauss-Newton Method for Convex Composite Models

The Gauss-Newton method can be used to solve problems of the form (8.21), and it uses a very simple strategy: **iteratively linearizing the smooth function F** [15]. More specifically, the Gauss-Newton method is an iterative method of the form

$$x^{\nu+1} = x^\nu + \gamma^\nu d^\nu, \tag{8.23}$$

where d^ν is the Gauss-Newton search direction, and γ^ν is a scalar that guarantees

$$f(x^{\nu+1}) < f(x^\nu). \tag{8.24}$$

The direction d^ν is obtained by solving the subproblem

$$d^\nu = \arg\min_d \tilde{f}(d) := \rho\left(F(x^\nu) + \nabla F(x^\nu)^\top d\right). \tag{8.25}$$

We then set

$$\tilde{\Delta} f(x^\nu) = \tilde{f}(d^\nu) - f(x^\nu).$$

By [15, Lemma 2.3, Theorem 3.6],

$$f'(x^\nu; d^\nu) \leq \tilde{\Delta} f(x^\nu) \leq 0, \tag{8.26}$$

with equality if and only if x^ν is a first-order stationary point for f. This implies that a suitable stopping criteria for the algorithm is the condition $\Delta f(x^\nu) \sim 0$. Moreover, x^ν is not a first-order stationary point for f, then the direction d^ν is a direction of strict descent for f at x^ν.

Once the direction d^ν is obtained with $\tilde{\Delta} f(x^\nu) < 0$, a step-size γ^ν is obtained by a standard backtracking line-search procedure: pick a values $0 < \lambda < 1$ and $0 < \kappa < 1$ (e.g., $\lambda = 0.5$ and $\kappa = 0.001$) and evaluate $f(x^\nu + \lambda^s d^\nu)$, $s = 0, 1, 2, \ldots$, until

$$f(x^\nu + \lambda^s d^\nu) \leq f(x^\nu) + \kappa \lambda^s \tilde{\Delta} f(x^\nu) \tag{8.27}$$

is satisfied for some $\bar{s}$, then set $\gamma^\nu = \lambda^{\bar{s}}$ and make the GN update (8.23). The fact that there is a finite value of s for which (8.27) is satisfied follows from inequality $f'(x^\nu; d^\nu) \leq \tilde{\Delta} f(x^\nu) < 0$. The inequality (8.27) is called the Armijo inequality. A general convergence theory for this algorithm as well as a wide range of others is found in [15]. For the NLLS case, the situation is simple, since ρ is a quadratic, and standard convergence theory is given for example in [27]. However, the more general theory is essential in the later sections.

8.3.3 Details for Kalman Smoothing

To implement the Gauss-Newton method described above, one must compute the solution d^ν to the Gauss-Newton subproblem (8.25) for (8.20). That is, one must compute

$$d^\nu = \arg\min_d \tilde{f}(d) = \frac{1}{2}\|G^\nu d - \underbrace{w - g(x^\nu)}_{w^\nu}\|^2_{Q^{-1}} + \frac{1}{2}\|H^\nu d - \underbrace{z - h(x^\nu)}_{z^\nu}\|^2_{R^{-1}}, \tag{8.28}$$

where

$$G^\nu = \begin{bmatrix} \mathrm{I} & 0 & & \\ -g_2^{(1)}(x_1^\nu) & \mathrm{I} & \ddots & \\ & \ddots & \ddots & 0 \\ & & -g_N^{(1)}(x_{N-1}^\nu) & \mathrm{I} \end{bmatrix}, \quad H^\nu = \mathrm{diag}\{h_1^{(1)}(x_1), \ldots, h_N^{(1)}(x_N)\}. \tag{8.29}$$

However, the problem (8.28) has exactly the same structure as (8.6); a fact that we have emphasized by defining

$$w^{\nu} := w - g(x^{\nu}), \quad z^{\nu} = z - h(x^{\nu}). \tag{8.30}$$

Therefore, we can solve it efficiently by using Algorithm 1.

The linearization step in (8.28) should remind the reader of the EKF. Note, however, that the Gauss-Newton method is iterative, and we iterate until convergence to a local minimum of (8.20). We also linearize along the entire state space sequence x^{ν} at once in (8.28), rather than re-linearizing as we make our way through the x_k^{ν}'s.

8.3.4 Numerical Example: Van Der Pol Oscillator

The Van der Pol oscillator is a popular nonlinear process for comparing Kalman filters, see [24] and [30, Sect. 4.1]. The oscillator is governed by a nonlinear ODE model

$$\dot{X}_1(t) = X_2(t) \text{ and } \dot{X}_2(t) = \mu[1 - X_1(t)^2]X_2(t) - X_1(t). \tag{8.31}$$

In contrast to the linear model (8.16), which was a *generic* process for a smooth signal, we now take the Euler discretization of (8.31) to be the specific process model for this situation.

Given $X(t_{k-1}) = x_{k-1}$ the Euler approximation for $X(t_{k-1} + \Delta t)$ is

$$g_k(x_{k-1}) = \begin{pmatrix} x_{1,k-1} + x_{2,k-1}\Delta t \\ x_{2,k-1} + \{\mu[1 - x_{1,k}^2]x_{2,k} - x_{1,k}\}\Delta t \end{pmatrix}. \tag{8.32}$$

For the simulation, the 'ground truth' is obtained from a stochastic Euler approximation of the Van der Pol oscillator. To be specific, with $\mu = 2$, $N = 80$ and $\Delta t = 30/N$, the ground truth state vector x_k at time $t_k = k\Delta t$ is given by $x_0 = (0, -0.5)^{\mathrm{T}}$ and for $k = 1, \ldots, N$,

$$x_k = g_k(x_{k-1}) + w_k, \tag{8.33}$$

where $\{w_k\}$ is a realization of independent Gaussian noise with variance 0.01 and g_k is given in (8.32). Our process model for state transitions is also (8.33), with $Q_k = 0.01\, I$ for $k > 1$, and so is identical to the model used to simulate the ground truth $\{x_k\}$. Thus, we have precise knowledge of the process that generated the ground truth $\{x_k\}$. The initial state x_0 is imprecisely specified by setting $g_1(x_0) = (0.1, -0.4)^{\mathrm{T}} \neq x_0$ with corresponding variance $Q_1 = 0.1\, I$. For $k = 1, \ldots, N$ noisy measurements z_k direct measurements of *the first component only* were used

$$z_k = x_{1,k} + v_k, \tag{8.34}$$

with $v_k \sim N(0, 1)$.

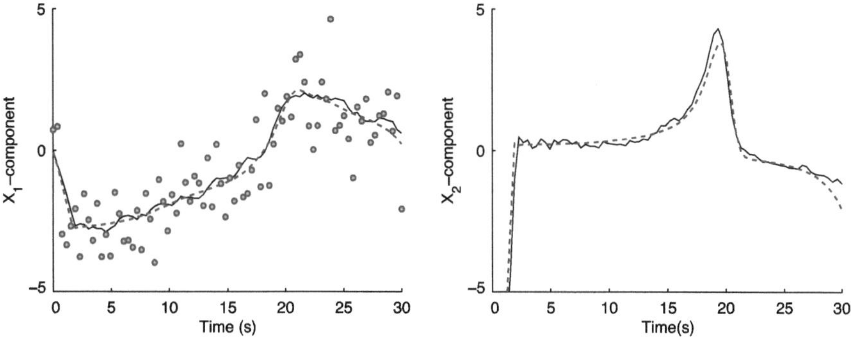

Fig. 8.3 Tracking the Van Der Pol Osciallator using a nonlinear process model (8.32) and direct (noisy) measurements (8.34) of X_1**-component only**. *Black solid line* is true signal, *blue dashed line* is nonlinear Kalman smoother estimate. Measurements are displayed as *circles*

The resulting fit is shown in Fig. 8.3. Despite the noisy measurements of only X_1, we are able to get a good fit for both components. The figure was generated using the `ckbs` package [6], see the file `vanderpol_experiment_simple.m`. The program and example are available for download from COIN-OR.

8.4 State Space Constraints

In almost every real-world problem, additional prior information is known about the state. In many cases, this information can be represented using *state space constraints*. For example, in tracking physical bodies, we often know (roughly or approximately) the topography of the terrain; this information can be encoded as a simple box constraint on the state. We may also know physical limitations (e.g., maximum acceleration or velocity) of objects being tracked, or hard bounds set by biological or financial systems. These and many other examples can be formulated using state space constraints. The ability to incorporate this information is particularly useful when measurements are inaccurate or far between.

In this section, we first show how to add affine inequality constraints to the affine smoother formulation in Sect. 8.2. This requires a novel methodology: *interior point (IP) methods*, an important topic in optimization [32, 37, 49]. IP methods work directly with optimality conditions, so we derive these conditions for the smoothing problem. Rather than review theoretical results about IP methods, we give a general overview and show how they specialize to the linear constrained smoother. The constrained Kalman smoother was originally proposed in [11], but we improve on that work here, and present a simplified algorithm, which is also faster and more numerically stable. We illustrate the algorithm using a numerical example, building on the example in Sect. 8.2.

Once the *linear* smoother with *linear* inequality constraints is understood, we review the constrained *nonlinear* smoother (which can have nonlinear process, measurement, and constraint functions). Using [11] and references therein, we show that the constrained nonlinear smoother is iteratively solved using linear constrained smoothing subproblems, analogously to how the nonlinear smoother in Sect. 8.3 is iteratively solved using linear smoothing subproblems from Sect. 8.2. Because of this hierarchy, the improvements to the affine algorithm immediately carry over to the nonlinear case. We end with a nonlinear constrained numerical example.

8.4.1 Linear Constrained Formulation

We start with the linear smoothing problem (8.6), and impose linear inequality constraints on the state space x:

$$B_k x_k \le b_k. \tag{8.35}$$

By choosing the matrix B_k and b_k appropriately, one can ensure x_k lies in any polyhedral set, since such a set is defined by a finite intersection of hyperplanes. Box constraints, one of the simplest and useful tools for modeling ($l_k \le x_k \le u_k$) can be imposed via

$$\begin{bmatrix} I \\ -I \end{bmatrix} x_k \le \begin{bmatrix} u_k \\ -l_k \end{bmatrix}.$$

In order to formulate the problem for the entire state space sequence, we define

$$B = \mathrm{diag}(\{B_k\}), \quad b = \mathrm{vec}(\{b_k\}), \tag{8.36}$$

and all of the constraints can be written simultaneously as $Bx \le b$. The constrained optimization problem is now given by

$$\begin{aligned} \min_x f(x) = \; & \frac{1}{2}\|Hx - z\|^2_{R^{-1}} + \frac{1}{2}\|Gx - w\|^2_{Q^{-1}} \\ \text{subject to} \; & Bx + s = b, \quad s \ge 0. \end{aligned} \tag{8.37}$$

Note that we have rewritten the inequality constraint as an equality constraint by introducing a new 'slack' variable s.

We derive the Karush-Kuhn-Tucker (KKT) conditions using the Lagrangian formulation. The Lagrangian corresponding to (8.36) is given by

$$\mathcal{L}(x, u, s) = \frac{1}{2}\|Hx - z\|^2_{R^{-1}} + \frac{1}{2}\|Gx - w\|^2_{Q^{-1}} + u^\top(Bx + s - b). \tag{8.38}$$

The KKT conditions are now obtained by differentiating $\mathcal{L}$ with respect to its arguments. Recall that the gradient of (8.6) is given by

$$(H^\top R^{-1}H + G^\top Q^{-1}G)x - H^\top R^{-1}z - G^\top Q^{-1}w.$$

As in (8.8) set $C = H^\top R^{-1}H + G^\top Q^{-1}G$, and for convenience set

$$c = H^\top R^{-1}z + G^\top Q^{-1}w \tag{8.39}$$

The KKT necessary and sufficient conditions for optimality are given by

$$\begin{aligned} \nabla_x \mathcal{L} &= Cx + c + B^\top u = 0 \\ \nabla_q \mathcal{L} &= Bx + s - b = 0 \\ u_i s_i &= 0 \quad \forall i \,;\, u_i, s_i \geq 0. \end{aligned} \tag{8.40}$$

The last set of nonlinear equations is known as *complementarity* conditions. In primal-dual interior point methods, the key idea for solving (8.37) is to successively solve relaxations of the system (8.40) that converge to a triplet $(\bar{x}, \bar{u}, \bar{s})$ which satisfy (8.40).

8.4.2 Interior Point Approach

IP methods work directly to find solutions of (8.40). They do so by iteratively relaxing the complementarity conditions $u_i s_i = 0$ to $u_i s_i = \mu$ as they drive the relaxation parameter μ to 0. The *relaxed* KKT system is defined by

$$F_\mu(s, u, x) = \begin{bmatrix} s + Bx - b \\ SU\mathbf{1} - \mu\mathbf{1} \\ Cx + B^T u - c \end{bmatrix}. \tag{8.41}$$

where S and U are diagonal matrices with s and u on the diagonal, and so the second equation in F_μ implements the relaxation $u_i s_i = \mu$ of (8.40). Note that the relaxation requires that $\mu_i, s_i > 0$ for all i. Since the solution to (8.37) is found by driving the KKT system to 0, at every iteration IP methods attempt to drive F_μ to 0 by Newton's method for root finding.

Newton's root finding method solves the linear system

$$F_\mu^{(1)}(s, u, x) \begin{bmatrix} \Delta s \\ \Delta u \\ \Delta x \end{bmatrix} = -F_\mu(s, u, x). \tag{8.42}$$

It is important to see the full details of solving (8.42) in order to see why it is so effective for constrained Kalman smoothing. The full system is given by

$$\begin{bmatrix} I & 0 & B \\ U & S & 0 \\ 0 & B^T & C \end{bmatrix} \begin{bmatrix} \Delta s \\ \Delta u \\ \Delta x \end{bmatrix} = - \begin{bmatrix} s + Bx - b \\ SU\mathbf{1} - \mu\mathbf{1} \\ Cx + B^T u - c \end{bmatrix}. \tag{8.43}$$

Applying the row operations

$$\begin{aligned} \text{row}_2 &\leftarrow \quad \text{row}_2 - U\text{row}_1 \\ \text{row}_3 &\leftarrow \text{row}_3 - B^T S^{-1}\text{row}_2 \end{aligned},$$

we obtain the equivalent system

$$\begin{bmatrix} I \; 0 \; B \\ 0 \; S \; -UB \\ 0 \; 0 \; C + B^T S^{-1} U B \end{bmatrix} \begin{bmatrix} \Delta s \\ \Delta u \\ \Delta x \end{bmatrix} = \\ - \begin{bmatrix} s + Bx - b \\ -U(Bx - b) - \mu\mathbf{1} \\ Cx + B^T u - c + B^T S^{-1}(U(Bx - b) + \mu\mathbf{1}) \end{bmatrix}. \tag{8.44}$$

In order to find the update for Δx, we have to solve the system

$$\left(C + B^T S^{-1} U B\right) \Delta x = Cx + B^T u - c + B^T S^{-1}(U(Bx - b) + \mu\mathbf{1}) \tag{8.45}$$

Note the structure of the matrix in the LHS of (8.45). The matrix C is the same as in (8.6), so it is positive definite symmetric block tridiagonal. The matrices S^{-1} and U are diagonal, and we always ensure they have only positive elements. The matrices B and $B^\top$ are both block diagonal. Therefore, $C + B^T S^{-1} U B$ has the same structure as C, and we can solve (8.45) using Algorithm 1.

Once we have Δx, the remaining two updates are obtained by back-solving:

$$\Delta u = U S^{-1}(B(x + \Delta x) - b) + \frac{\mu}{s} \tag{8.46}$$

and

$$\Delta s = -s + b - B(x + \Delta x). \tag{8.47}$$

This approach improves the algorithm presented in [11] solely by changing the order of variables and equations in (8.41). This approach simplifies the derivation while also improving speed and numerical stability.

It remains to explain how μ is taken to 0. There are several strategies, see [32, 37, 49]. For the Kalman smoothing application, we use one of the simplest: for two out of every three iterations μ is aggressively taken to 0 by the update $\mu = \mu/10$; while in the remaining iterations, μ is unchanged. In practice, one seldom needs more than 10 interior point iterations; therefore the constrained linear smoother performs at a constant multiple of work of the linear smoother.

8.4.3 Two Linear Numerical Examples

In this section, we present to simple examples, both with linear constraints.

8.4.3.1 Constant Box Constraints

In the first example, we impose box constraints in the example of Sect. 8.2.6. Specifically, we take advantage of the fact the state is bounded:

$$\begin{bmatrix}-1\end{bmatrix} \leq \begin{bmatrix}x\end{bmatrix}\begin{bmatrix}1\end{bmatrix} \tag{8.48}$$

We can encode this information in form (8.35) with

$$B_k = \begin{bmatrix}1 & 0\\ 0 & -1\end{bmatrix}, \quad b_k = \begin{bmatrix}1\\ 1\end{bmatrix}. \tag{8.49}$$

We contrast the performance of the constrained linear smoother with that of the linear smoother without constraints. To show the advantages of modeling with constraints, we increase the measurement noise in both situations to $\sigma^2 = 1$. The results are show in Fig. 8.4. The constrained smoother avoids some of the problems encountered by the unconstrained smoother. Of particular interest are the middle and end parts of the track, where the unconstrained smoother goes far afield because of bad measurement. The constrained smoother is able track portions of the track extremely well, having avoided the bad measurements with the aid of the bound constraints. The figure was generated using the `ckbs` package [6], specifically using the example file `affine_ok_boxC.m`.

8.4.3.2 Variable Box Constraints

In the second example, we impose time-varying constraints on the state. Specifically, we track an exponentially bounded signal with a linear trend:

$$\exp(-\alpha t)\sin(\beta t) + 0.1t$$

using the 'smooth signal' process model and direct measurements, as in Sect. 8.2.6. The challenge here is that as the oscillations start to die down because of the exponential damping, the variance of the measurements remains the same. We can improve the performance by giving the smoother the exponential damping terms as constraints.

We included the second example to emphasize that 'linearity' of constraints means 'with respect to the state'; in fact, the constraints in the second example are simply box constraints which are time dependent. The second example is no more complicated than the first one for the constrained smoother.

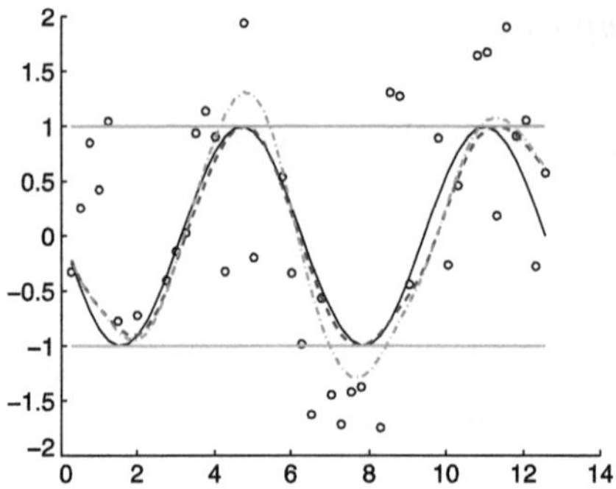

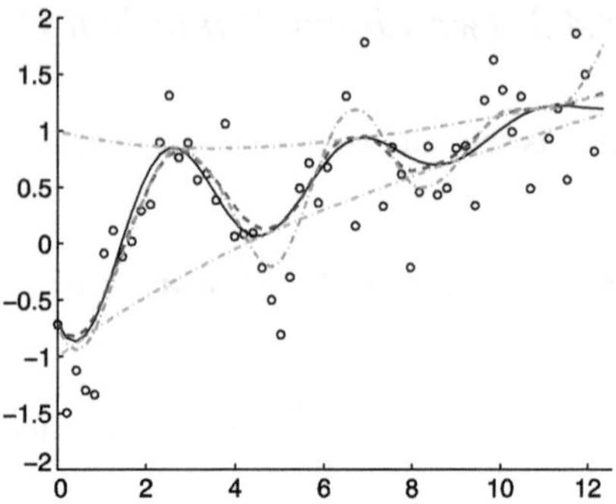

Fig. 8.4 Two examples of linear constraints. Black *solid line* is true signal, *magenta dash-dot lines* is unconstrained Kalman smoother, and *blue dashed line* is the constrained Kalman smoother. Measurements are displayed as *circles*, and bounds are shown as green horizontal lines. In the *left panel*, note that performance of the bounded smoother is significantly better around time 4–10—the unconstrained is fooled by the measurements at times 4 and 8. In the *right panel*, as the oscillations die down due to damping, the measurement variance remains unchanged, so it becomes much more difficult to track the signal without the bound constraints

8.4.4 Nonlinear Constrained Smoother

We now consider the nonlinear constrained smoother, where we allow process functions g_k, measurement functions h_k to be nonlinear, and also allow nonlinear smooth constraints $\xi_k(x_k) \le b_k$. To be consistent with the notation we use throughout the paper, we define a new function

$$\xi(x) = \begin{bmatrix} \xi_1(x_1) \\ \xi_2(x_2) \\ \vdots \\ \xi_N(x_N) \end{bmatrix}, \tag{8.50}$$

so that all the constraints can be written simultaneously as $\xi(x) \le b$.

The problem we would like to solve now is a constrained reformulation of (8.20)

$$\begin{aligned} \min_x \quad & f(x) = \frac{1}{2}\|g(x) - w\|^2_{Q^{-1}} + \frac{1}{2}\|h(x) - z\|^2_{R^{-1}} \\ \text{subject to} \quad & \xi(x) - b \le 0. \end{aligned} \tag{8.51}$$

At this point, we come back to the convex-composite representation described in Sect. 8.3.1. The constraint $\xi(x) - b \le 0$ may be represented using an additional term in the objective function:

$$\delta\left(\xi(x) - b \mid \mathbb{R}_-\right), \tag{8.52}$$

where $\delta(x \mid C)$ is the convex indicator function:

$$\delta(x \mid C) = \begin{cases} 0 & x \in C \\ \infty & x \notin C \end{cases}. \tag{8.53}$$

Therefore, the objective (8.51) can be represented as follows:

$$\begin{aligned} f(x) &= \rho(F(x)) \\ \rho\begin{pmatrix} y_1 \\ y_2 \\ y_3 \end{pmatrix} &= \frac{1}{2}\|y_1\|^2_{Q^{-1}} + \frac{1}{2}\|y_2\|^2_{R^{-1}} + \delta(y_3 \mid \mathbb{R}_-) \\ F(x) &= \begin{bmatrix} g(x) - w \\ h(x) - z \\ \xi(x) - b \end{bmatrix}. \end{aligned} \tag{8.54}$$

The approach to nonlinear smoothing in [11] is essentially the Gauss-Newton method described in Sect. 8.3.2, applied to (8.54). In other words, at each iteration ν, the function F is linearized, and the direction finding subproblem is obtained by solving

$$\begin{aligned} \min_d \quad & \frac{1}{2}\|G^\nu d - \underbrace{w - g(x^\nu)}_{w^\nu}\|^2_{Q^{-1}} + \frac{1}{2}\|H^\nu d - \underbrace{z - h(x^\nu)}_{z^\nu}\|^2_{R^{-1}}, \\ \text{subject to} \quad & B^\nu d \le \underbrace{b - \xi(x^\nu)}_{b^\nu}, \end{aligned} \tag{8.55}$$

where G^ν and H^ν are exactly as in (8.28), $B^\nu = \nabla_x \xi(x^\nu)$ is a block diagonal matrix because of the structure of ξ (8.50), and we have written the indicator function in (8.54) as an explicit constraint to emphasize the structure of the subproblem.

Note that (8.55) has exactly the same structure as the linear constrained smoothing problem (8.37), and therefore can be solved using the interior point approach in the previous section. Because the convex-composite objective (8.54) is not finite valued (due to the indicator function of the feasible set), to prove convergence of the nonlinear smoother, [11] uses results from [14]. We refer the interested reader to [11, Lemma 8, Theorem 9] for theoretical convergence results, and to [11, Algorithm 6] for the full algorithm, including line search details.

Because of the hierarchical dependence of the nonlinear constrained smoother on the linear constrained smoother, the simplified improved approach we presented in Sect. 8.4.2 pays off even more in the nonlinear case, where it is used repeatedly as a subroutine.

8.4.5 Nonlinear Constrained Example

The example in this section is reproduced from [11]. Consider the problem of tracking a ship traveling close to shore where we are given distance measurements from two fixed stations to the ship as well as the location of the shoreline. Distance to fixed stations is a nonlinear function, so the measurement model here is nonlinear.

In addition, the corresponding constraint functions $\{f_k\}$ are not affine because the shoreline is not a straight line. For the purpose of simulating the measurements $\{z_k\}$, the ship velocity $[X_1(t), X_3(t)]$ and the ship position $[X_2(t), X_4(t)]$ are given by

$$X(t) = [\ 1,\ t,\ \ -\cos(t),\ 1.3 - \sin(t)\ ]^\top$$

Both components of the ship's position are modeled using the smooth signal model in Sect. 8.2.6. Therefore we introduce two velocity components, and the process model is given by

$$G_k = \begin{bmatrix} 1 & 0 & 0 & 0 \\ \Delta t & 1 & 0 & 0 \\ 0 & 0 & 1 & 0 \\ 0 & 0 & \Delta t & 0 \end{bmatrix}, \quad Q_k = \begin{bmatrix} \Delta t & \Delta t^2/2 & 0 & 0 \\ \Delta t^2/2 & \Delta t^3/3 & 0 & 0 \\ 0 & 0 & \Delta t & \Delta t^2/2 \\ 0 & 0 & \Delta t^2/2 & \Delta t^3/3 \end{bmatrix}.$$

The initial state estimate is given by $g_1(x_0) = X(t_1)$ and $Q_1 = 100 I_4$ where I_4 is the four by four identity matrix. The measurement variance is constant for this example and is denoted by σ^2. The distance measurements are made from two stationary locations on shore. One is located at $(0, 0)$ and the other is located at $(2\pi, 0)$. The measurement model is given by

$$h_k(x_k) = \begin{pmatrix} \sqrt{x_{2,k}^2 + x_{4,k}^2} \\ \sqrt{(x_{2,k} - 2\pi)^2 + x_{4,k}^2} \end{pmatrix}, \quad R_k = \begin{pmatrix} \sigma^2 & 0 \\ 0 & \sigma^2 \end{pmatrix}.$$

We know that the ship does not cross land, so $X_4(t) \geq 1.25 - \sin[X_2(t)]$. This information is encoded by the constraints

$$\xi_k(x_k) = 1.25 - \sin(x_{2,k}) - x_{4,k} \leq 0.$$

The initial point for the smoother is $[0, 0, 0, 1]^\top$, which is not feasible. The results are plotted in Fig. 8.5. The constrained smoother performs significantly better than the unconstrained smoother in this example. The experiment was done using the `ckbs` program, specifically see `sine_wave_example.m`.

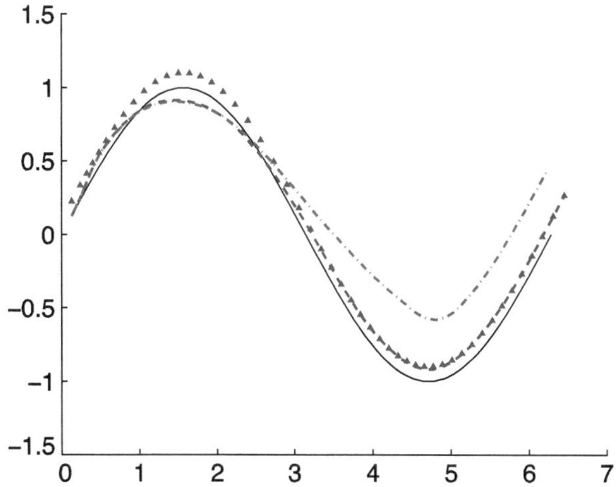

Fig. 8.5 Smoother results for ship tracking example with linear process model, nonlinear measurement model, and nonlinear constraints (with respect to the state). *Black solid line* is true state, *red triangles* denote the constraint, *magenta dash-dot line* is the unconstrained estimate, and *blue dashed line* gives the constrained nonlinear smoothed estimate

8.5 Robust Kalman Smoothing

In many applications, the probalistic model for the dynamics and/or the observations (8.1) is not well described by a Gaussian distribution. This occurs in the model for the observations when they are contaminated by outliers, or more generally, when the measurement noise v_k is heavy tailed [44], and it occurs in the model for the dynamics when tracking systems with rapidly changing dynamics, or jumps in the state values [31]. A *robust* Kalman filter or smoother is one that can obtain an acceptable estimate of the state when Gaussian assumptions are violated, and which continues to perform well when they are not violated.

We show how to accommodate non-Gaussian densities by starting with a simple case of non-Gaussian heavy tailed measurement noise v_k [7]. However, this general approach can be extended to w_k as well. Heavy tailed measurement noise occurs in applications related to glint noise [25], turbulence, asset returns, and sensor failure or machine malfunction. It can also occur in the presence of secondary noise sources or other kinds of data anomalies. Although it is possible to estimate a minimum variance estimate of the state using stochastic simulation methods such as Markov chain Monte-Carlo (MCMC) or particle filters [24, 35], these methods are very computationally intensive, and convergence often relies on heuristic techniques and is highly variable. The approach taken here is very different. It is based on the optimization perspective presented in the previous sections. We develop a method for computing the MAP estimate of the state sequence under the assumption that the observation noise comes from the ℓ_1-Laplace density often used in robust estimation, e.g., see [23, Eq. 2.3]. As we will see, the resulting optimization problem will again be one of convex composite type allowing us to apply a Gauss-Newton strategy for computing the MAP estimate. Again, the key to a successful computational strategy is the preservation of the underlying tri-diagonal structure.

8.5.1 An ℓ_1-Laplace Smoother

For $u \in \mathbb{R}^m$ we use the notation $\|u\|_1$ for the ℓ_1 norm of u; i.e., $\|u\|_1 = |u_1| + \ldots + |u_m|$. The multivariate ℓ_1-Laplace distribution with mean μ and covariance R has the following density:

$$\mathbf{p}(v_k) = \det(2R)^{-1/2} \exp\left[-\sqrt{2}\left\|R^{-1/2}(v_k - \mu)\right\|_1\right], \qquad (8.56)$$

where $R^{1/2}$ denotes a Cholesky factor of the positive definite matrix R; i.e., $R^{1/2}(R^{1/2})^{\mathrm{T}} = R$. One can verify that this is a probability distribution with covariance R using the change of variables $u = R^{-1/2}(v_k - \mu)$. A comparison of the Gaussian and Laplace distributions is displayed in Fig. 8.6. This comparison includes the densities, negative log densities, and influence functions, for both distributions.

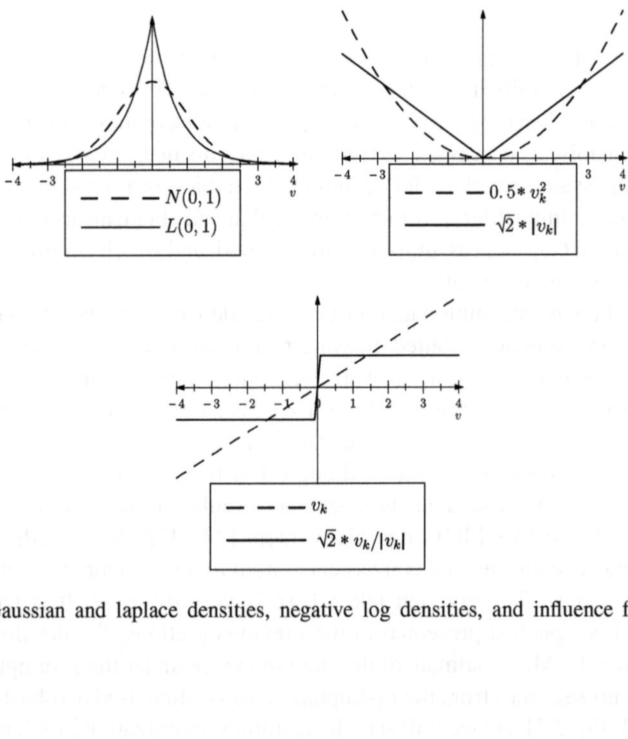

Fig. 8.6 Gaussian and laplace densities, negative log densities, and influence functions (for scalar v_k)

8.5.1.1 Maximum a Posteriori Formulation

Assume that the model for the dynamics and the observations is given by (8.1), where w_k is assumed to be Gaussian and v_k is modeled by the ℓ_1-Laplace density (8.56). Under these assumptions, the MAP objective function is given by

$$\begin{aligned} P\left(\{x_k\}\big|\{z_k\}\right) &\propto P\left(\{z_k\}\big|\{x_k\}\right) P\left(\{x_k\}\right) \\ &= \prod_{k=1}^{N} P(\{v_k\})P(\{w_k\}) \\ &\propto \prod_{k=1}^{N} \exp\Bigg(-\sqrt{2}\left\| R^{-1/2}(z_k - h_k(x_k))\right\|_1 \\ &\quad -\frac{1}{2}(x_k - g_k(x_{k-1}))^{\top} Q_k^{-1}(x_k - g_k(x_{k-1}))\Bigg). \end{aligned} \tag{8.57}$$

Dropping terms that do not depend on $\{x_k\}$, minimizing this MAP objective with respect to $\{x_k\}$ is equivalent to minimizing

$$f(\{x_k\}) :=$$
$$\sqrt{2}\sum_{k=1}^{N}\left\| R_k^{-1/2}[z_k - h_k(x_k)]\right\|_1 + \frac{1}{2}\sum_{k=1}^{N}[x_k - g_k(x_{k-1})]^{\mathrm{T}} Q_k^{-1}[x_k - g_k(x_{k-1})],$$

where, as in (8.1), x_0 is known and $g_0 = g_1(x_0)$. Setting

$$\begin{aligned} R &= \mathrm{diag}(\{R_k\}) \\ Q &= \mathrm{diag}(\{Q_k\}) \\ x &= \mathrm{vec}(\{x_k\}) \\ w &= \mathrm{vec}(\{g_0, 0, \dots, 0\}) \\ z &= \mathrm{vec}(\{z_1, z_2, \dots, z_N\}) \end{aligned}, \quad g(x) = \begin{bmatrix} x_1 \\ x_2 - g_2(x_1) \\ \vdots \\ x_N - g_N(x_{N-1}) \end{bmatrix}, \quad h(x) = \begin{bmatrix} h_1(x_1) \\ h_2(x_2) \\ \vdots \\ h_N(x_N) \end{bmatrix}, \tag{8.58}$$

as in (8.5) and (8.19), the MAP estimation problem is equivalent to

$$\underset{x \in \mathbf{R}^{Nn}}{\text{minimize}} \; f(x) = \frac{1}{2}\|g(x) - w\|_{Q^{-1}} + \sqrt{2}\left\|R^{-1/2}(h(x) - z)\right\|_1. \tag{8.59}$$

8.5.1.2 The Convex Composite Structure

The objective in (8.59) can again be written as a the composition of a convex function ρ with a smooth function F:

$$f(x) = \rho(F(x)), \tag{8.60}$$

where

$$\rho\begin{pmatrix}y_1\\y_2\end{pmatrix} = \frac{1}{2}\|y_1\|^2_{Q^{-1}} + \sqrt{2}\|R^{-1/2}y_2\|_1, \quad F(x) = \begin{bmatrix} g(x) - w \\ h(x) - z \end{bmatrix}. \tag{8.61}$$

Consequently, the generalized Gauss-Newton methodology described in Sect. 8.3.2 again applies. That is, given an approximate solution x^ν to (8.59), we compute a new approximate solution of the form

$$x^{\nu+1} = x^\nu + \gamma^\nu d^\nu,$$

where d^ν solves the subproblem

$$\underset{d\in\mathbb{R}^n}{\text{minimize}}\ \rho(F(x^\nu) + F'(x^\nu)d), \tag{8.62}$$

and γ^ν is computed using the backtracking line-search procedure described in Sect. 8.3.2. Following the pattern described in (8.28), the subproblem (8.62), where ρ and F are given in (8.61), has the form

$$d^\nu = \arg\min_d \tilde{f}(d) = \frac{1}{2}\|G^\nu d - \underbrace{w - g(x^\nu)}_{w^\nu}\|^2_{Q^{-1}} + \sqrt{2}\|R^{-1/2}(H^\nu d - \underbrace{z - h(x^\nu)}_{z^\nu})\|_1, \tag{8.63}$$

where

$$G^\nu = \begin{bmatrix} \mathrm{I} & 0 & & \\ -g_2^{(1)}(x_1^\nu) & \mathrm{I} & \ddots & \\ & \ddots & \ddots & 0 \\ & & -g_N^{(1)}(x_{N-1}^\nu) & \mathrm{I} \end{bmatrix}, \quad H^\nu = \operatorname{diag}\{h_1^{(1)}(x_1), \ldots, h_N^{(1)}(x_N)\}. \tag{8.64}$$

8.5.1.3 Solving the Subproblem by Interior Point Methods

By (8.63), the basic subproblem that must be solved takes the form

$$\min_d \frac{1}{2}\|Gd - w\|^2_{Q^{-1}} + \sqrt{2}\|R^{-1/2}(Hd - z)\|_1, \tag{8.65}$$

where, as in (8.5),

$$
\begin{aligned}
R &= \text{diag}(\{R_k\}) & x &= \text{vec}(\{x_k\}) \\
Q &= \text{diag}(\{Q_k\}) & w &= \text{vec}(\{w_1, w_2, \ldots, w_N\}) \\
H &= \text{diag}(\{H_k\}) & z &= \text{vec}(\{z_1, z_2, \ldots, z_N\})
\end{aligned}
\qquad
G = \begin{bmatrix} \mathrm{I} & 0 & & \\ -G_2 & \mathrm{I} & \ddots & \\ & \ddots & \ddots & 0 \\ & & -G_N & \mathrm{I} \end{bmatrix}.
\tag{8.66}
$$

Using standard optimization techniques, one can introduce a pair of auxiliary non-negative variables $p^+, p^- \in \mathbb{R}^M$ ($M = \sum_{k=1}^N m(k)$) so that this problem can be rewritten as

$$
\begin{array}{ll}
\text{minimize} & \frac{1}{2} d^\top C d + c^\top d + \sqrt{\mathbf{2}}^\top (p^+ + p^-) \\
\text{w.r.t.} & d \in \mathbb{R}^{nN},\ p^+, p^- \in \mathbb{R}^M \\
\text{subject to} & Bd + b = p^+ - p^-,
\end{array}
\tag{8.67}
$$

where

$$
C = G^\top Q^{-1} G = \begin{bmatrix} C_1 & A_2^\top & 0 & \\ A_2 & C_2 & A_3^\top & 0 \\ 0 & \ddots & \ddots & \ddots \\ & 0 & A_N & C_N \end{bmatrix},
\qquad
\begin{aligned}
A_k &= -Q_k^{-1} G_k \\
C_k &= Q_k^{-1} + G_{k+1}^\top Q_{k+1}^{-1} G_{k+1} \\
c &= G^\top w \\
B &= R^{-1/2} H \\
b &= -R^{-1/2} z
\end{aligned}
.
$$

The problem (8.67) is a convex quadratic program. If we define

$$
F_\mu(p^+, p^-, s^+, s^-, d) = \begin{bmatrix} p^+ - p^- - b - Bd \\ \text{diag}(p^-)\text{diag}(s^-)\mathbf{1} - \mu\mathbf{1} \\ s^+ + s^- - 2\sqrt{\mathbf{2}} \\ \text{diag}(p^+)\text{diag}(s^+)\mathbf{1} - \mu\mathbf{1} \\ Cd + c + B^\top(s^- - s^+)/2 \end{bmatrix},
\tag{8.68}
$$

for $\mu \geq 0$, then the KKT conditions for (8.67) can be written as

$$
F_0(p^+, p^-, s^+, s^-, d) = 0.
$$

The set of solutions to $F_\mu(p^+, p^-, s^+, s^-, d) = 0$ for $\mu > 0$ is called the central path. We solve the system for $\mu = 0$ by an interior point strategy which, as described earlier, is a Newton based predictor-corrector method for following the central path as $\mu \downarrow 0$. At each iteration of the interior point method we need to solve a system of the form

$$
F_\mu(p^+, p^-, s^+, s^-, d) + F_\mu'(p^+, p^-, s^+, s^-, d) \begin{bmatrix} \Delta p^+ \\ \Delta p^- \\ \Delta s^+ \\ \Delta s^- \\ \Delta y \end{bmatrix} = 0,
$$

where the vectors p^+, p^-, s^+, and s^- are componentwise strictly positive. Using standard methods of Gaussian elimination (as in Sect. 8.4.2), we obtain the solution

$$\begin{aligned}
\Delta y &= [C + B^{\mathrm{T}} T^{-1} B]^{-1} (\bar{e} + B^{\mathrm{T}} T^{-1} \bar{f}) \\
\Delta s^- &= T^{-1} B \Delta y - T^{-1} \bar{f} \\
\Delta s^+ &= -\Delta s^- + 2\sqrt{\mathbf{2}} - s^+ - s^- \\
\Delta p^- &= \mathrm{diag}(s^-)^{-1} [\tau \mathbf{1} - \mathrm{diag}(p^-) \Delta s^-] - p^- \\
\Delta p^+ &= \Delta p^- + B \Delta y + b + By - p^+ + p^-,
\end{aligned}$$

where

$$\begin{aligned}
\bar{d} &= \tau \mathbf{1}/s^+ - \tau \mathbf{1}/s^- - b - By + p^+ \\
\bar{e} &= B^{\mathrm{T}} (\sqrt{\mathbf{2}} - s^-) - Cy - c \\
\bar{f} &= \bar{d} - \mathrm{diag}(s^+)^{-1} \mathrm{diag}(p^+)(2\sqrt{\mathbf{2}} - s^-) \\
T &= \mathrm{diag}(s^+)^{-1} \mathrm{diag}(p^+) + \mathrm{diag}(s^-)^{-1} \mathrm{diag}(p^-).
\end{aligned}$$

Since the matrices T and B are block diagonal, the matrix $B^\top T B$ is also block diagonal. Consequently, the key matrix $C + B^{\mathrm{T}} T^{-1} B$ has exactly the same form as the block tri-diagonal matrix in (8.10) with

$$\begin{aligned}
c_k &= Q_k^{-1} + G_{k+1}^\top Q_{k+1}^{-1} G_{k+1} + H_k^\top T_k^{-1} H_k \quad & k = 1, \ldots, N, \\
a_k &= -Q_k^{-1} G_k & k = 2, \ldots, N,
\end{aligned}$$

where $T_k = \mathrm{diag}(s_k^+)^{-1} \mathrm{diag}(p_k^+) + \mathrm{diag}(s_k^-)^{-1} \mathrm{diag}(p_k^-)$. Algorithm 1 can be applied to solve this system accurately and stably with $O(n^3 N)$ floating point operations which preserves the efficiency of the classical Kalman Filter algorithm.

Further discussion on how to incorporate approximate solutions to the quadratic programming subproblems can be found in [7, Sect. V].

8.5.1.4 A Linear Example

In the linear case, the functions g_k and h_k is (8.1) are affine so that they equal their linearizations. In this case, the problems (8.59) and (8.62) are equivalent and only one subproblem of the form (8.65), or equivalently (8.67), needs to be solved. We illustrate the ℓ_1-Laplace smoother described in Sect. 8.5.1.1 by applying it to the example studied in Sect. 8.2.6, except now the noise term v_k is modeled using the ℓ_1-Laplace density. The numerical experiment described below is take from [7, Sect. VI].

The numerical experiment uses two full periods of $X(t)$ generated with $N = 100$ and $\Delta t = 4\pi/N$; i.e., discrete time points equally spaced over the interval $[0, 4\pi]$. For $k = 1, \ldots, N$ the measurements z_k were simulated by $z_k = X_2(t_k) + v_k$. In order to test the robustness of the ℓ_1 model to measurement noise containing outlier data, we generate v_k as a mixture of two normals with p denoting the fraction of outlier contamination; i.e.,

$$v_k \sim (1-p)\mathbf{N}(0, 0.25) + p\mathbf{N}(0, \phi). \tag{8.69}$$

This was done for $p \in \{0,\ 0.1\}$ and $\phi \in \{1, 4, 10, 100\}$. The model for the mean of z_k given x_k is $h_k(x_k) = (0, 1)x_k = x_{2,k}$. Here $x_{2,k}$ denotes the second component of x_k. The model for the variance of z_k given x_k is $R_k = 0.25$. This simulates a lack of knowledge of the distribution for the outliers; i.e., $p\mathbf{N}(0, \phi)$. Note that we are recovering estimates for the smooth function $-\sin(t)$ and its derivative $-\cos(t)$ using noisy measurements (with outliers) of the function values.

We simulated 1000 realizations of the sequence $\{z_k\}$ keeping the ground truth fixed, and for each realization, and each estimation method, we computed the corresponding state sequence estimate $\{\hat{x}_k\}$. The Mean Square Error (MSE) corresponding to such an estimate is defined by

$$\text{MSE} = \frac{1}{N}\sum_{k=1}^{N}[x_{1,k} - \hat{x}_{1,k}]^2 + [x_{2,k} - \hat{x}_{2,k}]^2, \tag{8.70}$$

where $x_k = X(t_k)$. In Table 8.1, the Gaussian Kalman Filter is denoted by (GKF), the Iterated Gaussian Smoother (IGS), and the Iterated ℓ_1-Laplace Smoother (ILS). For each of these estimation techniques, each value of p, and each value of ϕ, the corresponding table entry is the median MSE followed by the centralized 95 % confidence interval for the MSE. For this problem, the model functions $\{g_k(x_{k-1})\}$ and $\{h_k(x_k)\}$ are linear so the iterated smoothers IGS and ILS only require one iteration to estimate the sequence $\{\hat{x}_k\}$.

Note the ℓ_1-Laplace smoother performs nearly as well as the Gaussian smoother at the nominal conditions ($p = 0$). The ℓ_1-Laplace smoother performs better and more consistently in cases with data contamination ($p \geq 0.1$ and $\phi \geq 1$). It is also apparent that the smoothers perform better than the filters.

Outlier detection and removal followed by refitting is a simple approach to robust estimation and can be applied to the smoothing problem. An inherent weakness of this approach is that the outlier detection is done using an initial fit which assumes outliers are not present. This can lead to good data being classified as outliers and

Table 8.1 Median MSE and 95 % confidence intervals for the different estimation methods

p	ϕ	GKF	IGS	ILS
0	–	0.34 (0.24, 0.47)	0.04(0.02, 0.1)	0.04(0.01, 0.1)
0.1	1	0.41(0.26, 0.60)	0.06(0.02, 0.12)	0.04(0.02, 0.10)
0.1	4	0.59(0.32, 1.1)	0.09(0.04, 0.29)	0.05(0.02, 0.12)
0.1	10	1.0(0.42, 2.3)	0.17(0.05, 0.55)	0.05(0.02, 0.13)
0.1	100	6.8(1.7, 17.9)	1.3(0.30, 5.0)	0.05(0.02, 0.14)

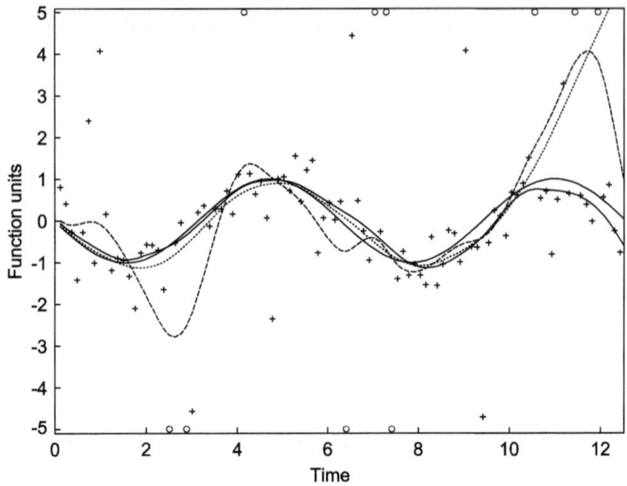

Fig. 8.7 Simulation: measurements (+), outliers (o) (absolute residuals more than three standard deviations), true function (*thick line*), ℓ_1-Laplace estimate (*thin line*), Gaussian estimate (*dashed line*), Gaussian outlier removal estimate (*dotted line*)

result in over fitting the remaining data. An example of this is illustrated in Fig. 8.7 which plots the estimation results for a realization of $\{z_k\}$ where $p = 0.1$ and $\phi = 100$. Outlier removal also makes critical review of the model more difficult. A robust smoothing method with a consistent model, such as the ℓ_1-Laplace smoother, does not suffer from these difficulties.

8.5.1.5 Stochastic Nonlinear Process Example

We now illustrate the behavior of the ℓ_1-Laplace smoother on the Van Der Pol Oscillator described in Sect. 8.3.4. The numerical experiment we describe is taken from [7, Sect. VI]. The corresponding nonlinear differential equation is

$$\dot{X}_1(t) = X_2(t) \text{ and } \dot{X}_2(t) = \mu[1 - X_1(t)^2]X_2(t) - X_1(t).$$

Given $X(t_{k-1}) = x_{k-1}$ the Euler approximation for $X(t_{k-1} + \Delta t)$ is

$$g_k(x_{k-1}) = \begin{pmatrix} x_{1,k-1} + x_{2,k-1}\Delta t \\ x_{2,k-1} + \{\mu[1 - x_{1,k}^2]x_{2,k} - x_{1,k}\}\Delta t \end{pmatrix}.$$

For this simulation, the 'ground truth' is obtained from a stochastic Euler approximation of the Van der Pol oscillator. To be specific, with $\mu = 2$, $N = 164$ and $\Delta t = 16/N$, the ground truth state vector x_k at time $t_k = k\Delta t$ is given by $x_0 = (0, -0.5)^{\mathrm{T}}$ and for $k = 1, \ldots, N$,

$$x_k = g_k(x_{k-1}) + w_k, \tag{8.71}$$

where $\{w_k\}$ is a realization of independent Gaussian noise with variance 0.01. Our model for state transitions (8.1) uses $Q_k = 0.01\, I$ for $k > 1$, and so is identical to the model used to simulate the ground truth $\{x_k\}$. Thus, we have precise knowledge of the process that generated the ground truth $\{x_k\}$. The initial state x_0 is imprecisely specified by setting $g_1(x_0) = (0.1, -0.4)^{\mathrm{T}} \neq x_0$ with corresponding variance $Q_1 = 0.1\, I$.

For $k = 1, \ldots, N$ the measurements z_k were simulated by $z_k = x_{1,k} + v_k$. The measurement noise v_k was generated as follows:

$$v_k \sim (1 - p)\mathbf{N}(0, 1.0) + p\mathbf{N}(0, \phi). \tag{8.72}$$

This was done for $p \in \{0, 0.1, 0.2, 0.3\}$ and $\phi \in \{10, 100, 1000\}$. The model for the mean of z_k given x_k is $h_k(x_k) = (1, 0)x_k = x_{1,k}$. As in the previous simulation, we simulated a lack of knowledge of the distribution for the outliers; i.e., $p\mathbf{N}(0, \phi)$. In (8.1), the model for the variance of z_k given x_k is $R_k = 1.0$.

We simulated 1,000 realizations of the ground truth state sequence $\{x_k\}$ and the corresponding measurement sequence $\{z_k\}$. For each realization, we computed the corresponding state sequence estimate $\{\hat{x}_k\}$ using both the IGS and IKS procedures. The Mean Square Error (MSE) corresponding to such an estimate is defined by equation (8.70), where x_k is given by equation (8.71). The results of the simulation appear in Table 8.2. As the proportion and variance of the outliers increase, the Gaussian smoother degrades, but the ℓ_1-Laplace smoother is not affected.

Figure 8.8 provides a visual illustration of one realization $\{x_k\}$ and its corresponding estimates $\{\hat{x}_k\}$. The left two panels demonstrate that, when no outliers are present, both the IGS and ILS generate accurate estimates. Note that we only observe the first component of the state and that the variance of the observation is relatively large (see top two panels). The right two panels show what can go wrong when outliers are present. The Van der Pol oscillator can have sharp peaks as a result of the nonlinearity in its process model, and outliers in the measurements can 'trick' the IGS into

Table 8.2 Median MSE over 1,000 runs and confidence intervals containing 95 % of MSE results

p	ϕ	IGS	ILS
0	–	0.07 (0.06, 0.08)	0.07 (0.06, 0.09)
0.1	10	0.07 (0.06, 0.10)	0.07 (0.06, 0.09)
0.2	10	0.08 (0.06, 0.11)	0.08 (0.06, 0.11)
0.3	10	0.08 (0.06, 0.11)	0.08 (0.06, 0.11)
0.1	100	0.10 (0.07, 0.14)	0.07 (0.06, 0.10)
0.2	100	0.12 (0.07, 0.40)	0.08 (0.06, 0.11)
0.3	100	0.13 (0.09, 0.64)	0.08 (0.07, 0.10)
0.1	1000	0.17 (0.11, 1.50)	0.08 (0.06, 0.11)
0.2	1000	0.21 (0.14, 2.03)	0.08 (0.06, 0.11)
0.3	1000	0.25 (0.17, 2.66)	0.09 (0.07, 0.12)

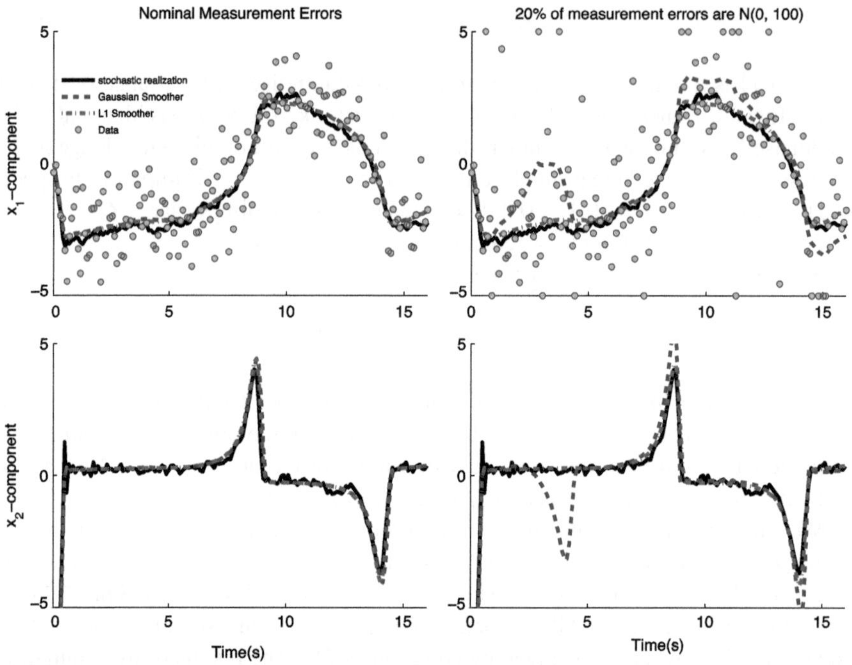

Fig. 8.8 The *left* two panels show estimation of x_1, (*top*) and x_2 (*bottom*) with errors from the nominal model. The stochastic realization is represented by a *thick black line*; the Gaussian smoother is the *blue dashed line*, and the ℓ_1-smoother is the *magenta dash-dotted* line. *Right* two panels show the same stochastic realization but with measurement errors now from $(p, \phi) = (0.2, 100)$. Outliers appear on the *top* and *bottom* boundary in the *top right panel*

these modes when they are not really present. In contrast, the Iterated ℓ_1-Laplace Smoother avoids this problem.

8.5.2 Further Extensions with Log-Concave Densities

Let us step back for a moment and examine a theme common to all of the variations on the Kalman smoother that we have examined thus far and compare the objective functions in (8.6, 8.20, 8.37, 8.51, 8.59). In all cases, the objective function takes the form

$$\sum_{k=1}^{N} V_k\left(h(x_k) - z_k; R_k\right) + J_k\left(x_k - g(x_{k-1}); Q_k\right), \tag{8.73}$$

where the mappings V_k and J_k are associated with log-concave densities of the form

$$p_{v,k}(z) \propto \exp\left(-V_k(z : R_k)\right)) \quad \text{and} \quad p_{w,k}(x) \propto \exp\left(-J_k(x; Q_k)\right)$$

with $p_{v,k}$ and $p_{w,k}$ having covariance matrices R_k and Q_k, respectively. The choice of the penalty functions V_k and J_k reflect the underlying model for distribution of the observations and the state, respectively. In many applications, the functions V_k and J_k are a members of the class of extended piecewise linear-quadratic penalty functions.

8.5.2.1 Extended Linear-Quadratic Penalties

Definition 1 For a nonempty polyhedral set $U \subset \mathbb{R}^m$ and a symmetric positive-semidefinite matrix $M \in \mathbb{R}^{m\times m}$ (possibly $M = 0$), define the function $\theta_{U,M} : \mathbb{R}^m \to \{\mathbb{R} \cup \infty\} := \overline{\mathbb{R}}$ by

$$\theta_{U,M}(w) := \sup_{u\in U} \left\{ \langle u, w\rangle - \frac{1}{2}\langle u, Mu\rangle \right\}. \tag{8.74}$$

Given and injective matrix $B \in \mathbb{R}^{m\times n}$ and a vector $b \in \mathbb{R}^m$, define $\rho : \mathbb{R}^n \to \overline{\mathbb{R}}$ as $\theta_{U,M}(b + By)$:

$$\rho_{U,M,b,B}(y) := \sup_{u\in U} \left\{\langle u, b + By\rangle - \tfrac{1}{2}\langle u, Mu\rangle\right\}. \tag{8.75}$$

All functions of the type specified in (8.74) are called *piecewise linear-quadratic* (PLQ) penalty functions, and those of the form (8.75) are called *extended piecewise linear-quadratic* (EPLQ) penalty functions.

Remarks 1 PLQ penalty functions are extensively studied by Rockafellar and Wets in [43]. In particular, they present a full duality theory for optimizations problems based on these functions.

It is easily seen that the penalty functions arising from both the Gaussian and ℓ_1-Laplace distributions come from this EPLQ class. But so do other important densities such as the Huber and Vapnik densities.

Example 1 : The ℓ_2, ℓ_1, Huber, and Vapnik penalties are representable in the notation of Definition 1.

1. L_2: Take $U = \mathbb{R}$, $M = 1$, $b = 0$, and $B = 1$. We obtain $\rho(y) = \sup_{u\in\mathbf{R}} \left\langle uy - \frac{1}{2}u^2 \right\rangle$. The function inside the sup is maximized at $u = y$, whence $\rho(y) = \frac{1}{2}y^2$.
2. ℓ_1: Take $U = [-1, 1]$, $M = 0$, $b = 0$, and $B = 1$. We obtain $\rho(y) = \sup_{u\in[-1,1]} \langle uy\rangle$. The function inside the sup is maximized by taking $u = \mathrm{sign}(y)$, whence $\rho(y) = |y|$.

3. Huber: Take $U = [-K, K]$, $M = 1$, $b = 0$, and $B = 1$. We obtain $\rho(y) = \sup_{u\in[-K,K]} \left\langle uy - \frac{1}{2}u^2 \right\rangle$. Take the derivative with respect to u and consider the following cases:

   a. If $y < -K$, take $u = -K$ to obtain $-Ky - \frac{1}{2}K^2$.
   b. If $-K \leq y \leq K$, take $u = y$ to obtain $\frac{1}{2}y^2$.
   c. If $y > K$, take $u = K$ to obtain a contribution of $Ky - \frac{1}{2}K^2$.

   This is the Huber penalty with parameter K, shown in the upper panel of Fig. 8.9.

4. Vapnik: take $U = [0, 1] \times [0, 1]$, $M = \begin{bmatrix} 0 & 0 \\ 0 & 0 \end{bmatrix}$, $B = \begin{bmatrix} 1 \\ -1 \end{bmatrix}$, and $b = \begin{bmatrix} -\epsilon \\ -\epsilon \end{bmatrix}$, for some $\epsilon > 0$. We obtain $\rho(y) = \sup_{u_1,u_2\in[0,1]} \left\langle \begin{bmatrix} y - \epsilon \\ -y - \epsilon \end{bmatrix}, \begin{bmatrix} u_1 \\ u_2 \end{bmatrix} \right\rangle$. We can obtain an explicit representation by considering three cases:

   a. If $|y| < \epsilon$, take $u_1 = u_2 = 0$. Then $\rho(y) = 0$.
   b. If $y > \epsilon$, take $u_1 = 1$ and $u_2 = 0$. Then $\rho(y) = y - \epsilon$.
   c. If $y < -\epsilon$, take $u_1 = 0$ and $u_2 = 1$. Then $\rho(y) = -y - \epsilon$.

   This is the Vapnik penalty with parameter ϵ, shown in the lower panel of Fig. 8.9.

8.5.2.2 PLQ Densities

We caution that not every EPLQ function is the negative log of a density function. For an ELQP function ρ to be associated with a density, the function $\exp(-\rho(x))$ must be integrable on $\mathbb{R}^n$. The integrability of $\exp(-\rho(x))$ can be established under a coercivity hypothesis.

Definition 2 A function $\rho : \mathbb{R}^n \to \mathbb{R} \cup \{+\infty\} = \overline{\mathbb{R}}$ is said to be coercive (or 0-coercive) if $\lim_{\|x\|\to\infty} \rho(x) = +\infty$.

Since the functions $\rho_{U,M,b,B}$ defined in (8.75) are not necessarily finite-valued, their calculus must be treated with care. An important tool in this regard is the essential dominion. The essential domain of $\rho : \mathbb{R}^n \to \overline{\mathbb{R}}$ is the set

$$\mathrm{dom}(\rho) := \{x : \rho(x) < +\infty\}.$$

The affine hull of $\mathrm{dom}(\rho)$ is the smallest affine set containing $\mathrm{dom}(\rho)$, where a set is affine if it is the translate of a subspace.

Theorem 1 [4, Theorem 6] (PLQ Integrability). *Let $\rho := \rho_{U,M,b,B}$ be defined as in* (8.75). *Suppose $\rho(y)$ is coercive, and let n_{aff} denote the dimension of* $\mathrm{aff}(\mathrm{dom}\,\rho)$. *Then the function $f(y) = \exp(-\rho(y))$ is integrable on* $\mathrm{aff}(\mathrm{dom}\,\rho)$ *with the n_{aff}-dimensional Lebesgue measure.* ■

Theorem 2 [4, Theorem 7] (Coercivity of ρ). *The function $\rho_{U,M,b,B}$ defined in* (8.75) *is coercive if and only if* $[B^{\mathrm{T}}\mathrm{cone}(U)]^\circ = \{0\}$. ■

If $\rho := \rho_{U,M,b,B}$ is coercive, then, by Theorem 1, then the function $f(y) = \exp(-\rho(y))$ is integrable on aff(dom ρ) with the n_{aff}-dimensional Lebesgue measure. If we define

$$\mathbf{p}(y) = \begin{cases} c_1^{-1}\exp(-\rho(y)) & y \in \mathrm{dom}\,\rho \\ 0 & \text{else,} \end{cases} \tag{8.76}$$

where

$$c_1 = \left(\int_{y\in\mathrm{dom}\,\rho} \exp(-\rho(y))dy\right),$$

and the integral is with respect to the Lebesgue measure with dimension n_{aff}, then $\mathbf{p}$ is a probability density on dom(ρ). We call these PLQ densities.

8.5.2.3 PLQ Densities and Kalman Smoothing

We now show how to build up the penalty functions V_k and J_k in (8.73) using PLQ densities. We will do this for the linear model (8.1–8.2) for simplicity. The nonlinear case can be handled as before by applying the Gauss-Newton strategy to the underlying convex composite function.

Using the notion given in 8.5, the linear model (8.1–8.2) can be written as

$$\begin{aligned} w &= Gx + \mathbf{w} \\ z &= Hx + \mathbf{v}. \end{aligned} \tag{8.77}$$

A general Kalman smoothing problem can be specified by assuming that the noises $\mathbf{w}$ and $\mathbf{v}$ in the model (8.77) have PLQ densities with means 0, variances Q and R (8.5). Then, for suitable $\{U_k^w, M_k^w, b_k^w, B_k^w\}$ and $\{U_k^v, M_k^v, b_k^v, B_k^v\}$, we have

$$\begin{aligned} \mathbf{p}(w) &\propto \exp(-\theta_{U^w,M^w}(b^w + B^w Q^{-1/2}w)) \\ \mathbf{p}(v) &\propto \exp(-\theta_{U^v,M^v}(b^v + B^v R^{-1/2}v)), \end{aligned} \tag{8.78}$$

where

$$\begin{aligned} U^w &= \prod_{k=1}^N U_k^w \subset \mathbb{R}^{nN} \\ U^v &= \prod_{k=1}^N U_k^v \subset \mathbb{R}^{M} \end{aligned}, \quad \begin{aligned} M^w &= \mathrm{diag}(\{M_k^w\}) \\ M^v &= \mathrm{diag}(\{M_k^v\}) \end{aligned}, \quad \begin{aligned} B^w &= \mathrm{diag}(\{B_k^w\}) \\ B^v &= \mathrm{diag}(\{B_k^v\}) \\ b^w &= \mathrm{vec}(\{b_k^w\}) \\ b^v &= \mathrm{vec}(\{b_k^v\}) \end{aligned}.$$

Then the MAP estimator for x in the model (8.77) is

$$\arg\min_{x\in\mathbb{R}^{nN}} \left\{ \begin{array}{l} \theta_{U^w,M^w}(b^w + B^w Q^{-1/2}(Gx - w)) \\ + \theta_{U^v,M^v}(b^v + B^v R^{-1/2}(Hx - z)) \end{array} \right\}. \tag{8.79}$$

Note that since w_k and v_k are independent, problem (8.79) is decomposable into a sum of terms analogous to (8.73). This special structure follows from the block diagonal structure of H, Q, R, B^v, B^w, the bidiagonal structure of G, and the product structure of sets U^w and U^v, and is key in proving the linear complexity of the solution method we propose.

8.5.2.4 Solving the Kalman Smoother Problem with PLQ Densities

Recall that, when the sets U^w and U^v are polyhedral, (8.79) is an Extended Linear Quadratic program (ELQP), described in [43, Example 11.43]. We solve (8.79) by working directly with its associated Karush-Kuhn-Tucker (KKT) system.

Lemma 1 [4, Lemma 3.1] *Suppose that the sets U_k^w and U_k^v are polyhedral, that is, they can be given the representation*

$$U_k^w = \{u|(A_k^w)^T u \le a_k^w\}, \quad U_k^v = \{u|(A_k^v)^T u \le a_k^v\}.$$

Then the first-order necessary and sufficient conditions for optimality in (8.79) *are given by*

$$\begin{array}{ll} 0 = (A^w)^{\mathrm{T}} u^w + s^w - a^w; & 0 = (A^v)^{\mathrm{T}} u^v + s^v - a^v \\ 0 = (s^w)^{\mathrm{T}} q^w; & 0 = (s^v)^{\mathrm{T}} q^v \\ & \\ 0 = \tilde{b}^w + B^w Q^{-1/2} G x - M^w u^w - A^w q^w & \\ 0 = \tilde{b}^v - B^v R^{-1/2} H x - M^v u^v - A^v q^v & \\ 0 = G^{\mathrm{T}} Q^{-\mathrm{T}/2} (B^w)^{\mathrm{T}} u^w - H^{\mathrm{T}} R^{-\mathrm{T}/2} (B^v)^{\mathrm{T}} u^v & \\ 0 \le s^w, s^v, q^w, q^v. & \end{array} , \tag{8.80}$$

where $\tilde{b}^w = b^w - B^w Q^{-1/2} w$ and $\tilde{b}^v = b^v - B^v R^{-1/2} z$. ■

We propose solving the KKT conditions (8.80) by an Interior Point (IP) method. IP methods work by applying a damped Newton iteration to a relaxed version of (8.80) where the relaxation is to the complementarity conditions. Specifically, we replace the complementarity conditions by

$$\begin{array}{l} (s^w)^{\mathrm{T}} q^w = 0 \to Q^w S^w \mathbf{1} - \mu \mathbf{1} = 0 \\ (s^v)^{\mathrm{T}} q^v = 0 \to Q^v S^v \mathbf{1} - \mu \mathbf{1} = 0, \end{array}$$

where Q^w, S^w, Q^v, S^v are diagonal matrices with diagonals q^w, s^w, q^v, s^v respectively. The parameter μ is aggressively decreased to 0 as the IP iterations proceed. Typically, no more than 10 or 20 iterations of the relaxed system are required to obtain a solution of (8.80), and hence an optimal solution to (8.79). The following theorem shows that the computational effort required (per IP iteration) is linear in the number of time steps whatever PLQ density enters the state space model.

Theorem 3 [4, Theorem 3.2] (PLQ Kalman Smoother Theorem) *Suppose that all w_k and v_k in the Kalman smoothing model* (8.1–8.2) *come from PLQ densities that satisfy* $\text{Null}(M) \cap U^\infty = \{0\}$. *Then an IP method can be applied to solve* (8.79) *with a per iteration computational complexity of $O(Nn^3 + Nm)$.* ■

The proof, which can be found in [4], shows that IP methods for solving (8.79) preserve the key block tridiagonal structure of the standard smoother. General smoothing estimates can therefore be computed in $O(Nn^3)$ time, as long as the number of IP iterations is fixed (as it usually is in practice, to 10 or 20).

It is important to observe that the motivating examples all satisfy the conditions of Theorem 3.

Corollary 1 [4, Corollary 3.3] *The densities corresponding to L^1, L^2, Huber, and Vapnik penalties all satisfy the hypotheses of Theorem 3.*

Proof We verify that $\text{Null}(M) \cap \text{Null}(A^\text{T}) = 0$ for each of the four penalties. In the L^2 case, M has full rank. For the L^1, Huber, and Vapnik penalties, the respective sets U are bounded, so $U^\infty = \{0\}$.

8.5.2.5 Numerical Example: Vapnik Penalty and Functional Recovery

In this section we present a numerical example to illustrate the use of the Vapnik penalty (see Fig. 8.9) in the Kalman smoothing context, for a functional recovery application.

We consider the following function

$$f(t) = \exp\left[\sin(8t)\right]$$

taken from [19]. Our aim is to reconstruct f starting from 2000 noisy samples collected uniformly over the unit interval. The measurement noise v_k was generated using a mixture of two normals with $p = 0.1$ denoting the fraction from each normal; i.e.,

$$v_k \sim (1-p)\mathbf{N}(0, 0.25) + p\mathbf{N}(0, 25),$$

where $\mathbf{N}$ refers to the Normal distribution. Data are displayed as dots in Fig. 8.10. Note that the purpose of the second component of the normal mixture is to simulate outliers in the output data and that all the measurements exceeding vertical axis limits are plotted on upper and lower axis limits (4 and -2) to improve readability.

The initial condition $f(0) = 1$ is assumed to be known, while the difference of the unknown function from the initial condition (i.e., $f(\cdot) - 1$) is modeled as a Gaussian process given by an integrated Wiener process. This model captures the Bayesian interpretation of cubic smoothing splines [48], and admits a 2-dimensional

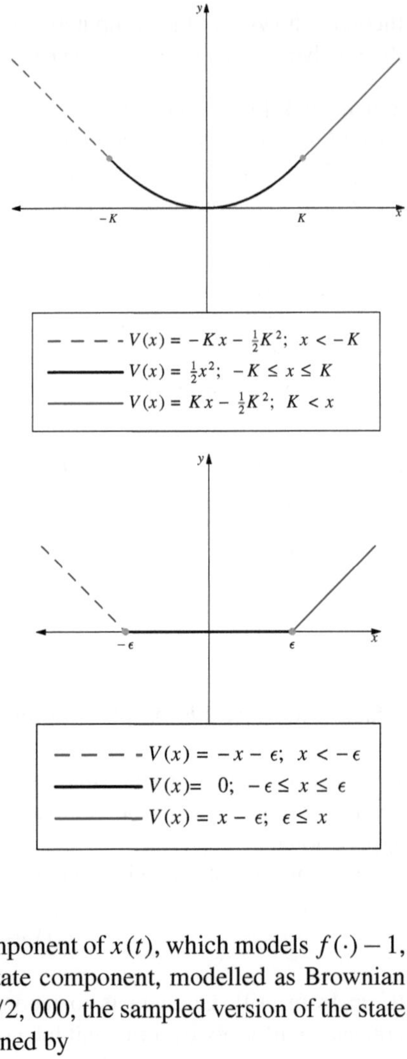

Fig. 8.9 Huber (*upper*) and Vapnik (*lower*) penalties

state space representation where the first component of $x(t)$, which models $f(\cdot) - 1$, corresponds to the integral of the second state component, modelled as Brownian motion. To be more specific, letting $\Delta t = 1/2,000$, the sampled version of the state space model (see [26, 38] for details) is defined by

$$G_k = \begin{bmatrix} 1 & 0 \\ \Delta t & 1 \end{bmatrix}, \qquad k = 2, 3, \ldots, 2,000$$
$$H_k = \begin{bmatrix} 0 & 1 \end{bmatrix}, \qquad k = 1, 2, \ldots, 2,000$$

with the autocovariance of w_k given by

$$Q_k = \lambda^2 \begin{bmatrix} \Delta t & \frac{\Delta t^2}{2} \\ \frac{\Delta t^2}{2} & \frac{\Delta t^3}{3} \end{bmatrix}, \qquad k = 1, 2, \ldots, 2,000,$$

where λ^2 is an unknown scale factor to be estimated from the data.

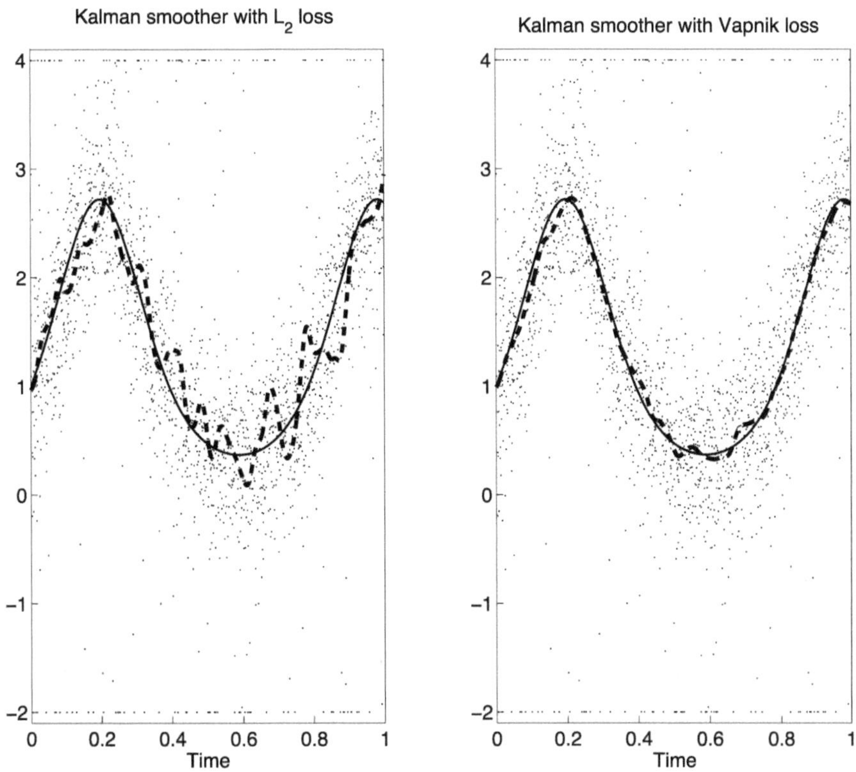

Fig. 8.10 Simulation: measurements ($\cdot$) with outliers plotted on axis limits (4 and -2), true function (*continuous line*), smoothed estimate using either the quadratic loss (*dashed line, left panel*) or the Vapnik's ϵ-insensitive loss (*dashed line, right panel*)

The performance of two different Kalman smoothers are compared. The first (classical) estimator uses a quadratic loss function to describe the negative log of the measurement noise density and contains only λ^2 as unknown parameter. The second estimator is a Vapnik smoother relying on the ϵ-insensitive loss, and so depends on two unknown parameters λ^2 and ϵ. In both of the cases, the unknown parameters are estimated by means of a cross validation strategy where the 2,000 measurements are randomly split into a training and a validation set of 1,300 and 700 data points, respectively. The Vapnik smoother was implemented by exploiting the efficient computational strategy described in the previous section, see [8] for specific implementation details. In this way, for each value of λ^2 and ϵ contained in a 10×20 grid on $[0.01, 10, 000] \times [0, 1]$, with λ^2 logarithmically spaced, the function estimate was rapidly obtained by the new smoother applied to the training set. Then, the relative average prediction error on the validation set was computed, see Fig. 8.11. The parameters leading to the best prediction were $\lambda^2 = 2.15 \times 10^3$ and $\epsilon = 0.45$, which give a sparse solution defined by fewer than 400 support vectors.

The value of λ^2 for the classical Kalman smoother was then estimated following the same strategy described above. In contrast to the Vapnik penalty, the quadratic loss does not induce any sparsity, so that, in this case, the number of support vectors equals the size of the training set.

The left and right panels of Fig. 8.10 display the function estimate obtained using the quadratic and the Vapnik losses, respectively. It is clear that the Gaussian estimate is heavily affected by the outliers. In contrast, as expected, the estimate coming from the Vapnik based smoother performs well over the entire time period, and is virtually unaffected by the presence of large outliers.

8.6 Sparse Kalman Smoothing

In recent years, sparsity promoting formulations and algorithms have made a tremendous impact in signal processing, reconstruction algorithms, statistics, and inverse problems (see e.g., [13] and the references therein). In some contexts, rigorous mathematical theory is available that can guarantee recovery from under-sampled sparse signals [20]. In addition, for many inverse problems, sparsity promoting optimization provides a way to exploit prior knowledge of the signal class as a way to improve the solution to an ill-posed problem, but conditions for recoverability have not yet been derived [36].

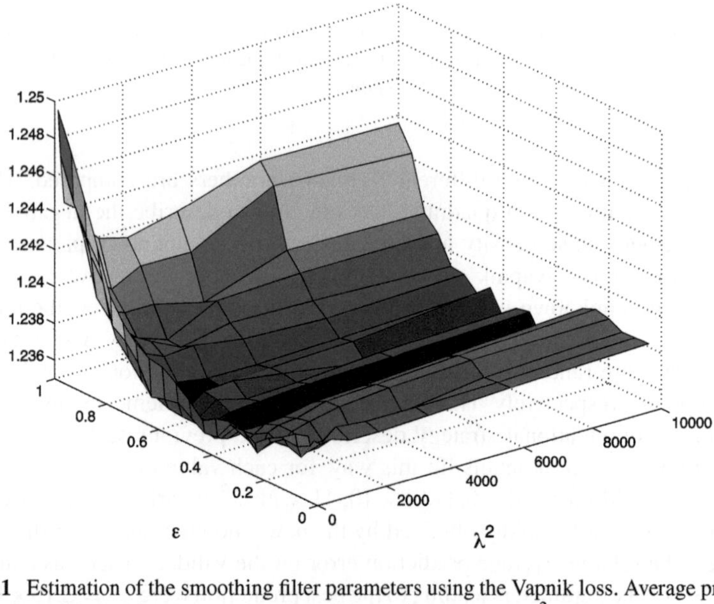

Fig. 8.11 Estimation of the smoothing filter parameters using the Vapnik loss. Average prediction error on the validation data set as a function of the variance process λ^2 and ϵ

In the context of dynamic models, several sparse Kalman filters have been recently proposed [1, 16, 17, 47]. In the applications considered, in addition to process and measurement models, the state space is also known to be sparse. The aim is to improve recovery by incorporating sparse optimization techniques. Reference [1] is very close to the work presented in this section, since they formulate a sparsity promoting optimization problem over the whole measurement sequence and solve it with an optimization technique shown to preserve computational efficiency.

In this section, we formulate the sparse Kalman smoothing problem as an optimization problem over the entire state space sequence, and suggest two new approaches for the solution of such problems. The first approach is based on the interior point methodology, and is a natural extension of the mathematics presented in earlier sections.

The second approach is geared towards problems where the dimension n (state at a single time point) is large. For this case, we propose a matrix free approach, using a different (constrained) Kalman smoothing formulation, together with the projected gradient method. In both methods, the structure of the Kalman smoothing problem is exploited to achieve computational efficiency.

We present theoretical development for the two approaches, leaving applications and numerical results to future work.

8.6.1 Penalized Formulation and Interior Point Approach

We consider only the linear smoother (8.6). A straight forward way to impose sparsity on the state is to augment this formulation with a 1-norm penalty:

$$\min_x f(x) := \frac{1}{2}\|Hx - z\|^2_{R^{-1}} + \frac{1}{2}\|Gx - w\|^2_{Q^{-1}} + \lambda\|Wx\|_1, \tag{8.81}$$

where W is a diagonal weighting matrix included for modeling convenience. For example, the elements of W can be set to 0 to exclude certain parts of the state dimension from the sparse penalty. A straightforward constrained reformulation of (8.81) is

$$\begin{aligned} \min_x \quad & \frac{1}{2}\|Hx - z\|^2_{R^{-1}} + \frac{1}{2}\|Gx - w\|^2_{Q^{-1}} + \lambda \mathbf{1}^T y \\ \text{s.t.} \quad & -y \le Wx \le y. \end{aligned} \tag{8.82}$$

Note that this is different from the constrained problem (8.37), because we have introduced a new variable y, with constraints in x and y. Nonetheless, an interior point approach may still be used to solve the resulting problem. We rewrite the constraint in (8.88) using non-negative slack variables s, r:

$$\begin{aligned} Wx - y + s &= 0 \\ -Wx - y + r &= 0, \end{aligned} \tag{8.83}$$

and form the Lagrangian for the corresponding system:

$$L(s, r, q, p, y, x) = x^T Cx + c^T x + \lambda \mathbf{1}^T y + q^T (Wx - y + s) + p^T (-Wx - y + r), \tag{8.84}$$

with C as in (8.8) and c as in (8.39)., and where q and p are the dual variables corresponding to the inequality constraints $Wx \leq y$ and $-Wx \leq -y$, respectively. The (relaxed) KKT system is therefore given by

$$F_\mu(s, r, q, p, y, x) := \begin{pmatrix} s - y + Wx \\ r - y - Wx \\ D(s)D(q)\mathbf{1} - \mu\mathbf{1} \\ D(r)D(p)\mathbf{1} - \mu\mathbf{1} \\ \lambda\mathbf{1} - q - p \\ Wq - Wp + Cx + c \end{pmatrix} = 0. \tag{8.85}$$

The derivative matrix $F_\mu^{(1)}$ is given by

$$F_\mu^{(1)} = \begin{bmatrix} I & 0 & 0 & 0 & -I & W \\ 0 & I & 0 & 0 & -I & -W \\ D(q) & 0 & D(s) & 0 & 0 & 0 \\ 0 & D(p) & 0 & D(r) & 0 & 0 \\ 0 & 0 & -I & -I & 0 & 0 \\ 0 & 0 & W & -W & 0 & C \end{bmatrix}, \tag{8.86}$$

and it is row equivalent to the system

$$\begin{bmatrix} I & 0 & 0 & 0 & -I & W \\ 0 & I & 0 & 0 & -I & -W \\ 0 & 0 & D(s) & 0 & D(q) & -D(q)W \\ 0 & 0 & 0 & D(r) & D(p) & D(p)W \\ 0 & 0 & 0 & 0 & \Phi & -\Psi W \\ 0 & 0 & 0 & 0 & 0 & C + W\Phi^{-1}\left(\Phi^2 - \Psi^2\right)W \end{bmatrix}$$

where

$$\begin{aligned} \Phi &= D(s)^{-1}D(q) + D(r)^{-1}D(p) \\ \Psi &= D(s)^{-1}D(q) - D(r)^{-1}D(p), \end{aligned} \tag{8.87}$$

and the matrix $\Phi^2 - \Psi^2$ is diagonal, with the iith entry given by $4q_i r_i$. Therefore, the modified system preserves the structure of C; specifically it is symmetric, block tridiagonal, and positive definite. The Newton iterations required by the interior point method can therefore be carried out, with each iteration having complexity $O(n^3 N)$.

8.6.2 Constrained Formulation and Projected Gradient Approach

Consider again the linear smoother (8.6), but now impose a 1-norm constraint rather than a penalty:

$$\begin{aligned} \min_x f(x) :=& \frac{1}{2}\|Hx - z\|^2_{R^{-1}} + \frac{1}{2}\|Gx - w\|^2_{Q^{-1}} \\ \text{s.t.} \quad & \|Wx\|_1 \le \tau. \end{aligned} \tag{8.88}$$

This problem, which equivalent to (8.81) for certain values of λ and τ, is precisely the LASSO problem [45], and can be written

$$\min \frac{1}{2}x^T Cx + c^T x \quad \text{s.t.} \quad \|Wx\|_1 \le \tau. \tag{8.89}$$

with $C \in \mathbb{R}^{nN \times nN}$ as in (8.8) and $c \in \mathbb{R}^{nN}$ as in (8.39). When n is large, the interior point method proposed in the previous section may not be feasible, since it requires exact solutions of the system

$$(C + W\Phi^{-1}\left(\Phi^2 - \Psi^2\right) W)x = r,$$

and the block-tridiagonal Algorithm 1 requires the inversion of $n \times n$ systems.

The problem (8.89) can be solved without inverting such systems, using the spectral projected gradient method, see e.g., [46, Algorithm 1]. Specifically, the gradient $Cx + c$ must be repeatedly computed, and then $x^\nu - (Cx^\nu + c)$ is projected onto the set $\|Wx\|_1 \le \tau$. (the word 'spectral' refers to the fact that the Barzilai-Borwein line search is used to get the step length).

In the case of the Kalman smoother, the gradient $Cx + c$ can be computed in $O(n^2 N)$ time, because of the special structure of C. Thus for large systems, the projected gradient method that exploits the structure of C affords significant savings per iteration relative to the interior point approach, $O(n^2N)$ vs. $O(n^3N)$, and relative to a method agnostic to the structure of C, $O(n^2N)$ vs. $O(n^2N^2)$. The projection onto the feasible set $\|Wx\|_1 \le \tau$ can be done in $O(nN \log(nN))$ time.

8.7 Conclusions

In this chapter, we have presented an optimization approach to Kalman smoothing, together with a survey of applications and extensions. In Sect. 8.2.5, we showed that the recursive Kalman filtering and smoothing algorithm is equivalent to Algorithm 1, an efficient method to solve block tridiagonal positive definite systems. In the following sections, we used this algorithm as a subroutine, allowing us to present new ideas on a high level, without needing to explicitly write down modified Kalman filtering and smoothing equations.

We have presented extensions to nonlinear process and measurement models in Sect. 8.3, described constrained Kalman smoothing (both the linear and nonlinear cases) in Sect. 8.4, and presented an entire class of robust Kalman smoothers (derived by considering log-linear-quadratic densities) in Sect. 8.5. For all of these applications, nonlinearity in the process, measurements, and constraints can be handled by a generalized Gauss-Newton method that exploits the convex composite structure discussed in Sects. 8.3.1 and 8.4.4. The GN subproblem can be solved either in closed form or via an interior point approach; in both cases Algorithm 1 was used. For all of these extensions, numerical illustrations have also been presented, and most are available for public release through the `ckbs` package [6].

In the case of the robust smoothers, it is possible to extend the density modeling approach by considering densities outside the log-concave class [3], but we do not discuss this work here.

We ended the survey of extensions by considering two novel approaches to Kalman smoothing of sparse systems, for applications where modeling the sparsity of the state space sequence improves recovery. The first method built on the readers' familiarity with the interior point approach as a tool for the constrained extension in Sect. 8.4. The second method is suitable for large systems, where exact solution of the linear systems is not possible. Numerical illustrations of the methods have been left to future work.

References

1. Angelosante D, Roumeliotis SI, Giannakis GB (2009) Lasso-kalman smoother for tracking sparse signals. In: 2009 conference record of the 43rd Asilomar conference on signals, systems and computers, pp 181–185
2. Ansley CF, Kohn R (1982) A geometric derivation of the fixed interval smoothing algorithm. Biometrika 69:486–487
3. Aravkin A, Burke J, Pillonetto G (2011) Robust and trend-following Kalman smoothers using students t. In: International federation of automaic control (IFAC), 16th symposium of system identification, Oct 2011
4. Aravkin A, Burke J, Pillonetto G (2011) A statistical and computational theory for robust and sparse Kalman smoothing. In: International federation of automaic control (IFAC), 16th symposium of system identification, Oct 2011
5. Aravkin AY (2010) Robust methods with applications to Kalman smoothing and bundle adjustment. Ph.D. Thesis, University of Washington, Seattle, June 2010
6. Aravkin AY, Bell BM, Burke JV, and Pillonetto G, (2007–2011) Matlab/Octave package for constrained and robust Kalman smoothing
7. Aravkin AY, Bell BM, Burke JV, Pillonetto G (2011) An ℓ_1-laplace robust kalman smoother. IEEE Trans Autom Control 56(12):2898–2911
8. Aravkin AY, Bell BM, Burke JV, Pillonetto G (2011) Learning using state space kernel machines. In: Proceedings of IFAC World congress 2011, Milan
9. Bell BM (1994) The iterated Kalman smoother as a Gauss-Newton method. SIAM J Opt 4(3):626–636

10. Bell BM (2000) The marginal likelihood for parameters in a discrete Gauss-Markov process. IEEE Trans Signal Process 48(3):626–636
11. Bell BM, Burke JV, Pillonetto G (2009) An inequality constrained nonlinear kalman-bucy smoother by interior point likelihood maximization. Automatica 45(1):25–33
12. Bell BM, Cathey F (1993) The iterated Kalman filter update as a Gauss-Newton method. IEEE Trans Autom Control 38(2):294–297
13. Bruckstein Alfred M, Donoho David L, Elad Michael (2009) From sparse solutions of systems of equations to sparse modeling of signals and images. SIAM Rev 51(1):34–81
14. Burke JV, Han SP (1989) A robust sequential quadratic programming method. Math Program 43:277–303. doi:10.1007/BF01582294
15. Burke James V (1985) Descent methods for composite nondifferentiable optimization problems. Math Program 33:260–279
16. Carmi A, Gurfil P, Kanevsky D (2010) Methods for sparse signal recovery using kalman filtering with embedded pseudo-measurement norms and quasi-norms. IEEE Trans Signal Process 58:2405–2409
17. Carmi A, Gurfil P, Kanevsky D (2008) A simple method for sparse signal recovery from noisy observations using kalman filtering. Technical report RC24709, Human Language Technologies, IBM
18. Van der Merwe R (2004) Sigma-point Kalman filters for probabilistic inference in dynamic state-space models. Ph.D. Thesis, OGI School of Science and Engineering, Oregon Health and Science University, April 2004
19. Dinuzzo F, Neve M, De Nicolao G, Gianazza UP (2007) On the representer theorem and equivalent degrees of freedom of SVR. J Mach Learn Res 8:2467–2495
20. Donoho DL (2006) Compressed sensing. IEEE Trans Inf Theory 52(4):1289–1306
21. Fahrmeir L, Kaufmann V (1991) On Kalman filtering, posterior mode estimation, and Fisher scoring in dynamic exponential family regression. Metrika 38: 37–60
22. Fahrmeir Ludwig, Kunstler Rita (1998) Penalized likelihood smoothing in robust state space models. Metrika 49:173–191
23. Gao Junbin (2008) Robust L1 principal component analysis and its Bayesian variational inference. Neural Comput 20(2):555–572
24. Gillijns V, Mendoza OB, Chandrasekar V, De Moor BLR, Bernstein DS, Ridley A (2006) What is the ensemble Kalman filter and how well does it work? In: Proceedings of the American control conference (IEEE 2006), pp 4448–4453
25. Hewer GA, Martin RD, Judith Zeh (1987) Robust preprocessing for Kalman filtering of glint noise. IEEE Trans Aerosp Electron Syst AES-23(1):120–128
26. Jazwinski A (1970) Stochastic processes and filtering theory. Dover Publications, Inc.
27. Dennis JE Jr, Schnabel. RB (1983) Numerical methods for unconstrained optimiation and nonlinear equations. Computational mathematics, Prentice-Hall, Englewood Cliffs
28. Julier Simon, Uhlmann Jeffrey, Durrant-White Hugh (2000) A new method for the nonlinear transformation of means and covariances in filters and estimators. IEEE Trans Autom Control 45(3):477–482
29. Kalman RE (1960) A new approach to linear filtering and prediction problems. Trans AMSE J Basic Eng 82(D):35–45
30. Kandepu R, Foss B, Imsland L (2008) Applying the unscented Kalman filter for nonlinear state estimation. J Process Control 18:753–768
31. Kim S-J, Koh K, Boyd S, Gorinevsky D (2009) ℓ_1 trend filtering. Siam Rev 51(2):339–360
32. Kojima M, Megiddo N, Noma T, Yoshise A (1991) A unified approach to interior point algorithms for linear complementarity problems. Lecture notes in computer science, vol 538. Springer Verlag, Berlin
33. Kourouklis S, Paige CC (1981) A constrained least squares approach to the general Gauss-Markov linear model. J Am Stat Assoc 76(375):620–625
34. Lefebvre T, Bruyninckx H, De Schutter J (2004) Kalman filters for nonlinear systems: A comparison of performance. Intl J Control 77(7):639–653

35. Liu Jun S, Chen Rong (1998) Sequential Monte Carlo methods for dynamic systems. J Am Stat Assoc 93:1032–1044
36. Mansour H, Wason H, Lin TTY, Herrmann FJ (2012) Randomized marine acquisition with compressive sampling matrices. Geophys Prospect 60(4):648–662
37. Nemirovskii A, Nesterov Y (1994) Interior-point polynomial algorithms in convex programming. Studies in applied mathematics, vol 13. SIAM, Philadelphia
38. Oksendal B (2005) Stochastic differential equations, 6th edn. Springer, Berlin
39. Paige CC, Saunders MA (1977) Least squares estimation of discrete linear dynamic systems using orthogonal transformations. Siam J Numer Anal 14(2):180–193
40. Paige CC (1985) Covariance matrix representation in linear filtering. Contemp Math 47:309–321
41. Pillonetto G, Aravkin AY, Carpin S (2010) The unconstrained and inequality constrained moving horizon approach to robot localization. In: 2010 IEEE/RSJ international conference on intelligent robots and systems, Taipei, pp 3830–3835
42. Rauch HE, Tung F, Striebel CT (1965) Maximum likelihood estimates of linear dynamic systems. AIAA J 3(8):1145–1150
43. Rockafellar RT, Wets RJ-B (1998) Variational analysis. A series of comprehensive studies in mathematics, vol 317. Springer, Berlin
44. Schick Irvin C, Mitter Sanjoy K (1994) Robust recursive estimation in the presence of heavy-tailed observation noise. Annal Stat 22(2):1045–1080
45. Tibshirani R (1996) Regression shrinkage and selection via the LASSO. J R Stat Soc Ser B 58(1):267–288
46. van den Berg E, Friedlander MP (2008) Probing the pareto frontier for basis pursuit solutions. SIAM J Sci Comput 31(2):890–912
47. Vaswani N (2008) Kalman filtered compressed sensing. In: Proceedings of the IEEE international conference on image processing (ICIP)
48. Wahba G (1990) Spline models for observational data. SIAM, Philadelphia
49. Wright SJ (1997) Primal-dual interior-point methods. Siam, Englewood Cliffs

Chapter 9
Compressive System Identification

Avishy Y. Carmi

Abstract The first part of this chapter presents a novel Kalman filtering-based method for estimating the coefficients of sparse, or more broadly, compressible autoregressive models using fewer observations than normally required. By virtue of its (unscented) Kalman filter mechanism, the derived method essentially addresses the main difficulties attributed to the underlying estimation problem. In particular, it facilitates sequential processing of observations and is shown to attain a good recovery performance, particularly under substantial deviations from ideal conditions, those which are assumed to hold true by the theory of compressive sensing. In the remaining part of this chapter we derive a few information-theoretic bounds pertaining to the problem at hand. The obtained bounds establish the relation between the complexity of the autoregressive process and the attainable estimation accuracy through the use of a novel measure of complexity. This measure is suggested herein as a substitute to the generally incomputable restricted isometric property.

9.1 Introduction

The common practice in engineering and science is to describe systems and processes by means of parametric models. In many cases the values of the parameters cannot be directly measured but rather they should be estimated by merely observing the system behaviour. The art and science of doing so is sometimes referred to as system identification. Within this discipline, autoregressive (AR) models/processes are perhaps one of the most valuable tools for time-series analysis. These models are widely used for change detection [7], dynamical system modeling [37], forecasting (e.g., the Box-Jenkins framework [1]) and causal reasoning (e.g., Granger causality [30]).

A. Y. Carmi (✉)
School of Mechanical and Aerospace Engineering, Nanyang Technological University, Singapore 639798, Singapore
e-mail: acarmi@ntu.edu.sg

A. Y. Carmi et al. (eds.), *Compressed Sensing & Sparse Filtering*,
Signals and Communication Technology, DOI: 10.1007/978-3-642-38398-4_9,

The underlying formalism has to do with non Markovian dynamics (e.g., future outcomes depend on the process lagged values) and is best known for its ability to adequately capture complex behaviours and nonlinearities such as those underlying natural phenomena [1].

By virtue of their linear formulation, AR models allow simple inference methods to be employed. The archetypical procedure is aimed at learning the process coefficients which best explain the available time-series data by means of maximum likelihood (ML) or least squares (LS) techniques. If the time-series data was indeed produced using an AR model then, under mild conditions, this procedure is guaranteed to yield the actual coefficients to within accuracy which primarily depends on the amount of observations. Notwithstanding, the causal generating mechanism underlying the observed data is inaccessible in virtually many real-world applications and in that case fitting an AR model would be considered as an approximation at most. In this respect, a fitted AR model is deemed plausible if it facilitates the prediction of future patterns and behaviours exhibited by the actual process.

9.1.1 Sparse and Compressible AR Models

Sparse and, more broadly, compressible AR models are those which consist of only a few elementary components with the rest being negligible in terms of their contribution to the observed phenomena. As we are essentially dealing with linear expressions this characterisation translates into having a significant number of nearly vanishing parameters. This configuration may be conceptualised by portraying the AR model as a graph for which the nodes and edges are, respectively, the underlying random variables and their statistical dependency. In this view, a sparse AR model corresponds to a graph with only a few connected nodes, which is otherwise known as a sparse graphical model (see Fig. 9.1). This kind of model, which is a well-studied concept in machine learning and statistics, is known to promote reduced-order descriptions and plausible inference [29]. Over the past years, sparse AR models have been used for fMRI analysis and causal discovery [9, 31], and more recently also for system identification [45].

One of the tasks in fitting an AR model is to determine its order, that is, the number of parameters necessary for describing the generating mechanism underlying the time-series data [10]. Conventional methods for doing so mostly rely on the Akaike information criterion (AIC) [32], or its counterpart, the Bayesian information

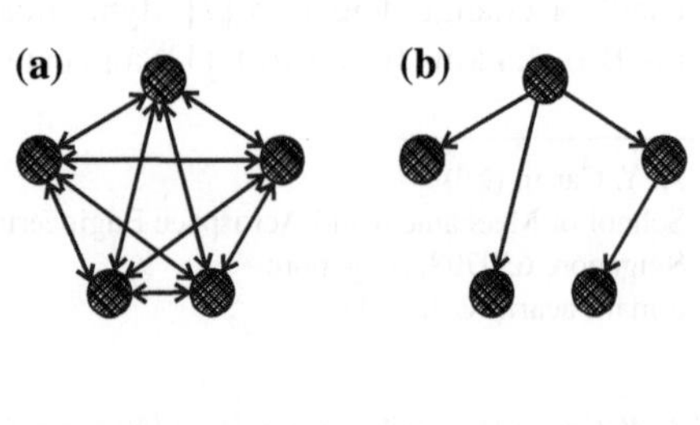

Fig. 9.1 Fully connected and sparsely connected graphical models

criterion (BIC) [46], both which penalise the underlying likelihood function with respect to the number of parameters. There are at least two reasons why this is vital for obtaining interpretable models. The first reason is that normally we are limited by the amount of available data which necessarily restricts the number of parameters that can be used in practice. The other, equally important, reason has to do with the principle of Occam's Razor which essentially attests in favor of simplified descriptions. In that sense, expendable parameters would increase the chance of departure from the factual generating mechanism underlying the data. As an example, the undesired effect referred to as overfitting is one of the best known manifestations of using extra parameters.

An autoregressive model may be considered sparse if it involves much more parameters than what is necessary for adequately describing the observed phenomenon. In this setting we would normally be interested in recovering only the few significant (non-vanishing) parameters that underly the time-series data. This is exactly what compressive sensing techniques are made for (see below). Furthermore, since not all parameters are of the same importance it turns out that fewer observations would be required to estimate only those which are truly necessary. This approach, which can be viewed as simultaneous parameter estimation and variable selection, completely does away with peripheral procedures for model determination such as AIC and BIC.

Complex systems are yet another example in which compressive AR models may be useful. Occasionally, in such cases there are multiple time-series each pertaining to an individual component or a group of components within the system. The standard AR formalism would naturally embody the interrelations among various system components, and in this respect the vector of parameters would be deemed sparse whenever only a small subset of entities affects the majority of constituents. This sort of hierarchical mechanism underlies a wide range of emergent behaviours in group dynamics and social networks.

In this work we suggest estimating the parameters of sparse and compressible AR models by means of compressive sensing techniques. We provide a brief overview of this paradigm in which we highlight a few implementation issues in the context of this problem. This part is rather crucial for understanding the aims and scope of this work.

9.1.2 Compressive Sensing

In recent years compressive sensing (CS) has drawn enormous amount of attention in the signal processing community. The key concept of reconstructing signals from fewer observations than what is normally considered to be sufficient, is the primary reason popularising CS in a wide variety of scientific domains where large-scale problems naturally arise. Compressive sensing has its origins in group testing paradigms and computational harmonic analysis [51]. Similar concepts have been around for more than two decades. The prevalent formalism is that of recovering sparse signals, i.e., those which are constituted by a relatively small number of non-vanishing components, using as few as possible linear measurements.

In computational harmonics this concept has come into being largely due to the pervasiveness of orthonormal basis functions underlying the Fourier and wavelet transforms. Here, the incoherence among the basis functions facilitates highly accurate recovery of sparse signals using an amount of measurements which may be drastically smaller than the one predicted by the Nyquist-Shannon sampling theorem. The signal (unknown) support and the corresponding spike magnitudes can be obtained by means of convex programming [11, 15, 16, 22, 23], and greedy algorithms [8, 24, 40, 42] (see also [50]).

The recovery of sparse signals is in general NP-hard [22]. State-of-the-art methods for addressing this problem commonly utilise convex relaxations, non-convex local optimisation steps and greedy search mechanisms. Convex relaxations are used in various methods such as the infamous homotopy-based LASSO and LARS [11, 28], as well as in the Dantzig selector [11], basis pursuit and basis pursuit de-noising [23]. Non-convex optimization approaches include Bayesian methodologies such as the relevance vector machine, otherwise known as sparse Bayesian learning [49] and the Bayesian compressive sensing (BCS) [32]. Notable greedy search algorithms are the matching pursuit (MP) [40], the orthogonal MP (OMP) [42], iterative hard thresholding [8], and the orthogonal least squares [22].

The extent to which we are able to materialise the concepts of CS has been progressively refined over the past years [15, 16, 44, 51] until the appearance of [12, 22, 26] in which it assumes a definite form. Roughly, there are two major influencing factors: the number of prominent basis components composing the underlying signal (i.e., its support size), and the incoherence between the columns of the sensing matrix. The latter factor has to do with the condition widely known as the restricted isometric property, or RIP, in short [12, 22]. Possibly the best way to understand this is by considering a typical CS problem. Thus,

$$x = H\alpha$$

where $x \in \mathbb{R}^N$ and $\alpha \in \mathbb{R}^n$ denote, respectively, a vector of measurements and a vector of unknowns. For being consistent with the standard terminology, the matrix H, which has been given different names in various disciplines (e.g., design matrix, measurement matrix, dictionary), would be referred to as simply a sensing matrix. In this formulation, N is considerably smaller than n, and hence the linear system is underdetermined. Nevertheless, a unique solution can yet be guaranteed assuming α is sufficiently sparse. In particular, the number of non-zero entries in α should be less than $\frac{1}{2}\text{spark}(H)$ [12], where the spark of H is defined as follows.

Definition 1 *(The spark of a matrix).* The spark of a matrix $H \in \mathbb{R}^{N \times n}$, denoted as $\text{spark}(H)$, is the smallest number of columns that constitute a linearly dependent set.

A solution to the CS problem can be efficiently computed (using any one of the methods mentioned above) provided that H obeys the RIP [22, 26]. To be more precise, the necessary and sufficient conditions for exact recovery are provided in terms of the RIP [12]. Essentially, it is required that the Gramian $H^T H$ would be nearly equal to the identity matrix.

9.1.3 Challenges

In spite of its popularity, the ideas brought forth by CS cannot be easily implemented in virtually many real-world applications. Exceptions are to be made in the fields of image and video processing where incoherent basis transforms are abundant and obey most of the necessary conditions for ensuring a significant gain in recovery performance [17]. In the context of our problem one concern is that the underlying sensing matrix does not obey the RIP to the extent maintained by the theory of CS. This fact is explained in detail in the next section. To get the gist of it, illustrations of two Gramian matrices are provided in Fig. 9.2. The first one is associated with an ideal RIP sensing matrix (random with Gaussian entries [22]) whereas the other is typical to the problem at hand. This depiction doubtlessly indicates a departure from the ideal RIP settings in our case as manifested by the prominent off-diagonal patterns in the Gramian matrix. This essentially reflects relatively high coherence among the columns of the sensing matrix.

The success of CS as a new emerging paradigm in signal processing has convinced many that sensing devices should be redesigned for fully exploiting the potential benefits. It turns out that the RIP property of a sensing matrix cannot be guaranteed or assessed in virtually many practical scenarios and hence most of the elegant theory of CS do not apply in such cases. This caveat has paved the way to new hardware design practices that inherently produce ideal RIP matrices which thereby enhance the performance of CS algorithms, sometimes even to the extent of meeting the theoretical bounds [27]. In general, however, we are not privileged to redesign the system nor can we guarantee its RIP properties. The question is whether something can be done in such cases.

Another difficulty is related to the nature of existing CS methods. The fundamentals of CS build upon convex optimisation and greedy search perspectives and as such it

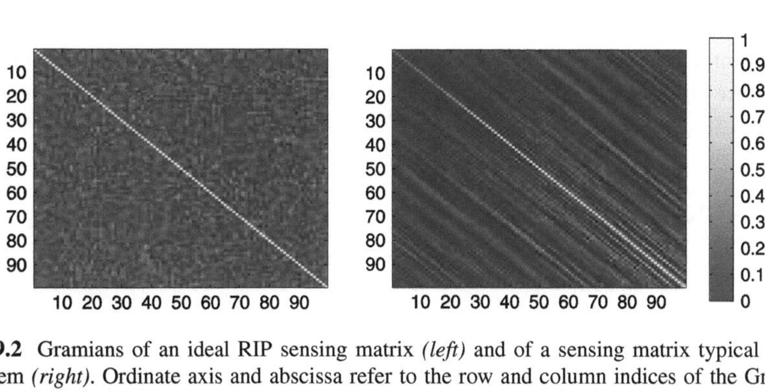

Fig. 9.2 Gramians of an ideal RIP sensing matrix *(left)* and of a sensing matrix typical to our problem *(right)*. Ordinate axis and abscissa refer to the row and column indices of the Gramian matrix, respectively

is conventionally assumed that the measurements are available in a batch form. As far as dynamical systems and time-series data are concerned this premise brings about serious limitations. In many applications observations are required to be sequentially processed which renders most of the existing CS approaches inadequate.

9.2 Contributions

In the first part of this chapter, a novel Kalman filtering-based method is introduced for estimating the coefficients of sparse, or more broadly, compressible AR models using fewer observations than normally required. By virtue of its Kalman filter (KF) mechanism, the derived method essentially addresses both of the previously mentioned difficulties. In particular, it facilitates sequential processing of observations and is shown to attain a good recovery performance, particularly under substantial deviations from ideal RIP conditions. In the remaining part of this chapter we derive a few information-theoretic bounds pertaining to the estimation problem at hand. The obtained bounds establish the relation between the complexity of the AR process and the attainable estimation accuracy through the use of a novel measure of complexity. This measure is used in this work as a substitute to the generally incomputable RIP. For the sake of readability we shall take the time to further elucidate the main concepts introduced in this work (not necessarily in their order of appearance).

9.2.1 Compressive Identification in Non-RIP Settings

The standard AR formulation poses a major challenge when adopting the prevalent views brought forth by CS. In order to better understand the underlying difficulty let us consider a typical compressive identification problem. Suppose that a vector of unknown coefficients,

$$\alpha^i := [\alpha^{i,1}(1), \dots, \alpha^{i,n}(1), \dots, \alpha^{i,1}(p), \dots, \alpha^{i,n}(p)]^T \in \mathbb{R}^{np}$$

(this notation is made clear below), consists of only a few dominant entries while all others are comparably small. These coefficients underly linear relationships among multiple $\mathbb{R}$-valued random processes, $\{x_k^j, \ k > p, \ j = 1, \dots, n\}$, that is

$$x_k^i = \sum_{t=1}^{p} \sum_{j=1}^{n} \alpha^{i,j}(t) x_{k-t}^j + \omega_k^i = \underbrace{\left[x_{k-1}^1, \dots, x_{k-1}^n, \dots, x_{k-p}^1, \dots, x_{k-p}^n\right]}_{\bar{x}_k^T} \alpha^i + \omega_k^i$$

for $i \in [1, n]$, where $\{\omega_k^i, \ k > p\}$ is a white sequence. From that point onwards we shall assume, without loss of generality, that $\{x_k^j, \ j = 1, \dots, n\}$ are zero-mean.

Usually, we are provided with N realisations of $\bar{x}_k$ and x_k^i which, respectively, constitute a sensing matrix and an observation vector. Based on these we seek to reconstruct α^i to within a satisfactory accuracy using fewer observations than its ambient dimension, np. In general the underlying sensing matrix cannot be expected to obey the RIP to the extent required by CS. This immediately follows upon noting that the columns of such a sensing matrix are not likely to be sufficiently incoherent. As this is a key detail in our developments throughout this work, we shall take the time to further explain it.

The rows of our sensing matrix are, in fact, independent samples from the (multivariate) distribution of $\bar{x}_k^T$. The covariance associated with this distribution naturally conveys the statistical coherence, or the correlation, between the underlying random variables (the samples of which constitute the columns of the sensing matrix). The entries of the corresponding $(np) \times (np)$ correlation matrix are given by

$$\mathcal{C}^{l,j} = \frac{E\left\{\bar{x}_k^j \bar{x}_k^l\right\}}{\sqrt{E\{\|\bar{x}_k^j\|_2^2\}E\{\|\bar{x}_k^l\|_2^2\}}}, \quad l, j \in [1, np] \tag{9.1}$$

where $E\{\cdot\}$ and and $\|\cdot\|_2$ denote, respectively, the expectation operator and the Euclidean norm. If, in addition, we let $H \in \mathbb{R}^{N\times(np)}$ be a normalised version of our sensing matrix, i.e., with all its columns having a unit magnitude, then a Monte Carlo estimate of the correlation matrix $\mathcal{C}$ is simply $\hat{\mathcal{C}} := H^T H$, which is otherwise known as the Gramian matrix. Note that, in the context of CS, $\hat{\mathcal{C}}$ is necessarily rank-deficient as $N < np$. Notwithstanding, the entries of $\hat{\mathcal{C}}$ approaches those of the actual correlation matrix with an increasing N, and in that sense they may adequately represent the statistical coherence. Hence, an approximation associated with (9.1) is the inner product between two columns in H, namely

$$\hat{\mathcal{C}}^{l,j} = \langle H^l, H^j \rangle, \quad l, j \in [1, np] \tag{9.2}$$

Having said that, it is rather clear that we can associate ideal RIP settings with nearly diagonal $\hat{\mathcal{C}}$ and $\mathcal{C}$. This can immediately be recognised by explicitly writing the RIP [12]

$$\left|\|H\alpha^i\|_2^2 - \|\alpha^i\|_2^2\right| = \left|(\alpha^i)^T \left(\hat{\mathcal{C}} - I_{((np)\times(np))}\right)\alpha^i\right| \le \delta_{2s}\|\alpha^i\|_2^2 \tag{9.3}$$

where the inequality holds for a fixed $\delta_{2s} \in (0, 1)$ and any sparse vector α^i having no more than s non vanishing entries, *which would henceforth be referred to as s-sparse*. The restriction on δ_{2s} to be fixed irrespective of α^i is known in the literature as a uniform recovery condition. Classical and contemporary results in the theory of CS guarantee perfect and highly accurate recovery of sparse parameter vectors when this property is satisfied for small values of δ_{2s} (see for example [6, 12, 14]). A set of concomitant results, known largely as concentration of measure, provide further evidence that random sensing matrices are most likely to obey this property with overwhelming probability under some restrictions on their dimensions [44, 51].

Further letting $\tilde{x}_k := [x_k^1, \ldots, x_k^n]$, it can be recognised that C is merely composed of the autocorrelations

$$\frac{E\left\{\tilde{x}_{k-r}\tilde{x}_{k-t}^T\right\}}{\sqrt{E\{\|\tilde{x}_{k-r}\|_2^2\}E\{\|\tilde{x}_{k-t}\|_2^2\}}}, \quad r, t \in [1, p]$$

and hence the overall structure of both C and its empirical approximation $\hat{C}$ has to do with the mixing properties of the underlying process $\{\tilde{x}_k\}_{k>p}$. As mentioned earlier, ideal RIP settings are associated with a nearly diagonal C which would only be the case when $\{\tilde{x}_k\}_{k>p}$ is strongly mixing. It is well known that this trait entails ergodicity and that some convergent processes become strongly mixing after reaching stationarity (e.g., stable linear stochastic systems). Generally, however, this cannot be expected to hold true and hence ideal RIP conditions are not guaranteed.

A notable attempt to alleviate this problem is made in [13] where the basic theory of CS is extended for accommodating highly coherent sensing matrices. This gives rise to an adapted and much stringent version of the RIP which is referred to in [13] as D-RIP. The D-RIP applies to not necessarily sparse vectors which are obtained as linear combination of no more than s columns in some coherent overcomplete dictionary (i.e., a matrix with morc columns than rows). Accurate recovery of such signals is likely, assuming the D-RIP coefficient δ_{2s} is an extremely small number. This premise obviously restricts the viability of the proposed recovery scheme in [13] to those applications where the sensing matrix satisfies this generally imperceptible condition. As demonstrated in [13], this difficulty may be alleviated by employing conventional random matrices which are known to obey the RIP to the desired extent.

Definition 2 *(Non-RIP Settings).* From this point onwards we shall refer to substantial departure from ideal RIP settings as simply non-RIP. In essence, this term indicates that the underlying sensing matrix is coherent to the extent where any of the following occurs:

- The RIP itself is not sufficiently strict (i.e., δ_{2s} approaches 1)
- The RIP is not satisfied with high probability [12, 22]
- Uniform recovery is not guaranteed (i.e., δ_s in (9.3) depends on the vector of unknowns α^i) and hence the RIP becomes an inadequate measure of recovery performance

Being a measure of uniform recovery, the RIP is possibly too strict for most real-world applications. The inability to assess its strictness (i.e., the RIP constant δ_{2s} [12]) for some sensing matrix which has not been specifically designed or selected from a pool of known RIP-preserving constructions is another major limitation. On the other hand, CS methods can be successfully applied in non-RIP settings. This discrepancy between theory and practice can be somewhat resolved by directly scrutinizing the correlation matrix C or the covariance of $\bar{x}_k$.

9.2.2 Measure of Sensing Complexity

The above concept is further materialised in this work, giving rise to an upper bound on the worst attainable estimation accuracy in compressive settings (i.e., where $N < np$). As distinct from conventional results in CS, our proposed bound, which is expressed in terms of (differential) information entropy, relaxes the requirement to specify the underlying RIP constant. This is achieved by introducing a novel complexity metric which is termed here *measure of sensing complexity*, or in an abbreviated form, MSC (see Sect. 9.7). The MSC relies on the correlation matrix $\mathcal{C}$ for quantifying the complexity underlying any sensing matrix H that may be composed from independent samples of $\bar{x}_k$. It indicates the amount of information required for encoding np statistically dependent random variables, the samples of which constitute the columns of the underlying sensing matrix. The measure itself is normalised with respect to the maximal complexity attained only when the random variables are statistically independent, or, in other words, when $\mathcal{C}$ is the identity matrix. As already noted, this case corresponds to an ideal RIP sensing matrix H. Designated as ρ, the MSC assumes values between 0 and 1 where the upper limit is reserved for ideal RIP matrices. On the opposite extreme $\rho \to 0$ is associated with highly coherent (non-RIP) sensing matrices, those which do not promote compressive reconstruction (see Fig. 9.3).

Getting a bit ahead of us, the MSC suggests a new perspective on the sparse recovery problem. This is evinced by the upper bounds in Sect. 9.7 and is further explained therein. In short, our bounds consist of two components, the first of which is rather familiar and concurs with the estimation error statistics of ML estimators.

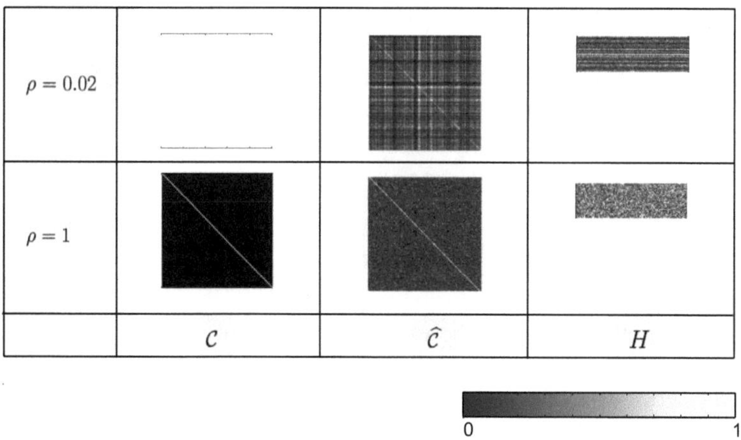

Fig. 9.3 Measure of sensing complexity (MSC) in non-RIP (*upper* row, $\rho = 0.02$) and ideal RIP (*lower* row, $\rho = 1$) settings. Showing the corresponding correlation matrix (second column from the *left*), the Gramian matrix (third column from the *left*), and a sample sensing matrix (*rightmost* column)

The second component relies on the MSC and essentially represents an additional uncertainty associated with the signal unknown support.

$$\text{Estimation error entropy} \leq \underbrace{\text{Observation uncertainty}}_{\text{Due to observation noise}} + \underbrace{\text{Combinatorial uncertainty (MSC)}}_{\text{Due to unknown support}}$$

The rightmost term in the above expression stems from the combinatorial nature of the recovery problem. This suggests that the MSC merely quantifies the information needed to encode all possible subsets of columns from H.

9.2.3 Unscented Kalman Filtering for Sequential Compressive Identification

The proposed CS algorithm in this work exclusively relies on the unscented Kalman filter (UKF) [36]. A preliminary version of this algorithm has been recently presented by the author and his colleagues in [20]. As demonstrated in the numerical study section, the newly derived method tends to exhibit a recovery performance which is comparable to that of the classical BCS and LASSO in standard settings. Apart from its other benefits which are detailed in the ensuing, our approach shows to possess the following properties: (1) it attains the best recovery accuracy among the examined CS algorithms under large deviations from ideal RIP settings, and (2) it manages to sustain a reasonable performance when the amount of observations is less than the threshold set by the theory of CS. The latter advantage is especially evident when estimating the coefficients of compressible rather than sparse AR models. Lastly, by virtue of its KF mechanism, the new method is one of the few existing techniques allowing sequential processing of observations in dynamic CS scenarios, i.e., where time-varying compressible signals are of interest.

9.2.4 Organization of this Chapter

Section 9.3 briefly introduces the essentials of (vector) autoregressive processes together with some of the commonly used techniques for estimating their parameters from time-series data. The problem of compressive identification is then discussed in Sect. 9.4. The Kalman filtering approach to compressed sensing is introduced and discussed in Sect. 9.5. The novel UKF-based compressed sensing method is derived in Sect. 9.6. Section 9.7 provides a detailed derivation of entropy bounds and other measures of recovery performance (i.e., MSC) pertaining to the compressive identification problem. The concepts introduced throughout this work are numerically demonstrated in Sect. 9.8. Finally, conclusions are offered in Sect. 9.9.

9.3 Preliminaries

9.3.1 Vector Autoregressive Processes

We shall begin our discussion with a standard AR model. Denote $k \in \mathbb{N}$ a discrete-time index and let $\{x_k,\ k \geq 0\}$ be a random process assuming values on the real line and obeying

$$x_k = \sum_{t=1}^{p} \alpha(t) x_{k-t} + \omega_k \tag{9.4}$$

where the scalars $\alpha(t),\ t = 1, \ldots, p$ are known as the AR coefficients, and $\{\omega_k,\ k \geq 0\}$ is a zero-mean white sequence. The formulation (9.4) can generally be viewed as a non-stationary output of a dynamic system driven by a white noise. In this respect, a convenient way of representing (9.4) is obtained by letting $\bar{x}_k := [x_k, \ldots, x_{k-p+1}]^T$ and recasting

$$\bar{x}_k = A\bar{x}_{k-1} + B\omega_k \tag{9.5}$$

where the matrices $A \in \mathbb{R}^{p \times p}$ and $B \in \mathbb{R}^p$ are given by

$$A = \begin{bmatrix} \alpha(1), \ldots, \alpha(p) \\ I_{(p-1)\times(p-1)},\ \mathbf{0}_{(p-1)\times 1} \end{bmatrix}, \quad B = \begin{bmatrix} 1 \\ \mathbf{0}_{(p-1)\times 1} \end{bmatrix} \tag{9.6}$$

with I and $\mathbf{0}$ being the identity matrix of an appropriate dimension and a null vector, respectively. The practice underlying (9.5) can, to some extent, be interpreted as the trading of temporal and spatial complexities, both which are indicated by the model order parameter p. Hence, what signifies the process memory in (9.4) translates into the dimension of the Markovian system (9.5).

A natural extension of the basic Eq. (9.4) is the vector AR model which encompasses several possibly interacting processes. Denote $\{x_k^i,\ k \geq 0\}$ the i-th process and write

$$x_k^i = \sum_{j=1}^{n} \sum_{t=1}^{p} \alpha^{i,j}(t) x_{k-t}^j + \omega_k^i, \quad i = 1, \ldots, n \tag{9.7}$$

Hence, the entire system can be described in a fashion similar to (9.5) with

$$A = \begin{bmatrix} A(1), \ldots, A(p) \\ I_{[n(p-1)]\times[n(p-1)]},\ \mathbf{0}_{[n(p-1)]\times n} \end{bmatrix}, \quad B = \begin{bmatrix} I_{n\times n} \\ \mathbf{0}_{[n(p-1)]\times n} \end{bmatrix} \tag{9.8}$$

where this time

$$\bar{x}_k = [z_k^T, \ldots, z_{k-p+1}^T]^T, \quad \bar{\omega}_k = [\omega_k^1, \ldots, \omega_k^n]^T \tag{9.9}$$

with

$$z_k := [x_k^1, \ldots, x_k^n]^T \tag{9.10}$$

The $n \times n$ sub-matrices $A(t)$, $t = 1, \ldots, p$ comprise of the processes coefficients, that is $A(t) = [\alpha^{i,j}(t)]$. An analogous representation of (9.4) in the multivariate case is accordingly

$$z_k = \sum_{t=1}^{p} A(t) z_{k-t} + \bar{\omega}_k \tag{9.11}$$

It is worthwhile noting that by setting $p = 1$, the model (9.11) reduces to a simple linear time invariant (Markov) system.

9.3.2 The Yule-Walker Equations

The Yule-Walker (YW) equations resolve the relations among the process autocorrelation functions. It is a system of linear equations in which the AR coefficients constitute the vector of unknowns. These equations are readily obtained by multiplying both sides of (9.11) with z_{k-l} and taking the expectation with respect to the underlying variables. Repeating this procedure for $l = 0, \ldots, p$ yields the following

$$C(k, k-l) = \sum_{t=1}^{p} A(t) C(k-t, k-l) + Q_k \delta(l), \quad l = 0, \ldots, p \tag{9.12}$$

where $C(k-t, k-l) = E\left\{z_{k-t} z_{k-l}^T\right\}$, $Q_k = E\left\{\bar{\omega}_k \bar{\omega}_k^T\right\}$ and $\delta(\cdot)$ denotes the Kronecker delta. It can be recognized that (9.12) is in fact a set of $n^2 p + n$ equations that essentially allow solving for the unknown matrix coefficients $A(1), \ldots, A(p)$ and n diagonal entries of the noise covariance matrix Q_k, which are, for the sake of simplicity, assumed fixed and denoted henceforth q^i, $i = 1, \ldots, n$. This can be more easily verified by decomposing (9.12) into n independent subsets of $np + 1$ equations

$$C^i(k-l, k) = \sum_{t=1}^{p} C(k-l, k-t) A_i(t)^T, \quad l = 1, \ldots, p \tag{9.13}$$

$$C^{i,i}(k, k) = \sum_{t=1}^{p} (C^i(k-t, k))^T A_i(t)^T + q^i \tag{9.14}$$

where $C^i(k-t, k-l) = E\left\{z_{k-t} x_{k-l}^i\right\}$ and $C^{i,j}(k-t, k-l)$ denote, respectively, the ith column and the (i, j) entry of $C(k-t, k-l)$. The designation $A_i(t)$ represents the i row of $A(t)$, namely, $[\alpha^{i,1}(t), \ldots, \alpha^{i,n}(t)]$. We note that the last Eq. (9.14) is used for resolving the noise variance in a rather standalone manner (i.e., independently of (9.13), albeit based on its solution).

The above argument considers an ideal situation where the autocorrelation matrices, $C(k-t, k-l)$, are perfectly known, which is rarely the case. In practice, these quantities are substituted with appropriate empirical approximations, i.e., sample autocorrelations. This would generally require having multiple realisations of z_k which would substantially complicate things. One of the prevalent approaches for alleviating this issue assumes that the underlying process is both ergodic and stationary. These restrictions, respectively, ensure that the underlying statistical quantities can be derived from merely computing time averages and that the samples are identically distributed. This in turn renders the autocorrelations independent of k, i.e., $C(k-t, k-l) = C(0, t-l)$, and hence an empirical estimate $\hat{\alpha}^i$ of

$$\alpha^i := [A_i(1), \ldots, A_i(p)]^T = \left[\alpha^{i,1}(1), \ldots, \alpha^{i,n}(1), \ldots, \alpha^{i,1}(p), \ldots, \alpha^{i,n}(p)\right]^T \tag{9.15}$$

can be obtained based on the sample autocorrelations $\hat{C}(0, t-l)$. Let $\bar{X}_k$ and X_k^i denote, respectively, the realisations of $\bar{x}_k$ and x_k^i. Then, for every $i \in [1, n]$

$$\hat{\alpha}^i = \left[\sum_{k=p+1}^{N+p} \bar{X}_{k-1}\bar{X}_{k-1}^T\right]^{-1} \left[\sum_{k=p+1}^{N+p} X_k^i \bar{X}_{k-1}\right] \tag{9.16}$$

assuming the ensemble matrix within the brackets is not rank-deficient. It is worth noting that, at least theoretically, the estimate (9.16) almost surely approaches the actual value with an increasing N, where its standard deviation is roughly $\mathcal{O}(1/\sqrt{N})$. It is also rather obvious that (9.16) is unique only when there are at least np linearly independent samples $\bar{X}_k$, which necessarily entails $N \geq np$.

9.3.3 Relaxing the Requirements: LS and ML Estimation

A closer look at (9.16) unfolds its identity as a LS solution subject to the following observation model

$$x_k^i = \bar{x}_{k-1}^T \alpha^i + \omega_k^i, \qquad k > p, \;\; i \in [1, n] \tag{9.17}$$

where, as before, ω_k^i represents a zero-mean white noise. Consequently, we conclude that both requirements of stationarity and ergodicity are superfluous when approximating the autoregressive coefficients in (9.16). This readily follows from the fact that the LS approach does not impose any of these restrictions and yet yields a solution which is identical to (9.16). We point out that these assumptions were made in the first place for the mere reason of substituting intractable statistical moments with time averages. The LS approach asserts that time averages can yet be used, however, without the need to interpret them as ergodic estimates of the underlying autocorrelations.

Recalling the Gauss-Markov theorem, the LS solution (9.16) coincides with the best linear unbiased estimator (BLUE) assuming the observation noise ω_k^i in (9.17) is zero-mean and white, i.e., whenever $E\left\{\omega_k^i\right\} = 0$ and $E\left\{\omega_k^i \omega_{k+t}^i\right\} = 0, \ \forall t \neq 0$ (see [41]). It also coincides with the ML solution under the additional restriction of normally distributed noise.

9.3.4 Ill-Posed Problems, Regularisation and Priors

The LS solution (9.16) is unique only when there are at least np linearly independent realisations $\bar{X}_{k-1}$, which necessarily entails $N \geq np$. Yet, there are many intriguing cases that either violate this assumption or end up with an ill-conditioned information matrix in (9.16). This limitation can be addressed by means of a well-known technique known as regularisation. The idea is fairly simple and consists of adding a positive definite term to the information matrix in (9.16). Hence,

$$\hat{\alpha}^i = \left[P_0^{-1} + \sum_{k=p+1}^{N+p} \bar{X}_{k-1} \bar{X}_{k-1}^T \right]^{-1} \left[\sum_{k=p+1}^{N+p} X_k^i \bar{X}_{k-1} \right] \tag{9.18}$$

where $P_0^{-1} \in \mathbb{R}^{(np)\times(np)}$ is a positive definite matrix. In the statistical literature, (9.18) is normally referred to as Tikhonov regularisation or ridge regression. This expression is merely the solution of the l_2-penalised LS problem

$$\min_{\hat{\alpha}^i} \left\| P_0^{-1/2} \hat{\alpha}^i \right\|_2^2 + \sum_{k=p+1}^{N+p} \left\| X_k^i - \bar{X}_{k-1}^T \hat{\alpha}^i \right\|_2^2 \tag{9.19}$$

where $P_0^{-1/2}$ denotes the matrix square root of P_0^{-1}.

The regularised LS solution (9.18) can be shown to coincide with the maximum a-posteriori (MAP) estimator of a random parameter vector α^i for which the prior is a zero-mean Gaussian with covariance $q^i P_0$. At least conceptually both these approaches are substantially different as the LS is applicable for deterministic parameters whereas the MAP assumes a random α^i.

9.3.5 Estimation Error Statistics

To some extent, the observation Eq. (9.17) is a departure from the classical linear model appearing in many text books [41]. This follows from the fact that the sensing matrix, which is composed of $\{\bar{x}_{k-1}^T,\ k > p\}$, is essentially random. Consequently, any statistics computed based on this observation model is conditioned upon

$$\mathcal{X}_N := \{\bar{x}_p, \ldots, \bar{x}_{N+p-1}\} \tag{9.20}$$

This obviously applies to the estimation error covariance of an unbiased estimator of α^i, denoted here as $\hat{\alpha}^i$. Hence,

$$P_N(\hat{\alpha}^i) := E\left\{(\alpha^i - \hat{\alpha}^i)(\alpha^i - \hat{\alpha}^i)^T \mid \mathcal{X}_N\right\} \tag{9.21}$$

which is, by itself, a random quantity.

9.4 Problem Statement

The compressive identification problem can be summarised as follows. Given the time-series data $\mathcal{X}_N$ we wish to estimate the presumably compressible vector of AR coefficients α^i using fewer observations than its ambient dimension, that is $N < np$. We require that the sought-after estimator $\hat{\alpha}^i$ would be optimal in the sense

$$\min_{\hat{\alpha}^i} \operatorname{Tr}\left\{P_N(\hat{\alpha}^i)\right\} \tag{9.22}$$

subject to the observation model (9.17), where $\operatorname{Tr}(\cdot)$ denotes the trace operator.

Conventionally, a solution to (9.22) in the non compressive settings may be given by the Tikhonov regularisation (9.18), which essentially coincides with the LS solution for $P_0^{-1} = 0$. As already mentioned, this solution also coincides with the MAP estimate assuming α^i and ω_k^i in (9.17) are normally distributed. It is worth noting that in this approach α^i is considered as random rather than deterministic. The key concept which allows solving (9.22) for the compressive case is explained next.

9.4.1 Compressive Identification

Let us decompose the set $\mathcal{X}_N$ (9.20) into a $N \times (np)$ sensing matrix H. Hence

$$H = \begin{bmatrix} \bar{x}_p^T \\ \vdots \\ \bar{x}_{N+p-1}^T \end{bmatrix} \tag{9.23}$$

Using this notation, the observation model (9.17) yields

$$y^i = H\alpha^i + \xi^i \tag{9.24}$$

where $y^i := [x^i_{p+1}, \ldots, x^i_{N+p}]^T$ and $\xi^i := [\omega^i_{p+1}, \ldots, \omega^i_{N+p}]^T$. In what follows the sensing matrix does not appear explicitly, yet the conditions pertaining to H applies indirectly. Having this in mind, we mention the following identity which is used in the ensuing

$$\|y^i - H\alpha^i\|_2^2 = \sum_{k=p+1}^{N+p} \left(x^i_k - \bar{x}^T_{k-1}\hat{\alpha}^i\right)^2 \tag{9.25}$$

It is also worth noting that $\mathcal{X}_N$ and H are two random entities that can be interchangeably used in conditional expectations, e.g., $P_N(\hat{\alpha}^i)$.

Suppose that α^i is s-sparse (i.e., it consists of no more than s non-vanishing entries) and that $\text{spark}(H) > 2s$. It has been shown that under these conditions, α^i can be accurately recovered by solving the following problem [22, 26]

$$\min \| \hat{\alpha}^i \|_0 \text{ s.t. } \sum_{k=p+1}^{N+p} \left(x^i_k - \bar{x}^T_{k-1}\hat{\alpha}^i\right)^2 \leq \epsilon \tag{9.26}$$

for a sufficiently small ϵ, where $\|\alpha^i\|_0$ denotes the support size of α^i (i.e., the number of its non-vanishing entries). Following a similar rationale, for a random α^i we may write

$$\min \| \hat{\alpha}^i \|_0 \text{ s.t. } \text{Tr}\left\{P_N(\hat{\alpha}^i)\right\} \leq \epsilon \tag{9.27}$$

where we have implicitly used the fact that $E\left\{\|\hat{\alpha}^i\|_0 \mid \mathcal{X}_N\right\} = \|\hat{\alpha}^i\|_0$ (because $\hat{\alpha}^i$ is a function of the observation set $\mathcal{X}_N$). Unfortunately, both (9.26) and (9.27) are generally NP-hard and cannot be solved efficiently. The remarkable result of CS asserts that both these problems can be solved by means of a simple convex program assuming the sensing matrix H obeys the RIP (9.3) to within a certain tolerance. In particular, accurate recovery of α^i is guaranteed for $\delta_{2s} < \sqrt{2}-1$ via solving [12, 22]

$$\min \| \hat{\alpha}^i \|_1 \text{ s.t. } \sum_{k=p+1}^{N+p} \left(x^i_k - \bar{x}^T_{k-1}\hat{\alpha}^i\right)^2 \leq \epsilon \tag{9.28}$$

or

$$\min \| \hat{\alpha}^i \|_1 \text{ s.t. } \text{Tr}\left\{P_N(\hat{\alpha}^i)\right\} \leq \epsilon \tag{9.29}$$

Under this additional assumption, the solutions of (9.28) and (9.29) coincide, respectively, with those of the original problems (9.26) and (9.27). The key idea is that as opposed to the generally intractable problems (9.26) and (9.27), the convex relaxations can be efficiently solved using a myriad of existing methods. In this work we employ a novel Kalman filtering technique for solving a dual problem of (9.29). This approach is explained next.

9.5 Kalman Filtering Approach to Compressive Sensing

Followed by the pioneering works [18, 19, 52], in which the KF has shown a remarkable success in estimating sparse and compressible signals, several dynamic CS schemes have been proposed over the last two years [2, 5, 3]. The KF algorithm constitutes a vital part also in the works of [4, 21, 38]. Indeed, the KF is elegant and simple and above all is the linear optimal minimum mean square error (MMSE) estimator irrespective of noise statistics. Despite its appealing features, rarely it is used in its standard formulation which is primarily designed for linear time-varying models. Modifying the KF structure and extending its capabilities have already become a common practice in many engineering and scientific fields. The resulting KF-based methods are vastly used for nonlinear filtering, constrained state estimation, distributed estimation, learning in neural networks, and fault-tolerant filtering.

The KF-based methodologies for dynamic CS can be divided into two broad classes: hybrid, and self-reliant. Whereas the former class refers to KF-based approaches involving the utilisation of peripheral optimisation schemes for handling sparseness and support variations, the latter class refers to methods that are entirely independent of any such scheme. Hybrid KF-based approaches refer to works such as [4, 21, 38, 52]. The only self-reliant KF method available to that end is the one of [18, 19].

The self-reliant KF method in [19] benefits from ease of implementation. It avoids intervening in the KF process which thereby maintains the filtering statistics as adequate as possible. The key idea behind it is to apply the KF in constrained filtering settings using the so-called pseudo-measurement technique. It may, however, exhibit an inferior performance when improperly tuned or when insufficient number of iterations had been carried out. In this work, we improve over [19] by employing the UKF [36] for the pseudo-measurement update stage.

The resulting UKF-based CS algorithm has the following benefits: (1) Self-reliant and easy to implement, (2) Recursively updates the mean and covariance of the filtering probability density function (pdf), (3) Facilitates sequential processing of measurements, (4) Non iterative—as opposed to [19] no reiterations are needed at any stage, (5) Its computational complexity is nearly equal to that of a standard UKF.

9.5.1 The Pseudo-Measurement Technique

The derivation of the UKF-based CS algorithm in this work is based on the notion of pseudo-measurement (PM) from [19]. The key idea is fairly simple and has been vastly employed for constrained state estimation [25, 35]. Thus, instead of solving the l_1-relaxation (9.29), the unconstrained minimisation (9.22) is considered with the observation set $\mathcal{X}_N$ augmented by an additional fictitious measurement satisfying [19]

$$0 = \|\alpha^i\|_1 - v_k \tag{9.30}$$

where v_k is a Gaussian random variable with some predetermined mean and variance, μ_k and r_k, respectively. The above PM is in essence the stochastic analogous of the l_1 constraint in the dual problem [33]

$$\min_{\hat{\alpha}^i} \operatorname{Tr}\left\{P_N(\hat{\alpha}^i)\right\} \text{ s.t. } \|\hat{\alpha}^i\|_1 \leq \epsilon' \tag{9.31}$$

The role of (9.30) can be better apprehended by noting that v_k is aimed to capture the first two statistical moments of the random variable regulating the sparseness degree, $\|\alpha^i\|_1$. The distribution of $\|\alpha^i\|_1$ is, in general, analytically intractable and consequently either approximations or tuning procedures should be utilised for determining appropriate values for μ_k and r_k. We note, however, that the resulting method is rather robust to the underlying parameters as demonstrated in [19].

9.5.2 *Adaptive Pseudo-Measurement Approximation*

Equation (9.30) cannot be straightforwardly processed in the framework of Kalman filtering as it is nonlinear. In practice, this equation is substituted with the following approximation [19]

$$0 = \operatorname{sign}(\hat{\alpha}^i_k)^T \alpha^i - \bar{v}_k \tag{9.32}$$

Here, $\hat{\alpha}^i_k$ and $\operatorname{sign}(\hat{\alpha}^i_k)$ denote, respectively, the estimator of α^i based on k measurements, and a vector composed of either 1 or -1 corresponding to the entries of $\hat{\alpha}^i_k$. The second moment $\bar{r}_k$ of the effective measurement noise, $\bar{v}_k$, obeys

$$\bar{r}_k = \mathcal{O}\left(\|\hat{\alpha}^i_k\|_2^2 + g^T P_k g\right) + r_k \tag{9.33}$$

where $g \in \mathbb{R}^{np}$ is some (tunable) constant vector, and P_k is the estimation error covariance of the supposedly unbiased estimator $\hat{\alpha}^i_k$. For improved readability, the proof of (9.33) is deferred to the last section of this work.

The practical implementation of the approximate PM technique in the context of our problem is demonstrated by the pseudo-code in Algorithm 1. This scheme is in fact the standard CS-embedded KF of [19] (CSKF in short).

Algorithm 1 The CSKF for estimating the coefficients of compressible AR models

1. *Initialisation*

$$\hat{\alpha}_0^i = E\left\{\alpha^i\right\} \tag{9.34a}$$

$$P_0 = E\left\{(\alpha^i - \hat{\alpha}_0^i)(\alpha^i - \hat{\alpha}_0^i)^T\right\} \tag{9.34b}$$

2. *Measurement Update*

$$K_{k-1} = \frac{P_{k-1}\bar{X}_{k-1}}{\bar{X}_{k-1}^T P_{k-1}\bar{X}_{k-1} + q^i} \tag{9.35a}$$

$$\hat{\alpha}_k^i = \hat{\alpha}_k^i + K_{k-1}\left(X_k^i - \bar{X}_{k-1}^T \hat{\alpha}_{k-1}^i\right) \tag{9.35b}$$

$$P_k = (I - K_{k-1}\bar{X}_{k-1}^T)P_{k-1} \tag{9.35c}$$

3. *CS Pseudo Measurement:* Let $P^1 = P_k$ and $\gamma^1 = \hat{\alpha}_k^i$.
4. for $m = 1, 2, \ldots, N_m - 1$ iterations

$$\gamma^{m+1} = \gamma^m - \frac{P^m \operatorname{sign}(\gamma^m)\|\gamma^m\|_1}{\operatorname{sign}(\gamma^m)^T P^m \operatorname{sign}(\gamma^m) + r_k} \tag{9.36a}$$

$$P^{m+1} = P^m - \frac{P^m \operatorname{sign}(\gamma^m)\operatorname{sign}(\gamma^m)^T P^m}{\operatorname{sign}(\gamma^m)^T P^m \operatorname{sign}(\gamma^m) + r_k} \tag{9.36b}$$

5. end for
6. Set $P_k = P^{N_m}$ and $\hat{\alpha}_k^i = \gamma^{N_m}$.

9.6 Sigma Point Filtering for Compressive Sensing

The UKF and its variants, which are broadly referred to as Sigma point filters, parameterise the filtering pdf via the first two statistical moments, namely the mean and covariance, thus providing an approximation to the optimiser of (9.22) (i.e., the conditional mean). These methods amend the KF algorithm for handling generalised nonlinear process and measurement models. As distinct from the extended KF (EKF) which employs infamous linearisation techniques, the UKF relies on the unscented transformation (UT), which is otherwise known as statistical linearisation. This approach is acclaimed for its ease of implementation and its improved estimation performance owing to a rather adequate computation of the underlying covariance matrix. By virtue of its mechanism, the UKF alleviates filtering inconsistencies which in most cases results in improved robustness to model nonlinearities and initial conditions.

The UT can be readily understood by considering a simple example. Let $z \sim \mathcal{N}(\mu, \Sigma)$ be a random vector of dimension n, and let also $f(\cdot) : \mathbb{R}^n \to \mathbb{R}^m$ be some function. Suppose that we are interested in computing the mean and covariance of $f(z)$ to within a certain accuracy. It turns out that a fairly reasonable approximation of these quantities can be made by carefully choosing a finite set of L instrumental vectors $Z^j \in \mathbb{R}^n$, $j = 0, \ldots, L-1$, and corresponding weights w^j. The UT essentially provides a convenient deterministic mechanism for generating $2n+1$ such points which are known by the name Sigma points. As Σ is a symmetric matrix it can be decomposed as $\Sigma = DD^T$ (e.g., Cholesky decomposition). The Sigma points are then given as

$$\begin{aligned} Z^j &= \mu + \sqrt{L} D^j \\ Z^{j+n} &= \mu - \sqrt{L} D^j, \quad j = 1, \ldots, n \end{aligned} \tag{9.37}$$

where D^j denotes the jth column of D, and $Z^0 = \mu$. Note that the sample mean and sample covariance of Z^j, $j = 0, \ldots, 2n$, are μ and Σ, respectively (i.e., this set of points captures the statistics of z). Now, the mean and covariance of $f(z)$ can be approximated by

$$\hat{\mu}_f = \sum_{j=0}^{2n} w^j f(Z^j) \tag{9.38a}$$

$$\hat{\Sigma}_f = \sum_{j=0}^{2n} w^j f(Z^j) f(Z^j)^T - \hat{\mu}_f \hat{\mu}_f^T. \tag{9.38b}$$

9.6.1 CS-UKF: A Compressive Sigma Point Filter

In this work, we amend the UKF for handling sparse and compressible signals. The resulting algorithm, the CS-UKF as we termed it, is a Bayesian CS algorithm that is capable of estimating compressible signals sequentially in time. The sparseness constraint is imposed in a manner similar to [19] via the use of the PM approximation (9.32), nevertheless, without the need for reiterating the PM update. This in turn maintains a computational overhead similar to that of a standard UKF.

The CS-UKF consists of the two traditional UKF stages, the prediction and update, along with an additional refinement stage during which the sigma points gradually become compressible. In particular, after a single standard UKF cycle the sigma points are individually updated in a manner similar to the PM update stage in [19].

Let P_k and Z_k^j be the updated covariance and the jth sigma point at time k, respectively (i.e., after the measurement update). A set of compressible Sigma points at time k is thus given by

$$\beta_k^j = Z_k^j - \frac{P_k \text{sign}(Z_k^j) \|Z_k^j\|_1}{\text{sign}(Z_k^j)^T P_k \text{sign}(Z_k^j) + \bar{r}_k^j} \tag{9.39}$$

with

$$\bar{r}_k^j := c(\|Z_k^j\|_2^2 + g^T P_k g) + r_k \tag{9.40}$$

for $j = 0, \ldots, 2n$, where c is some positive tuning parameter. Once the set $\{\beta_k^j\}_{j=0}^{2n}$ is obtained, its sample mean and sample covariance (see (9.38)) substitutes the updated mean and covariance of the UKF at time k. Note that if the process and measurement models are linear then the prediction and update stages of the UKF can be substituted with those of the standard KF. In this case, the resulting CS-UKF algorithm would consist of the KF prediction and update stages together with a Sigma point-based PM refinement phase.

In the context of our problem the CS-UKF is employed for estimating the compressible vector of AR coefficients, α^i, based on the observation model (9.17). This scheme is summarised in Algorithm 2.

Algorithm 2 The CS-UKF for estimating the coefficients of compressible AR models

1. *Initialisation*

$$\hat{\alpha}_0^i = E\left\{\alpha^i\right\} \tag{9.41a}$$

$$P_0 = E\left\{(\alpha^i - \hat{\alpha}_0^i)(\alpha^i - \hat{\alpha}_0^i)^T\right\} \tag{9.41b}$$

2. *Measurement Update*

$$K_{k-1} = \frac{P_{k-1}\bar{X}_{k-1}}{\bar{X}_{k-1}^T P_{k-1}\bar{X}_{k-1} + q^i} \tag{9.42a}$$

$$\hat{\alpha}_k^i = \hat{\alpha}_k^i + K_{k-1}\left(X_k^i - \bar{X}_{k-1}^T \hat{\alpha}_{k-1}^i\right) \tag{9.42b}$$

$$P_k = (I - K_{k-1}\bar{X}_{k-1}^T)P_{k-1} \tag{9.42c}$$

3. *CS Pseudo Measurement:* Generate $2np + 1$ Sigma points

$$\begin{aligned} Z_k^0 &= \hat{\alpha}_k^i \\ Z_k^j &= \hat{\alpha}_k^i + \sqrt{L}D_k^j \\ Z_k^{j+np} &= \hat{\alpha}_k^i - \sqrt{L}D_k^j, \quad j = 1, \ldots, np \end{aligned} \tag{9.43}$$

where $P_k = D_k D_k^T$.

Compute the compressive Sigma points

$$\beta_k^j = Z_k^j - \frac{P_k \operatorname{sign}(Z_k^j)\|Z_k^j\|_1}{\operatorname{sign}(Z_k^j)^T P_k \operatorname{sign}(Z_k^j) + \bar{r}_k^j}, \quad \bar{r}_k^j = c(\|Z_k^j\|_2^2 + g^T P_k g) + r_k \tag{9.44}$$

for $j = 0, \ldots, 2np$

4. Set

$$\hat{\alpha}_k^i = \sum_{j=0}^{2np} w^j \beta_k^j, \quad P_k = \sum_{j=0}^{2np} w^j \beta_k^j \left(\beta_k^j\right)^T - \hat{\alpha}_k^i \left(\hat{\alpha}_k^i\right)^T \tag{9.45}$$

9.7 Information Entropy Bounds

This section provides tools for assessing the performance of compressive identification schemes in diverse, possibly non-RIP, settings. This is substantiated by introducing a few upper bounds on the estimation error entropy. In contrast with the classical theoretical guarantees in the literature, which essentially rely on the RIP, our bounds involve an alternative and generally perceptible information-theoretic measure, the MSC. Let h_N be the (differential) entropy of a multivariate Gaussian distribution associated with the estimation error, that is

$$h_N = \frac{1}{2} \log \left\{ (2\pi e)^{np} \det \left(P_N(\hat{\alpha}^i) \right) \right\} \tag{9.46}$$

where $P_N(\hat{\alpha}^i)$ is the underlying estimation error covariance based on the N measurements, $\mathcal{X}_N$ (see (9.20)). In what follows, h_N is sometimes referred to as simply the estimation error entropy. This, in general, should not be interpreted as if the actual estimation error is normally distributed. The two entropies, the one corresponding to the distribution of $\alpha^i - \hat{\alpha}^i$, and h_N, do coincide whenever the AR driving noise, ω_k^i, is normally distributed.

As far as compressive identification is considered, $P_N(\hat{\alpha}^i)$ may be ill-conditioned or undefined due to the fact that the number of observations N is smaller than the dimension of α^i, namely $N < np$. Obviously, this could have been resolved had the support of α^i was given to us in advance. Suppose for a moment that this is indeed the case which in turn renders the entropy (9.46) feasible. Here, the estimation error covariance is no longer a $(np) \times (np)$ matrix but rather is $s \times s$ sub-matrix composed out of columns and rows from $P_N(\hat{\alpha}^i)$. In particular,

$$h_N^s = \frac{1}{2} \log \left\{ (2\pi e)^s \det \left(P_N[j_1, \ldots, j_s] \right) \right\} \tag{9.47}$$

where $P_N[j_1, \ldots, j_s]$ denotes a sub-matrix composed of the entries from $P_N(\hat{\alpha}^i)$ for which the indices are given by the set $\{[j_1, \ldots, j_s] \times [j_1, \ldots, j_s]\}$. Having said that we are now ready to state our theorem.

Theorem 1 (Upper Bound for Compressive Identification). *Suppose that the subset of N successive random variables, $\bar{x}_{k_0+1}, \ldots, \bar{x}_{k_0+N}$, constitute an ergodic wide-sense stationary process where, without any loss of generality, $E\left\{\bar{x}_{k_0+j}\right\} = 0$, $j = 1, \ldots, N$. Assume also that α^i is a deterministic s-sparse vector and that the number N is smaller than* $\dim(\alpha^i)$, *i.e., $N < np$. Define $C_k := E\left\{\bar{x}_k \bar{x}_k^T\right\}$ and let $C = C_{k_0+1} = \ldots = C_{k_0+N}$ be the covariance matrix corresponding to the underlying stationary distribution. Decompose the process outcome into a $N \times (np)$ sensing matrix, namely*

$$H = \begin{bmatrix} \bar{x}_{k_0+1}^T \\ \vdots \\ \bar{x}_{k_0+N}^T \end{bmatrix} \tag{9.48}$$

and assume that $\operatorname{spark}(H) > 2s$ *almost surely. If, in addition,*

$$s = \mathcal{O}\left(\bar{c}(\epsilon)^2 N / \log(np)\right) \tag{9.49}$$

for some positive constant $\epsilon < 1$, then any estimation error entropy, computed based on a support consisting of not more than s entries, is bounded above by

$$h_N^s \leq \frac{1}{2} s \log\left(\frac{2\pi e q^i}{(1-\epsilon)N}\right) + \frac{np}{2}(1-\rho)\log\left(1 + \max_j \mathcal{C}^{j,j}\right) \tag{9.50}$$

with probability exceeding $1 - \delta(\epsilon)$. In particular, the probability approaches 1 as the problem dimensions, np and N, increase while yet maintaining

$$N \geq \bar{c}(\epsilon)^2 \|\mathcal{C}\|^2 s \log(np) \tag{9.51}$$

The exact expressions for $\bar{c}(\epsilon)$ and $\delta(\epsilon)$ are provided in Lemma 2 as part of the proof. Finally, ρ in (9.50) denotes a unique measure of complexity taking values between 0 and 1. This quantity, which is referred to as the measure of sensing complexity, or MSC in short, is defined in the next theorem.

Corollary 1 *If the covariance matrix $\mathcal{C}$ in Theorem 1 is a correlation matrix, then the upper bound* (9.50) *assumes the form*

$$h_N^s \leq \frac{1}{2} s \log\left(\frac{2\pi e q^i}{(1-\epsilon)N}\right) + \frac{np}{2}(1-\rho)\log 2 \tag{9.52}$$

Theorem 2 (Measure of Sensing Complexity). *The measure of sensing complexity, which is defined as*

$$\rho = \frac{\log\det\left(\mathcal{C} + I_{(np)\times(np)}\right)}{np \log\left(1 + \max_j \mathcal{C}^{j,j}\right)} \tag{9.53}$$

satisfies the following conditions:

1. $\rho \in (0, 1]$
2. $\rho = 1$ if $\mathcal{C} = cI_{(np)\times(np)}$ *for some positive constant c*
3. $\lim_{np\to\infty} \rho = 0$ *for* $\mathcal{C} = E\left\{[x, \ldots, x]^T [x, \ldots, x]\right\}$, *i.e., multiple copies of the same random variable*
4. *If the condition number of $\mathcal{C}$ approaches 1 then so is ρ.*

9.7.1 Discussion

The proposed entropy bounds (9.50) and (9.52) consist of two (right-hand-side) ingredients. The first one can readily be identified as the information entropy associated with the estimation error covariance of a ML or LS estimator working in ideal settings. It yet accounts for a correction term $\lambda_{\min}(\mathcal{C})$ regulating the signal to noise ratio due to the observation model (9.17). By saying "ideal settings" we essentially refer to an hypothetical situation in which the support of the sparse parameter vector α^i is perfectly known. As this is not truly the case, there comes the second term which represents the information loss due to the indefinite support. Whereas the first term may be negative the second one is always non-negative

$$\frac{np}{2}(1-\rho)\log\left(1+\max_j C^{j,j}\right)\geq 0 \tag{9.54}$$

reflecting the underlying information loss. Notwithstanding, as ρ approaches its upper limit of 1, this undesired effect is reduced, which is the case in ideal RIP settings.

9.7.2 Democracy and Dictatorship

It has been previously pointed out that RIP matrices are democratic in some sense [39]. Our measure of complexity, the MSC, suggests a nice interpretation in this respect. According to Theorem 2, ρ attains its upper limit of 1 whenever the covariance C equals $cI_{(np)\times(np)}$, i.e., when the underlying random variables are statistically independent. As much as it is the case with democratic ballots, the balanced nature of this configuration, in which every random variable has its own "mindset" and is by no means affected by its counterparts, entails the least prognostic situation. This may, alternatively, be viewed as the problem of selecting an arbitrary set of individuals from np candidates with no preference whatsoever. A fast calculation shows that the information required to encode this setting attains the maximum value of

$$-\frac{1}{2^{np}}\sum_{j=1}^{2^{np}}\log\left(\frac{1}{2^{np}}\right)=np\log 2\ \ [\text{nats}]$$

Unsurprisingly, this is exactly the number we would get in the denominator in (9.53) assuming C is a correlation matrix (i.e., $c=1$) (see also the rightmost term in (9.52). As ideal RIP matrices are associated with $\rho\rightarrow 1$ we may regard them as "democratic" in the above sense.

On the other extreme, a dictatorship is the situation where we have an indecisive preference for only one individual from the pool of candidates. Following a rationale similar to the one above yields an entropy of 0. This essentially concurs with the case where multiple copy of the same random variable are present (i.e., the third item in Theorem 2). Having this in mind and recalling some of the points in Sect. 9.7.1, we conclude the following.

Corollary 2 *The dependency upon the problem dimension, np, relaxes in* (9.50) *and* (9.52) *as the MSC approaches the democratic limit of 1. In that case the bound* (9.52) *attains an ideal limit*

$$\lim_{\rho \to 1} h_N^s \leq \frac{1}{2} s \log \left(\frac{2\pi e q^i}{(1-\epsilon)N} \right). \tag{9.55}$$

9.7.3 Practical Considerations and Extensions

Our derivations above assume the knowledge of the covariance matrix $\mathcal{C}$. In the conventional settings of CS, where sensing matrices are designed or chosen from a prescribed pool of random constructions, the computation of $\mathcal{C}$ may be straightforward. In general, however, this might not be the case. It turns out that for the problem at hand, the covariance matrix depends on the unknown parameters α^i, $i = 1, \ldots, n$ as evinced by the observation model (9.17). If α^i is deterministic then the MSC and likewise the probabilistic bounds (9.50) and (9.52) can be computed in practice via solving the following discrete Lyapunov equation for $\mathcal{C}$

$$A\mathcal{C}A^T - \mathcal{C} + BQB^T = 0 \tag{9.56}$$

where $Q = E\left\{\bar{\omega}_k \bar{\omega}_k^T\right\}$ is the driving noise covariance, and the matrices A and B are defined in (9.8). A solution to (9.56) exists only if A is a stable matrix which also implies that the process $\bar{x}_k$ becomes stationary in the wide-sense with an increasing k.

Extending the upper bounds for accommodating random parameters α^i requires that the solution of (9.56) almost surely exists. Here, both the obtained covariance $\mathcal{C}$ and the MSC are random quantities, essentially depending on α^i.

$$\mathcal{C}_A = \mathcal{C}(\alpha^1, \ldots, \alpha^n), \quad \rho_A = \rho(\alpha^1, \ldots, \alpha^n) \tag{9.57}$$

An amended version of the bound (9.50) for the random case is hence suggested by the following theorem.

Theorem 3 (Upper Bound Assuming Random Parameters). *Suppose that the subset of N successive random variables, $\bar{x}_{k_0+1}, \ldots, \bar{x}_{k_0+N}$, constitute an ergodic wide-sense stationary process where, without any loss of generality, $E\left\{\bar{x}_{k_0+j}\right\} = 0$, $j = 1, \ldots, N$. Assume also that α^i is an s-sparse random vector and that the number N is smaller than* $\dim(\alpha^i)$, *i.e., $N < np$. Define $\mathcal{C}_k := E\left\{\bar{x}_k \bar{x}_k^T\right\}$ and let $\mathcal{C}_A = \mathcal{C}_{k_0+1} = \ldots = \mathcal{C}_{k_0+N}$ be the covariance matrix corresponding to the underlying stationary distribution. Decompose the process outcomes into a $N \times (np)$ sensing matrix H, as in Theorem 1, and assume that* $\text{spark}(H) > 2s$ *almost surely. If, in addition,*

$$s = \mathcal{O}\left(\bar{c}(\epsilon)^2 N / \log(np)\right) \tag{9.58}$$

almost surely holds for some positive $\epsilon < 1$, then

$$E\left\{h_N^s \mid \mathcal{X}_N\right\} \leq E\left\{\frac{1}{2}s\log\left(\frac{2\pi e q^i}{(1-\epsilon)N}\right)+\frac{np}{2}(1-\rho_A)\log\left(1+\max_j C_A^{j,j}\right)\middle|\mathcal{X}_N\right\} \tag{9.59}$$

with probability exceeding $1-\delta(\epsilon)$*, and where the expectation is with respect to the random variables* $(\alpha^1,\dots,\alpha^n \mid \mathcal{X}_N)$*, namely*

$$E\left\{f(\alpha^1,\dots,\alpha^n) \mid \mathcal{X}_N\right\} = \int_{a^1}\cdots\int_{a^n} f(a^1,\dots,a^n)p_{\alpha^1,\dots,\alpha^n|\mathcal{X}_N}(a^1,\dots,a^n \mid \mathcal{X}_N)da^1\cdots da^n. \tag{9.60}$$

9.8 Numerical Study

The concepts introduced throughout this chapter are numerically demonstrated in this section. We compare the estimation performance of the newly derived scheme from Sect. 9.6.1, the CS-UKF, with that of the CSKF [19], BCS [32], OMP [42] and LARS [28]. As the latter methods, the BCS, OMP and LARS, are non sequential, in our experiments they are fed at any given time with the whole batch of measurements available up to the specific instance. In contrast, both KF variants process a single measurement at a time. The CS methods are employed for estimating the coefficients of a vector AR process for which the time evolution is

$$\bar{x}_k = A\bar{x}_{k-1} + B\bar{\omega}_k \tag{9.61}$$

where the matrices $A \in \mathbb{R}^{(np)\times(np)}$ and $B \in \mathbb{R}^{(np)\times n}$ are given in (9.8). At the beginning of each run the non-trivial coefficients of A, namely, $\{\alpha^{i,j}(t),\ \ i,j = 1,\dots,n,\ \ t=1,\dots,p\}$, are randomly sampled from a uniform distribution $[-d,d]$, where d is selected such that the resulting A is stable. That is, we maintain

$$\|\lambda_m(A)\| < 1,\quad m = 1,\dots,np$$

We consider two types of parameter vectors, $\alpha^i \in \mathbb{R}^{np},\ \ i=1,\dots,n$, which are either sparse or compressible. In our examples, compressible vectors consist of many relatively small entries which are uniformly sampled over the interval $[-0.03, 0.03]$. On the other hand, the significant entries of α^i are uniformly sampled over $[-1,1]$. The initial distribution of $\bar{x}_k$ is chosen as zero-mean Gaussian with covariance matrix $I_{(np)\times(np)}$. The driving noise $\bar{\omega}_k$ is also assumed zero-mean Gaussian with covariance matrix $Q = qI_{n\times n}$, where $q = 10^{-4}$. At every run, the underlying algorithms are seeded by the set of realisations $\{\bar{X}_1,\dots,\bar{X}_N\}$, where $N < np$.

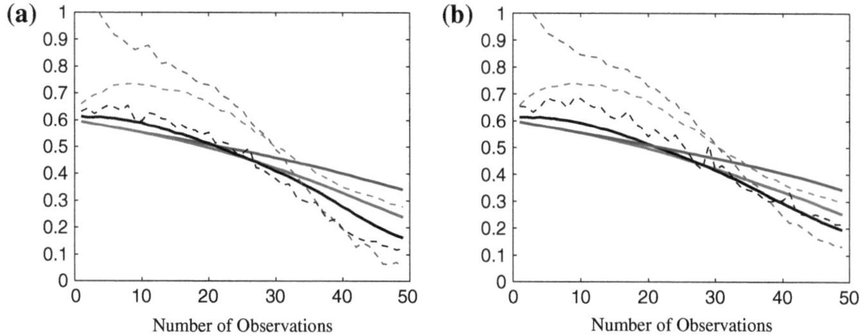

Fig. 9.4 Recovery performance of the various methods for a system dimension of $np = 70$ with $s = 14$ non-vanishing/significant parameters. Batch methods: BCS *(blue dashed line)*, OMP *(red dashed line)*, and LARS *(black dashed line)*. Sequential methods: CSKF (red line), CS-UKF *(black line)*, and KF/ridge regression *(blue line)*. **a** Normalised RMSE (Sparse Parameters). **b** Normalised RMSE (Compressible Parameters)

9.8.1 Sequential Processing of AR Measurements

The recovery performance based on 100 Monte Carlo (MC) runs of all methods with respect to an increasing number of observations is depicted in Fig. 9.4. In this figures and the ones that follow we use the mean normalised RMSE as a measure of estimation performance. This metric, which is given by

$$\sqrt{\sum_{i=1}^{n} \frac{\|\alpha^i - \hat{\alpha}^i\|_2^2}{\|\alpha^i\|_2^2}} \tag{9.62}$$

is averaged over all MC runs.

As seen from Fig. 9.4, in this example, the mean performance of the CS-UKF approaches that of the LARS and the BCS as the number of observations, N, increases, particularly in the compressible case. The KF-based CS methods ,however, process a single measurement at a time and thereby maintain around the same computational overhead irrespective of N. The rest of the methods, on the other hand, deal with an increased complexity as N grows. As opposed to the CSKF which employs 40 PM iterations and covariance updates per measurement, the CS-UKF uses only one iteration per sigma point and two covariance updates per measurement (which accounts for measurement update and PM refinement). Moreover, in contrast with the OMP and BCS algorithms, it seems that both KF-based CS methods maintain a reasonable accuracy over the entire range of N, and even when the number of observations is considerably small. Finally, the performance of a standard KF is also provided which clearly demonstrates the advantage of using the PM stage in compressive settings.

9.8.2 Large Deviations from Ideal RIP Conditions

We further examine the performance of the CS methods in different settings. In this example the state dimension np varies between 40 and 70 while the number of observations remains fixed with $N = 30$. The sparseness degree (i.e., the number of significant entries in α^i) is set as $s = \text{int}[c \cdot (np)]$ where $\text{int}[\cdot]$ denotes the integer part, and c, which is referred to as the sparseness index, may be between 0.1 and 0.5. As c approaches its upper limit of 0.5 the associated sensing matrix may severely deviate from ideal RIP conditions. This follows from the fact that as s increases there are more statistically dependent entries in $\bar{x}_k$ which would ultimately be reflected by a less "democratic" ρ, i.e., the MSC would become smaller.

The normalised RMSE for all possible problem dimensions is shown for the underlying methods in Fig. 9.5. The advantage of using the KF-based CS approaches is rather decisive in this figure. This observation is further evinced in Fig. 9.6 where the normalised RMSE is averaged along one of the dimensions, either the state dimension or the sparseness degree. In both these figures the CS-UKF exhibits the best recovery performance over almost the entire range of values.

The MSC corresponding to the various settings in this example is illustrated via a level plot in Fig. 9.7. This figure was obtained following the approach described in Sect. 9.7.3. Thus, our conjecture from above can immediately be validated. Indeed, as shown in this figure the MSC tends to decrease with a growing sparseness index. This is further manifested in Fig. 9.7b where the MSC is averaged over the entire range of state dimensions. Finally, as the average MSC drops the attained recovery performance of all methods conclusively deteriorates as seen from Fig. 9.6b.

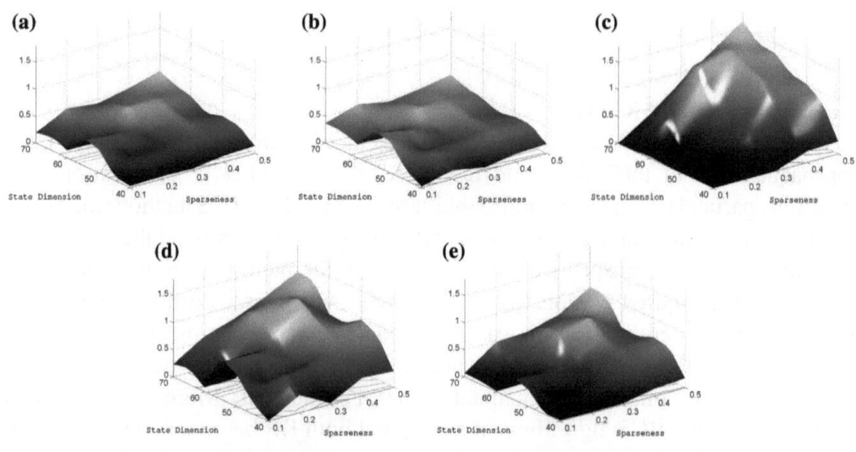

Fig. 9.5 **a** CS-UKF. **b** BCS. **c** OMP. **d** LARS Normalised RMSE with respect to state dimension and sparseness index

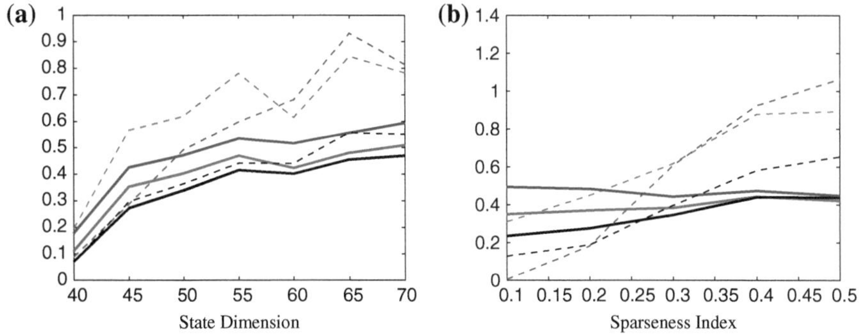

Fig. 9.6 **a** Varying State Dimension. **b** Varying Sparseness Index Normalised RMSE with respect to state dimension np (averaged over the entire range of sparseness indices), and the normalised RMSE with respect to sparseness index (averaged over the entire range of state dimensions). Batch methods: BCS *(blue dashed line)*, OMP *(red dashed line)*, and LARS *(black dashed line)*. Sequential methods: CSKF *(red line)*, CS-UKF *(black line)*, and KF/ridge regression *(blue line)*

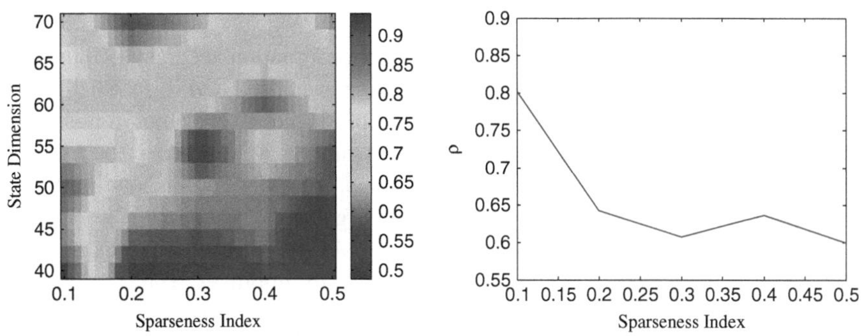

Fig. 9.7 The MSC with respect to the state dimension and sparseness degree. **a** MSC. **b** Averaged MSC versus sparseness index

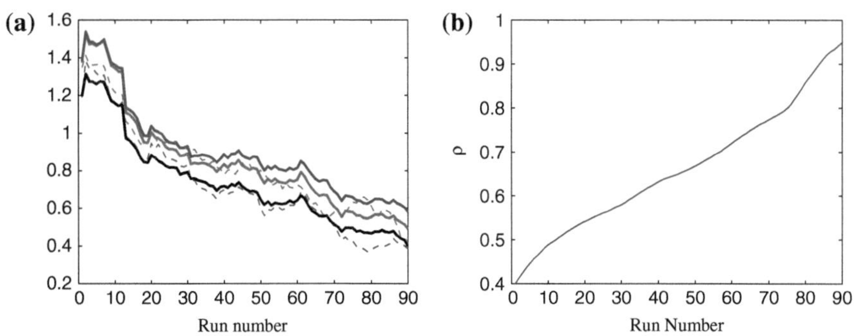

Fig. 9.8 Normalised RMSE and the corresponding MSC. Batch methods: BCS *(blue dashed line)*, OMP *(red dashed line)*. Sequential methods: CSKF *(red line)*, CS-UKF *(black line)*, and KF/ridge regression *(blue line)*. LARS is out of scale. **a** Normalised RMSE. **b** MSC

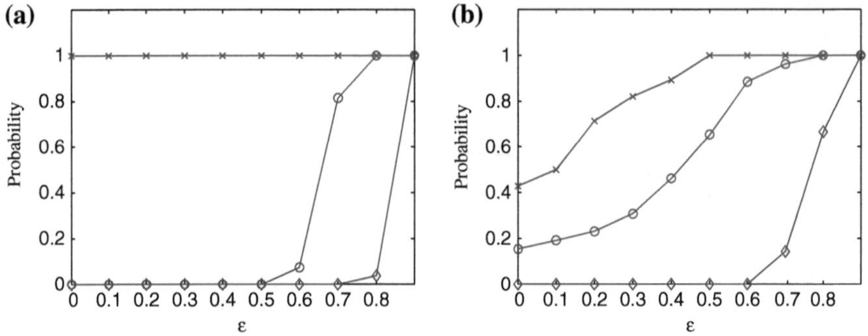

Fig. 9.9 An approximation of the probabilistic upper bound (9.50) for sparseness degrees of $s = 4$ *(crosses)*, $s = 6$ *(circles)*, and $s = 8$ *(diamonds)*. **a** CS-UKF. **b** BCS

9.8.3 MSC and Upper Bounds

The relation between the MSC and the attainable estimation accuracy is illustrated in Fig. 9.8. The left panel in this figure shows the normalised RMSE for 90 different cases in which a new set of parameters α^i was produced. Every point in this figure represents the performance of the corresponding method averaged over 100 MC runs in which the underlying parameters remain fixed. The MSC corresponding to any set of parameters is shown in the right panel in Fig. 9.8.

Based on Fig. 9.8 we conclude the following. As the MSC approaches its upper limit the recovery performance almost monotonically improves. The CS-UKF and the BCS exhibit almost identical recovery performance in this example. Yet, the CS-UKF recursively updates its estimates by processing a single observation at a time.

The upper bound in Theorem 1 is numerically assessed in Fig. 9.9. Thus, the probability of (9.50) to hold true is approximated based on 100 MC runs assuming various sparseness degrees and values of ϵ. In this experiment only two methods, the CS-UKF and the BCS, admit the bound which may indicate that, in this case, both CS schemes attain an optimal recovery performance.

9.8.4 Detecting Interactions in a Multi-Agent System

The following example demonstrates how compressible AR models can be used for efficiently detecting interactions among agents in a dynamic complex system. Consider a multi-agent system where each individual agent is free to move in the 2-dimensional plane. Denoting $(x_k^1)_i$, $(x_k^2)_i$ and $(x_k^3)_i$, $(x_k^4)_i$ as, respectively, the position and velocity of the ith agent, the motion is generally governed by the following discrete-time Markovian evolution

$$\begin{bmatrix} (x_k^1)_i \\ (x_k^2)_i \\ (x_k^3)_i \\ (x_k^4)_i \end{bmatrix} = \begin{bmatrix} 1 & 0 & \Delta t & 0 \\ 0 & 1 & 0 & \Delta t \\ 0 & 0 & 1 & 0 \\ 0 & 0 & 0 & 1 \end{bmatrix} \begin{bmatrix} (x_{k-1}^1)_i \\ (x_{k-1}^2)_i \\ (x_{k-1}^3)_i \\ (x_{k-1}^4)_i \end{bmatrix} + \begin{bmatrix} 0 \\ 0 \\ (\zeta_k)_i \end{bmatrix} \tag{9.63}$$

where $\{(\zeta_k)_i\}_{k \geq 0}$ is a zero-mean white Gaussian sequence with covariance $E\left\{(\zeta_k)_i(\zeta_k)_i^T\right\} = 0.1^2 I_{2\times 2}$, and Δt is a sampling time interval. Nevertheless, within this system there are some agents which interact with their counterparts by way of attraction, that is, they aware of the other agents' position and they moderate their velocity for approaching an arbitrarily chosen individual. The resulting motion of these agents essentially consists of adding nonlinear terms to (9.63). Hence, assuming the ith agent is attracted to the jth one, yields

$$\begin{bmatrix} (x_k^1)_i \\ (x_k^2)_i \\ (x_k^3)_i \\ (x_k^4)_i \end{bmatrix} = \begin{bmatrix} 1 & 0 & \Delta t & 0 \\ 0 & 1 & 0 & \Delta t \\ 0 & 0 & 0 & 0 \\ 0 & 0 & 0 & 0 \end{bmatrix} \begin{bmatrix} (x_{k-1}^1)_i \\ (x_{k-1}^2)_i \\ (x_{k-1}^3)_i \\ (x_{k-1}^4)_i \end{bmatrix} + \frac{\nu}{d_{k-1}^{j,i}} \begin{bmatrix} 0 \\ 0 \\ (x_{k-1}^1)_j - (x_{k-1}^1)_i \\ (x_{k-1}^2)_j - (x_{k-1}^2)_i \end{bmatrix} + \begin{bmatrix} 0 \\ 0 \\ (\zeta_k)_i \end{bmatrix} \tag{9.64}$$

where ν and $d_{k-1}^{j,i}$ denote, respectively, a positive scalar and the distance between the ith and jth agents at time $k-1$, that is

$$d_{k-1}^{j,i} = \sqrt{\left[(x_{k-1}^1)_j - (x_{k-1}^1)_i\right]^2 + \left[(x_{k-1}^2)_j - (x_{k-1}^2)_i\right]^2} \tag{9.65}$$

The objective here is to reconstruct the interaction pattern among agents based on their positions and velocities without being aware of the actual state dynamics, i.e., the actual models (9.63) and (9.64) are not provided to the detection algorithm. Our approach relies on the AR formulation (9.17) where this time the augmented state vector $\bar{x}_k$ is defined as

$$\bar{x}_k = \left[z_k^T, \ldots, z_{k-p+1}^T\right]^T, \quad z_k = \left[(x_k^1)_1, \ldots, (x_k^4)_1, \ldots, (x_k^1)_{N_a}, \ldots, (x_k^4)_{N_a}\right]^T \tag{9.66}$$

with N_a being the total number of agents.

The rationale underlying this formulation is the following. Let $\mathcal{G}_i$ and $\mathcal{G}_j$ be the set of indices pertaining to the ith and jth agents. If agent i is indeed affected by agent j then this is expected to be reflected by the estimated AR coefficients, and particularly,

$$\sum_{l \in \mathcal{G}_j} \sum_{t=1}^{p} |\hat{\alpha}^{m,l}(t)| \gg \sum_{l \notin \{\mathcal{G}_j \cup \mathcal{G}_i\}} \sum_{t=1}^{p} |\hat{\alpha}^{m,l}(t)|, \quad m \in \mathcal{G}_i \tag{9.67}$$

In other words, the coefficients associated with the influence of the jth agent on the behaviour of the ith agent are conjectured to be considerably larger in magnitude compared with the rest of the coefficients, excluding those which correspond to

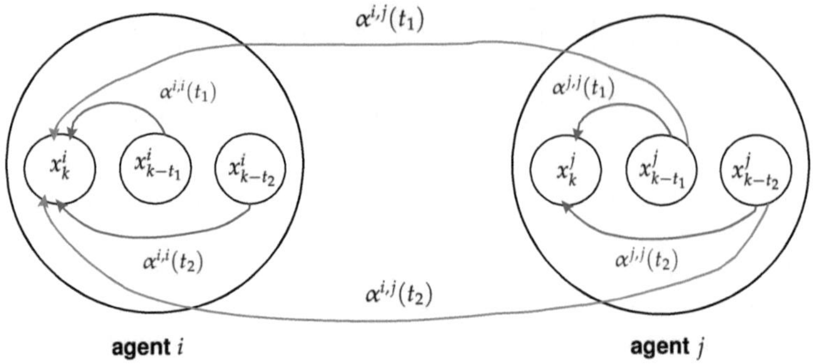

Fig. 9.10 Autoregressive coefficients representing interactions in a multi-agent system

the ith agent own states (i.e., for which the indices are in $\mathcal{G}_i$). Consequently, the vectors of AR coefficients, α^l, $l = 1, \ldots, 4N_a$, are assumed to be compressible. Figure 9.10 further illustrates how AR coefficients are used to represent interactions among agents.

We simulate a system with $N_a = 4$ agents for which the initial state is independently sampled from a uniform distribution, $(x_0^1)_i \sim U[50, 50]$, $(x_0^2)_i \sim U[50, 50]$, $(x_0^3)_i \sim U[5, 5]$, $(x_0^4)_i \sim U[5, 5]$. All agents travel freely in the 2D plane except of the second agent, which is attracted to the first one. Following the aforementioned rationale, we expect that in this case the most dominant entries in the AR vectors of the first agent would belong to both $\mathcal{G}_1$ and $\mathcal{G}_2$. On the other hand, the most dominant entries in the AR vectors of any other agent, $i \neq 1$, would belong exclusively to $\mathcal{G}_i$.

The AR model is applied with a time-lag parameter $p = 10$, which consequently yields 160 coefficients in each AR vector. For each agent, only two coefficient vectors are considered which correspond to its coordinates $(x_k^1)_i, (x_k^2)_i$. Any one of these coefficient vectors is independently estimated using either a regular KF or the proposed CS-UKF, both which employ the observation model (9.17). The total number of observations used in this example is $N = 80$.

The estimated coefficients in this experiment are shown for all agents in Fig. 9.11. For brevity, the two coefficient vectors associated with each agent are summed up and shown as a single line in any one of the panels in this figure. The blue and red lines correspond to the coefficients obtained using, respectively, a regular KF (i.e., non compressive estimation) and the CS-UKF (i.e., compressive estimation). There are four panels in Fig. 9.11, one for each agent. Each panel is further divided into four segments which put together coefficients representing the influence of the same agent, i.e., the first segment in the upper left panel (agent 1) consists of the coefficients $\alpha^{m,l}(1), \ldots, \alpha^{m,l}(p), \quad m \in \mathcal{G}_1, \quad l \in \mathcal{G}_1$, the second segment in the same panel consists of $\alpha^{m,l}(1), \ldots, \alpha^{m,l}(p), \quad m \in \mathcal{G}_1, \quad l \in \mathcal{G}_2$, the third segment consists of $\alpha^{m,l}(1), \ldots, \alpha^{m,l}(p), \quad m \in \mathcal{G}_1, \quad l \in \mathcal{G}_3$, and so forth.

Figure 9.11 clearly shows that both the KF and the CS-UKF correctly identify the dominant entries influencing the dynamics of each agent, and also the (nonlinear)

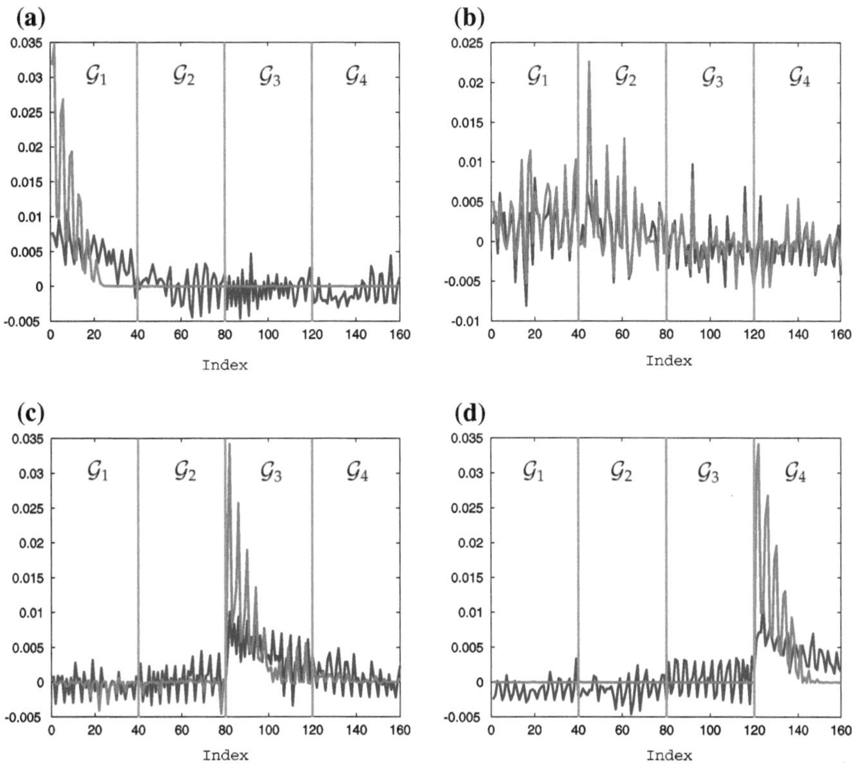

Fig. 9.11 Compressive *(red line)* and non compressive *(blue line)* estimation of the interaction coefficients, $\alpha^{i,j}(t)$, in a multi-agent system. **a** Agent 1. **b** Agent 2. **c** Agent 3. **d** Agent 4

influence of $\mathcal{G}_1$ on $\mathcal{G}_2$. Nevertheless, the CS-UKF exhibits a decisive superiority in this respect, as its estimates are far more adequate. In particular, the CS-UKF manages to accurately recover the agents' own dynamic coefficients whereas the unrelated entries almost completely vanish.

9.9 Concluding Remarks

We consider the problem of estimating the coefficients of sparse and compressible AR models using less observations than conventionally needed. This setting constitutes a challenge to standard compressive sensing schemes as it generally entails a substantial deviation from ideal RIP conditions (which is referred to here as non-RIP). The derived unscented KF-based compressive sensing method, the CS-UKF, exploits the KF machinery for processing the underlying observations sequentially in time, which is also one of its distinctive features. Other benefits of this algorithm account for its ability to maintain a good recovery performance even under substantial deviations from ideal RIP conditions, which thereby renders this method the most successful compared with the other compressive sensing algorithms considered here (i.e., OMP, LARS, BCS, and CSKF). This is evinced by its recovery performance, which is on the average, more accurate than that of the other methods (see for example Fig. 9.6). In other cases, its performance is comparable to that of classical methods such as the LARS and BCS. Notwithstanding, its recursive nature facilitates a nearly fixed computational overhead at every iteration when a new observation becomes available. This is a desired aspect especially when dealing with dynamic systems.

A feature of the CS-UKF which has not been explored in this work refers to its ability to estimate dynamic compressible signals, i.e., signals with varying support and/or entries. Thus, the CS-UKF and likewise its counterpart, the CSKF, have the potential of detecting structural changes in compressible AR models, where one or more parameters may vary with time. In this respect, the CS-UKF falls within the category of dynamic CS schemes, and hence shares common objectives with methods such as the dynamic LASSO and l_1-regularised recursive least squares (see discussion in Sect. 9.5).

9.9.1 Entropy Bounds and Sensing Complexity

The information bounds derived in Sect. 9.7 completely avoid RIP constants. This is achieved by introducing a novel complexity metric, which is referred to as MSC. The MSC relies on the AR process correlation matrix for quantifying the complexity underlying the sensing matrix, H. In essence, it indicates the amount of information required for encoding the set of statistically dependent random variables forming the columns of H. The measure itself is normalised with respect to the maximal complexity attained only when these random variables are statistically independent. As already noted, this case corresponds to an ideal RIP sensing matrix H. Designated as ρ, the MSC assumes values between 0 and 1 where the upper limit is reserved for ideal RIP matrices. On the opposite extreme $\rho \to 0$ is associated with highly coherent (non-RIP) sensing matrices, those which do not promote compressive reconstruction.

The MSC suggests a new perspective on the sparse recovery problem which is substantiated by the information upper bounds in Sect. 9.7. Theese bounds consist of two components, the first of which is rather familiar and concurs with the estimation error statistics of ML estimators. The second component relies on the MSC and essentially represents an additional uncertainty associated with the signal unknown support.

$$\text{Estimation error entropy} \leq \underbrace{\text{Observation uncertainty}}_{\text{Due to observation noise}} + \underbrace{\text{Combinatorial uncertainty (MSC)}}_{\text{Due to unknown support}}$$

The rightmost term in the above expression stems from the combinatorial nature of the recovery problem. This suggests that the MSC merely quantifies the information needed to encode all possible subsets of columns from H.

9.10 Proofs of Theorems

9.10.1 Proof of Equation (9.33)

Consider the PM observation

$$0 = \|\alpha^i\|_1 - v_k = \text{sign}(\alpha^i)^T \alpha^i - v_k \tag{9.68}$$

Because α^i is unknown we make use of the relation $\alpha^i = \hat{\alpha}_k^i + \tilde{\alpha}_k^i$ to get

$$0 = \text{sign}\left(\hat{\alpha}_k^i + \tilde{\alpha}_k^i\right)^T \alpha^i - v_k \tag{9.69}$$

where $\tilde{\alpha}_k^i$ is the estimation error after processing k observations. As sign$(\cdot)$ is a bounded function, we may write

$$0 = \left[\text{sign}(\hat{\alpha}_k^i) + g\right]^T \alpha^i - v_k = \text{sign}(\hat{\alpha}_k^i)^T \alpha^i + \underbrace{g^T \alpha^i - v_k}_{\bar{v}_k} \tag{9.70}$$

where g is $\mathcal{X}_k$-measurable, and $\|g\| \leq c$ almost surely. The expression (9.70) is the approximate PM with an effective observation noise $\bar{v}_k$. A the mean of $\bar{v}_k$ cannot be easily obtained we approximate the non centralised second moment

$$E\left\{\bar{v}_k^2 \mid \mathcal{X}_k\right\} = E\left\{g^T \left(\hat{\alpha}_k^i + \tilde{\alpha}_k^i\right)\left(\hat{\alpha}_k^i + \tilde{\alpha}_k^i\right)^T g | \mathcal{X}_k\right\} + r_k \tag{9.71}$$

which follows from the fact that $\hat{\alpha}_k^i$ and v_k are statistically independent, where $E\{v_k\} = 0$ and $E\{v_k^2\} = r_k$. Substituting the estimation error covariance with the KF-computed one, P_k, in (9.71), yields

$$\begin{aligned}\bar{r}_k = E\left\{\bar{v}_k^2 \mid \mathcal{X}_k\right\} &= g^T \hat{\alpha}_k^i (\hat{\alpha}_k^i)^T g + g^T E\left\{\tilde{\alpha}_k^i \left(\tilde{\alpha}_k^i\right)^T |\mathcal{X}_k\right\} g + r_k \\ &\approx \mathcal{O}(\|\hat{\alpha}_k^i\|^2) + g^T P_k g + r_k\end{aligned} \tag{9.72}$$

where it was implicitly assumed that $E\{\tilde{\alpha}_k^i \mid \mathcal{X}_k\} = 0$, i.e., $\hat{\alpha}_k^i$ is unbiased. QED.

9.10.2 Proof of Theorem 1

The proof of this theorem consists of two subsequent parts. The first part establishes a relation between the estimation error entropy (9.47) and the corresponding covariance sub-matrix of the stationary process, $\mathcal{C}[j_1, \ldots, j_s]$. The second part extends the result of the first part to account for the $(np) \times (np)$ covariance matrix, $\mathcal{C}$.

Part one: We begin with the main result in [43], where it is proved that

$$E\left\{\left\|\frac{1}{N}\sum_{i=1}^{N}\bar{x}_{k_0+i}\bar{x}_{k_0+i}^T - I_{(np)\times(np)}\right\|\right\} \le c\sqrt{\frac{\log np}{N}}E\left\{\|\bar{x}_{k_0}\|^{\log N}\right\}^{1/\log N} \tag{9.73}$$

holds for independent and identically distributed (iid) vectors $\bar{x}_{k_0+i}, i = 0, \ldots, N$, having an identity covariance matrix, namely $E\left\{\bar{x}_{k_0}\bar{x}_{k_0}^T\right\} = I_{(np)\times(np)}$.

Two distinctions are made here. Firstly, in our case the vectors $\bar{x}_{k_0+i}$ are not truly independent, as they are generated by a Markovian system (see Sect. 9.3.1). Nevertheless, under the ergodicity and stationarity assumptions, from a certain time point onwards, these vectors become identically distributed and may be considered, to some extent, as statistically independent. By saying this, we mean that any two distinct vectors, $\bar{x}_{k_0+i}$ and $\bar{x}_{k_0+j}$, are nearly independent providing that $|i - j| \ge \tau$, where the positive integer τ has to do with the mixing properties of the process. Using statistically dependent vectors for computing the ensemble average in (9.73) introduces a Monte Carlo error which diminishes with the number of samples N. It is shown in [47] that for a small number of samples the following holds

$$\frac{1}{N}\sum_{i=1}^{N}\bar{x}_{k_0+i}\bar{x}_{k_0+i}^T = c_\tau\frac{1}{N}\sum_{i=1}^{N}\bar{y}_i\bar{y}_i^T \tag{9.74}$$

where $c_\tau > 0$, and $\bar{y}_i, i = 0, \ldots, N$ are iid with covariance $E\left\{\bar{y}_0\bar{y}_0^T\right\} = E\left\{\bar{x}_{k_0}\bar{x}_{k_0}^T\right\}$. Furthermore, as N increases then $c_\tau \to 1$. Equation (9.74) allows us to substitute the statistically dependent vectors in (9.73) with statistically independent ones. Thus, combining both (9.74) and (9.73) yields

$$E\left\{\left\|\frac{1}{N}\sum_{i=1}^{N}\bar{x}_{k_0+i}\bar{x}_{k_0+i}^T - I_{(np)\times(np)}\right\|\right\} \le c'\sqrt{\frac{\log np}{N}}E\left\{\|\bar{x}_{k_0}\|^{\log N}\right\}^{1/\log N} \tag{9.75}$$

where $c' = c_\tau^{1/2}c$.

The second distinction is the following: we note that with only a slight modification, (9.75) applies to vectors $\bar{x}_{k_0+i}$ with arbitrary covariance, $\mathcal{C}$.

Lemma 1 *Let* $\mathcal{C} = E\left\{\bar{x}_{k_0+i}\bar{x}_{k_0+i}^T\right\}$, $i = 0, \ldots, N$. *Then,*

$$E\left\{\left\|\frac{1}{N}\sum_{i=1}^{N}\bar{x}_{k_0+i}\bar{x}_{k_0+i}^T - C\right\|\right\} \leq c'\|C\|\sqrt{\frac{\log np}{N}}E\left\{\|\bar{y}_0\|^{\log N}\right\}^{1/\log N} \tag{9.76}$$

where $\bar{y}_0$ is a zero-mean random vector with a unit covariance.

Proof Decompose $C = U\Lambda U^T$, where U and Λ denote, respectively, an orthogonal matrix and a non-negative diagonal matrix. Hence, (9.75) yields

$$E\left\{\left\|\frac{1}{N}\sum_{i=1}^{N}\bar{x}_{k_0+i}\bar{x}_{k_0+i}^T - C\right\|\right\} = E\left\{\left\|U\Lambda^{1/2}\left(\frac{1}{N}\sum_{i=1}^{N}\bar{y}_i\bar{y}_i^T - I\right)\Lambda^{1/2}U^T\right\|\right\} =$$
$$E\left\{\left\|\Lambda^{1/2}\left(\frac{1}{N}\sum_{i=1}^{N}\bar{y}_i\bar{y}_i^T - I\right)\Lambda^{1/2}\right\|\right\} \leq \|\Lambda\|E\left\{\left\|\frac{1}{N}\sum_{i=1}^{N}\bar{y}_i\bar{y}_i^T - I\right\|\right\} \leq$$
$$c'\|\Lambda\|\sqrt{\frac{\log np}{N}}E\left\{\|\bar{y}_0\|^{\log N}\right\}^{1/\log N} \tag{9.77}$$

where $\bar{y}_i = \Lambda^{-1/2}U^T\bar{x}_{k_0+i}$, $i = 0, \ldots, N$. Finally, recognising that $\|\Lambda\| = \|C\|$, and $E\left\{\bar{y}_i\bar{y}_i^T\right\} = \Lambda^{-1/2}U^TCU\Lambda^{-1/2} = I_{(np)\times(np)}$, yields the lemma. QED.

In what follows, we make use of (9.76) in the context of the information entropy (9.47), where the estimation error covariance sub-matrix, $P_N[j_1, \ldots, j_s]$, appears. For that reason, we slightly rephrase (9.76) for encompassing this case. Let $\bar{x}_{k_0+i}^s$, $i \in [0, N]$ be a vector comprised of s entries from $\bar{x}_{k_0+i}$, for which the indices are $\{j_1, \ldots, j_s\}$. Then, (9.76) implies

$$E\left\{\left\|\frac{1}{N}\sum_{i=1}^{N}\bar{x}_{k_0+i}^s(\bar{x}_{k_0+i}^s)^T - C[j_1, \ldots, j_s]\right\|\right\}$$
$$\leq c'\|C\|\sqrt{\frac{\log np}{N}}E\left\{\|\bar{y}_0^s\|^{\log N}\right\}^{1/\log N} \tag{9.78}$$

where $E\left\{\bar{y}_0^s(\bar{y}_0^s)^T\right\} = I_{s\times s}$. Before proceeding any further we note that the right hand side in (9.78) satisfies

$$c'\|C\|\sqrt{\frac{\log np}{N}}E\left\{\|\bar{y}_0^s\|^{\log N}\right\}^{1/\log N} \geq c'\|C\|\sqrt{\frac{\log np}{N}}E\left\{\|\bar{y}_0^s\|^2\right\}^{1/2} \tag{9.79}$$

for $\log N > 2$. As, $E\left\{\|\bar{y}_0^s\|^2\right\} = s$, this further implies

$$c'\|C\|\sqrt{\frac{\log np}{N}}E\left\{\|\bar{y}_0^s\|^{\log N}\right\}^{1/\log N} \geq c'\|C\|\sqrt{\frac{s\log np}{N}} \tag{9.80}$$

Therefore, a necessary condition for the right hand side in (9.78) to be smaller than 1, is

$$N > (c' \|\mathcal{C}\|)^2 s \log(np) \quad \text{or} \quad s = \mathcal{O}\left(N/\log(np)\right) \tag{9.81}$$

Equation (9.78) is related to the estimation error entropy (9.47) in the following way. The ensemble average on the left hand side in (9.78) is, in fact, a scaled version of the optimal estimation error covariance (in the MSE sense) assuming the support of α^i is provided. In other words, we have

$$P_N[j_1, \ldots, j_s] = q^i \left(\sum_{i=1}^{N} \bar{x}^s_{k_0+i} (\bar{x}^s_{k_0+i})^T \right)^{-1} \tag{9.82}$$

which is no other than the estimation error covariance of a least squares estimator (which almost surely exists owing to the condition $\text{spark}(H) > 2s$). The above expression has to do with the process covariance sub-matrix $\mathcal{C}[j_1, \ldots, j_s]$ as manifested by (9.78). Unfortunately, this premise cannot be directly used in the information entropy (9.47). Getting around with this requires an intermediate stage in which the formulation assumes a probabilistic twist. In detail, we invoke the Markov inequality for relating the determinants of $P_N[j_1, \ldots, j_s]$ and $\mathcal{C}[j_1, \ldots, j_s]$. The argument proceeds as follows.

Equations (9.78) and (9.82) together with the Markov inequality, yield

$$\begin{aligned} \Pr\left(\left\| \frac{q^i}{N} P_N[j_1, \ldots, j_s]^{-1} - \mathcal{C}[j_1, \ldots, j_s] \right\| > \epsilon' \right) \leq \\ \frac{1}{\epsilon'} E\left\{ \left\| \frac{q^i}{N} P_N[j_1, \ldots, j_s]^{-1} - \mathcal{C}[j_1, \ldots, j_s] \right\| \right\} \leq \\ \frac{c'}{\epsilon'} \|\mathcal{C}\| \sqrt{\frac{\log np}{N}} E\left\{ \|\bar{y}^s_0\|^{\log N} \right\}^{1/\log N} \end{aligned} \tag{9.83}$$

for which the complementary inequality is

$$\begin{aligned} \Pr\left(\left\| \frac{q^i}{N} P_N[j_1, \ldots, j_s]^{-1} - \mathcal{C}[j_1, \ldots, j_s] \right\| \leq \epsilon' \right) \geq \\ 1 - \frac{c'}{\epsilon'} \|\mathcal{C}\| \sqrt{\frac{\log np}{N}} E\left\{ \|\bar{y}^s_0\|^{\log N} \right\}^{1/\log N} \end{aligned} \tag{9.84}$$

Recalling Weyl inequalities, (9.84) further implies the following relations among the eigenvalues

$$\lambda_1(P_N[j_1, \ldots, j_s]^{-1}) \geq \cdots \geq \lambda_s(P_N[j_1, \ldots, j_s]^{-1})$$

and

$$\lambda_1(\mathcal{C}[j_1, \ldots, j_s]) \geq \cdots \geq \lambda_s(\mathcal{C}[j_1, \ldots, j_s])$$

of the two symmetric matrices $P_N[j_1, \ldots, j_s]^{-1}$ and $\mathcal{C}[j_1, \ldots, j_s]$. Thus,

$$\Pr\left(\lambda_m\left(\frac{q^i}{N}P_N[j_1,\ldots,j_s]^{-1}\right) - \lambda_{l-m+1}\left(\mathcal{C}[j_1,\ldots,j_s]\right) \geq -\epsilon'\right) \geq$$
$$\Pr\left(\lambda_m\left(\frac{q^i}{N}P_N[j_1,\ldots,j_s]^{-1} - \mathcal{C}[j_1,\ldots,j_s]\right) \geq -\epsilon'\right) \geq$$
$$\Pr\left(\left\{\lambda_m\left(\frac{q^i}{N}P_N[j_1,\ldots,j_s]^{-1} - \mathcal{C}[j_1,\ldots,j_s]\right) \geq -\epsilon'\right\}\bigcap\right.$$
$$\left.\left\{\lambda_m\left(\frac{q^i}{N}P_N[j_1,\ldots,j_s]^{-1} - \mathcal{C}[j_1,\ldots,j_s]\right) \leq \epsilon'\right\}\right) =$$
$$\Pr\left(\left|\lambda_m\left(\frac{q^i}{N}P_N[j_1,\ldots,j_s]^{-1} - \mathcal{C}[j_1,\ldots,j_s]\right)\right| \leq \epsilon'\right) \geq$$
$$1 - \frac{c'}{\epsilon'}\|\mathcal{C}\|\sqrt{\frac{\log np}{N}}E\left\{\|\bar{y}_0^s\|^{\log N}\right\}^{1/\log N} \tag{9.85}$$

for any $1 \leq l - m + 1 \leq s$.

Let $\epsilon' = (1-d)\lambda_{l-m+1}(\mathcal{C}[j_1,\ldots,j_s]) - d$ and assume that $d/(1-d) < \lambda_{l-m+1}(\mathcal{C}[j_1,\ldots,j_s])$ so as to ensure $\epsilon' > 0$. Substituting ϵ' into (9.85) and slightly rearranging, yields

$$\Pr\left(\lambda_m\left(\frac{q^i}{N}P_N[j_1,\ldots,j_s]^{-1}\right) \geq d[\lambda_{l-m+1}(\mathcal{C}[j_1,\ldots,j_s]) + 1]\right) \geq$$
$$1 - \underbrace{c(d)\|\mathcal{C}\|\sqrt{\frac{\log np}{N}}E\left\{\|\bar{y}_0^s\|^{\log N}\right\}^{1/\log N}}_{\delta(d)} \tag{9.86}$$

from which it is clear that d is necessarily positive. The positive constant $c(d)$ in (9.86) is obtained as

$$c(d) = c'/[(1-d)\lambda_{\min}(\mathcal{C}) - d] \tag{9.87}$$

owing to Cauchy interlacing theorem which essentially implies $\lambda_{\min}(\mathcal{C}) \leq \lambda_{\min}(\mathcal{C}[j_1,\ldots,j_s]) \leq \lambda_{l-m+1}(\mathcal{C}[j_1,\ldots,j_s])$. The inequality (9.86) gives rise to the following lemma.

Lemma 2 *Let ϵ be a positive constant satisfying*

$$\min\left\{0, \frac{1-\lambda_{\min}(\mathcal{C})\max_j \mathcal{C}^{j,j}}{1+\lambda_{\min}(\mathcal{C})}\right\} < \epsilon < 1 \tag{9.88}$$

Suppose that

$$N > \bar{c}(\epsilon)^2\|\mathcal{C}\|^2 s\log(np)$$

or equivalently

$$s = \mathcal{O}\left(\bar{c}(\epsilon)^2 N / \log(np)\right)$$

Then the following holds with probability exceeding $1 - \delta(\epsilon)$

$$\det\left(P_N[j_1, \ldots, j_s]\right) \leq \left(\frac{q^i}{N} \frac{1 + \max_j \mathcal{C}^{j,j}}{1 - \epsilon}\right)^s \det\left(I + \mathcal{C}[j_1, \ldots, j_s]\right)^{-1} \quad (9.89)$$

Proof From (9.86) it follows that

$$\det\left(P_N[j_1, \ldots, j_s]\right) = \left(\frac{q^i}{N}\right)^s \prod_{m=1}^{s} \frac{1}{\lambda_m\left(\frac{q^i}{N} P_N[j_1, \ldots, j_s]^{-1}\right)} \leq$$
$$\left(\frac{q^i}{d \cdot N}\right)^s \prod_{1 \leq l-m+1 \leq s} \lambda_{l-m+1}\left(\mathcal{C}[j_1, \ldots, j_s] + I_{s \times s}\right)^{-1} =$$
$$\left(\frac{q^i}{d \cdot N}\right)^s \det\left(\mathcal{C}[j_1, \ldots, j_s] + I_{s \times s}\right)^{-1} \quad (9.90)$$

with probability of at least $1 - \delta(d)$. Set $d = (1 - \epsilon)/(1 + \max_j \mathcal{C}^{j,j})$, and assume that ϵ obeys (9.88), which in turn fulfills the conditions underlying (9.86), i.e., $d > 0$, and $d/(1 - d) < \lambda_{l-m+1}(\mathcal{C}[j_1, \ldots, j_s])$. Finally, substituting d into (9.90) yields the lemma with the constants $\bar{c}(\epsilon)$ and $\delta(\epsilon)$ given by

$$\bar{c}(\epsilon) = c\left((1 - \epsilon)/(1 + \max_j \mathcal{C}^{j,j})\right) = c' \frac{1 + \max_j \mathcal{C}^{j,j}}{\lambda_{\min}(\mathcal{C})(\max_j \mathcal{C}^{j,j}) - 1 + \epsilon(1 + \lambda_{\min}(\mathcal{C}))} \quad (9.91a)$$

$$\delta(\epsilon) = \bar{c}(\epsilon)\|\mathcal{C}\|\sqrt{\frac{\log np}{N}} E\left\{\|\bar{y}_0^s\|^{\log N}\right\}^{1/\log N} \quad (9.91b)$$

where, under the conditions of the lemma, $\delta(\epsilon) < 1$. It is worthwhile noting that as $\epsilon \to 1$ then $\bar{c}(\epsilon) \to c'/\lambda_{\min}(\mathcal{C})$, and hence

$$\lim_{\epsilon \to 1} \delta(\epsilon) = c' \text{cond}(\mathcal{C})\sqrt{\frac{\log np}{N}} E\left\{\|\bar{y}_0^s\|^{\log N}\right\}^{1/\log N} \quad (9.92)$$

where $\text{cond}(\mathcal{C})$ denotes the condition number of $\mathcal{C}$. QED.

Lemma 2 entails

$$h_N^s = \frac{1}{2} \log\left\{(2\pi e)^s \det\left(P_N[j_1, \ldots, j_s]\right)\right\} \leq$$
$$\frac{1}{2} \log\left\{\left(\frac{2\pi e q^i (1 + \max_j \mathcal{C}^{j,j})}{(1 - \epsilon)N}\right)^s \det\left(I + \mathcal{C}[j_1, \ldots, j_s]\right)^{-1}\right\} \quad (9.93)$$

which concludes the first part of the proof.

Part two: Bearing in mind that the underlying support $\{j_1, \ldots, j_s\}$ is normally unknown, the second part of the proof is concerned with finding a suitable replacement for the covariance sub-matrix, $\mathcal{C}[j_1, \ldots, j_s]$, in (9.93). This is carried out by first permuting the covariance matrix $\mathcal{C}$ such that the sub-matrix $\mathcal{C}[j_1, \ldots, j_s]$ appears as the uppermost block in the resulting permuted matrix $\mathcal{C}'$. Because permutations preserve the eigenvalues of the original matrix, it follows that $\det(\mathcal{C}) = \det(\mathcal{C}')$. Using a well-known property of determinants of non-negative matrices gives

$$\det(\mathcal{C}) = \det(\mathcal{C}') \leq \det(\mathcal{C}[j_1, \ldots, j_s]) \det(\mathcal{C}[j_{s+1}, \ldots, j_{np}]) \tag{9.94}$$

where $j_{s+1}, \ldots, j_{np}$ are the remaining indices which do not belong to the support $\{j_1, \ldots, j_s\}$. Therefore,

$$\begin{aligned}
&\det(I_{(np)\times(np)} + \mathcal{C}) \leq \\
&\quad \det(I_{s\times s} + \mathcal{C}[j_1, \ldots, j_s]) \det(I_{(np-s)\times(np-s)} + \mathcal{C}[j_{s+1}, \ldots, j_{np}]) \leq \\
&\qquad \det(I_{s\times s} + \mathcal{C}[j_1, \ldots, j_s]) \prod_{m=s+1}^{np} (1 + \mathcal{C}^{j_m, j_m}) \leq \\
&\qquad\quad \det(I_{s\times s} + \mathcal{C}[j_1, \ldots, j_s]) \left(1 + \max_j \mathcal{C}^{j,j}\right)^{np-s}
\end{aligned} \tag{9.95}$$

which thereby yields

$$\det(I_{s\times s} + \mathcal{C}[j_1, \ldots, j_s])^{-1} \leq \left(1 + \max_j \mathcal{C}^{j,j}\right)^{np-s} \det(I_{(np)\times(np)} + \mathcal{C})^{-1} \tag{9.96}$$

Further, substituting (9.96) into (9.93), reads

$$h_N^s \leq \frac{s}{2} \log\left(\frac{2\pi e q^i}{(1-\epsilon)N}\right) + \frac{np}{2} \log\left(1 + \max_j \mathcal{C}^{j,j}\right) - \frac{1}{2} \log\det(I + \mathcal{C}) \tag{9.97}$$

which essentially coincides with (9.50) assuming ρ is as defined in (9.53). QED.

9.10.3 Proof of Theorem 2

We prove the following properties of ρ:

1. $\rho \in (0, 1]$
2. $\rho = 1$ if $\mathcal{C} = cI_{(np)\times(np)}$ for some positive constant c
3. $\lim_{np\to\infty} \rho = 0$ for $\mathcal{C} = E\left\{[x, \ldots, x]^T [x, \ldots, x]\right\}$, i.e., multiple copies of the same random variable
4. If the condition number of $\mathcal{C}$ approaches 1 then so is ρ

Proof

1. The first property follows straightforwardly from the fact that $\mathcal{C}$ is neither negative-definite nor is it a null matrix. Thus,

$$\log\det(I + \mathcal{C}) = \sum_{i=1}^{np} \log(1 + \lambda_i(\mathcal{C})) > 0 \tag{9.98}$$

as all eigenvalues of $\mathcal{C}$ are non negative and at least one eigenvalue $\lambda_i(\mathcal{C})$ is strictly larger than 0. Similarly,

$$np \log(1 + \max_j \mathcal{C}^{j,j}) \geq np \log\left(1 + (np)^{-1}\mathrm{tr}(\mathcal{C})\right) \geq$$
$$np \log\left(1 + (np)^{-1}\sum_i \lambda_i(\mathcal{C})\right) > 0 \tag{9.99}$$

from which we conclude that $\rho > 0$. Finally, because $\mathcal{C}$ is nonnegative-definite and symmetric, it follows that

$$\log\det(I + \mathcal{C}) \leq \log\prod_j (1 + \mathcal{C}^{j,j}) \leq np \log(1 + \max_j \mathcal{C}^{j,j}) \tag{9.100}$$

which implies $\rho \leq 1$.

2. Assuming $\mathcal{C} = cI$, $c > 0$,

$$\rho = \frac{\log\det((1 + c)I_{(np)\times(np)})}{np \log(1 + c)} = \frac{\log((1 + c)^{np})}{np \log(1 + c)} = 1 \tag{9.101}$$

3. It readily follows that $\mathcal{C}$ in this case is rank-deficient with only a single non-vanishing eigenvalue equals to $\lambda_{\max}(\mathcal{C}) = \mathrm{tr}(\mathcal{C})$. Therefore,

$$\rho = \frac{\log\det(I + \mathcal{C})}{np \log(1 + \max_j \mathcal{C}^{j,j})} = \frac{\log(1 + \mathrm{tr}(\mathcal{C}))}{np \log(1 + \mathcal{C}^{1,1})} = \frac{\log(1 + np\mathcal{C}^{1,1})}{np \log(1 + \mathcal{C}^{1,1})} \tag{9.102}$$

which implies $\lim_{np\to\infty} \rho = 0$.

4. This immediately follows upon recognising that

$$\frac{\log(1 + \lambda_{\min}(\mathcal{C}))}{\log(1 + \lambda_{\max}(\mathcal{C}))} \leq \rho \leq \frac{\log(1 + \lambda_{\max}(\mathcal{C}))}{\log(1 + \lambda_{\min}(\mathcal{C}))} \tag{9.103}$$

References

1. Alan P (1983) Forecasting with univariate Box-Jenkins models: concepts and cases. Wiley, New York
2. Angelosante D, Bazerque JA, Giannakis GB (2010) Online adaptive estimation of sparse signals: where RLS meets the l_1-norm. IEEE Trans Signal Process 58:3436–3447
3. Angelosante D, Giannakis GB, Grossi E (2009) Compressed sensing of time-varying signals. Proceedings of the 16th international conference on digital signal processing
4. Asif MS, Charles A, Romberg J, Rozell C (2011)Estimation and dynamic updating of time-varying signals with sparse variations. In: International conference on acoustics, speech and signal processing (ICASSP), pp 3908–3911
5. Asif MS, Romberg J (2009) Dynamic updating for sparse time varying signals. In: Proceedings of the conference on information sciences and systems, pp 3–8
6. Baraniuk RG, Davenport MA, Ronald D, Wakin MB (2008) A simple proof of the restricted isometry property for random matrices. Constr Approx 28:253–263
7. Benveniste A, Basseville M, Moustakides GV (1987) The asymptotic local approach to change detection and model validation. IEEE Trans Autom Control 32:583–592
8. Blumensath T, Davies M (2009) Iterative hard thresholding for compressed sensing. Appl Comput Harmon Anal 27:265–274
9. Bosch-Bayard J. et al (2005) Estimating brain functional connectivity with sparse multivariate autoregression. Philos Trans R Soc 360:969–981
10. Brockwell PJ, Davis RA (2009) Time Series: theory and methods, Springer, New York
11. Candes E, Tao T (2007) The Dantzig selector: statistical estimation when p is much larger than n. Ann Stat 35:2313–2351
12. Candes EJ (2008) The restricted isometry property and its implications for compressed sensing. C R Math 346:589–592
13. Candes EJ, Eldar YC, Needell D, Randall P (2011) Compressed sensing with coherent and redundant dictionaries. Appl Comput Harmon Anal 31:59–73
14. Candes EJ, Romberg J, Tao T (2006) Robust uncertainty principles: exact signal reconstruction from highly incomplete frequency information. IEEE Trans Inf Theory 52:489–509
15. Candes EJ, Tao T (2005) Decoding by linear programming. IEEE Trans Inf Theory 51:4203–4215
16. Candes EJ, Tao T (2006) Near-optimal signal recovery from random projections: universal encoding strategies? IEEE Trans Inf Theory 52:5406–5425
17. Candes EJ, Wakin MB (2008) An introduction to compressive sampling. IEEE Signal Process Mag 25:21–31
18. Carmi A, Gurfil P, Kanevsky D (2008) A simple method for sparse signal recovery from noisy observations using Kalman filtering. Technical Report RC24709, Human Language Technologies, IBM
19. Carmi A, Gurfil P, Kanevsky D (2010) Methods for sparse signal recovery using Kalman filtering with embedded pseudo-measurement norms and quasi-norms. IEEE Trans Signal Process 58:2405–2409
20. Carmi A, Mihaylova L, Kanevsky D (2012) Unscented compressed sensing. In: Proceedings of the IEEE international conference on acoustics, speech and signal processing (ICASSP)
21. Charles A, Asif MS, Romberg J, Rozell C (2011) Sparsity penalties in dynamical system estimation. In: Proceedings of the conference on information sciences and systems, pp 1–6
22. Chen S, Billings SA, Luo W (1989) Orthogonal least squares methods and their application to non-linear system identification. Int J Contro 50:1873–1896
23. Chen SS, Donoho DL, Saunders MA (1998) Atomic decomposition by basis pursuit. SIAM J Sci Comput 20:33–61
24. Davis G, Mallat S, Avellaneda M (1997) Greedy adaptive approximation. Constr Approx 13:57–98
25. Deurschmann J, Bar-Itzhack I, Ken G (1992) Quaternion normalization in spacecraft attitude determination. In: Proceedings of the AIAA/AAS astrodynamics conference, pp 27–37

26. Donoho DL (2006) Compressed sensing. IEEE Trans Inf Theory 52:1289–1306
27. Durate MF, Davenport MA, Takhar D, Laska JN, Sun T, Kelly KF, Baraniuk RG (2008) Single pixel imaging via compressive sampling, IEEE Signal Process Mag
28. Efron B, Hastie T, Johnstone I, Tibshirani R (2004) Least angle regression. Ann Stat 32:407–499
29. Friedman N, Nachman I, Peer D (1999) Learning Bayesian network structure from massive datasets: The sparse candidate algorithm. In: Proceedings of the fifteenth conference annual conference on uncertainty in, artificial intelligence (UAI-99), pp 206–215.
30. Granger CWJ (1969) Investigating causal relations by econometric models and cross-spectral methods. Econometrica 37:424–438
31. Haufe S, Muller K, Nolte G, Kramer N (2008) Sparse causal discovery in multivariate time series, NIPS Workshop on causality
32. Hirotugu A (1974) A new look at the statistical model identification. IEEE Trans Autom Control 19:716–723
33. James GM, Radchenko P, Lv J (2009) DASSO: connections between the Dantzig selector and LASSO. J Roy Stat Soc 71:127–142
34. Ji S, Xue Y, Carin L (June 2008) Bayesian compressive sensing. IEEE Trans Signal Process 56:2346–2356
35. Julier SJ, LaViola JJ (2007) On Kalman filtering with nonlinear equality constraints. IEEE Trans Signal Process 55:2774–2784
36. Julier SJ, Uhlmann JK (1997) A new extension of the Kalman filter to nonlinear systems. In: Proceedings of the international symposium on aerospace/defense sensing, simulation and controls, pp 182–193
37. Kailath T (1980) Linear Systems. Prentice Hall,Englewood Cliffs
38. Kalouptsidis N, Mileounis G, Babadi B, Tarokh V (2011) Adaptive algorithms for sparse system identification. Signal Proc 91:1910–1919
39. Laska JN, Boufounos PT, Davenport MA, Baraniuk RG (2011) Democracy in action: quantization, saturation, and compressive sensing. Appl Comput Harmon Anal 31:429–443
40. Mallat S, Zhang Z (1993) Matching pursuits with time-frequency dictionaries. IEEE Trans Signal Process 4:3397–3415
41. Mendel JM (1995) Lessons in estimation theory for signal processing, communications, and control. Prentice Hall, Englewood-Cliffs
42. Pati YC, Rezifar R, Krishnaprasad PS (1993) Orthogonal matching pursuit: recursive function approximation with applications to wavelet decomposition. In: Proceedings of the 27th asilomar conf. on signals, systems and comput., pp 40–44
43. Rudelson M (1999) Random vectors in the isotropic position. J Funct Anal 164:60–72
44. Rudelson M, Vershynin R (2005) Geometric approach to error correcting codes and reconstruction of signals. Int Math Res Not 64:4019–4041
45. Sanadaji BM, Vincent TL, Wakin MB, Toth, Poola K (2011) Compressive System Identification of LTI and LTV ARX models. In: Proceedings of the IEEE conference on decision and control and european control conference (CDC-ECC), pp 791–798
46. Schwarz GE (1978) Estimating the dimension of a model. Ann Stat 6:461–464
47. Sokal AD (1989) Monte carlo methods in statistical mechanics: foundations and new algorithms. Cours de Troisieme Cycle de la Physique en Suisse Romande, Laussane
48. Tibshirani R (1996) Regression shrinkage and selection via the LASSO. J Roy Stat Soc B Method, 58:267–288
49. Tipping ME (2001) Sparse Bayesian learning and the relevance vector machine. Int J Mach Learn Res 1:211–244
50. Tropp JA (2004) Greed is good: Algorithmic results for sparse approximation. IEEE Trans Inf Theory 50:2231–2242
51. Tropp JA, Gilbert AC (2007) Signal recovery from random measurements via orthogonal matching pursuit. IEEE Trans Inf Theory 53:4655–4666
52. Vaswani N (2008) Kalman filtered compressed sensing. In: Proceedings of the IEEE international conference on image processing (ICIP) pp 893–896

Chapter 10
Distributed Approximation and Tracking Using Selective Gossip

Deniz Üstebay, Rui Castro, Mark Coates and Michael Rabbat

Abstract This chapter presents selective gossip which is an algorithm that applies the idea of iterative information exchange to vectors of data. Instead of communicating the entire vector and wasting network resources, our method adaptively focuses communication on the most significant entries of the vector. We prove that nodes running selective gossip asymptotically reach consensus on these significant entries, and they simultaneously reach an agreement on the indices of entries which are insignificant. The results demonstrate that selective gossip provides significant communication savings in terms of the number of scalars transmitted. In the second part of the chapter we propose a distributed particle filter employing selective gossip. We show that distributed particle filters employing selective gossip provide comparable results to the centralized bootstrap particle filter while decreasing the communication overhead compared to using randomized gossip to distribute the filter computations.

10.1 Introduction

Many applications of wireless sensor networks require collection and processing of large amounts of data. The main challenge in fulfilling these tasks is preserving network resources such as lifetime and bandwidth. One approach to fuse and

D. Üstebay (✉) · M. Coates · M. Rabbat
Department of Electrical and Computer Engineering, McGill University, Montreal, QC, Canada
e-mail: deniz.ustebay@mail.mcgill.ca

M. Coates
e-mail: coates@ece.mcgill.ca

M. Rabbat
e-mail: michael.rabbat@mcgill.ca

R. Castro
Department of Mathematics, Eindhoven University of Technology, Eindhoven, The Netherlands
e-mail: rmcastro@tue.nl

A. Y. Carmi et al. (eds.), *Compressed Sensing & Sparse Filtering*,
Signals and Communication Technology, DOI: 10.1007/978-3-642-38398-4_10,

process large amounts of data without draining network resources is to reduce the data dimensionality. We present an algorithm called selective gossip to approximate high dimensional vectors of network data in an efficient manner. Our method is based on gossip algorithms which are decentralized methods studied extensively for scalar network data. In essence, gossip algorithms utilize iterative information exchange between pairs of nodes, and asymptotically all nodes reach consensus on a network aggregate. Selective gossip applies the idea of iterative information exchange to vectors of data. Instead of communicating the entire vector and wasting network resources, our method adaptively focuses communication on the most significant entries of the vector. We prove that nodes running selective gossip asymptotically reach consensus on these significant entries, and they simultaneously reach an agreement on the indices of entries which are insignificant.

Selective gossip can be taken as a building block and used in various distributed signal processing algorithms. Here we study the distributed target tracking problem where the nodes of a sensor network collaboratively track a moving object. For problems involving nonlinear dynamics, nonlinear measurements, and non-Gaussian noise, particle filtering is the current state-of-the-art-estimation method. We propose a distributed particle filter implementation using selective gossip. In this setting, nodes maintain a shared particle filter to sequentially estimate the state of the target. The measurements taken by sensors are fused by reaching a consensus on the likelihood associated with the each particle. Selective gossip efficiently identifies particles with large weights and focuses communication resources on computing these important weights. Through a simulation study we demonstrate that selective gossip requires lower communication overhead while achieving similar accuracy as compared to the state-of-the-art distributed particle filtering approaches on a scenario involving bearings-only measurements of a maneuvering target.

This chapter is organized as follows. Section 10.2 reviews gossip algorithms. Section 10.3 discusses the distributed averaging problem for vectors. Section 10.4 proposes the selective gossip algorithm in its three versions and also provides the convergence results. Section 10.5 introduces distributed tracking problem and proposes the distributed particle filter using selective gossip. A distributed target tracking scenario is presented to illustrate the performance of this algorithm. Section 10.6 concludes the chapter with a discussion of the results.

10.2 Gossip Algorithms

Operating under energy and bandwidth constraints, wireless sensor networks require efficient and reliable methods for processing. The traditional approach of centralized processing has several drawbacks. It introduces a single point of failure to the network. Furthermore, in dense networks, the links close to the central authority can become bottlenecks. To avoid congestion and, also, to exploit processing capabilities of sensor nodes, in-network processing algorithms are proposed. In-network processing can be performed using spanning trees or Hamiltonian cycles. These are effective methods when the network topology does not change over time. However, since they require forming and maintaining routes, these methods also have significant communication overhead when nodes are mobile or wireless networking conditions are not reliable. Gossip algorithms, on the other hand, are decentralized methods which do not require specialized routes. They are known to provide robust and scalable solutions for in-network processing.

Gossip algorithms have been widely studied as solutions to distributed consensus, a problem which dates back to early work of Tsitsiklis et al. [32, 33]. This problem requires nodes to reach an agreement by using only local exchanges. It is acknowledged as a canonical problem in distributed control and signal processing (see, e.g., the surveys [7, 23]). Some example applications are cooperative control of multiple autonomous vehicles [18], parameter estimation [30], distributed optimization [22], and source localization [26].

The standard example of distributed consensus is the *average consensus* problem, where in a network of n nodes, each node v has a scalar value $x^v \in \mathbb{R}$, and the goal is to compute the average

$$\overline{x} = \frac{1}{n}\sum_{v=1}^{n} x^v, \tag{10.1}$$

at every node. Although averaging of scalars is a basic problem, it can be generalized to computation of any linear function of the node values and to averaging of vectors. Due to this capacity for generalization, algorithms that solve average consensus are attractive for a wide range of wireless sensor network applications.

Gossip algorithms can be synchronous or asynchronous. The synchronous version requires that at each iteration all nodes broadcast their values [37]. Having received the values of its neighbors, each node then updates its value with a weighted average of its value and the values it received. The asynchronous version, on the other hand, does not require synchronization and only one pair of nodes update at each iteration. In the remainder of this chapter when we refer to gossip algorithms, we refer to asynchronous gossip.

Randomized gossip algorithm describes a randomized and asynchronous version of gossip [3]. This algorithm restricts information exchange at each iteration to only a pair of neighboring nodes. Below we summarize the randomized gossip algorithm.

For a network of n nodes, let the undirected graph $\mathcal{G} = (V, E)$ represent the network connectivity where $V = \{1, \dots, n\}$ is the set of nodes, and $E \subseteq V \times V$ is the set of edges such that $(u, v) \in E$ if and only if nodes u and v can perform bidirectional wireless communication. The set of neighbors of node u (not including u itself) is denoted by $\mathcal{N}_u = \{v\colon (u, v) \in E\}$. The gossip iterations are indexed using $k = 1, 2, \dots,$ where $k = 0$ corresponds to the initial state. Each node $v \in V$ maintains a gossip value $x^v(k)$ which is initialized with $x^v(0) = x^v$.

Asynchronous time model [2]. A clock ticks at each node according to an independent rate-1 Poisson process. Since there are $|V| = n$ nodes, this is equivalent to there being a network coordinator running a Poisson clock with rate n, and when the coordinator's clock ticks, it assigns the tick to a node drawn uniformly from V. Each tick of the coordinator's clock corresponds to one iteration and we assume that the communication and update steps involved in each iteration occur instantaneously so that no two iterations overlap.

In a practical setting, the updates take some non-trivial amount of time. One could either tune the rate of the Poisson clocks at each node so that two updates overlap (e.g., leading to interference) with probability zero, or one could adopt a more complex scheduling mechanism to avoid interference. These issues are beyond the scope of this work.

Communication model. There is a pre-defined communication matrix P with entries $P_{u,v} \geq 0$ and $\sum_{v \in V} P_{u,v} = 1$. In addition, $P_{u,v} > 0$ if and only if $(u, v) \in E$. Suppose the kth clock tick occurs at node u. Then u contacts a random neighbor v which is drawn according to the distribution $\{P_{u,v}\}_{v \in V}$, and nodes u and v perform an update.

Update rule. When nodes u and v gossip, they update their values with the average,

$$x^u(k+1) = x^v(k+1) = \frac{1}{2}\big(x^u(k) + x^v(k)\big). \tag{10.2}$$

All other nodes $v' \in V \setminus \{u, v\}$ remain unchanged; i.e., $x^{v'}(k+1) = x^{v'}(k)$.

Intuitively, the convergence of randomized gossip is guaranteed if there exists a path between each pair of nodes so that information can flow between each pair infinitely often. Hence, one can show that, for a connected graph $\mathcal{G}$, under mild conditions on the way a random neighbor, v, is chosen, the values $x^u(k)$ converge to $\overline{x}$ at every node u as $k \to \infty$ [37]. The number of randomized gossip iterations required to achieve consensus scales with the number of nodes in the network; the rate of scaling depends on the network topology. For topologies that are generally used to model wireless sensor networks such as grids and random geometric graphs, randomized gossip converges slowly [3]. Motivated by this fact, there has been a body of work studying faster versions of gossip, e.g., [1, 6, 19, 24, 35].

Another research direction involves using gossip algorithms as a building box in complex signal processing applications (see [7] and references therein). Motivated by applications in distributed estimation, we study gossiping on vectors of data. Below we state this problem and propose selective gossip for efficient distributed approximation of vectors.

10.3 Gossiping on Vectors

The scalar average consensus problem described in the previous section can be immediately generalized to distributed averaging of vectors where, initially, each node $v \in V$ has a vector $\mathbf{x}^v \in \mathbb{R}^M$ and the aim is to compute the average

$$\bar{\mathbf{x}} = \frac{1}{n}\sum_{v=1}^{n} \mathbf{x}^v, \tag{10.3}$$

at each node v.

The basic solution to this problem is to run one scalar gossip algorithm for each dimension of the vector in parallel such that the entire average vector is computed at all nodes. Parallel gossip sessions can be implemented using the standard gossip setup with a modified update rule which involves exchanging and averaging vectors instead of scalars. Note that in practical sensor network scenarios, each wireless packet can carry only a certain amount of data and, consequently, exchanging long vectors may require several packets to be transmitted. Since energy consumption is proportional to the number of packets transmitted, exchange of long vectors instead of scalars increases the energy consumption of wireless communication. Increased number of packet transmissions also increases the bandwidth consumption of gossip updates.

However, often we only care about computing the largest entries of the average vector and not the entire vector. One example is decentralized field estimation where sensor nodes are deployed in an area to take scalar measurements [34]. Starting with local measurements, the goal is to reach a network state where each node has an approximation of the field. Transform coding is based on the idea that many natural signals are sparse (or nearly sparse) when they are transformed into a suitable domain. Hence, the signal representing the field can be well-approximated using only a few transform coefficients (those with the large magnitude). Assuming that a suitable transformation is available, one can use gossip algorithms to reach a consensus on the transform coefficients in a decentralized manner. Since only a few of the transform coefficients have large magnitudes, reaching a consensus only on these coefficients is satisfactory. The problem is that we do not know which coefficients have large magnitudes before actually computing them. Thus, any gossip algorithm that aims to decrease the communication cost by computing only these coefficients needs to also identify their locations.

Another example is the scenario where nodes must collectively decide among one of a large number of hypotheses. Initially, each node has its own data. Under the assumption that the likelihood of the data at different nodes is conditionally independent given the hypothesis, the network-wide log-likelihood of any hypothesis is simply the sum of the log-likelihoods at each node. However, if the number of hypotheses is very large, then it is more efficient for the nodes to focus their resources on computing the log-likelihood of only the most likely hypothesis or hypotheses, rather than all of them. In Sect. 10.5, we will consider the related setting of distributed particle filtering, where nodes gossip on the weights of particles which can be viewed as hypotheses.

Motivated by these applications in distributed signal processing and decision making, we study a method which adaptively identifies the largest elements of a vector while computing their values. The next section describes this method and provides results related to its performance.

10.4 Selective Gossip

To address the average consensus problem in multi-dimensional setting, we propose an efficient distributed averaging algorithm called *selective gossip* [34, 36]. Selective gossip conceptually builds on the randomized gossip algorithm described in [3]. In particular, we adopt the asynchronous time model and the communication model, explained in Sect. 10.2. The update rule, on the other hand, is where selective gossip differs from randomized gossip.

Each node $v \in V$ maintains a gossip vector $\mathbf{x}^v(k) \in \mathbb{R}^M$ at iteration k and this vector is initialized with $\mathbf{x}^v(0) = \mathbf{x}^v$. Let $x_j^v(k)$ represent the jth entry of $\mathbf{x}^v(k)$. The gossip vectors in the entire network at iteration k are denoted by $X(k) = \{\mathbf{x}^v(k)\}_{v \in V}$. Let $\bar{x}_{(i)}$ denote the ith highest entry of $\bar{\mathbf{x}}$, so that $\bar{x}_{(1)} \geq \bar{x}_{(2)} \geq \cdots \geq \bar{x}_{(M)}$.

We aim to reach a consensus on the locations and values of the largest entries of $\bar{\mathbf{x}}$. Depending on how the concept of *largest entries* is defined, the problem statement and the solution changes. Here we consider two possibilities:

1. **Threshold**. Given a non-negative threshold τ, let H_τ be the set of entries larger than the threshold, i.e.,
$$H_\tau = \{j : \bar{x}_j \geq \tau\}. \tag{10.4}$$
The goal is to have the iterates $X(k)$ approach $\mathcal{X}_\tau^*$ as efficiently as possible, where
$$\mathcal{X}_\tau^* = \left\{\{\mathbf{x}^v\} :\ \text{for all } v \in V,\ \begin{matrix} x_j^v = \bar{x}_j \text{ if } j \in H_\tau \\ x_j^v < \tau \ \text{ if } j \notin H_\tau \end{matrix}\right\}. \tag{10.5}$$
If $X(k) \in \mathcal{X}_\tau^*$ then we say that the network has *reached consensus on* $\bar{\mathbf{x}}$ *for entries larger than the threshold* τ.
2. **Top-m**. Given a non-negative integer $m < M$, let $H_{\text{top-}m}$ be the set of entries of the highest m entries of $\bar{\mathbf{x}}$, i.e.,
$$H_{\text{top-}m} = \{j : \bar{x}_j \geq \bar{x}_{(m)}\}. \tag{10.6}$$
Note that the cardinality of $H_{\text{top-}m}$, denoted $|H_{\text{top-}m}|$, may in fact be larger than m, e.g., if $\bar{x}_{(m+1)} = \bar{x}_{(m)}$. Similar to above, the goal is to have the iterates $X(k)$ approach $\mathcal{X}_{\text{top-}m}^*$ as efficiently as possible, where
$$\mathcal{X}_{\text{top-}m}^* = \left\{\{\mathbf{x}^v\} :\ \text{for all } v \in V,\ \begin{matrix} x_j^v = \bar{x}_j \quad \text{if } j \in H_{\text{top-}m} \\ x_j^v < \bar{x}_{(m)} \text{ if } j \notin H_{\text{top-}m} \end{matrix}\right\}. \tag{10.7}$$
If $X(k) \in \mathcal{X}_{\text{top-}m}^*$ then we say that the network has *reached consensus on* $\bar{\mathbf{x}}$ *for the largest m entries*.

Our goal is to *efficiently* reach a state $X(k) \in \mathcal{X}_\tau^*$ or $X(k) \in \mathcal{X}_{\text{top-}m}^*$. Our measure of efficiency aims to capture the amount of data communicated between nodes over the network. Specifically, we count the total number of scalar values transmitted. Of course, in order to obtain $X(k) \in \mathcal{X}_\tau^*$ or $X(k) \in \mathcal{X}_{\text{top-}m}^*$, one could run a standard distributed averaging algorithm [2, 3, 23] on each dimension, in which case standard results guarantee that $\mathbf{x}^v(k) \to \bar{\mathbf{x}}$ as $k \to \infty$ for all $v \in V$. Since $\bar{\mathbf{x}} \in \mathcal{X}_\tau^*$ and $\bar{\mathbf{x}} \in \mathcal{X}_{\text{top-}m}^*$, this achieves our objective in both cases. However, if $|H_\tau| \ll M$ or $m \ll M$, then this is wasteful since the nodes expend communication resources calculating entries which are not relevant. Selective gossip aims to achieve a network state in $\mathcal{X}_\tau^*$ or $\mathcal{X}_{\text{top-}m}^*$, but not necessarily one where any node computes the entire vector $\bar{\mathbf{x}}$. The main challenge is that the nodes do not know, a priori, the index set (H_τ or $H_{\text{top-}m}$) as it depends on the initial values, $X(0)$, and so it must also be estimated.

Below we present three versions of selective gossip; the first version addresses the threshold-based problem and the next two versions address the top-m problem.

10.4.1 Threshold Selective Gossip

Threshold selective gossip algorithm employs a threshold τ, which is fixed and known by all nodes, to determine which entries to communicate and update at each iteration. For a node $v \in V$, let $H_\tau^v(k)$ represent the entries with values higher than τ, i.e.,

$$H_\tau^v(k) = \{j : x_j^v(k) \geq \tau\}. \tag{10.8}$$

When nodes u and v wake up according to the asynchronous time model and communication model described in Sect. 10.2, they update entries that at least one of them believes to be one of the largest. Namely, they update only the entries $j \in H_\tau^u(k-1) \cup H_\tau^v(k-1)$ by setting

$$x_j^u(k) = x_j^v(k) = \frac{1}{2}\big(x_j^u(k-1) + x_j^v(k-1)\big). \tag{10.9}$$

No change is made to entries $j \notin H_\tau^u(k-1) \cup H_\tau^v(k-1)$, and these values are not transmitted in order to save energy. Also, all other nodes $v' \in V \setminus \{u, v\}$ keep their gossip vectors unchanged.

Threshold selective gossip asymptotically converges to the correct values for entries $j \in H_\tau$. Since there is no coupling between the different entries of the vector $\bar{\mathbf{x}}$, we treat each entry individually and focus on analyzing the behavior of the algorithm for a single scalar entry. Without loss of generality, let $x^v(0)$ denote the initial value for this entry at node v, let $\bar{x}$ denote the average, and let $\tau > 0$ be the given threshold. It is well known that, under the assumptions stated above, randomized gossip converges asymptotically to the average consensus [3]. Selective gossip differs from randomized gossip in that, at some iterations, two nodes may not update a particular entry. Thus, intuitively, to show convergence when $\bar{x} \geq \tau$ we just need to show that nodes gossip sufficiently often so that eventually they all have $x_v(k) \geq \tau$; at that point selective gossip is identical to randomized gossip.

Theorem 1 [34]. *Let* $S(k) = \sum_{v=1}^{n}(x^v(k) - \bar{x})^2$ *and suppose* $\bar{x} \geq \tau$. *Then*

$$\mathbb{E}[S(k)|S(0)] \leq \left(1 - \frac{1}{n^4 \operatorname{diam}(\mathcal{G})^2 \Delta_{\max}}\right)^k S(0), \tag{10.10}$$

where $\operatorname{diam}(\mathcal{G})$ *is the diameter of the network* $\mathcal{G}$ *and* $\Delta_{\max} = \max_v |\mathcal{N}_v|$ *is the maximum degree.*

Sketch of proof When a pair of neighboring nodes (u, v) decide to gossip at the kth iteration, $S(k)$ decreases such that $S(k+1) = S(k) - \frac{1}{2}\left(x^u(k) - x^t(k)\right)^2$. Taking the expectation over all pairs of neighboring nodes with non-zero probability of gossiping at iteration k, we get

$$\mathbb{E}[S(k+1)|S(k)] \leq S(k) - \frac{1}{n\Delta_{\max}}\left(x^u(k) - x^v(k)\right)^2. \tag{10.11}$$

Since consensus is not reached yet, there exists at least one node a with $x^a(k) \geq \bar{x} + \frac{1}{n}\sqrt{\frac{S(k)}{n}}$. Constructing a path from node a to any node b with $x^b(k) < \bar{x}$, we find that there exists a pair of neighboring nodes (a', b') on this path for which

$$(x^{a'}(k) - x^{b'}(k))^2 > \frac{S(k)}{n^3 \operatorname{diam}(\mathcal{G})^2},$$

and with (10.11) the statement of the theorem follows. □

Theorem 1 shows that for entries $j \in H_\tau$, selective gossip always computes the correct value in expectation. Furthermore, since $\mathbb{E}[S(k+1)|S(k)] \leq S(k)$ and $S(k) \geq 0$ for all k, the sequence $\{S(k)\colon k \geq 0\}$ is a non-negative supermartingale with respect to itself. Using the Martingale convergence theorem, one can show that the limit $S_\infty = \lim_{k\to\infty} S(k)$ exists almost surely [12]. Moreover, standard arguments [3] based on Markov's inequality can be applied to this result to show convergence in probability. Next we give the result for entries $j \notin H_\tau$.

Theorem 2 [34]. *Let* $\mathcal{G} = \mathcal{K}_n$ *be the complete graph. Suppose that* $\bar{x} < \tau$ *and* $\tau - \bar{x} = c > 0$. *If* $S(0) > 0$ *and there exists at least one node with non-zero probability of gossiping, then there exists a finite constant* $K < \infty$ *such that after* $k \geq K$ *iterations,* $x^v(k) < \tau$ *for all nodes* v *with probability 1.*

Sketch of proof In this case, one can find two nodes (u, v) such that $(x^u(k) - x^v(k))^2 \geq c^2$. Since $\Delta_{\max} = n - 1$ for the complete graph and using the bound (10.11), we get $\mathbb{E}[S(k)|S(0)] \leq S(0) - \frac{kc^2}{n(n-1)}$. Applying Markov's inequality yields

$$\Pr\left(S(k) \geq c^2 | S(0)\right) \leq \frac{S(0)}{c^2} - \frac{k}{n(n-1)}.$$

Therefore, if $k \geq K = \frac{n(n-1)}{c^2} S(0)$, then $x^v(k) < \tau$, for all v with probability 1. □

Theorem 2 addresses the case where $\bar{x} < \tau$ only for the complete graph. This approach does not directly extend to general connected topologies. In particular, in the proof of Theorem 2, one cannot guarantee that the nodes u and v will be neighbors in a general topology. However, the convergence can be shown using an approach similar to that presented below for the proof of Theorem 3.

It is also worth noting that the bounds given in Theorems 1 and 2 are extremely loose since we only consider the gossiping of one pair of nodes instead of all pairs, and hence these bounds should not be taken as an indicator of the rate of convergence. In fact, it is easy to see that once all nodes agree that an entry j is in H_τ, threshold selective gossip behaves identically to randomized gossip, and so asymptotically the rates of convergence are the same as reported in [3] for randomized gossip. As illustrated in the simulations presented below, the error decay rate of threshold selective gossip, as a function of the number of scalar values transmitted, is in fact substantially faster than running randomized gossip in parallel for all entries.

10.4.2 Adaptive Threshold Selective Gossip

Threshold selective gossip requires a fixed preset threshold, τ, to determine the entries to be computed. However, having a fixed threshold is typically not practical since we may not have accurate prior knowledge of the distribution of values in the average vector. To address this problem, we describe a heuristic called *adaptive threshold selective gossip* which aims to find the appropriate threshold at each node in a decentralized way. By appropriate threshold, we mean $\tau \in (\bar{x}_{(m)}, \bar{x}_{(m+1)})$, where m is given as input to the algorithm. In other words, our heuristic deals with the top-m problem and tries to reach the index set $H_{\text{top-}m}$ by adaptively changing the threshold at every node. For this, each node keeps an estimate of the threshold as well as the gossip vectors $\mathbf{x}^v(k)$.

Let the threshold estimate of each node v be denoted by $\tau^v(k)$ at time k. The threshold estimate of each node is initialized with the mth largest entry of its gossip vector, i.e., $\tau^v(0) = x^v_{(m)}(0)$. When two nodes u and v perform a gossip update, they modify the entries $j \in H^u_{\tau^u(k-1)}(k-1) \cup H^v_{\tau^v(k-1)}(k-1)$ by setting

$$x^u_j(k) = x^v_j(k) = \frac{1}{2}\left(x^u_j(k-1) + x^v_j(k-1)\right). \tag{10.12}$$

All other entries remain unchanged and all other nodes keep their gossip vectors unchanged. After the update, nodes u and v reassess their approximation quality. If the current threshold of node v provides fewer than m entries in $H^v_\tau(k)$, then the node decreases its threshold. If the node has more than m entries in $H^v_\tau(k)$, then the threshold value is increased. Specifically, node v updates its threshold according to the following rule

$$\tau^v(k+1) = \begin{cases} (1+c_1)\tau^v(k) & |H^v_\tau(k)| > m \\ (1-c_2)\tau^v(k) & |H^v_\tau(k)| < m \\ \tau^v(k) & |H^v_\tau(k)| = m \end{cases} \tag{10.13}$$

where $c_1, c_2 > 0$ are predefined constants. Note that we choose $c_1 \neq c_2$ as having $c_1 = c_2$ may cause undesirable oscillations in the threshold estimates.

The adaptive threshold heuristic does not have any convergence guarantees but intuitively should be more efficient than randomized gossip since it aims to compute only the largest entries of the average vector. We present simulation results in the upcoming sections to illustrate the performance of this method.

10.4.3 Top-m Selective Gossip

Since the adaptive threshold version of selective gossip is a heuristic without convergence guarantees, we propose another variation of gossip that solves the top-m problem and also has provable guarantees. Top-m selective gossip takes a positive integer m as an input and adaptively focuses communication on the largest m entries of the gossip vectors.

Let $x^v_{(m)}(k)$ denote the mth largest value in the gossip vector $\mathbf{x}^v(k)$ at node v, and let $H^v_{\text{top-}m}(k)$ denote the set of largest m indices of node v, i.e.,

$$H^v_{\text{top-}m}(k) = \{j : x^v_j(k) \geq x^v_{(m)}(k)\}. \tag{10.14}$$

When nodes u and v perform an update, they first exchange those entries of their gossip vectors which at least one of them believes to be among the m largest; i.e., they exchange values for entries $j \in H^u_{\text{top-}m}(k-1) \cup H^v_{\text{top-}m}(k-1)$. Then, they update

$$x^u_j(k) = x^v_j(k) = \frac{1}{2}\big(x^u_j(k-1) + x^v_j(k)(k-1)\big), \tag{10.15}$$

for entries $j \in H^u_{\text{top-}m}(k-1) \cup H^v_{\text{top-}m}(k-1)$, and they set $x^u_j(k) = x^u_j(k-1)$ and $x^v_j(k) = x^v_j(k-1)$ for entries $j \notin H^u_{\text{top-}m}(k-1) \cup H^v_{\text{top-}m}(k-1)$. Likewise, the gossip vectors of all nodes $v' \in V \setminus \{u, v\}$ who do not participate in the update remain unchanged; i.e., $\mathbf{x}^{v'}(k) = \mathbf{x}^{v'}(k-1)$.

Although the threshold and top-m approaches appear similar at first glance, there are subtle differences which make top-m selective gossip considerably more challenging to analyze. When the aim is to compute all entries which exceed a threshold, the updates applied to each entry of the vector can be decoupled, since the final result only depends on whether the average for that entry does or does not exceed the threshold. On the other hand, when the aim is to compute the largest m entries of the average vector, all entries are coupled since the final result depends on the rank ordering. Subsequently, a different approach is required to show convergence.

The following theorem shows that this algorithm converges asymptotically on any connected graph to a state where all nodes agree on the indices and values of the m largest entries, where m is a given parameter.

Theorem 3 [36]. *The gossip vectors generated by top-m selective gossip converge to a limit* $\{\mathbf{x}^v(k)\}_{v\in V} \to \{\tilde{\mathbf{x}}^v\}_{v\in V}$ *as* $k \to \infty$*, where*

$$\tilde{x}_j^v = \bar{x}_j, \quad for\ j \in H_{top\text{-}m}, \quad v \in V,$$
$$\tilde{x}_j^v < \bar{x}_{(m)}, \quad for\ j \notin H_{top\text{-}m}, \quad v \in V.$$

Sketch of proof Let $\mathbf{x}_j(k) \in \mathbb{R}^n$ denote the jth entry at each node, $x_j^v(k)$, stacked into a vector. Observe that the update Eqs. (10.14) and (10.15) for top-m selective gossip, can be written as a collection of linear updates,

$$\mathbf{x}_j(k) = \mathbf{W}_j(k)\mathbf{x}_j(k-1), \quad \text{for } j = 1, 2, \ldots, M \tag{10.16}$$

where $\mathbf{W}_j(k)$ is time-varying and depends on the entire state $X(k)$ through the sets $H_{\text{top-}m}^v$ as described next. Let $[\mathbf{W}]_{u,v}$ the (u, v)th entry of the matrix $\mathbf{W}$. Suppose that nodes u and v perform the kth gossip update. If $j \in H_{\text{top-}m}^u(k-1) \cup H_{\text{top-}m}^v(k-1)$, then

$$[\mathbf{W}_j(k)]_{u,u} = [\mathbf{W}_j(k)]_{u,v} = [\mathbf{W}_j(k)]_{v,u} = [\mathbf{W}_j(k)]_{v,v} = \frac{1}{2}, \tag{10.17}$$

and

$$[\mathbf{W}_j(k)]_{u',u'} = 1, \quad [\mathbf{W}_j(k)]_{u',v'} = 0, \tag{10.18}$$

for all $u', v' \notin \{u, v\}$, since only nodes u and v update their gossip vector, and all other nodes make no changes. If $j \notin H_{\text{top-}m}^u(k-1) \cup H_{\text{top-}m}^v(k-1)$, then no node updates this entry of the gossip vector, and $\mathbf{W}_j(k) = I$. In particular, note that every matrix $\mathbf{W}_j(k)$ is symmetric and doubly stochastic with non-zero entries at least 1/2.

Recent theory [16, 31] for time-varying linear systems of the form (10.16) makes it possible to characterize the behavior of the limit $\lim_{k\to\infty} \mathbf{x}_j(k)$. Specifically, for matrices such as $\mathbf{W}_j(k)$ satisfying the properties mentioned above, the limit $\tilde{\mathbf{x}}_j = \lim_{k\to\infty} \mathbf{x}_j(k)$ exists. In addition, consider the graph $G_j = (V, E_j)$ with $(u, v) \in E_j$ if $[\mathbf{W}_j(k)]_{u,v}$ infinitely often. If G_j is connected (i.e., if there is a path in G_j connecting every pair of nodes), then $\tilde{x}_j^u = \tilde{x}_j^v$ for all u, v; i.e., all nodes asymptotically reach a consensus on the jth entry of the gossip vector. Moreover, since every $\mathbf{W}_j(k)$ is doubly stochastic, $\tilde{x}_j^v = \bar{x}_j = \frac{1}{n}\sum_u x_j^u(0)$, and the nodes reach a consensus on the average. Thus, to determine which entries of the gossip vectors converge to the average (if any), we need to characterize which entries are updated infinitely often as $k \to \infty$.

From the definition of the asynchronous time model, since all nodes initiate updates according to a rate-1 Poisson process, it follows that as $k \to \infty$, every node will participate in an infinite number of updates. Each time an update is performed, the nodes u and v update those entries in the set $H_{\text{top-}m}^u(k-1) \cup H_{\text{top-}m}^v(k-1)$, which contains at least m elements (more if the two sets are not identical). Thus, there exists a set of indices $\mathcal{J}$ which are updated infinitely often, and thus for $j \in \mathcal{J}$, the limit $\tilde{\mathbf{x}}_j$ is the consensus vector with all elements equal to $\bar{x}_j$. Moreover, those indices $i \notin \mathcal{J}$ are only updated a finite number of times. It remains to be shown that $\mathcal{J} \equiv H_{\text{top-}m}$.

Suppose that $j \in H_{\text{top-}m}$. It follows that at every iteration k there exists a node u_k such that $x_j^{u_k}(k) \geq \bar{x}_j \geq \bar{x}_{(m)}$, and so $j \in H_{\text{top-}m}^{u_k}(k)$. Thus there is a non-zero probability of element j being updated at every iteration (since there is a non-zero probability that node u_k will participate in the update), and it follows that $j \in \mathcal{J}$ and $\tilde{x}_j^v = \bar{x}_j$ for all $v \in V$ and $j \in H_{\text{top-}m}$.

Next, suppose that $j \notin H_{\text{top-}m}$ and $j \in \mathcal{J}$. Since $j \notin H_{\text{top-}m}$, we have $\bar{x}_j < \bar{x}_{(m)}$. As more updates are performed on entry j, the entries $x_j^v(k)$ for all nodes v approach $\bar{x}_j$. At some time k', it is necessarily true that $\max_v x_j^v(k') < \min_v \min_{i \in H_{\text{top-}m}} x_i^v(k') \leq \bar{x}_{(m)}$. But then $j \notin H_{\text{top-}m}^u(k')$ for any node u, and so entry j will no longer be updated. Thus, the values of entries $j \notin H_{\text{top-}m}$ converge to a limit $\tilde{x}_j^v < \bar{x}_{(m)}$ which may be different at every node. □

Note that the goals stated above can also be generalized to cases where one aims to reach a consensus on the largest entries in absolute value; i.e., entries with $|\bar{x}_j| \geq \tau$, or sorting the entries according to magnitude, $|\bar{x}_{(1)}| \geq |\bar{x}_{(2)}| \geq \dots$, when defining which are the m most significant. For example, in the decentralized field estimation application where transform coefficients can be computed via selective gossip, it may be more meaningful to compute the m entries (transform coefficients) with largest magnitude, rather than simply the largest m coefficients. All three versions of selective gossip can be modified to address this formulation, at the expense of more cumbersome notation. This extension has been performed for threshold selective gossip and the results, along with a comparison to the corresponding version of adaptive threshold selective gossip, are reported in [34]. We expect that a similar extension for top-m selective gossip should be possible using similar techniques.

10.4.4 Simulation Results

In this section we demonstrate the performance of selective gossip through numerical experiments. The simulation setup consists of a network of $n = 50$ nodes which are distributed uniformly at random in the unit square. The communication topology is a random geometric graph, i.e., there is an edge between two nodes that are within a distance r from each other. This distance is set to $r = \sqrt{2 \log n / n}$ so that the graph is connected with high probability [14]. The dimension of the gossip vectors is $M = 25$.

To generate an initial network state, $X(0)$, we first determine the average vector, $\bar{\mathbf{x}}$. The top panels of Figs. 10.1 and 10.2 show two different vectors $\bar{\mathbf{x}}$ used in our experiments. The first one has a clear separation between the averages of the first 5

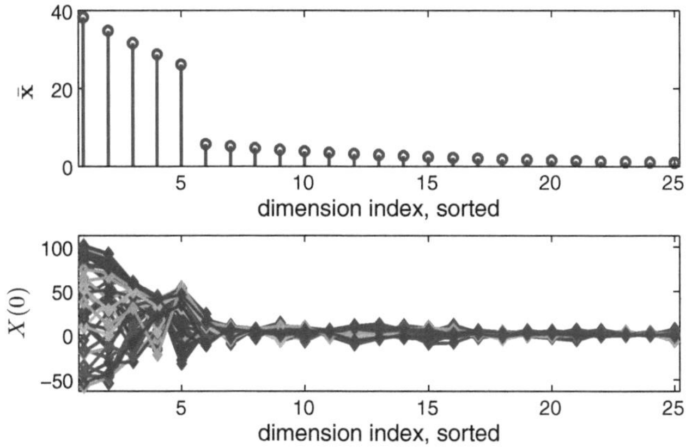

Fig. 10.1 *Top*: The average vector $\bar{\mathbf{x}}$ in descending order for initialization 1. *Bottom*: The initial state of the network with indices in the same order as $\bar{\mathbf{x}}$ above. *Diamonds* represent $x_j^v(0)$ and those that belong to the same node are connected with a *solid line*

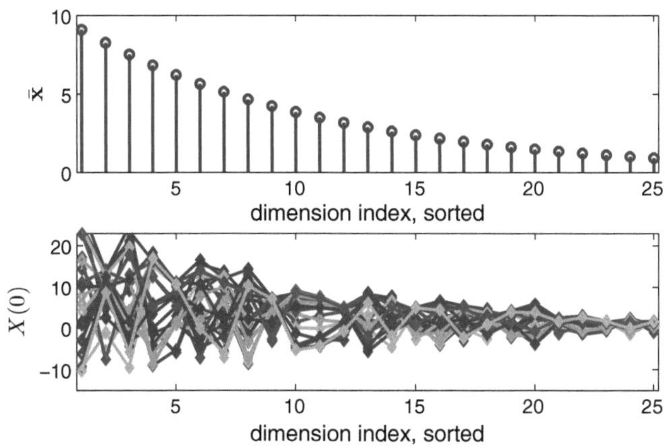

Fig. 10.2 *Top*: The average vector $\bar{\mathbf{x}}$ in descending order for initialization 2. *Bottom*: The initial state of the network with indices in the same order as $\bar{\mathbf{x}}$ above. *Diamonds* represent $x_j^v(0)$ and those that belong to the same node are connected with a *solid line*

indices and the rest, making $m = 5$ a natural choice. The second average vector is more smoothly distributed across its dimensions.

Motivated by applications in sensor networks, we assume that the node values represent measurements of natural phenomena. For each index j, we select a point μ_j uniformly at random in the unit square. Then for each node v we generate $x_j^v(0)$ such that the nodes geographically closer to μ_j will have higher values and the average over all nodes is equal to $\bar{x}_j$. The initial values are distributed such that the highest m indices at each node are not necessarily the same as $H_{\text{top-}m}$. The bottom panels of Figs. 10.1 and 10.2 illustrate the vectors $\{x_j^v(0)\}_{v\in V}$ and how they are distributed over the network.

We compare the performance of adaptive threshold selective gossip and top-m selective gossip with the performance of randomized gossip [3]. Randomized gossip is guaranteed to converge to $\bar{\mathbf{x}}$, but it is wasteful since the nodes gossip on every entry of the gossip vector at every iteration. Since randomized gossip computes every entry of $\bar{\mathbf{x}}$ it is equivalent to running top-m selective gossip with $m = M$.

The performance is measured with the mean squared error which is defined as

$$MSE(k) = \frac{1}{n}\sum_{v\in V}\sum_{j\in H_{\text{top-}m}} \left(x_j^v(k) - \bar{x}_j\right)^2.$$

Since we are interested in the amount of data that is communicated during the course of gossip, we plot the error against the number of scalars that are transmitted instead of the iterations k. Figures 10.3 and 10.4 compare the MSE of the three algorithms for different values of m. The results show the average performance over 500 different realizations of gossip.

Since randomized gossip updates all entries of gossip vectors at every iteration, its performance is the same as the performance of selective gossip for $m = 25$. In fact, for $m = 25$ all three methods perform the same, and hence their MSE curves overlap. For other values, we can see that the performance of top-m selective gossip is always better than that of adaptive threshold selective gossip.

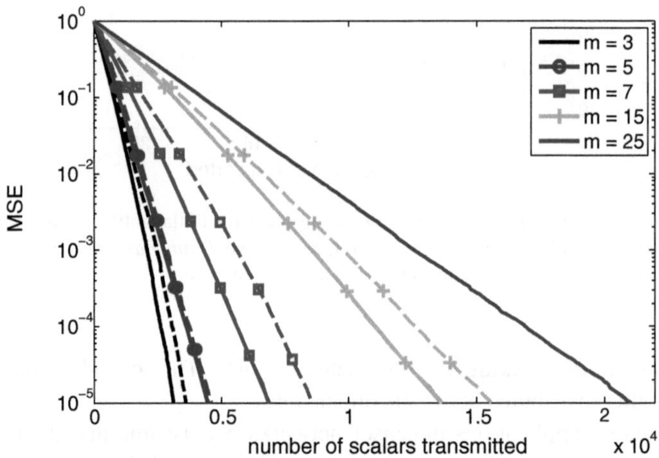

Fig. 10.3 A comparison of error performances for initialization given in Fig. 10.1. The algorithms that are compared are top-m selective gossip *(solid)* and adaptive threshold selective gossip *(dashed)* for varying m values and randomized gossip (corresponds to $m = 25$ in the plot as it updates all entries at each iteration). The plot illustrates the performance averaged over 500 realizations of gossip

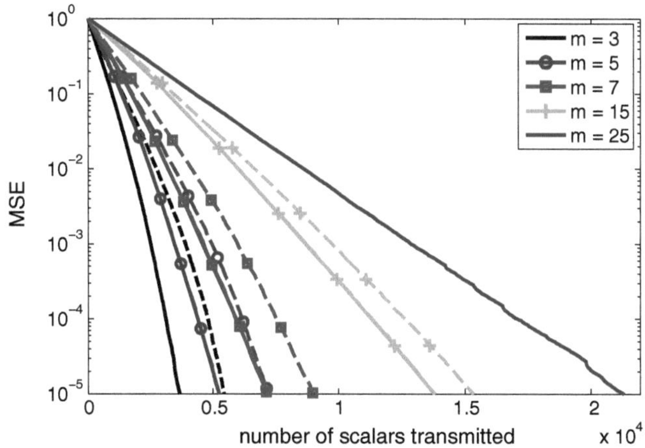

Fig. 10.4 A comparison of error performances for initialization given in Fig. 10.2. The algorithms that are compared are top-m selective gossip *(solid)* and adaptive threshold selective gossip *(dashed)* for varying m values and randomized gossip (corresponds to $m = 25$ in the plot as it updates all entries at each iteration). The plot illustrates the performance averaged over 500 realizations of gossip

The effects of varying m can be seen in Fig. 10.4 for the initialization shown in Fig. 10.1. The difference between the top-m and adaptive threshold versions of selective gossip is minimal when m is equal to the number of entries of $\bar{\mathbf{x}}$ that are significantly higher than the rest. For the initialization of Fig. 10.2, the adaptive threshold version performs worse for every m. In particular, for low values of m, top-m selective gossip computes more entries of the average vector with the same number of transmitted scalars compared to the adaptive threshold version.

To investigate how well one could hope to do using top-m selective gossip, we also implement a version of top-m selective gossip where every node clairvoyantly knows $H_{\text{top-}m}$ from the start and only updates entries $j \in H_{\text{top-}m}$ at each iteration. The corresponding results are shown in Figs. 10.5 and 10.6.

10.5 Distributed Tracking Using Selective Gossip

In this section we propose a distributed tracking algorithm that utilizes selective gossip. Before explaining the details of the algorithm, we provide some background on the problem.

Tracking is an important task in wireless sensor networks. The goal of tracking is to estimate the state of a dynamical system sequentially in time using measurements recorded by the sensors. For example, the state can be the position and the velocity of a moving target or, in the case of monitoring environmental conditions, it can represent the soil moisture and temperature. In these scenarios, we do not have direct

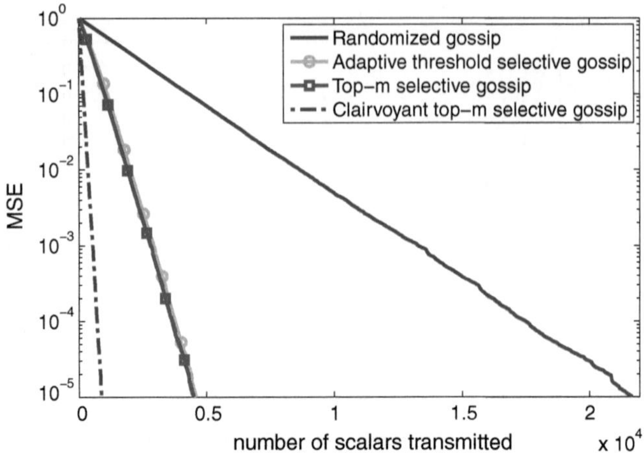

Fig. 10.5 A comparison of performances of randomized gossip, top-m selective gossip, adaptive threshold selective gossip for the initialization given in Fig. 10.1 and $m = 5$. The plot also includes clairvoyant top-m selective gossip which updates only the entries in $H_{\text{top-}m}$ at each iteration

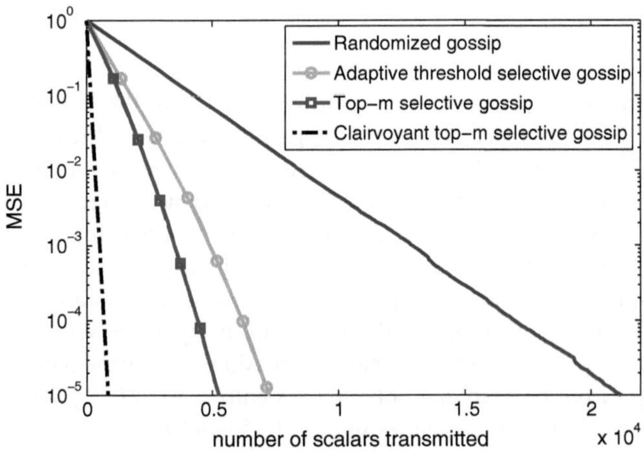

Fig. 10.6 A comparison of performances of randomized gossip, top-m selective gossip, adaptive threshold selective gossip for the initialization given in Fig. 10.2 and $m = 5$. The plot also includes clairvoyant top-m selective gossip which updates only the entries in $H_{\text{top-}m}$ at each iteration

access to the state of the dynamical system. Instead, the state can only be observed via the noise-corrupted measurements of the sensors.

The sequential estimation problem arises in many areas including robotics, tracking, financial econometrics and computer vision (see [4, 8, 28] and the references therein). The optimal estimator for this problem when the dynamics and observation models are linear and the noise distributions are Gaussian is the well-known Kalman filter. However, many practical scenarios (e.g., the tracking of a maneuvering target) involve nonlinearities and/or non-Gaussian noise, in which case the Kalman filter does not apply. Some popular approaches for more general settings are the extended Kalman filter, the Gaussian sum filter, the unscented Kalman filter, and particle filter methods (also known as sequential Monte Carlo methods) [28]. Due to their flexibility, ease of implementation, and performance, particle filter methods are widely accepted as the state-of-the-art approach to sequential estimation for the case of nonlinear dynamic models and non-Gaussian noise distributions [8, 9].

10.5.1 Sequential Estimation

In this section we review the sequential estimation problem, adopting definitions and terminology from [4, 5, 9, 28].

The state-space modeling framework describes the state of the system as an unobserved Markov process denoted by $\{\mathbf{y}_t\}_{t\in\mathbb{N}}$. The state evolution is determined by the initial distribution $p(\mathbf{y}_0)$ and the transition distribution $p(\mathbf{y}_t|\mathbf{y}_{t-1})$. The observations $\{\mathbf{z}_t\}_{t\in\mathbb{N}^+}$ are assumed to be conditionally independent given the state $\mathbf{y}_t$, and they are of marginal distribution $p(\mathbf{z}_t|\mathbf{y}_t)$. Such state-space models are also known as hidden Markov models.

The goal is to characterize the distribution of the state at the present time using the information provided by all observations received up to the present time. Let the sequence of states up to time t be denoted by $\mathbf{y}_{0:t}$ and let the sequence of observations up to time t be denoted by $\mathbf{z}_{1:t}$. We are interested in sequential estimation of the posterior distribution $p(\mathbf{y}_{0:t}|\mathbf{z}_{1:t})$ and the filtering distribution $p(\mathbf{y}_t|\mathbf{z}_{1:t})$.

The analytical solution is available as a two-stage recursion for both the posterior and filtering distributions. The stages of the recursion for the filtering distribution are termed prediction and update steps, and are presented in the following format:

$$\text{Prediction:} \quad p(\mathbf{y}_t|\mathbf{z}_{1:t-1}) = \int p(\mathbf{y}_t|\mathbf{y}_{t-1})p(\mathbf{y}_{t-1}|\mathbf{z}_{1:t-1})d\mathbf{y}_{t-1} \tag{10.19}$$

$$\text{Update:} \quad p(\mathbf{y}_t|\mathbf{z}_{1:t}) = \frac{p(\mathbf{z}_t|\mathbf{y}_t)p(\mathbf{y}_t|\mathbf{z}_{1:t-1})}{p(\mathbf{z}_t|\mathbf{z}_{1:t-1})} \tag{10.20}$$

where, assuming $p(\mathbf{y}_{t-1}|\mathbf{z}_{1:t-1})$ is available, the system model is used to predict the prior distribution at time t and the observation $\mathbf{z}_t$ is used in the second stage to update the prior via Bayes' rule.

10.5.2 Particle Filtering

Particle filters approximate the distributions $p(\mathbf{y}_{0:t}|\mathbf{z}_{1:t})$ and $p(\mathbf{y}_t|\mathbf{z}_{1:t})$ by a set of random samples termed particles. These particles are candidates for the state and their associated weights represent the accuracy of the estimate. Particle filters, also known as sequential Monte Carlo methods, have been around since the 1960s [15], but due to their computational complexity they were not widely used. The early implementations also suffered from particle degeneracy which is due to the increase in variance of weights over time. After some iterations, many particles have negligible weights and thus do not contribute to the estimation. This problem was solved in 1993 by Gordon et al. with the introduction of resampling [11].

The sequential importance resampling (SIR) particle filter maintains a weighted particle approximation $\{\mathbf{y}_{1:t}^{(i)}, w_t^{(i)}\}_{i=1}^{M}$ to estimate a posterior of interest $p(\mathbf{y}_{1:t}|\mathbf{z}_{1:t})$. The posterior is estimated by the distribution

$$\hat{p}_M(\mathbf{y}_{1:t}|\mathbf{z}_{1:t}) = \frac{1}{M}\sum_{i=1}^{M} w_t^{(i)}\delta(\mathbf{y}_{1:t} - \mathbf{y}_{1:t}^{(i)}), \tag{10.21}$$

where $\delta(\cdot)$ is the Dirac delta function.

Assuming it has a weighted particle approximation at time $t-1$, SIR propagates the particles to time t by sampling from an importance function q, evaluates the likelihoods of the extended particles, and updates the weights accordingly. A common approach is to use the prior as the importance function, i.e., $q = p(\mathbf{x}_t|\mathbf{z}_{t-1}^{(i)})$. Then there is an optional resampling step to construct a set of particles with more evenly distributed weights. Resampling replicates particles with high weights and discards particles with low weights. In [11] the prior is used as the importance function and resampling is done at every step. The authors call this implementation the bootstrap particle filter. Algorithm 1 provides the pseudo-code for the bootstrap particle filter algorithm.

10.5.3 Particle Filters in Wireless Sensor Networks

One approach to implement particle filters in networks is the leader node framework [38]. One node is selected as leader and all nodes send their measurements to this node. The leader node runs a centralized particle filter using all the information from the network. This leader node may change over time to distribute the responsibility of processing among the nodes. Being centralized, the leader node framework allows only the leader node to be queried and introduces a single point of failure. In addition, to be able to process the raw measurements of the sensors, the leader node needs to know the observation models, sensor locations, and calibration parameters of the sensors. Since only the leader node has access to the output of the particle filter, it must also make sensor management decisions such as which nodes take measurements next and with which modality.

Another approach is to distribute the computation. Each node calculates its local likelihood and the information is fused to form a global posterior. Virtually all such distributed filters rely on an assumption of conditional independence of the measure-

Algorithm 1 Bootstrap particle filter

// Initialization at time $t = 1$
1. For each particle $i = 1, \ldots, M$ do
   - Sample $\mathbf{y}_1^{(i)} \sim q_1(\cdot)$
   - Set $w_1^{(i)} = \frac{p(\mathbf{z}_1|\mathbf{y}_1^{(i)})p(\mathbf{y}_1^{(i)})}{q_1(\mathbf{y}_1^{(i)})}$
2. end
3. Normalize weights $w_1^{(i)}$ so that $\sum_{i=1}^{M} w_1^{(i)} = 1$
4. Resample $\left\{\mathbf{y}_1^{(i)}, w_1^{(i)}\right\}_{i=1}^{M}$ to obtain $\left\{\mathbf{y}_1'^{(i)}, \frac{1}{M}\right\}_{i=1}^{M}$
5. For times $t > 1$:
// For each particle $i = 1, \ldots, M$ do
   - Set $\mathbf{y}_{1:t-1}^{(i)} = \mathbf{y}_{1:t-1}'^{(i)}$
   - Sample $\mathbf{y}_t^{(i)} \sim q(\mathbf{y}_t|\mathbf{y}_{t-1}^{(i)})$
   - Set $w_t^{(i)} = \frac{p(\mathbf{z}_t|\mathbf{y}_t^{(i)})p(\mathbf{y}_t^{(i)}|\mathbf{y}_{t-1}^{(i)})}{q(\mathbf{y}_t|\mathbf{y}_{t-1}^{(i)})}$
6. end
7. Normalize weights $w_t^{(i)}$ so that $\sum_{i=1}^{M} w_t^{(i)} = 1$
8. Resample $\left\{\mathbf{y}_{1:t}^{(i)}, w_t^{(i)}\right\}_{i=1}^{M}$ to obtain $\left\{\mathbf{y}_{1:t}'^{(i)}, \frac{1}{M}\right\}_{i=1}^{M}$

ments made at each node given the target state. Several of these distributed particle filters require a spanning tree or Hamiltonian cycle for communication [5, 29]. Construction and maintenance of such routes can be very challenging when nodes are mobile or wireless conditions are adverse. Hence the algorithms are highly vulnerable to link and node failures.

Alternatively, gossip algorithms can be used for distributing the computation [10, 13, 17, 20, 21, 25]. The algorithm in [13] uses the expectation-maximization (EM) algorithm based on gossip to estimate the parameters of a mixture approximation to the global posterior, but it imposes significant constraints on the structure of the likelihood function. In the procedure in [25], each node forms a Gaussian approximation to a local posterior and then a gossip algorithm is used to fuse the means and covariance matrices to construct a Gaussian approximation of the global posterior. This algorithm has a much lower communication overhead, but its accuracy diminishes when posteriors cannot be adequately approximated by a Gaussian. The method presented in [17] constructs a polynomial approximation of the joint likelihood at each node using distributed averaging. Hence this algorithm also involves reduced communication overhead and is restricted to certain types of likelihood functions.

The algorithms in [10, 20] do not form parametric approximations to the posterior; instead they share particles among different nodes. In [20], particles undergo a random walk through the sensor network, and their weights are successively multiplied by a function of the local likelihood. The function is carefully chosen so that the particle weights converge to the same values that a centralized particle filter would calculate. This algorithm has attractive properties, but it only supports importance-sampling from the prior, which can lead to poor performance of a particle filtering algorithm [9]. The algorithm in [20] also has no mechanism for eliminating particles with small weights, leading to wasteful communication.

The algorithm in [10] was designed to allow sampling from a better importance distribution (one that better matches the posterior). It estimates regions of concentrated mass in the global posterior by calculating the intersection of the regions of concentration in the local posteriors. The importance sampling function is then constructed to focus on the calculated region, and the gossip procedure is used to calculate the global likelihoods and hence the particle weights. This procedure achieves high accuracy, but the communication cost (in terms of number of values exchanged) is high because the weights of all particles must be calculated, even if many are very small. Also, the computation of regions of concentration requires oversampling of particles at each node, increasing the local computation complexity.

To improve upon currently available algorithms, we propose using selective gossip in a distributed implementation of the bootstrap particle filter. The next section describes our problem statement.

10.5.4 Distributed Tracking Problem Statement

We consider a wireless sensor network consisting of n nodes and represent network connectivity as a graph, $\mathcal{G} = (V, E)$. We assume that the graph is connected, and that although nodes are unaware of the global topology, they do have the knowledge of their neighbors. The goal is to sequentially estimate a state, denoted by $\mathbf{y}_t$ at time index t. The state may represent a target's kinematics, typically position and velocity, or a set of environmental conditions, such as temperature, wind speed, or soil moisture. Let d be the dimension of the state, i.e., $\mathbf{y}_t \in \mathbb{R}^d$. At time t, node v makes a noisy measurement $\mathbf{z}_t^v$. The set of all measurements made by the network at time t is then $\mathbf{z}_t^V = \{\mathbf{z}_t^v : v \in V\}$ and the joint likelihood of these measurements is given by the function $p(\mathbf{z}_t^V|\mathbf{y}_t)$.

Nodes do not have access to the measurement modalities, noise models, or calibration parameters of other nodes in the network. Hence they cannot process raw measurements from other nodes. However we assume that the noise distributions at different nodes are conditionally independent given the state. Therefore the joint likelihood can be factorized into

$$p(\mathbf{z}_t^V|\mathbf{y}_t) = \prod_{v \in V} p(\mathbf{z}_t^v|\mathbf{y}_t), \tag{10.22}$$

where $p(\mathbf{z}_t^v|\mathbf{y}_t)$ is the likelihood of the observation made by node v.

Since the global likelihood is factorisable, its computation can be reduced to a set of local tasks at nodes followed by a final networked aggregation step. We are interested in a particle filter implementation that takes advantage of this factorization and achieves decentralized sequential estimation. Because wireless sensor networks have battery and bandwidth constraints, the distributed implementation needs to be efficient in terms of the number of values exchanged.

10.5.5 Distributed Particle Filter Using Selective Gossip

We now present our distributed particle filter algorithm which is based on the bootstrap particle filter. In this algorithm, every node in the network runs a copy of the same particle filter provided that the following two conditions hold. First, the measurements at nodes are synchronized so that measurements made at the same time index reflect the same state at all nodes. Second, the random number generators of the nodes are synchronized (e.g., the nodes use pseudo random generators initialized with the same seed). This ensures that nodes sample the same values when they are given the same set of weighted particles as input. These two conditions can be achieved via a decentralized routine that is executed before the sequential estimation.

The challenge in implementing distributed particle filters lies in the fact that the global weights depend on the measurements $\mathbf{z}_t^V$, but each node v only has access to its own measurement $\mathbf{z}_t^v$. We address this challenge by exploiting the factorization of the global likelihood and the fact that the computation of global weights is reduced to local computation tasks which need to be followed by a multiplication procedure. The local tasks can be performed independently at each node and do not require knowledge of the modality, noise, or calibration details of other nodes. Instead of multiplication, we use summation in the logarithm domain, which is suitable for distributed averaging.

We start by introducing local pre-weights $\{\phi_t^{v,(i)}\}_{i=1}^M$ where $\phi_t^{v,(i)} = n \log p(\mathbf{z}_t^v|\mathbf{y}_t^{(i)})$. Then the weight of particle i can be expressed using local pre-weights as

$$w_t^{(i)} = \frac{\exp(\frac{1}{n}\sum_{v\in V}\phi_t^{v,(i)})p(\mathbf{y}_t^{(i)}|\mathbf{y}_{t-1}^{(i)})}{q(\mathbf{y}_t|\mathbf{y}_{t-1}^{(i)})}. \tag{10.23}$$

Hence the weights can be calculated via averaging of an M−dimensional vector equation. Once the weights are computed, the bootstrap filter requires a normalization and resampling step so that particles are more evenly distributed. In particular, resampling discards particles with low weights and replicates the particles that have high weights. Figure 10.7 illustrates the distribution of particle weights for an example filter running with $M = 2000$ particles. Most of the particles have low weights and since particles with low weights are not to be kept, computing their values via distributed averaging wastes scarce network resources. Hence we are interested in computing only the weights that are high instead of computing all weights $\{w_t^{(i)}\}_{i=1}^M$.

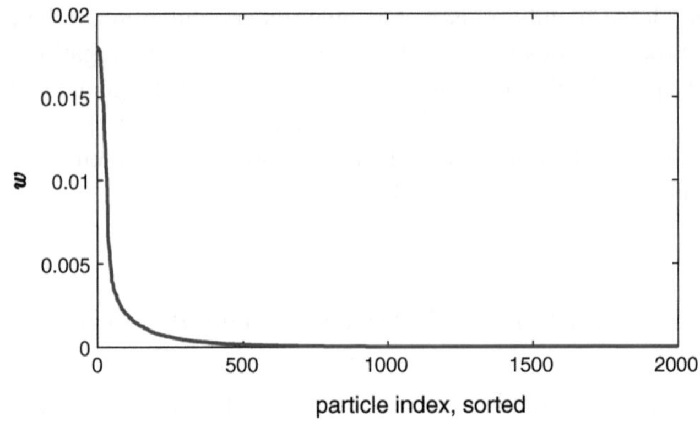

Fig. 10.7 The distribution of particle weights, sorted in descending order

Of course the challenge is that nodes do not know which weights are higher from only the local information that they have.

We propose to use selective gossip to focus communication on only the highest m weights. With the input of local pre-weights, $\{\phi^v\}_{v=1}^n$, at n nodes, and the given integer m, selective gossip identifies the set $H_{\text{top-}m}$ of the particles with the highest m weights and provides each node with the pre-weight estimates of these particles, $\{\widetilde{\phi}^{v,(H_{\text{top-}m})}\}_{v=1}^n$.

We then run a max gossip procedure to ensure that all nodes have exactly the same values, i.e., the same pre-weight vector $\widehat{\phi}^{(H_{\text{top-}m})}$. Similar to selective gossip, max gossip is based on the asynchronous time model and the communication model given in Sect. 10.2. When two nodes u and v perform a max gossip iteration, they identify the entries to update in the same way as selective gossip does. However, max gossip differs from selective gossip in that, instead of averaging, the nodes take the maximum of their previous values; i.e., nodes u and v update entries $j \in H_\tau^u(k-1) \cup H_\tau^v(k-1)$ by setting

$$x_j^u(k) = x_j^v(k) = \max\left(x_j^u(k-1), x_j^v(k-1)\right). \tag{10.24}$$

When all nodes have the exact same pre-weight values for particles in the set $H_{\text{top-}m}$, then they can compute the weights for these particles and proceed with the normalization and resampling. Since they have synchronized seeds, they will sample the same particles and reach the same set of weighted particles at the end of each step of the algorithm. The complete algorithm is described in Algorithm 2.

10.5.6 Numerical Example: Bearings Only Distributed Tracking of a Maneuvering Target

To evaluate the performance of our method, we study a distributed tracking scenario where a maneuvering target is monitored by a network of bearings sensors. Such a

Algorithm 2 Distributed Bootstrap Particle Filter with Selective Gossip

// Initialization at time $t = 1$

1. For each node $v = 1, \ldots, n$ do
   - For each particle $i = 1, \ldots, M$ do
     - Sample $\mathbf{y}_1^{(i)} \sim q_1(\cdot)$
     - Set $\phi^{v,(i)} = n \log p(\mathbf{z}_1^v | \mathbf{y}_1^{(i)})$
   - end
2. end
3. $\{\widetilde{\phi}^{v,(H_{\text{top-}m})}\}_{v=1}^n = \text{SelectiveGossip}(\{\phi^v\}_{v=1}^n, m)$
4. $\{\widehat{\phi}^{(H_{\text{top-}m})}\}_{v=1}^n = \text{MaxGossip}(\{\widetilde{\phi}^{v,(H_{\text{top-}m})}\}_{v=1}^n)$
5. For each node $v = 1, \ldots, n$ do
   - For each particle $i \in H_{\text{top-}m}$ do
     - Set $w_1^{(i)} = \frac{\exp(\widehat{\phi}^{(i)}) p(\mathbf{y}_1^{(i)})}{q_1(\mathbf{y}_1^{(i)})}$
   - end
   - Normalize weights $w_1^{(i)}$ so that $\sum_{i \in H_{\text{top-}m}} w_1^{(i)} = 1$
   - Resample $\left\{\mathbf{y}_1^{(i)}, w_1^{(i)}\right\}_{i \in H_{\text{top-}m}}$ to obtain $\left\{\mathbf{y}_1'^{(i)}, \frac{1}{M}\right\}_{i=1}^M$
6. end

// For times $t > 1$:

7. For each node $v = 1, \ldots, n$ do
   - For each particle $i = 1, \ldots, M$ do
     - Set $\mathbf{y}_{1:t-1}^{(i)} = \mathbf{y}_{1:t-1}'^{(i)}$
     - Sample $\mathbf{y}_t^{(i)} \sim q(\mathbf{y}_t | \mathbf{y}_{t-1}^{(i)})$
     - Set $\phi^{v,(i)} = n \log p(\mathbf{z}_t^v | \mathbf{y}_t^{(i)})$
   - end
8. end
9. $\{\widetilde{\phi}^{v,(H_{\text{top-}m})}\}_{v=1}^n = \text{SelectiveGossip}(\{\phi^v\}_{v=1}^n, m)$
10. $\{\widehat{\phi}^{v,(H_{\text{top-}m})}\}_{v=1}^n = \text{MaxGossip}(\{\widetilde{\phi}^{v,(H_{\text{top-}m})}\}_{v=1}^n)$
11. For each node $v = 1, \ldots, n$ do
    - For each particle $i \in H_{\text{top-}m}$ do
      - Set $w_t^{(i)} = \frac{\exp(\widehat{\phi}^{(i)}) p(\mathbf{y}_t^{(i)} | \mathbf{y}_{t-1}^{(i)})}{q(\mathbf{y}_t | \mathbf{y}_{t-1}^{(i)})}$
    - end
    - Normalize weights $w_t^{(i)}$ so that $\sum_{i=1}^M w_t^{(i)} = 1$
    - Resample $\left\{\mathbf{y}_{1:t}^{(i)}, w_t^{(i)}\right\}_{i=1}^M$ to obtain $\left\{\mathbf{y}_{1:t}'^{(i)}, \frac{1}{M}\right\}_{i=1}^M$
12. end

scenario of tracking based on only angle measurements is generally termed bearings-only tracking (sometimes also appearing in the literature under the names passive ranging and target motion analysis [27]).

We consider a two-dimensional setup where the bearing is defined as the angle from the vertical axis of Cartesian plane to the line of sight between the observer and the target. The bearing angle is measured positive in the clockwise direction. The state of the target at time t is

$$\mathbf{y}_t = \left[y_{t,1} \; y_{t,2} \; \dot{y}_{t,1} \; \dot{y}_{t,2} \right]^T, \tag{10.25}$$

where $y_{t,1}$ and $y_{t,2}$ correspond to the position in the X and Y coordinates in the Cartesian plane and $\dot{y}_{t,1}$ and $\dot{y}_{t,2}$ are the velocity values in these coordinates. The state of the observing sensor node $v \in V$ is similarly defined as $\mathbf{y}_t^v = \left[y_1^v \; y_2^v \; 0 \; 0 \right]^T$. Note that the velocity values are equal to zero because the sensor nodes are static. We assume that each node is aware of its state. The measurement made by node v at time t is denoted by z_t^v.

The dynamics of the maneuvering target are modeled using three different motion models [28]. We assume that at any time the target makes one of the following motions: (1) constant velocity (CV), (2) clockwise coordinated turn (CT), or (3) counter-clockwise coordinated turn (CCT). The target moves according to these three motion models with probabilities of p_{CV}, p_{CT} and, p_{CCT}, respectively. We also assume that the probability of both coordinated turns are equal, i.e., $p_{\text{CT}} = p_{\text{CCT}}$, and there are no other motions available, that is $p_{\text{CV}} + p_{\text{CT}} + p_{\text{CCT}} = 1$.

The state at time $t+1$ can be expressed as a function of the previous state $\mathbf{y}_t$ and process noise $\mathbf{v}_t$

$$\mathbf{y}_{t+1} = \mathbf{F}_t^j \mathbf{y}_t + \mathbf{G}\mathbf{v}_t, \tag{10.26}$$

where $\mathbf{F}_t^j$ is the transition matrix corresponding to the motion model $j \in \{1, 2, 3\}$ and

$$\mathbf{G} = \begin{bmatrix} T^2/2 & 0 \\ 0 & T^2/2 \\ T & 0 \\ 0 & T \end{bmatrix}. \tag{10.27}$$

Here T is the sampling interval and $\mathbf{v}_t \sim \mathcal{N}(0, \sigma_a I_{2\times 2})$ with scalar σ_a. The transition matrix corresponding to the constant velocity model is

$$\mathbf{F}_t^1 = \begin{bmatrix} 1 & 0 & T & 0 \\ 0 & 1 & 0 & T \\ 0 & 0 & 1 & 0 \\ 0 & 0 & 0 & 1 \end{bmatrix}, \tag{10.28}$$

whereas the coordinated turn models are governed by

$$\mathbf{F}_t^j = \begin{bmatrix} 1\,0 & \frac{\sin(\Omega_t^{(j)}T)}{\Omega_t^{(j)}} & -\frac{1-\cos(\Omega_t^{(j)}T)}{\Omega_t^{(j)}} \\ 0\,1 & \frac{1-\cos(\Omega_t^{(j)}T)}{\Omega_t^{(j)}} & \frac{\sin(\Omega_t^{(j)}T)}{\Omega_t^{(j)}} \\ 0\,0 & \cos(\Omega_t^{(j)}T) & -\sin(\Omega_t^{(j)}T) \\ 0\,0 & \sin(\Omega_t^{(j)}T) & \cos(\Omega_t^{(j)}T) \end{bmatrix}, \quad j = 2, 3. \tag{10.29}$$

The turning rates for clockwise and counter clockwise coordinated turn models are

$$\Omega_t^2 = \frac{a}{\sqrt{(\dot{y}_{t,1})^2 + (\dot{y}_{t,2})^2}}, \qquad \Omega_t^3 = -\frac{a}{\sqrt{(\dot{y}_{t,1})^2 + (\dot{y}_{t,2})^2}}, \tag{10.30}$$

where $a > 0$ is the maneuver acceleration parameter. Note that the turning rates are nonlinear functions of the state.

The angle measurements are also a nonlinear function of the state. The measurement taken by node v at time t is modeled as

$$z_t^v = \arctan\left(\frac{y_{t,1} - y_{t,1}^v}{y_{t,2} - y_{t,2}^v}\right) + w_t, \tag{10.31}$$

where $w_t \sim \mathcal{N}(0, \sigma_\theta^2)$ is the measurement noise.

We consider a network of $n = 49$ sensor nodes, forming a grid topology. The network spans an area of 1 km^2. The initial state of the target is

$$\mathbf{y}_1 = \left[\,702\,\text{m}\;\; 621\,\text{m}\;\; 10\,\text{m/min}\;\; 80\,\text{m/min}\,\right]^T. \tag{10.32}$$

The target follows a trajectory for a duration of $t_{max} = 20$ min. The sensor locations and the trajectory of the target are shown in Fig. 10.8.

The particle filter at each node is initialized with the same distribution centered at the initial state of the target [28]. In particular, we assume prior knowledge of the target's initial range, speed, and course (i.e., the angle with the vertical axis of the Cartesian plane). The position components of the state are initialized using the bearing measurement recorded by the closest sensor at time $t = 1$ and the initial range $\hat{r}_1$. We assume that $\hat{r}_1 \sim \mathcal{N}(r_1, \sigma_r^2)$ where r_1 is the initial true target range and $\sigma_r^2 = r_1/8$. Similarly, the velocity components of the state are initialized using the initial speed, $\hat{s}_1$, and initial course, $\hat{c}_1$. We assume that $\hat{s}_1 \sim \mathcal{N}(s_1, \sigma_s^2)$ where s_1 is the target's true initial speed and $\sigma_s^2 = s_1/8$. Likewise, $\hat{c}_1 \sim \mathcal{N}(c_1, \sigma_c^2)$ where c_1 is the true initial course and $\sigma_c^2 = \pi/\sqrt{6}$. Note that this initialization is suitable for many problems, but there may be cases where target acquisition also needs to be performed. This is beyond the scope of the current chapter.

We model the target motion using the following parameters: the process noise is $\sigma_a = 0.1$, the acceleration parameter is $a = 30$, the probability of constant velocity model is $p_{\text{CV}} = 0.6$ and the probabilities of turning clockwise or counter-clockwise are $p_{\text{CT}} = p_{\text{CCT}} = 0.2$.

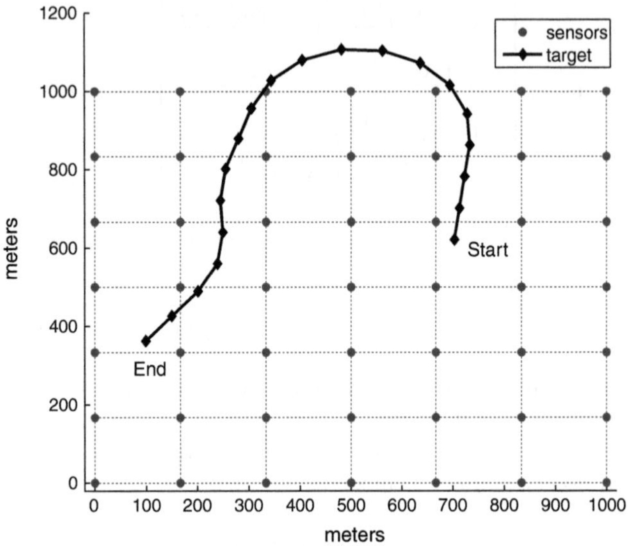

Fig. 10.8 Sensor network and the trajectory of the target. *Dashed lines* represent wireless communication links between sensors. The target makes a movement for 20 min and the markers show its location at the beginning of each minute. The start and end points of the trajectory are also marked

The nodes take measurements corrupted with additive Gaussian noise of standard deviation $\sigma_\theta = 3^\circ$. We assume that nodes have a limited sensing range, i.e., they can only provide bearing measurements for targets within their sensing range. The sensing range of each node is set to 200 m which is slightly longer than the distance between two horizontally or vertically adjacent nodes. Measurements are made only by the nodes that have the current estimate of target location within their sensing range. For the trajectory given in Fig. 10.8, up to 4 sensors take measurements at each time step. The sampling interval is $T = 1\text{min}$.

The experiment for each algorithm is repeated for 1,000 Monte Carlo trials. Let l denote the trial index. The position error for each trial l is calculated according to

$$E_t(l) = \sqrt{(\hat{y}_{t,1} - y_{t,1})^2 + (\hat{y}_{t,2} - y_{t,2})^2}, \tag{10.33}$$

where $\hat{y}_{t,1}$ and $\hat{y}_{t,2}$ are the estimated position of the target. Note that the error $E_t(l)$ is same at each node as the distributed particle filters are synchronized. The trials that exceed the error value of 250 m at any time t are considered as lost tracks. Then for the tracks that are not lost, we calculate the root-mean-squared (RMS) position error

$$RMSE(l) = \sqrt{\frac{1}{t_{max}} \sum_{t=1}^{t_{max}} E_t(l)^2}. \tag{10.34}$$

Table 10.1 A comparison of the performances of the centralized bootstrap particle filter and the distributed particle filters for $M = 2000$ and $m = 500$

Algorithm	Average RMSE	Track loss	Scalars
Centralized bootstrap	10.28 ± 6.4	0.1	–
Adaptive threshold selective gossip	11.05 ± 9.0	6.3	3.90e+06
Clairvoyant threshold selective gossip	11.09 ± 7.9	0.5	4.41e+06
Top-m selective gossip	11.01 ± 7.2	0.8	2.85e+06
Clairvoyant top-m selective gossip	10.92 ± 8.2	1.0	2.40e+06
Randomized gossip	11.17 ± 8.8	0.3	9.60e+06

For each filter the average RMS position error ± standard deviation, percentage of track loss, and the number of transmitted scalars are presented

Similarly, the communication overhead, which is represented with the number of scalars transmitted, does not include the trials that resulted in lost tracks.

We compare the performance of the distributed particle filter using the two versions of selective gossip: adaptive threshold and top-m selective gossip. To illustrate the decrease in communication cost compared to randomized gossip, we run the same algorithm with $m = N$ which corresponds to updating each entry at each gossip iteration, that is randomized gossip run in parallel for each particle weight. We also run a centralized bootstrap particle filter as a performance benchmark.

In addition, we simulate two clairvoyant versions of selective gossip. The first version, clairvoyant threshold selective gossip, represents the case where all nodes clairvoyantly know the threshold value corresponding to the largest mth entry of the average consensus vector. This is obtained by setting $\tau = \bar{\mathbf{x}}_{(m)}$ during the initialization of the algorithm. The second version, called clairvoyant top-m selective gossip, represents the case where each node clairvoyantly knows the indices of the set $H_{\text{top-}m}$ and only updates these entries. This is obtained by setting $H^{v}_{\text{top-}m}(k) = H_{\text{top-}m}$ at all nodes $v \in V$ and all iterations k. The distributed filter computations are performed with n^2 selective gossip iterations and $10n$ max gossip iterations.

For $M = 2,000$ particles and $m = 500$, Table 10.1 shows the average RMS position error, the track loss percentage and the number of scalars transmitted for each particle filter. The top-m version of selective gossip performs very close to the clairvoyant algorithms and better than the adaptive threshold selective gossip. Adaptive threshold selective gossip loses a high percentage of tracks while transmitting more scalars. Top-m selective gossip also provides performance similar to randomized gossip in terms of both error and track loss while decreasing the communication overhead more than three times. Figure 10.9 demonstrates sample tracks of the centralized and distributed particle filters, in particular the tracks with the median RMS performance for each filter.

Next, we investigate the effect of m on the performance of distributed particle filter with adaptive threshold and top-m selective gossip. Note that increasing m results in increased communication overhead. Figure 10.10 shows the percentage of lost tracks as a function of m. Figure 10.11 shows the RMS position error averaged over tracks that are not loss. We see that the distributed particle filter with top-m selective

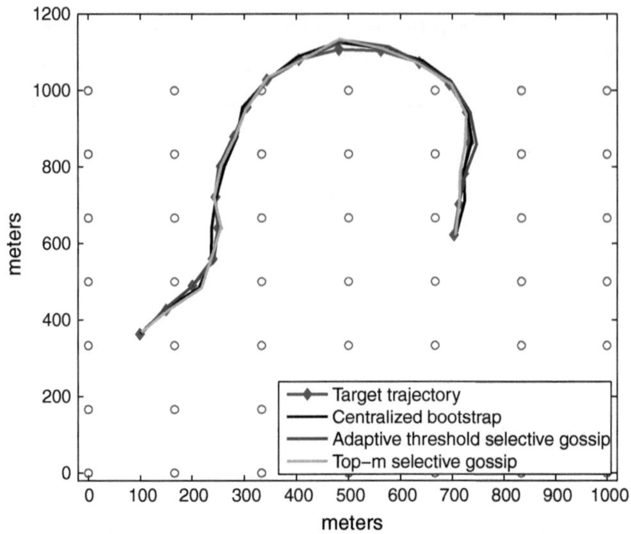

Fig. 10.9 Target trajectory and sample tracks corresponding to the median RMS position error for each filter. The unit of distance values on the axes is meter

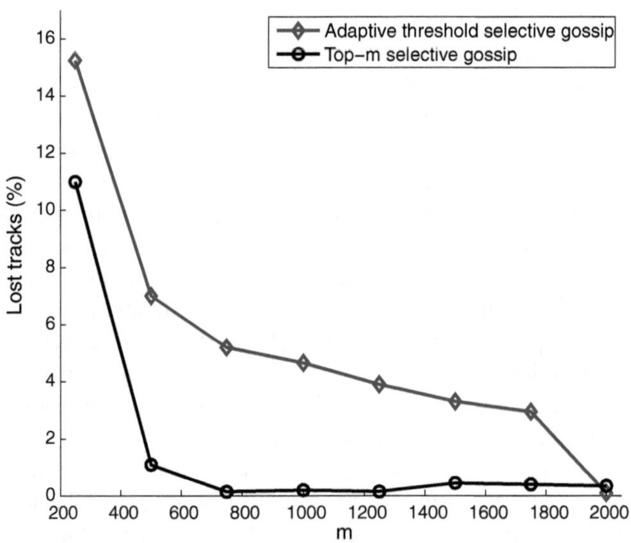

Fig. 10.10 Percentage of track loss as a function of m

gossip achieves good performance for m values 500 and more. Taken together, these results illustrate that the distributed particle filter with top-m selective gossip provides significantly better performance in terms of track loss and RMS error performance for non-divergent tracks compared to the adaptive threshold selective gossip.

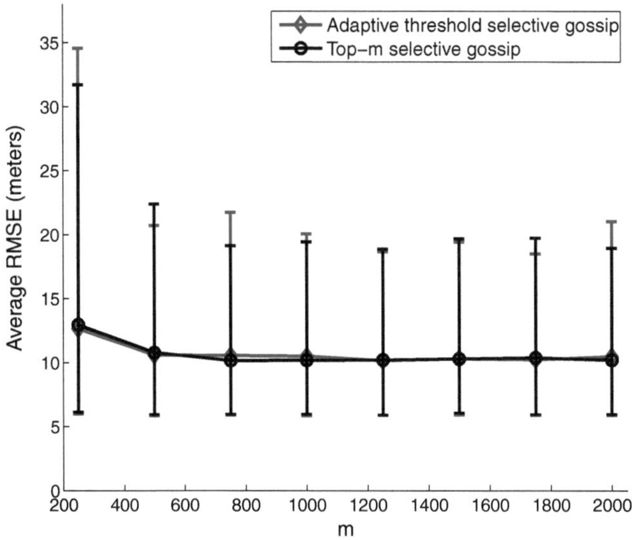

Fig. 10.11 Average RMS position error as a function of m. 95 % confidence bars are also depicted (the end points of these bars correspond to the 5 and 95 % percentiles)

10.6 Conclusions

Many complex signal processing tasks of wireless sensor networks can be formulated using distributed averaging of vector-valued network data where the vectors are possibly high-dimensional. Standard gossip algorithms, which are typically described for averaging scalar quantities, can easily be extended to the vector case by communicating all entries of the vectors. However, this is inefficient in applications where only a small percentage of the entries of the average vector is significant. This chapter presented selective gossip, an algorithm that reduces the dimension of the exchanged data by adaptively focusing communication resources on the entries which are significant for the nodes that are performing the exchange. We proved that focusing on locally significant data, nodes can asymptotically identify the locations of the significant entries of the average vector and also reach a consensus on the values of these entries. To investigate the communication overhead compared to randomized gossip, we presented a simulation study. The results demonstrate that selective gossip provides significant communication savings in terms of number of scalars transmitted. In the second part of the chapter, we proposed a distributed particle filter using selective gossip. In a target tracking scenario with bearings sensors, we showed that distributed particle filters implemented using our algorithm provide comparable results to the centralized bootstrap particle filter while decreasing the communication overhead compared to using randomized gossip to distribute the filter computations.

Our results demonstrate that selective gossip provides a decentralized and efficient building block for wireless sensor network applications. In particular, the top-m

version of selective gossip is potentially more interesting as it has convergence guarantees. This version also provides better tracking performance in the simulation setup we considered. Note that we presented selective gossip based on randomized gossip but it can be implemented with other gossip algorithms such as the synchronous gossip algorithm and faster pairwise gossip algorithms available in the literature.

The future work involves the investigation of the rates of convergence for selective gossip. Since the entries updated at each iteration depends on the vectors in the network at that iteration, the standard methods used for quantifying the convergence rate of randomized gossip do not apply.

References

1. Bénézit F, Dimakis A, Thiran P, Vetterli M (2007) Gossip along the way: Order-optimal consensus through randomized path averaging. In: Proceedings of the Allerton Conference on Communication, Control, and Computing, Monticello
2. Bertsekas DP, Tsitsiklis JN (1997) Parallel and distributed computation: Numerical methods. Athena Scientific, Belmont
3. Boyd S, Ghosh A, Prabhakar B, Shah D (2006) Randomized gossip algorithms. IEEE Trans Info Theory 52(6):2508–2530
4. Cappé O, Moulines E, Ryden T (2005) Inference in hidden Markov models. Springer-Verlag, New York
5. Coates M (2004) Distributed particle filters for sensor networks. In: Proceedings of the International Symposium on Information Processing in Sensor Networks (IPSN), Berkeley
6. Dimakis A, Sarwate A, Wainwright M (2006) Geographic gossip: Efficient aggregation for sensor networks. In: Proceedings of the International Conference on Information Processing in Sensor Networks (IPSN), Nashville
7. Dimakis AG, Kar S, Moura JMF, Rabbat MG, Scaglione A (2010) Gossip algorithms for distributed signal processing. Proc IEEE 98(11):1847–1864
8. Doucet A, de Freitas N, Gordon N (eds) (2001) Sequential Monte Carlo methods in practice. Springer-Verlag, New York
9. Doucet A, Johansen M (2010) Oxford handbook of nonlinear filtering, chapter A tutorial on particle filtering and smoothing: fifteen years later. Oxford University Press, to appear
10. Farahmand S, Roumeliotis SI, Giannakis GB (2011) Set-membership constrained particle filter: Distributed adaptation for sensor networks. IEEE Trans Signal Process 59(9):4122–4138
11. Gordon NJ, Salmond DJ, Smith AFM (1993) Novel approach to nonlinear/non-Gaussian Bayesian state estimation. IEE Proc-F 140(2):107–113
12. Grimmett GR, Stirzaker DR (2001) Probability and random processes. Oxford University Press, New York
13. Gu D (2007) Distributed particle filter for target tracking. In: Proceedings IEEE International Conference on Robotics and Automation, Rome
14. Gupta P, Kumar PR (2000) The capacity of wireless networks. IEEE Trans Info Theory 46(2):388–404
15. Handschin JE, Mayne DQ (1969) Monte Carlo techniques to estimate the conditional expectation in multi-stage non-linear filtering. Int J Control 9(5):547–559
16. Hendrickx JM, Tsitsiklis JN (2011) Convergence of type-symmetric and cut-balanced consensus seeking systems. Submitted; available at http://arxiv.org/abs/1102.2361
17. Hlinka O, Sluciak O, Hlawatsch F, Djurić PM, Rupp M (2010) Likelihood consensus: Principles and application to distributed particle filtering. In: The forty fourth Asilomar Conference on Signals, Systems and Computers (ASILOMAR)

18. Jadbabaie A, Lin J, Morse AS (2003) Coordination of groups of mobile autonomous agents using nearest neighbor rules. IEEE Trans Autom Control 48(6):988–1001
19. Kokiopoulou E, Frossard P (2009) Polynomial filtering for fast convergence in distributed consensus. IEEE Trans Signal Process 57(1):342–354
20. Lee SH, West M (2009) Markov chain distributed particle filters (MCDPF). In: Proceedings of the IEEE Conference on Decision and Control, Shanghai
21. Mohammadi A, Asif A (2011) Consensus-based distributed unscented particle filter.In: Proceedings of the IEEE Statistical Signal Processing Workshop (SSP), 237–240
22. Nedić A, Ozdaglar A (2009) Distributed subgradient methods for multi-agent optimization. IEEE Trans Autom Control 54(1):48–61
23. Olfati-Saber R, Fax JA, Murray RM (2007) Consensus and cooperation in networked multi-agent systems. Proc IEEE 95(1):215–233
24. Oreshkin BN, Coates MJ, Rabbat MG (2010) Optimization and analysis of distributed averaging with short node memory. IEEE Trans Signal Process 58(5):2850–2865
25. Oreshkin BN, Coates MJ (2010) Asynchronous distributed particle filter via decentralized evaluation of Gaussian products. In: Proceedings of the ISIF International Conference on Information Fusion, Edinburgh
26. Rabbat M, Nowak R, Bucklew J (2005) Robust decentralized source localization via averaging In: Proceedings of the IEEE International Conference on Acoustics, Speech, and Signal Processing (ICASSP). Philadelphia
27. Ristic B, Arulampalam MS (2003) Tracking a manoeuvring target using angle-only measurements: algorithms and performance. Signal Process 83(6):1223–1238
28. Ristic B, Arulampalam S, Gordon N (2004) Beyond the Kalman filter: particle filters for tracking applications. Artech House, Norwood, MA, USA
29. Sheng X, Hu Y-H, Ramanathan P (2005) Distributed particle filter with GMM approximation for multiple targets localization and tracking in wireless sensor network. In: Proceedings of the International Symposium on Information Processing in Sensor Networks (IPSN), Los Angeles
30. Sundhar Ram S, Veeravalli VV, Nedić A (2010) Distributed and recursive parameter estimation in parametrized linear state-space models. IEEE Trans Autom Control 55(2):488–492
31. Touri B (2011) Product of random stochastic matrices and distributed averaging. PhD thesis, Univeristy of Illinois at Urbana-Champaign
32. Tsitsiklis JN (1984) Problems in decentralized decision making and computation. PhD Thesis, MIT
33. Tsitsiklis JN, Bertsekas DP, Athans M (1986) Distributed asynchronous deterministic and stochastic gradient optimization algorithms. IEEE Trans Autom Control 31(9):803–812
34. Üstebay D, Castro R, Rabbat M (2011) Efficient decentralized approximation via selective gossip. IEEE J Sel Top Sign Proc 5(4):805–816
35. Üstebay D, Oreshkin B, Coates M, Rabbat M (2008) Rates of convergence for greedy gossip with eavesdropping. In: Proceedings of the Allerton Conference on Communication, Control, and Computing. Monticello, pp 367–374
36. Üstebay D, Rabbat M Efficiently reaching consensus on the largest entries of a vector. In: IEEE Conference on Decision and Control (CDC) '12, Maui, HI, USA
37. Xiao L, Boyd S (2004) Fast linear iterations for distributed averaging. Syst Control Lett 53(1):65–78
38. Zhao F, Shin J, Reich J (2002) Information-driven dynamic sensor collaboration. IEEE Signal Process Mag 19(2):61–72

Chapter 11
Recursive Reconstruction of Sparse Signal Sequences

Namrata Vaswani and Wei Lu

Abstract In this chapter we describe our recent work on the design and analysis of recursive algorithms for causally reconstructing a time sequence of (approximately) sparse signals from a greatly reduced number of linear projection measurements. The signals are sparse in some transform domain referred to as the sparsity basis and their sparsity patterns (support set of the sparsity basis coefficients) can change with time. By "recursive", we mean using only the previous signal's estimate and the current measurements to get the current signal's estimate. We also briefly summarize our exact reconstruction results for the noise-free case and our error bounds and error stability results (conditions under which a time-invariant and small bound on the reconstruction error holds at all times) for the noisy case. Connections with related work are also discussed. A key example application where the above problem occurs is dynamic magnetic resonance imaging (MRI) for real-time medical applications such as interventional radiology and MRI-guided surgery, or in functional MRI to track brain activation changes. Cross-sectional images of the brain, heart, larynx or other human organ images are piecewise smooth, and thus approximately sparse in the wavelet domain. In a time sequence, their sparsity pattern changes with time, but quite slowly. The same is also often true for the nonzero signal values. This simple fact, which was first observed in our work, is the key reason that our proposed recursive algorithms can achieve provably exact or accurate reconstruction from very few measurements.

N. Vaswani (✉) · W. Lu
Department of Electrical and Computer Engineering,
Iowa State University, Ames, IA 50011, USA
e-mail: namrata@iastate.edu

W. Lu
e-mail: greatjackylu@gmail.com

A. Y. Carmi et al. (eds.), *Compressed Sensing & Sparse Filtering*,
Signals and Communication Technology, DOI: 10.1007/978-3-642-38398-4_11,

11.1 Introduction

In this chapter, we describe our recent work on the design and analysis of recursive algorithms for causally reconstructing a time sequence of (approximately) sparse signals from a greatly reduced number of linear projection measurements. The signals are sparse in some transform domain referred to as the sparsity basis and their sparsity patterns (support set of the sparsity basis coefficients) can change with time. The most important example of the above problem occurs in dynamic magnetic resonance imaging (MRI) for real-time medical applications such as interventional radiology, MR image guided surgery, or functional MRI to track brain activation changes. MRI is a technique for cross-sectional imaging that sequentially captures the 2D Fourier projections of the cross-section to be reconstructed. Cross-sectional images of the brain, heart, larynx or other human organ images are usually piecewise smooth, e.g. see Fig. 11.1, and thus approximately sparse in the wavelet domain. In a time sequence, the sparsity pattern changes with time, but slowly. Often, the signal values also change gradually over time. We demonstrate this for a larynx and a cardiac MRI sequence in Fig. 11.1.

Since MR data acquisition is sequential, the ability to accurately reconstruct with fewer measurements directly translates to reduced scan times. Shorter scan times along with online (causal) and fast (recursive) reconstruction allow the possibility of real-time imaging of fast changing physiological phenomena. Other example applications where real-time imaging is needed include real-time single-pixel video imaging [1], real-time video compression/decompression, real-time sensor network based sensing of time-varying fields [2], or real-time extraction of the foreground image sequence (sparse image) from a slow changing background image sequence (well modeled as lying in a low-dimensional space [3]) using recursive projected compressive sensing (CS) [4, 5]. For other potential applications, see [6, 7].

Since the recent introduction of compressive sensing (CS) [8–10], the static sparse reconstruction problem has been thoroughly studied. But most existing algorithms for the dynamic problem just use CS to jointly reconstruct the entire time sequence in one go [11–13]. This is an offline and batch solution with very high complexity. The alternative — doing CS at each time separately (simple CS) — is online and fast but requires many more measurements. The question then is: *for a time sequence of sparse signals, how can we obtain a recursive solution that improves the accuracy of simple CS by using past observations, and does this will keep the computational complexity only as much as that of simple CS (and thus much lower than that of the batch methods)? In particular, how can we use slow or correlated sparsity pattern change, and in certain cases also slow signal value change, to do this?* By *"recursive"*, we mean a solution that uses only the previous signal estimate and the current observation vector at the current time.

This problem was first studied in [14] which proposed a solution called Kalman Filtered Compressed Sensing (KF-CS). In later work, a simpler special case of KF-CS, called Least Squares CS-residual (LS-CS) was analyzed in detail [15]; and more powerful approaches such as Modified-CS [16, 17], Modified-CS-residual [18, 19] and regularized modified-CS [20, 21] were introduced. Performance guarantees — exact recovery conditions in the noise-free case [16, 17, 21] and time-invariant error bounds (stability) in the noisy case [15, 22] — were also obtained. We describe all of these ideas in the next few sections. We first begin by providing a short background on sparse recovery and compressed sensing, followed by giving a formal problem definition for our problem and discussing related work.

11.2 Notation and Sparse Recovery Background

11.2.1 Notation

We use T^c to denote the complement of T w.r.t. $[1, m] := [1, 2, \dots m]$, i.e. $T^c := \{i \in [1, m] : i \notin T\}$. The notation $|T|$ denotes the size (cardinality) of the set T. The set operations $\cup$, $\cap$, and $\setminus$ have the usual meanings.

For a vector, v, and a set, T, v_T denotes the $|T|$ length sub-vector containing the elements of v corresponding to the indices in the set T. Also, $\|v\|_k$ denotes the ℓ_k norm of a vector v. When $k = 0$, $\|v\|_0$ counts the number of nonzero elements in the vector v. If just $\|v\|$ is used, it refers to $\|v\|_2$.

For a matrix M, $\|M\|_k$ denotes its induced k-norm, while just $\|M\|$ refers to $\|M\|_2$. M' denotes the transpose of M and $M^\dagger$ denotes its Moore-Penrose pseudo-inverse. For a tall matrix, M, $M^\dagger := (M'M)^{-1}M'$. For a fat matrix A, A_T denotes the sub-matrix obtained by extracting the columns of A corresponding to the indices in T.

The restricted isometry constant (RIC) [10], δ_S, for a matrix A, is the smallest real number satisfying

$$(1 - \delta_S)\|c\|^2 \leq \|A_T c\|^2 \leq (1 + \delta_S)\|c\|^2 \tag{11.1}$$

for all subsets $T \subseteq [1, m]$ of cardinality $|T| \leq S$ and all real vectors c of length $|T|$. It is easy to see that $\|A_T{}'A_T\| \leq (1 + \delta_S)$, $\|(A_T{}'A_T)^{-1}\| \leq 1/(1 - \delta_S)$ and $\|A_T{}^\dagger\| \leq 1/\sqrt{(1 - \delta_S)}$.

The restricted orthogonality constant (ROC) [10], $\theta_{S,S'}$, for a matrix A, is the smallest real number satisfying

$$|c_1{}'A_{T_1}{}'A_{T_2}c_2| \leq \theta_{S,S'} \; \|c_1\| \; \|c_2\| \tag{11.2}$$

for all disjoint sets $T_1, T_2 \subseteq [1, m]$ with $|T_1| \leq S$, $|T_2| \leq S'$, $S + S' \leq m$, and for all vectors c_1, c_2 of length $|T_1|$, $|T_2|$.

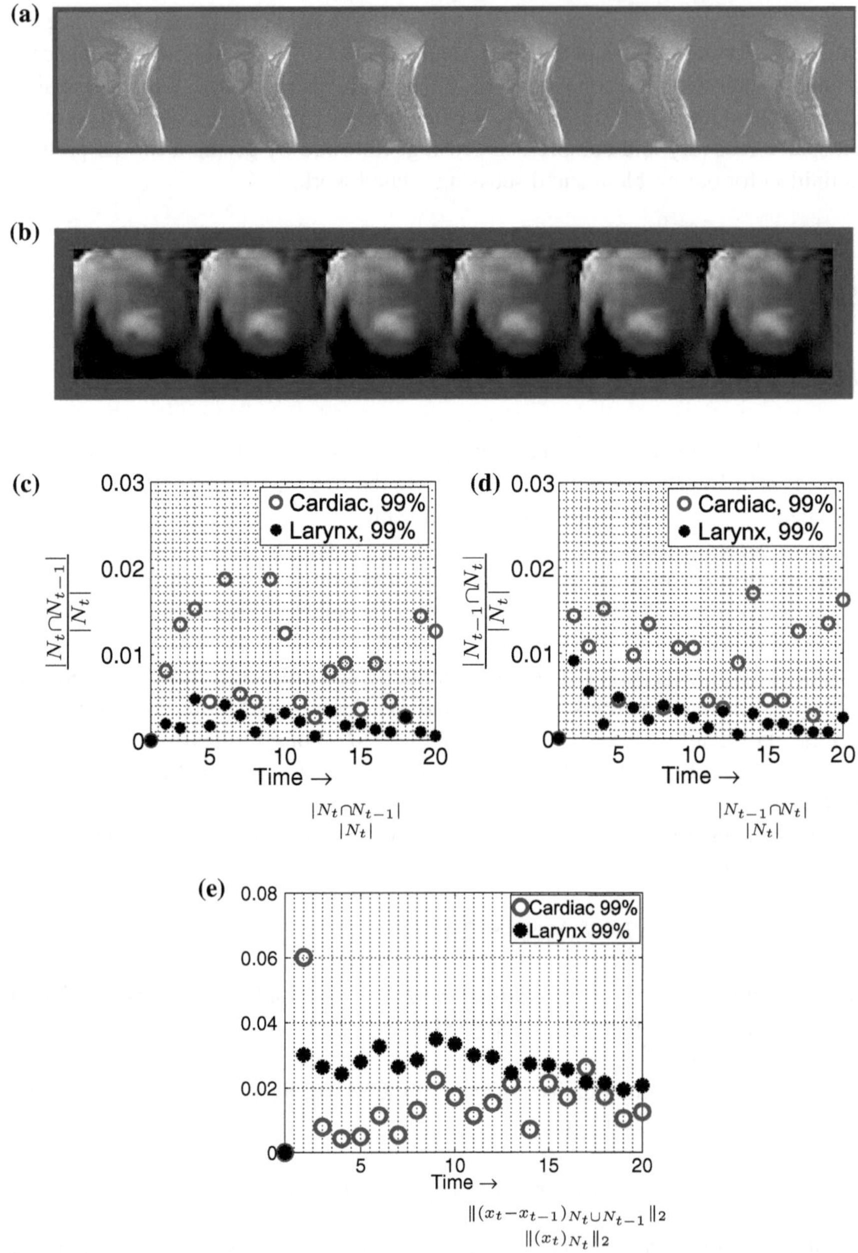

Fig. 11.1 In **a, b** we show two MRI image sequences: cardiac and a larynx sequence). In **c–e**, x_t is the two-level Daubechies-4 2D discrete wavelet transform (DWT) of the cardiac or the larynx image at time t and the set N_t is its 99 % energy support (the smallest set containing 99 % of the vector's energy). The support size was between 6–7 % of the image size. In **c** we plot signal value change.As can be seen from the plots, all support changes (both additions and removals) are less than 2 % of the support size. Also, almost all signal value changes are are less than 4 % of $\|(x_t)_{N_t}\|_2$

11.2.2 Background on Sparse Recovery

The sparse recovery problem has been studied for a very long time, e.g. see [23–25]. The goal in sparse recovery, or what is now interchangeably referred to as compressive sensing (CS), is to recover a sparse signal from a reduced number of its linear projection measurements. To be precise, we would like to recover an m length sparse vector, x, with support size, s, from $y := Ax$, or, in the noisy case, from $y := Ax + w$, when A is a fat matrix (a matrix with more columns than rows). Consider the noise-free case. The sparse recovery problem is solved if we can find the sparsest vector b among all vectors satisfying $y = Ab$, i.e. if we can solve

$$\min_b \|b\|_0 \text{ s.t. } y = Ab$$

and if A is such that every set of $2s$ columns of A are linearly independent [10, 24]. Finding the sparsest vector requires a combinatorial search and thus has complexity of the order of m^s [10]. The exponential complexity in s makes it impractical to directly solve this for any reasonable sized problem. Practical (polynomial complexity) approaches to this problem include (i) ℓ_1 minimization methods (replace the ℓ_0 norm by the ℓ_1 norm which is the closest norm to ℓ_0 that makes the problem convex) such as basis pursuit [24] and its noisy relaxations – basis pursuit denoising (BPDN) [24, 26, 27], Dantzig selector [28] and others; (ii) greedy methods such as matching pursuit [23], orthogonal matching pursuit [29] and many other recent works [30, 31]; and (iii) various other more recent approaches. While these approaches have been proposed and used since the 1990s, the recent work on compressed sensing provided strong performance guarantees for them: exact recovery conditions [8–10] and bounds on reconstruction error when exact recovery is not possible [26–28].

11.3 Problem Definition and Related Work

The recursive reconstruction problem explained here was first introduced in the ICIP 2008 paper on Kalman filtered compressed sensing (KF-CS) [14]. Let $(z_t)_{m\times 1}$ denote the spatial signal at time t and $(y_t)_{n\times 1}$, with $n < m$, denote its noise-corrupted measurements' vector at t, i.e. $y_t = Hz_t + w_t$ where w_t is measurement noise and H is the measurement matrix. The signal, z_t, is sparse in a given sparsity basis (e.g. wavelet) with orthonormal basis matrix, $\Phi_{m\times m}$, i.e. $x_t := \Phi' z_t$ is a sparse vector. Thus the observation model can be written as

$$y_t = Ax_t + w_t,\ A := H\Phi \tag{11.3}$$

We assume that A has unit norm columns. We study both the noise-free case, i.e. $w_t = 0$, and the bounded noise case, i.e. $\|w_t\|_2 \le \epsilon$. We use N_t to denote the support of x_t, i.e.

$$N_t := \text{supp}(x_t) = \{i : (x_t)_i \neq 0\}.$$

The goal is to recursively estimate x_t (or equivalently the signal, $z_t = \Phi x_t$) using $y_1, \dots y_t$. By *recursively*, we mean, use only y_t and the estimate from $t-1$, $\hat{x}_{t-1}$, to compute the estimate at t. This is done under one or both of the following two assumptions.

1. *Slow Support Change.* The support additions, $|N_t \setminus N_{t-1}| \leq S_a \ll |N_t|$ and the removals, $|N_{t-1} \setminus N_t| \leq S_a \ll |N_t|$ at all times t. This assumption is verified for MRI sequences in Fig. 11.1.
2. *Slow Signal Value Change.* The magnitude of the nonzero signal values also changes slowly with time, i.e. $\|(x_t - x_{t-1})_{N_t}\|_2 \ll \|(x_t)_{N_t}\|_2$. This assumption is also verified in Fig. 11.1.

Consider first the class of problems for which only the first assumption holds. Under this assumption, the above problem can be reformulated as one of sparse recovery in the presence of partial support knowledge. We can use the support estimate obtained from the previous time instant, $\hat{N}_{t-1}$, as the "partial support knowledge". We describe this problem and the proposed solutions for it in Sect. 11.4. If both assumptions hold, the above problem can be reformulated as one of sparse recovery with partial support and signal value knowledge. This is discussed in Sect. 11.5. Performance guarantees (exact reconstruction results, error bounds, and conditions for time-invariant error bounds) are briefly discussed in Sect. 11.6 and some interesting experimental results are shown in Sect. 11.7. Conclusions are given in Sect. 11.8.

11.3.1 Related Work

The recursive reconstruction problem was first studied in [14]. Before this, the only works that dealt with time sequences of sparse signals were either batch methods [11–13] or the work of Reddy et al [67] which applied CS on difference measurements to reconstruct only the difference signal (we refer to this as CS-diff). To reconstruct the original signal sequence, this will be unstable over time, except if the difference signal is much more sparse compared to the original signal. This assumption usually does not hold in practice [17].

A modification of KF-CS was studied in [32]. Recent work on Bayesian or other model-based approaches to sequential sparse estimation with time-varying supports includes [33–36, 50]. The work of [13] gives an approximate batch solution for dynamic MRI that is quite fast, but is offline. Related work on model-based and Bayesian approaches for a single signal includes [37–43].

The problem of sparse reconstruction with partial knowledge of the support was simultaneously addressed in [16, 17] and in [45, 46]. The work of [45] obtains exact recovery thresholds for weighted ℓ_1, similar to those in [47], for the case when a probabilistic prior on the signal support is available. Some later work motivated by modified-CS includes modified OMP [48], modified CoSaMP [49], modified block

CS [51], error bounds on modified BPDN [20, 22, 52, 53], better conditions for modified-CS based exact recovery [54], and exact support recovery conditions for multiple measurement vectors (MMV) based recursive recovery [50].

There is other recent work that may also be referred to as recursive sparse reconstruction, but whose goals are quite different from the problem that we discuss in this chapter. This includes (i) homotopy methods, e.g. [55, 56], whose goal is to only speed up the optimization algorithm using homotopy or warm starts and the previous reconstructed signal, but not to reduce the number of measurements required; (ii) [55, 57–59] which reconstruct a single signal from sequentially arriving measurements; and (iii) [60–62], which iteratively improve support estimation for a single sparse signal. Another recent work [63] proposes causal but batch methods, and does this only for signal sequences with the *same* support. This can be interpreted as a causal approach to solve the MMV problem.

We should note that none of the above works obtain conditions under which a time-invariant bound on the reconstruction error (i.e. stability) holds. Except [45] and [62], none of these obtain exact reconstruction conditions either.

11.4 Sparse Recovery with Partial Support Knowledge

This problem was first formulated in [16, 17]. The goal is to recover a sparse vector, x, with support set N, either from noise-free undersampled measurements, $y := Ax$, or from noisy measurements, $y := Ax + w$, when partial and possibly erroneous support knowledge, T, is available. The true support N can be rewritten as

$$N = T \cup \Delta \setminus \Delta_e \text{ where } \Delta := N \setminus T, \ \Delta_e := T \setminus N$$

It is easy to see that

$$|N| = |T| + |\Delta| - |\Delta_e|$$

Here we refer to the set Δ as the *misses* in the support knowledge and the set Δ_e is the *extras* in it. We say the *support knowledge is accurate if* $|\Delta| \ll |N|$ and $|\Delta_e| \ll |N|$.

Least Squares CS-residual (LS-CS) introduced in [15, 64] can be interpreted as the first solution to the above problem. We describe this next. The first solution that gives exact recovery under weaker conditions (using fewer measurements) than what simple CS needs was Modified-CS [16, 17]. We explain this in Sect. 11.4.2. For using either LS-CS or modified-CS for recursive reconstruction, we use the support estimate from the previous time instant as the partial knowledge set T. Support estimation approaches are discussed in Sect. 11.4.3 and LS-CS or modified-CS for recursive reconstruction is given in Sect. 11.4.4.

Algorithm 1 Dynamic LS-CS: LS-CS for recursive reconstruction

Simple CS. At $t = 0$, set $T = \emptyset$ and compute $\hat{x}_0$ as the solution of $\min_b \|b\|_1$ s.t. $\|y - Ab\|_2 \leq \epsilon$
For $t > 0$, do

1. Set $T = \hat{N}_{t-1}$.
2. *Initial LS.*
   a. Compute the initial LS estimate $(\hat{x}_{t,\text{init}})_T = (A_T{}'A_T)^{-1}A_T{}'y_t, \ (\hat{x}_{t,\text{init}})_{T^c} = 0$
3. *CS-residual.*
   a. Compute the observation residual, $\tilde{y}_t = y_t - A\hat{x}_{t,\text{init}}$
   b. Solve the ℓ_1 problem for the residual, i.e. compute $\hat{\beta}_t$ as the solution of $\min_b \|b\|_1$ s.t. $\|\tilde{y}_t - Ab\|_2 \leq \epsilon$
   c. Compute $\hat{x}_t = \hat{x}_{t,\text{init}} + \hat{\beta}_t$
4. *Support Estimation via Add-LS-Del.*

$$\begin{aligned} T_{\text{add}} &= T \cup \{i \in T^c : |(\hat{x}_t)_i| > \alpha_{\text{add}}\} \\ (\hat{x}_{\text{add}})_{T_{\text{add}}} &= A_{T_{\text{add}}}{}^{\dagger} y_t, \ (\hat{x}_{\text{add}})_{T_{\text{add}}^c} = 0 \\ \hat{N}_t &= T_{\text{add}} \setminus \{i \in T : |(\hat{x}_{\text{add}})_i| \leq \alpha_{\text{del}}\} \end{aligned} \tag{11.4}$$

5. *Final LS Estimate.*

$$(\hat{x}_{t,\text{final}})_{\hat{N}_t} = A_{\hat{N}_t}{}^{\dagger} y_t, \ (\hat{x}_{t,\text{final}})_{\hat{N}_t^c} = 0 \tag{11.5}$$

11.4.1 Least Squares CS-residual (LS-CS)

The key idea of LS-CS is as follows [15, 64]. Using T as the support set, compute an initial estimate of x by computing an LS estimate on T and setting all other elements to zero, i.e. compute

$$(\hat{x}_{\text{init}})_T = (A_T{}'A_T)^{-1}A_T{}'y_t, \ (\hat{x}_{\text{init}})_{T^c} = 0 \tag{11.6}$$

Compute the observation residual, $\tilde{y}$,

$$\tilde{y} = y - A\hat{x}_{\text{init}} \tag{11.7}$$

followed by solving the ℓ_1 minimization problem for this residual, i.e. compute $\hat{\beta}$ as the solution of

$$\arg\min_b \|b\|_1 \text{ s.t. } \|\tilde{y} - Ab\|_2 \leq \epsilon \tag{11.8}$$

Then, the estimate of x is computed as

$$\hat{x} = \hat{x}_{\text{init}} + \hat{\beta}. \tag{11.9}$$

This is followed by support estimation and computing a final LS estimate on the estimated support as described in Sect. 11.4.3.

Notice that, the signal residual, $\beta := x - \hat{x}_{\text{init}}$, is supported on $T \cup \Delta$ and satisfies

$$\begin{aligned}\beta_T &= (A_T{}'A_T)^{-1}A_T{}'(A_\Delta x_\Delta + w), \quad \|\beta_T\|_2 \leq \frac{\theta_{|T|,|\Delta|}}{1-\delta_{|T|}}\|x_\Delta\|_2 + \frac{1}{\sqrt{1-\delta_{|T|}}}\epsilon \\ \beta_{T^c} &= x_\Delta\end{aligned}$$

If $|\Delta|$ is small enough, θ is small. If $|\Delta_e|$ is small enough, $|T| \leq |N| + |\Delta_e|$ is not too large and so $1/(1-\delta_{|T|})$ is only a little more than one. Finally if the noise is also small, the above implies that $\|\beta_T\|_2 \ll \|x_\Delta\|_2$. Thus, if T is a good estimate of the true support, N; the measurement matrix A is incoherent enough; and the noise is small enough; then β is small on the set T. Or, in other words, β is approximately supported only on Δ. Since T is a good estimate of the true support, $|\Delta| \ll |N|$ and so the ℓ_1 problem that we need to solve in this case is much easier than in case of simple CS. As a result, it is possible to show that LS-CS results in small reconstruction error using much fewer measurements than what simple CS needs [15, Theorem 1]. We summarize LS-CS for recursive reconstruction in Algorithm 1.

However, notice that the exact sparsity size (total number of nonzero components) of the signal residual, β, is $|T| + |\Delta|$ and this is equal to or larger than that of the signal, $|N|$. Since the number of measurements required for exact reconstruction is governed by the exact sparsity size, LS-CS is not able to achieve exact reconstruction using fewer noiseless measurements than those needed by simple CS. The search for such a solution led us to our next and more powerful idea called Modified-CS.

11.4.2 Modified-CS

The key idea of Modified-CS is as follows [16, 17]. Suppose first that Δ_e is empty, i.e. $N = T \cup \Delta$. Thus, the sparse recovery problem now becomes one of trying to find the sparsest vector whose support contains T among all vectors that satisfy the data constraint. Or in other words, we would like to find the vector that is sparsest outside the set T among all vectors that satisfy the data constraint. In the noise-free case, this can be written as

$$\min_b \|b_{T^c}\|_0 \text{ s.t. } y = Ab$$

The above also works if Δ_e is not empty. It is easy to show that it can exactly recover x if $w = 0$ (noise-free case) and if every set of $|T| + 2|\Delta| = |N| + |\Delta| + |\Delta_e|$ columns of A are linearly independent [17, Proposition 1]. In comparison, the original ℓ_0 problem given in Sect. 11.2.2 requires every set of $2|N|$ columns of A to be linearly independent [10]. This is much stronger when $|\Delta| \approx |\Delta_e| \ll |N|$.

Algorithm 2 **Dynamic Modified-CS: Modified-CS for recursive reconstruction**

1. In Algorithm 1, replace steps 2 and 11.8 by the following *Modified-CS* step.
   a. Compute $\hat{x}_t$ as the solution of $\min_b \|b_{T^c}\|_1$ s.t. $\|y_t - Ab\|_2 \leq \epsilon$.

The above ℓ_0 problem also has exponential complexity, and hence as in case of CS, we replace it by the ℓ_1 problem (ℓ_1 norm is the closest norm to ℓ_0 that makes the optimization problem convex). Thus, *modified-CS* solves

$$\min_b \|b_{T^c}\|_1 \text{ s.t. } y = Ab \tag{11.10}$$

and we denote its solution by $\hat{x}$. Once again, this works, and can provably achieve exact recovery, even when Δ_e is not empty. We give the exact recovery conditions in Sect. 11.6.1. For noisy measurements, one can relax the data constraint as follows.

$$\min_b \|b_{T^c}\|_1 \text{ s.t. } \|y - Ab\|_2 \leq \epsilon \tag{11.11}$$

We summarize modified-CS for recursive reconstruction in Algorithm 2.

In practice, for large scale problems, one always adds the data term as a soft constraint and solves the following unconstrained problem (which is less expensive to solve and does not require knowledge of the noise bound). We refer to this as modified-BPDN [20, 53].

$$\min_b \gamma \|b_{T^c}\|_1 + 0.5\|y - Ab\|_2^2. \tag{11.12}$$

11.4.3 Support Estimation: Thresholding and add-LS-del

In order to use either LS-CS or modified-CS for recursive reconstruction, we use the support estimate from the previous time as the set T. Thus, we need to estimate the support of the signal at each time. The simplest way to do this is by thresholding, i.e. we compute

$$\hat{N} = \{i : |(\hat{x})_i| > \alpha\}$$

where $\alpha \geq 0$ is the zeroing threshold. In cases of exact reconstruction, i.e. if $\hat{x} = x$, we can use $\alpha = 0$. In other situations, we need a nonzero value. In cases of very accurate reconstruction, we can set α to be slightly smaller than the magnitude of the smallest nonzero element of x (assuming its rough estimate is available) [17]. This will ensure close to zero misses and few false additions. In general, α should depend on both the noise level and the magnitude of the smallest nonzero element of x.

For compressible signals, one should do the above but with "support" replaced by the b %-energy support. For a given number of measurements, b can be chosen to be the largest value so that all elements of the b %-energy support can be exactly reconstructed [17].

Single step thresholding as above means that the threshold, α, needs to be large enough to ensure that most missed elements from T are correctly deleted while ensuring that there are few false detections. However, notice that, in both modified-CS and LS-CS, $\hat{x}$ is a biased estimate of x. Consider modified-CS. Along $\Delta \subset T^c$, the values of $\hat{x}$ are biased towards zero (because we minimize $\|(\beta)_{T^c}\|_1$), while, along $\Delta_e \subset T$, they may be biased away from zero (since there is no constraint on $(\beta)_T$). The same also happens for LS-CS although the reasoning is a bit different [15, Sec II-A]. Since the estimates along Δ are biased towards zero, one needs a smaller threshold to detect them, whereas, since those along Δ_e may be biased away from zero, one may need a higher threshold to delete them. One partial solution to this problem is to use the following three step Add-LS-Del approach:

$$T_{\text{add}} = T \cup \{i : |(\hat{x})_i| > \alpha_{\text{add}}\} \tag{11.13}$$

$$(\hat{x}_{\text{add}})_{T_{\text{add}}} = A_{T_{\text{add}}}{}^{\dagger} y, \quad (\hat{x}_{\text{add}})_{T_{\text{add}}^c} = 0 \tag{11.14}$$

$$\hat{N} = T_{\text{add}} \setminus \{i : |(\hat{x}_{\text{add}})_i| \leq \alpha_{\text{del}}\} \tag{11.15}$$

The above add-LS-del procedure involves a support addition step, that uses a smaller threshold, α_{add}, as in (11.13); followed by LS estimation on the new support estimate, T_{add}, as in (11.14); and then a deletion step that thresholds the LS estimate, as in (11.15). The addition step threshold, α_{add}, needs to be just large enough to ensure that the matrix used for LS estimation, $A_{T_{\text{add}}}$ is well-conditioned. If α_{add} is chosen properly and if the number of measurements, n, is large enough, the LS estimate on T_{add} will have smaller error, and will be less biased, than $\hat{x}$ (modified-CS or LS-CS output). As a result, deletion will be more accurate when done using this estimate. This also means that one can use a larger deletion threshold, α_{del}, which will ensure deletion of more extras.

A similar issue for noisy CS, and a possible solution (Gauss-Dantzig selector), was first discussed in [28]. The add-LS-del idea was first introduced in the KF-CS and LS-CS papers [14, 15, 22] for recursive reconstruction and simultaneously also in [30, 31] for greedy algorithms for static sparse reconstruction.

Support estimation is usually followed by LS estimation on the final support estimate, in order to get a solution with reduced bias (Gauss-Dantzig selector idea), i.e. one computes

$$(\hat{x}_{\text{final}})_{\hat{N}} = A_{\hat{N}}{}^{\dagger} y, \quad (\hat{x}_{\text{final}})_{\hat{N}^c} = 0 \tag{11.16}$$

11.4.4 Recursive Recovery

For recursive recovery, in case of slow support change, one can use $T = \hat{N}_{t-1}$. We summarize the complete algorithm for LS-CS in Algorithm 1 and that for Modified-CS in Algorithm 2. Recent work [4] has introduced solutions for the more general case where the support change may not be slow, but is still highly correlated over time.

11.5 Sparse Recovery with Partial Support and Signal Value Knowledge

So far we only talked about the case where prior support information is available. In certain applications, one may also have partial signal value knowledge. In recursive recovery problems, it often happens that signal values also change slowly over time. In this case the problem can be formulated as follows. The goal is to recover a sparse vector x, with support set N, either from noise-free undersampled measurements, $y := Ax$, or from noisy measurements, $y := Ax + w$, when partial erroneous support knowledge, T, is available and partial erroneous signal value knowledge on T, $\hat{\mu}_T$, is available. The true support N can be written as

$$N = T \cup \Delta \setminus \Delta_e \text{ where } \Delta := N \setminus T, \ \Delta_e := T \setminus N$$

and the true signal x can be written as

$$\begin{aligned} (x)_{N \cup T} &= (\hat{\mu})_{N \cup T} + e \\ (x)_{N^c} &= 0, (\hat{\mu})_{T^c} = 0 \end{aligned} \tag{11.17}$$

The error e in the prior signal estimate is assumed to be small, i.e. $\|e\| \ll \|x\|$.

11.5.1 Regularized Modified-CS

Regularized modified-CS adds the slow signal value change constraint to modified-CS and solves the following [20, 21].

$$\min_b \|b_{T^c}\|_1 \text{ s.t. } \|y - Ab\|_2 \leq \epsilon, \text{ and } \|b_T - \hat{\mu}_T\|_\infty \leq \rho \tag{11.18}$$

As before, the following Lagrangian version (constraints added as weighted costs to get an unconstrained problem) is more useful in practice

$$\min_b \gamma \|b_{T^c}\|_1 + 0.5\|y - Ab\|_2^2 + 0.5\lambda\|b_T - \hat{\mu}_T\|_2^2 \tag{11.19}$$

Regularized modified-CS is analyzed in detail in [20] and [21].

11.5.2 Modified-CS-residual

The idea of modified-CS-residual is to combine the modified-CS idea with the CS-residual idea. One solves modified-CS on the observation residual computed using $\hat{x}_{\text{init}} = \hat{\mu}$. Once again the following unconstrained version is most useful:

$$\min_b \|b_{T^c}\|_1 + 0.5\alpha\|(y - A\hat{\mu} - Ab)\|_2^2 \tag{11.20}$$

For recursive reconstruction, one again uses $T = \hat{N}_{t-1}$. For $\hat{\mu}$, one can either use $\hat{\mu} = \hat{x}_{t-1}$, or, in certain situations where the signal values do not change much w.r.t. the first frame, using $\hat{\mu} = \hat{x}_0$ is a better idea. *For practical problems, e.g. real functional MRI sequences* [19], *modified-CS-residual with* $\hat{\mu} = \hat{x}_0$ *turns out to be the most promising approach to use.*

As we explain next, in recursive reconstruction problems, if a model on signal value change is available, one can also obtain $\hat{\mu}$ by using a Kalman filter.

11.5.3 Kalman Filtered CS-residual (KF-CS) and Kalman Filtered Modified-CS-residual (KalMoCS)

Kalman Filtered CS (KF-CS) was introduced in the context of recursive reconstruction in [14]. The key idea is to replace the initial LS step of LS-CS by a regularized LS step. One then computes the observation residual, followed by solving the ℓ_1 problem on this residual, exactly as in LS-CS. In KalMoCS, one replaces the ℓ_1 problem on this residual by the modified-ℓ_1.

Regularized LS becomes the KF in case of recursive recovery. The extra piece of information needed for KF-CS or KalMoCS is a model on signal value change. Typically, in most cases, one can assume a simple random walk model with equal change variance in all directions [14]. We summarize KF-CS and KalMoCS in Algorithm 3. This will outperform LS-CS and modified-CS when support changes occur every so often (allows the KF to stabilize to a small error before the next support change).

Algorithm 3 Kalman Filtered Modified-CS-residual (KalMoCS) and KF-CS

For $t > 0$ do,

1. Set $T = \hat{N}_{t-1}$.
2. *Initial KF.*

$$\begin{aligned} P_{t|t-1} &= P_{t-1} + \hat{Q}_t, \text{ where } \hat{Q}_t := \sigma_{sys}^2 I_T \\ (P_{t-1} + \hat{Q}_t) K_t &= P_{t|t-1} A' (A P_{t|t-1} A' + \sigma^2 I)^{-1} \\ P_t &= (I - K_t A) P_{t|t-1} \\ \hat{x}_{t,\text{init}} &= (I - K_t A) \hat{x}_{t-1} + K_t y_t s \end{aligned} \tag{11.21}$$

3. *CS-residual or Modified-CS-residual.*
   a. Compute the KF residual, $\tilde{y}_t$, using $\tilde{y}_t = y_t - A\hat{x}_{t,\text{init}}$
   b. KalMoCS: Solve modified-ℓ_1 on the residual: compute $\hat{\beta}_t$ as the solution of

$$\min_b \|b_{T^c}\|_1 \text{ s.t. } \|\tilde{y}_t - Ab\|_2 \le \epsilon$$

   - In case of KF-CS: replace $\|b_{T^c}\|_1$ by $\|b\|_1$.

   c. Compute $\hat{x}_t = \hat{x}_{t,\text{init}} + \hat{\beta}_t$
4. *Support Estimation via Add-LS-Del.*

$$\begin{aligned} T_{\text{add}} &= T \cup \{i \in T^c : |(\hat{x}_t)_i| > \alpha_{\text{add}}\} \\ (\hat{x}_{\text{add}})_{T_{\text{add}}} &= A_{T_{\text{add}}}{}^{\dagger} y, \ \ (\hat{x}_{\text{add}})_{T_{\text{add}}^c} = 0 \\ \hat{N}_t &= T_{\text{add}} \setminus \{i \in T : |(\hat{x}_{\text{add}})_i| \le \alpha_{\text{del}}\} \end{aligned} \tag{11.22}$$

5. *Final Estimate:* If $\hat{N}_t$ is equal to T, set

$$\hat{x}_{t,\text{final}} = \hat{x}_{t,\text{init}}$$

else, compute an LS estimate using $\hat{N}_t$ and update P_t as follows.

$$\begin{aligned} &(\hat{x}_{t,\text{final}})_{\hat{N}_t} = A_{\hat{N}_t}{}^{\dagger} y_t, \ \ (\hat{x}_{t,\text{final}})_{\hat{N}_t^c} = 0 \\ &(P_t)_{\hat{N}_t,\hat{N}_t} = (A_{\hat{N}_t}{}' A_{\hat{N}_t})^{-1} \sigma^2, \ \ (P_t)_{\hat{N}_t^c,[1,m]} = 0, \ \ (P_t)_{[1,m],\hat{N}_t^c} = 0 \end{aligned} \tag{11.23}$$

11.6 Theoretical Results

We first summarize the exact reconstruction results for modified-CS and regularized modified-CS and their implications. Next, we briefly discuss the error bounds for the noisy case. Finally, we address the most important question for recursive recovery: when is the algorithm "stable" over time, i.e. when can we get time-invariant bounds on its error over time?

11.6.1 Exact Reconstruction in Noise-free Case

As explained earlier, LS-CS and KF-CS cannot achieve exact recovery under weaker conditions than what is needed for simple CS. However, modified-CS [17] and regularized modified-CS can [21]. We give below the RIC based exact recovery conditions for modified-CS [17]:

Theorem 1 (Exact Recovery Conditions – Modified-CS). *[17, Theorem 1] Given a sparse vector, x, whose support, $N = T \cup \Delta \setminus \Delta_e$, where $\Delta = N \setminus T$ and $\Delta_e = T \setminus N$, consider reconstructing it from $y := Ax$ by solving (11.10). Let $k := |T|$, $u := |\Delta|$, $e := |\Delta_e|$ and $s := |N|$. Then, x is the unique minimizer of (11.10) if*

1. $\delta_{k+u} < 1$ *and* $\delta_{2u} + \delta_k + \theta_{k,2u}^2 < 1$ *and*
2. $a_k(2u, u) + a_k(u, u) < 1$ *where* $a_k(S, \check{S}) := \frac{\theta_{\check{S},S} + \frac{\theta_{\check{S},k}\,\theta_{S,k}}{1-\delta_k}}{1-\delta_S - \frac{\theta_{S,k}^2}{1-\delta_k}}$

The above conditions can also be rewritten in terms of s, e, u by substituting $k = s + e - u$.

A simpler sufficient condition for modified-CS that uses only the RIC is [17, Corollary 1]:

$$2\delta_{2u} + \delta_{3u} + \delta_{s+e-u} + \delta_{s+e}^2 + 2\delta_{s+e+u}^2 < 1.$$

Compare this with simple CS which requires [27, 28, 65]

$$\delta_{2s} < \sqrt{2} - 1 \text{ or } \delta_{2s} + \delta_{3s} < 1.$$

To compare these conditions numerically, we can use $u = e = 0.02s$ which is typical for time series applications (see Fig. 11.1). Using $\delta_{cr} \le c\delta_{2r}$ [31, Corollary 3.4], it can be show that modified-CS only requires $\delta_{2u} < 0.004$. On the other hand, simple CS requires $\delta_{2u} < 0.008$ which is clearly stronger.

Exact recovery conditions for regularized modified-CS in the noise-free case, i.e. for (11.18) with $\epsilon = 0$ are obtained in [21, Theorem 1]. These are weaker than those for modified-CS if $x_i - \hat{\mu}_i = \pm\rho$ for some $i \in T$ (some of the constraints $\|b_T - \hat{\mu}_T\|_\infty \le \rho$ are active for the true signal, x) and some elements of this active set satisfy the condition given in [21, Theorem 1]. One set of practical applications where $x_i - \hat{\mu}_i = \pm\rho$ with nonzero probability is when dealing with quantized signals and quantized signal estimates.

11.6.2 Error Bounds for the Noisy Case

When measurements are noisy, one cannot get exact recovery, but can only bound the reconstruction error. We give here the error bounds for both LS-CS [15] and modified-CS [22]. The LS-CS-residual step error can be bounded as follows. The proof follows in exactly the same way as that given in [15] where CS is done using the Dantzig selector instead of constrained BPDN as in (11.8).

Theorem 2 (LS-CS-residual error bound). *[15, Lemma 1] Let x be a sparse vector with support N and let $y := Ax + w$ with $\|w\| \leq \epsilon$. Also, let $\Delta := N \setminus T$ and $\Delta_e := T \setminus N$. Let $\hat{x}$ be computed as in (11.9). If $\delta_{2|\Delta|} < (\sqrt{2}-1)/2$ and $\delta_{|T|} < 1/2$,*

$$\|x - \hat{x}\| \leq C'(|T|, |\Delta|)\epsilon + \theta_{|T|,|\Delta|} C''(|T|, |\Delta|)\|x_\Delta\| \tag{11.24}$$

where $C'(|T|, |\Delta|) := C_1(2|\Delta|) + \sqrt{2}C_2(2|\Delta|)\sqrt{\frac{|T|}{|\Delta|}}$, $C''(|T|, |\Delta|) := 2C_2(2|\Delta|)$ $\sqrt{|T|}$, $C_1(S) := \frac{4\sqrt{1+\delta_S}}{1-(\sqrt{2}+1)\delta_S}$, and $C_2(S) := 2\frac{1+(\sqrt{2}-1)\delta_S}{1-(\sqrt{2}+1)\delta_S}$.

By adapting the approach of [27], the error of modified-CS can be bounded as a function of $|T| = |N| + |\Delta_e| - |\Delta|$ and $|\Delta|$. This was done by Jacques in [66] and by us in [22].

Theorem 3 (modified-CS error bound). *[22, Lemma 1] Let x be a sparse vector with support N and let $y := Ax + w$ with $\|w\| \leq \epsilon$. Also, let $\Delta := N \setminus T$ and $\Delta_e := T \setminus N$. Let $\hat{x}$ denote the solution of (11.11). If $\delta_{|T|+3|\Delta|} < (\sqrt{2} - 1)/2$, then*

$$\|x - \hat{x}\| C_1(|T| + 3|\Delta|)\epsilon \leq 9.8\epsilon, \text{ where } C_1(S) := \frac{4\sqrt{1 + \delta_S}}{1 - (\sqrt{2} + 1)\delta_S} \tag{11.25}$$

For both LS-CS and modified-CS, the error after the final LS step can be bounded in terms of $\tilde{T} := \hat{N}$ and $\tilde{\Delta} := \hat{N} \setminus N$ as follows.

$$\|x - \hat{x}_{\text{final}}\| \leq (1 + \frac{\theta_{|\tilde{T}|,|\tilde{\Delta}|}}{1 - \delta_{|\tilde{T}|}})\|x_{\tilde{\Delta}}\|_2 + \frac{1}{\sqrt{1 - \delta_{|\tilde{T}|}}}\epsilon \tag{11.26}$$

11.6.3 Recursive Reconstruction: Time-Invariant Error Bounds (Stability)

Let $\tilde{T} := \hat{N}$, $\tilde{\Delta} := \hat{N} \setminus N$ and $\tilde{\Delta}_e := N \setminus \hat{N}$. So far we bounded the LS-CS-residual error or the modified-CS error as a function of $|T|$, $|\Delta|$. The bound is small as long as $|\Delta_e|$ and $|\Delta|$ are small. Similarly the bound on the error of the final LS estimate, given in (11.26), is small if $|\tilde{\Delta}|$ and $|\tilde{\Delta}_e|$ are small. However for recursive reconstruction, what we need is conditions under which we can get a time invariant bound on $|\Delta_e|$ and $|\Delta|$ as well as on $|\tilde{\Delta}|$ and $|\tilde{\Delta}_e|$. Otherwise, it can happen that the support errors keep adding up and become large and the same will happen to the reconstruction errors.

The study of error stability over time requires a signal change model. We assume the following simple deterministic model [15, Signal Model 1]. (a) There is nonzero delay, d, between new coefficient addition and removal times; (b) at most S_a additions and removals occur at every change time; (c) new coefficients' magnitudes increase gradually from zero for sometime and finally reach a constant value; and (d) coefficients' magnitudes decrease gradually before becoming zero (getting removed from support). Under this model, one can show the following. The actual conditions in the final result are somewhat messy and so we skip them. We only state a qualitative version here.

Theorem 4 (Time-invariant error bounds). *[15, Theorem 2] Assume the above signal change model. If*

1. *the initial simple CS step is accurate enough,*
2. *the noise is bounded and the number of measurements, n, is large enough so that certain conditions on the RIC and ROC hold,*
3. *the addition and deletion thresholds are appropriately set,*
4. *for a given n and noise bound, a) the smallest constant coefficient magnitude is large enough, b) the rates of coefficient magnitude increase and decrease are large enough, and c) the delay between addition times, d, is larger than the "worst case detection delay" plus coefficient decrease time,*

then,

1. *the number of final misses $|\tilde{\Delta}_t|$ and extras $|\tilde{\Delta}_{e,t}|$ as well as the initial misses $|\Delta_t|$ and extras $|\Delta_{e,t}|$ are bounded by S_a or by a quantity slightly larger than S_a,*
2. *within a finite delay, all new additions get detected and not falsely deleted, i.e. $|\tilde{\Delta}_t| = 0$, and all the extras get deleted, i.e. $|\tilde{\Delta}_{e,t}| = 0$,*
3. *and the reconstruction error is bounded by a time-invariant and small values at all times.*

As long as the number of new additions or removals, $S_a \ll |N_t|$ (slow support change), the above result shows that the worst case number of misses or extras is also small compared to the support size. This makes it a meaningful result. Similarly, we can argue that the reconstruction error bound is small compared to the signal energy.

The above result was proved for LS-CS in [15, Theorem 2]. It is possible to prove an exactly analogous result for modified-CS as well. The key ideas in obtaining this result are as follows. (i) One needs to ensure that within a finite delay of a new addition time, all new additions definitely get detected and not false deleted (this delay is the "worst case detection delay"). (ii) This needs to be done while ensuring that there are no false deletions of the constant coefficients. (iii) Also, the deletion threshold needs to be high enough to definitely delete all the extras every-so-often (ensure $|\tilde{T}_t|$ is bounded). (iv) Finally, the "worst case detection delay" plus the coefficient decrease time need to be smaller than the delay between two addition times.

The above result assumes support change every d frames. One can also show stability under a more general signal model that allows support changes at every time. This has been done for both modified-CS and LS-CS in [22].

11.7 Experiments

We briefly describe three sets of experiments here. The first set consists of simulation experiments that demonstrate that modified-CS achieves exact reconstruction using significantly fewer measurements than what simple CS needs. The second set consists of simulation experiments that compare the reconstruction errors of LS-CS, KF-CS, modified-CS (actually modified-BPDN) and regularized modified-BPDN with each other and with other existing work in literature (CS-diff and weighted ℓ_1). The third set of experiments studies recursive recovery for a simulated dynamic MRI experiment. Here we took actual (not sparsified) larynx or cardiac image sequences and simulated MRI by taking a randomly selected set of their partial Fourier measurements. In this case, we did not add measurement noise, however since the signal sequence is not exact sparse, one could think of the compressible coefficients as "noise" (this noise is correlated with the signal, but none of our analysis uses any probability model, so the correlation does not matter). We demonstrate error stability over time of modified-CS and LS-CS and we also show that modified-CS has lower error than LS-CS.

11.7.1 Exact Reconstruction Probability Computation via Monte Carlo

In Sect. 11.6.1, we only compared sufficient conditions for CS and modified-CS. However, this does not mean that the required number of measurements, n, for CS is definitely smaller than what modified-CS needs. To actually compare this, we need to use Monte Carlo. We obtained a Monte Carlo estimate of the probability of exact reconstruction for CS and for modified-CS, for a given A (i.e. we averaged over the joint distribution of x and y given A) as follows [17]. Fix signal length, $m = 256$ and its support size, $s = 0.1m = 26$. In the experiment we describe here we also fixed $u = e = 0.08m$. We varied n. For each n, we generated a $n \times m$ random-Gaussian matrix, A once. We then repeated the following 500 times. (i) Generate the support, N, of size s, uniformly at random from $[1, m]$ and generate $(x)_N \sim \mathcal{N}(0, 100I)$. Set $(x)_{N^c} = 0$. (ii) Set $y := Ax$. (iii) Generate Δ of size u uniformly at random from the elements of N. (iv) Generate Δ_e of size e, uniformly at random from the elements of $[1, m] \setminus N$. (v) Set $T = N \cup \Delta_e \setminus \Delta$. (vi) Solve modified-CS, i.e. solve (11.10). Call the solution $\hat{x}_{modCS}$. (vii) Solve simple CS, i.e. solve (11.10) with T being the empty set. Call the solution $\hat{x}_{CS}$.

At the end, estimate the probability of exact reconstruction using modified-CS by counting the number of times $\hat{x}_{modCS}$ was equal to x ("equal" was defined as $\|\hat{x}_{modCS} - x\|_2/\|x\|_2 < 10^{-5}$) and dividing by 500. Do the same for CS using $\hat{x}_{CS}$. In this experiment, we observed the following.

1. With 19 % measurements, modified-CS gives exact recovery with probability (w.p.) 99.8 %, while CS does this w.p. zero.

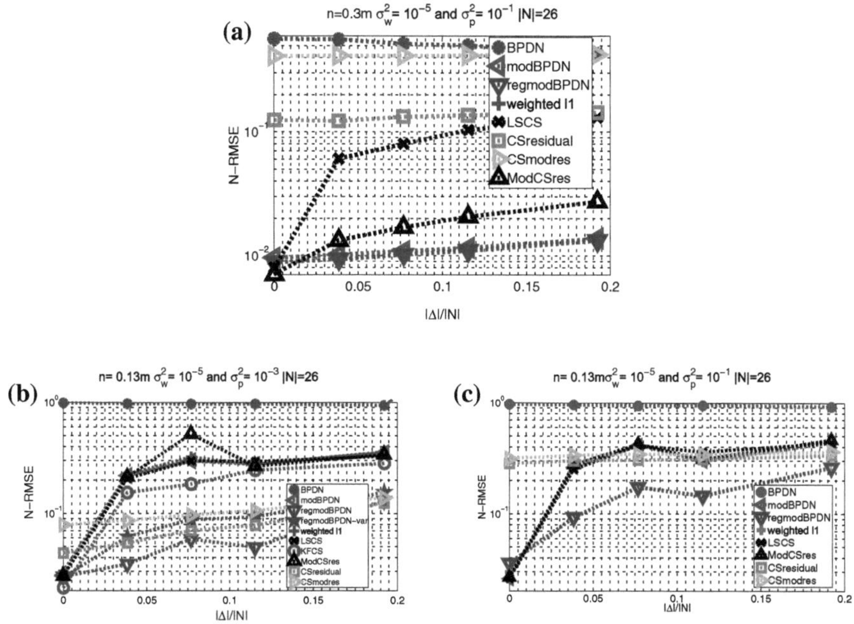

Fig. 11.2 The N-RMSE for reg-mod-BPDN, mod-BPDN, BPDN, LS-CS, KF-CS, weighted ℓ_1, CS-residual, CS-mod-residual and modified-CS-residual are plotted against $|\Delta|/|N|$. **a** $n = 0.2m$, $\sigma_p^2 = 10^{-1}$, $\sigma_w^2 = 10^{-5}$, **b** $n = 0.13m$, $\sigma_p^2 = 10^{-3}$, $\sigma_w^2 = 10^{-5}$, **c** $n = 0.13m$, $\sigma_p^2 = 10^{-1}$, $\sigma_w^2 = 10^{-5}$

2. With 25 % measurements, modified-CS gives exact recovery with probability (w.p.) 100 %, while CS does this w.p. 0.2 %.
3. CS requires 40 % measurements to work "reliably", i.e. to give exact recovery w.p. at least 98 %.

More detailed simulation results for various choices of u and e are summarized in [17, Table 1].

11.7.2 Reconstruction Error Comparisons

In Fig. 11.2, we compare the Monte Carlo average of the reconstruction error of reg-mod-BPDN given in (11.19) with that of modified-BPDN given in (11.12), modified-CS-residual given in (11.20), BPDN [24], weighted ℓ_1 [45], CS-residual [67] and CS-mod-residual. Weighted ℓ_1 solves $\min_b \gamma \|b_{T^c}\|_1 + \gamma' \|b_T\|_1 + \frac{1}{2}\|y - Ab\|_2^2$. CS-residual is an improved version of CS-diff [67]. It computes $\hat{x} = \hat{\mu} + \hat{b}$ where $\hat{b}$ solves $\min_b \gamma \|b\|_1 + \frac{1}{2}\|y - A\hat{\mu} - Ab\|_2^2$

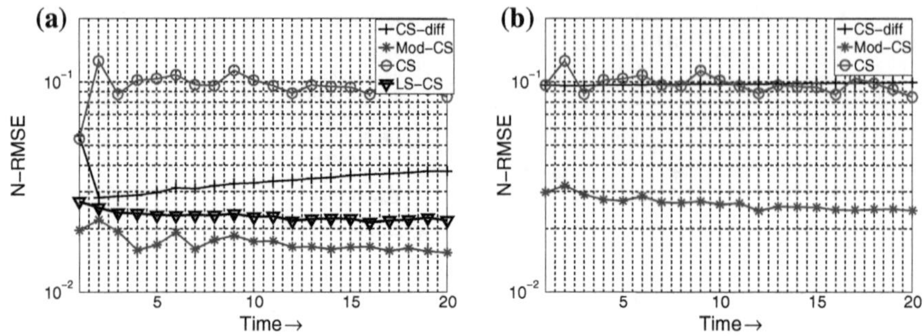

Fig. 11.3 Reconstructing a 256 × 256 *actual (compressible)* vocal tract (larynx) image sequence from *simulated MRI* measurements. Both figures used $n = 0.19m$ for $t > 0$ but used different values of n_0. Image size, $m = 256^2 = 65536$. 99 % energy support size, $|N_t| \approx 0.07m$; support change size $|N_t \setminus N_{t-1}| \approx 0.001m$, **a** $n_0 = 0.2m, n = 0.19m$, **b** $n_0 = 0.19m, n = 0.19m$

The simulation model used is as specified in [20]. The measurements were random-Gaussian projections corrupted by zero mean i.i.d. Gaussian noise with variance σ_w^2. We used $m = 256$, support size $|N| = 0.1m = 26$ and support extras size, $|\Delta_e| = 0.1|N| = 3$. We plot the errors against $|\Delta|/|N|$. The parameters, e.g. γ, λ, γ', used in each of the minimizations were selected as explained in [20]. Notice that with $n = 30\,\%$ measurements and a bad signal prior (large σ_p^2), reg-mod-BPDN, mod-BPDN and weighted ℓ_1 have similar performance. LS-CS is worse than either of these, but better than simple CS and CS-residual. With $n = 13\,\%$ in (b) and (c), reg-mod-BPDN significantly outperforms all the others. In (b), the signal prior is good (small σ_p^2) and so CS-residual is better than modified-CS or weighted ℓ_1 (which do not use signal value knowledge at all) whereas all three of them have similar performance in (c) when the signal prior is bad.

11.7.3 Recursive Reconstruction: Simulated Dynamic MRI

We now show comparisons for recursively reconstructing an actual (compressible) vocal tract image sequence from simulated dynamic MRI measurements [17]. The original image sequence is shown in Fig. 11.1. In Fig. 11.3, we show normalized root mean squared error (N-RMSE) comparisons of modified-CS and LS-CS with simple CS [10, 24] and CS-diff [67]. In the plot shown, the LS-CS error is close to that of modified-CS because we implemented LS estimation using conjugate gradient and did not allow the solution to converge (forcibly ran it with a reduced number of iterations). Without this, LS-CS error was much higher, since the computed initial LS estimate itself was inaccurate. Notice from the figure that modified-CS and LS-CS significantly outperform CS and CS-diff. Also, modified-CS has smaller error than LS-CS. In Fig. 11.3b, CS-diff performs so poorly, because the initial error at $t = 0$ is itself very large (since we use only $n_0 = 0.19m$). As a result the difference signal at $t = 1$ is not compressible enough, making its error large and so on. But even when n_0 is larger and the initial error is small, as in Fig. 11.3a, the CS-diff error is still unstable, i.e. it increases over time.

11.8 Conclusions

In this chapter, we summarized our recent work on algorithms for recursive reconstruction of sparse signal sequences. The key ideas we used are that in many such sequences, the sparsity pattern changes slowly over time, and, in certain cases, the same is true also for signal value change. Using just the first assumption, the recursive recovery problem can be reformulated as one of sparse recovery in the presence of partial support knowledge. We discussed two solutions to this problem, the first is called least squares CS-residual (LS-CS), and the second and more powerful one is called Modified-CS. Modified-CS achieves provably exact recovery under weaker conditions (using fewer measurements) than what simple CS needs whenever the support knowledge is accurate enough. When measurements are noisy, the errors are provably bounded. For recursive recovery with noisy measurements, the most important question is, when can we obtain time-invariant bounds on the reconstruction errors, i.e. when can we show error stability over time? We showed that this can be done under fairly mild assumptions for both LS-CS and modified-CS. For problems where both slow support and signal value change hold, we introduced Kalman filtered CS-residual (KF-CS) or its improved versions, Kalman filtered Modified-CS-residual (KalMoCS). Their performance analysis is still mostly a part of ongoing work. *Among all the ideas introduced in this chapter, we think Modified-CS, explained in* Sect. ,11.4.2, *and Modified-CS-residual, explained in* Sect. 11.5.2, *are the most promising approaches.*

Ongoing work is looking at how to utilize correlated, but not necessarily slow, support change to design recursive reconstruction algorithms [4]. Another line of work is exploring the problem of recursive reconstruction in the presence of (potentially) very large but correlated noise [5, 68].

References

1. Wakin M, Laska J, Duarte M, Baron D, Sarvotham S, Takhar D, Kelly K, Baraniuk R (2006) An architecture for compressive imaging. In: IEEE Intl Conf Image Proc (ICIP)
2. Haupt J, Nowak R (2006) Signal reconstruction from noisy random projections. IEEE Trans Inf Theory 52(9):4036–4048
3. Candès EJ, Li X, Ma Y, Wright J (2009) Robust principal component analysis? J ACM 58(1): 1–37
4. Qiu C, Vaswani N (2011) Support predicted modified-cs for recursive robust principal components' pursuit. In: IEEE Intl Symp Info Th (ISIT)
5. Qiu C, Vaswani N (2011) Recursive sparse recovery in large but correlated noise. Allerton Conf on Communication, Control, and, Computing

6. Carron I "Nuit blanche", in http://nuit-blanche.blogspot.com/
7. "Rice compressive sensing resources", in http://www-dsp.rice.edu/cs
8. Candes E, Romberg J, Tao T (2006) Robust uncertainty principles: Exact signal reconstruction from highly incomplete frequency information. IEEE Trans Inf Theory 52(2):489–509
9. Donoho D (2006) Compressed sensing. IEEE Trans Inf Theory 52(4):1289–1306
10. Candes E, Tao T (2005) Decoding by linear programming. IEEE Trans Inf Theory 51(12):4203–4215
11. Gamper U, Boesiger P, Kozerke S (2008) Compressed sensing in dynamic mri. Magn Reson Med 59(2):365–373
12. Wakin M, Laska J, Duarte M, Baron D, Sarvotham S, Takhar D, Kelly K, Baraniuk R (2006) Compressive imaging for video representation and coding. In: Proc April, Picture Coding Symposium (PCS), Beijing, China
13. Jung H, Sung KH, Nayak KS, Kim EY, Ye JC (2009) k-t focuss: a general compressed sensing framework for high resolution dynamic mri. Magn Reson Med 61:103–116
14. Vaswani N (2008) Kalman filtered compressed sensing. In: IEEE Intl Conf Image Proc (ICIP)
15. Vaswani N (2010) LS-CS-residual (LS-CS): Compressive Sensing on Least Squares residual. IEEE Trans Signal Process 58(8):4108–4120
16. Vaswani N, Lu W (2009) Modified-cs: Modifying compressive sensing for problems with partially known support. In: IEEE Intl Symp Info Th (ISIT)
17. Vaswani N, Lu W (2010) Modified-cs: Modifying compressive sensing for problems with partially known support. IEEE Trans Signal Process 58(9):4595–4607
18. Lu W, Vaswani N (2009) Modified Compressive Sensing for Real-time Dynamic MR Imaging. In IEEE Intl Conf Image Proc (ICIP)
19. Lu W, Li T, Atkinson I, Vaswani N (2011) Modified-cs-residual for recursive reconstruction of highly undersampled functional mri sequences. In, IEEE Intl Conf Image Proc (ICIP)
20. Lu W, Vaswani N (2012) Regularized modified bpdn for noisy sparse reconstruction with partial erroneous support and signal value knowledge. IEEE Trans Signal process 60(1):182–196
21. Lu W, Vaswani N (2012) Exact reconstruction conditions for regularized modified basis pursuit. IEEE Trans Signal Process 60(5):2634–2640
22. Vaswani N (2010) Stability (over time) of Modified-CS for Recursive Causal Sparse Reconstruction. In Allerton Conf Communication, Control, and, Computing
23. Mallat SG, Zhang Z (1993) Matching pursuits with time-frequency dictionaries. IEEE Trans Signal Process 41(12):3397–3415
24. Chen S, Donoho D, Saunders M (1998) Atomic decomposition by basis pursuit. SIAM J Sci Comput 20:33–61
25. Wipf DP, Rao BD (2004) Sparse bayesian learning for basis selection. IEEE Trans Signal Process 52:2153–2164
26. Tropp JA (2006) Just relax: Convex programming methods for identifying sparse signals. IEEE Trans Inf Theory 1030–1051
27. Candes E (2008) The restricted isometry property and its implications for compressed sensing. Compte Rendus de l'Academie des Sciences, Paris, Serie I:589–592
28. Candes E, Tao T (2007) The dantzig selector: statistical estimation when p is much larger than n. Ann Stat 35(6):2313–2351
29. Tropp J, Gilbert A (2007) Signal recovery from random measurements via orthogonal matching pursuit. IEEE Trans Inf Theory 53(12):4655–4666
30. Dai W, Milenkovic O (2009) Subspace pursuit for compressive sensing signal reconstruction. IEEE Trans Inf Theory 55(5):2230–2249
31. Needell D, Tropp JA (May 2009) Cosamp: Iterative signal recovery from incomplete and inaccurate samples. Appl Comp Harmonic Anal 26(3):301–321
32. Carmi A, Gurfil P, Kanevsky D (2010) Methods for sparse signal recovery using kalman filtering with embedded pseudo-measurement norms and quasi-norms. IEEE Trans Signal Process 2405–2409
33. Sejdinovic D, Andrieu C, Piechocki R (2010) Bayesian sequential compressed sensing in sparse dynamical systems. In Allerton Conf Communication, Control, and, Computing

34. Ziniel J, Potter LC, Schniter P (2010) Tracking and smoothing of time-varying sparse signals via approximate belief propagation. Asilomar Conf Sig Sys Comp
35. Zhang Z, Rao BD (2011) Sparse signal recovery with temporally correlated source vectors using sparse bayesian learning. IEEE J Sel Topics Sig Proc (Special Issue on Adaptive Sparse Representation of Data and Applications in Signal and Image Processing) 5(5):912–926
36. Charles A, Asif MS, Romberg J, Rozell C (2011) Sparsity penalties in dynamical system estimation. In Conf Info, Sciences and Systems
37. Garcia-Frias J, Esnaola I (2007) Exploiting prior knowledge in the recovery of signals from noisy random projections. In Data Comp Conf
38. Ji S, Xue Y, Carin L (2008) Bayesian compressive sensing. IEEE Trans Signal Process 56(6):2346–2356
39. Schniter P, Potter L, Ziniel J (2008) Fast bayesian matching pursuit: Model uncertainty and parameter estimation for sparse linear models. In: Information Theory and Applications (ITA)
40. La C, Do M (2005) "Signal reconstruction using sparse tree representations", in SPIE Wavelets XI. San Diego, California
41. Baraniuk R, Cevher V, Duarte M, Hegde C (2010) Model-based compressive sensing. IEEE Trans Inf Theory 56(4):1982–2001
42. Eldar Yonina C, Moshe Mishali (2009) Robust recovery of signals from a structured union of subspaces. IEEE Trans Inf Theory 55(11):5302–5316
43. Som S, Potter LC, Schniter P (2010) Compressive imaging using approximate message passing and a markov-tree prior. In Asilomar Conf Sig Sys Comp
44. Khajehnejad A, Xu W, Avestimehr A, Hassibi B (2009) Weighted l1 minimization for sparse recovery with prior information. In: IEEE Intl Symp Info Th (ISIT)
45. Khajehnejad A, Xu W, Avestimehr A, Hassibi B (2011) Weighted ℓ_1 minimization for sparse recovery with Nonuniform Sparse Models. IEEE Trans Signal Process 59(5):1985–2001
46. Miosso CJ, von Borries R, Argez M, Valazquez L, Quintero C, Potes C (2009) Compressive sensing reconstruction with prior information by iteratively reweighted least-squares. IEEE Trans Signal Process 57(6):2424–2431
47. Donoho D (2006) For most large underdetermined systems of linear equations, the minimal ell-1 norm solution is also the sparsest solution. Comm Pure App Math 59(6):797–829
48. Stankovic V, Stankovic L, Cheng S (2009) Compressive image sampling with side information. In: ICIP
49. Carrillo R, Polania LF, Barner K (2010) Iterative algorithms for compressed sensing with patially known support. In: ICASSP
50. Jongmin K, Ok Kyun L, Jong Chul Y (2012) Dynamic sparse support tracking with multiple measurement vectors using compressive MUSIC. In: ICASSP
51. Stojnic M (2010) Block-length dependent thresholds for ℓ_2/ℓ_1-optimization in block-sparse compressed sensing. In: ICASSP
52. Jacques L (2010) A short note on compressed sensing with partially known signal support, Signal Processing 90(12):3308–3312
53. Lu W, Vaswani N (2010) Modified bpdn for noisy compressive sensing with partially known support. In: IEEE Intl Conf Acoustics, Speech, Sig Proc (ICASSP)
54. Friedlander MP, Mansour H, Saab R, Yilmaz O (2012) Recovering compressively sampled signals using partial support information. IEEE Trans Inf Theory 58(2):1122–1134
55. Asif MS, Romberg J (2009) Dynamic updating for sparse time varying signals In: Conf. Info, Sciences and Systems
56. Yin W, Osher S, Goldfarb D, Darbon J (2008) Bregman iterative algorithms for l1-minimization with applications to compressed sensing. SIAM J Imaging Sci 1(1):143–168
57. Malioutov DM, Sanghavi S, Willsky AS (2008) Compressed sensing with sequential observations. In: IEEE Intl Conf Acoustics, Speech, Sig Proc (ICASSP)
58. Angelosante D, Giannakis GB (2009) Rls-weighted lasso for adaptive estimation of sparse signals. In: IEEE Intl Conf Acoustics, Speech, Sig Proc (ICASSP)
59. Pierre J. Garrigues and Laurent El Ghaoui (2008) An homotopy algorithm for the lasso with online observations. In: Adv Neural Info Proc Sys (NIPS)

60. Candes EJ, Wakin MB, Boyd SP (2008) Enhancing sparsity by reweighted l(1) minimization. J Fourier Anal Appl 14(5–6):877–905
61. Chartrand R, Yin W (2008) Iteratively reweighted algorithms for compressive sensing. In: IEEE Intl Conf Acoustics, Speech, Sig Proc (ICASSP)
62. Wang Y, Yin W (2010) Sparse signal reconstruction via iterative support detection. SIAM J Imaging Sci 3(3):462–491
63. Angelosante D, Giannakis GB, Grossi E (2009) Compressed sensing of time-varying signals. Dig Sig Proc Workshop, In
64. Vaswani N (2009) Analyzing least squares and kalman filtered compressed sensing. In: IEEE Intl Conf Acoustics, Speech, Sig Proc (ICASSP)
65. Foucart S, Lai MJ (2009) Sparsest solutions of underdetermined linear systems via ell-q-minimization for 0 <= q <= 1. Appl Comput Harmonic Anal 26:395–407
66. Jacques L (2009) A short note on compressed sensing with partially known signal support. ArXiv, preprint 0908.0660
67. Cevher V, Sankaranarayanan A, Duarte M, Reddy D, Baraniuk R, Chellappa R (2008) Compressive sensing for background subtraction. In: Eur Conf Comp Vis (ECCV)
68. Qiu C, Vaswani N (2010) Real-time robust principal components' pursuit. Allerton Conf on Communications, Control and, Computing

Chapter 12
Estimation of Time-Varying Sparse Signals in Sensor Networks

Manohar Shamaiah and Haris Vikalo

Abstract In this chapter, we consider the problem of reconstructing time-varying sparse signals in a sensor network with limited communication resources. In each time interval, the fusion center transmits the predicted signal estimate and its corresponding error covariance to a selected subset of sensors. The selected sensors compute quantized innovations and transmit them to the fusion center. We consider the situation where the signal is sparse, i.e., a large fraction of its components is zero-valued. We discuss algorithms for signal estimation in the described scenario, analyze their complexity, and demonstrate their near-optimal performance even in the case where sensors transmit a single bit (i.e., the sign of innovation) to the fusion center.

12.1 Introduction

In recent years, reconstruction of time-varying signals in sensor networks with limited communication resources received a lot of attention (see, e.g., [1–3] and the references therein). Due to limited bandwidth and power resources, sensors are often allowed to transmit only partial (e.g., quantized) information to a fusion center. Quantizing and transmitting sensor measurements is often prohibitive since a large number of quantization levels may be needed to ensure a satisfactory performance of signal reconstruction algorithms. However, as demonstrated in [1–3], schemes relying on quantized innovations provide performance which is comparable to that of the full information filtering schemes.

M. Shamaiah (✉) · H. Vikalo
University of Texas, Austin, TX, USA
e-mail: manohar.shamaiah@gmail.com

H. Vikalo
e-mail: hvikalo@ece.utexas.edu

A. Y. Carmi et al. (eds.), *Compressed Sensing & Sparse Filtering*,
Signals and Communication Technology, DOI: 10.1007/978-3-642-38398-4_12,

On another note, in many applications signals exhibit sparseness and hence may allow for an economic use of sensing resources. Recently developed compressed sensing techniques enable reconstruction of sparse signals from potentially far fewer measurements than unknowns [4, 5]. Hence, exploiting sparseness of signals in sensor networks may lessen the demands for communication between sensors and the fusion center, saving both bandwidth and power [6, 7]. Reconstruction of time-varying sparse signals in sensor networks was recently studied in [8] using group lasso and fused lasso techniques. The group-fused lasso assumes the time-invariant support but allows the non-zero components of the signal to vary over time. This is a batch algorithm which relies on quadratic programming to recover the unknown signal. A computationally efficient recursive lasso algorithm (R-lasso), introduced in [9], estimates the sparse signal at each point in time recursively. In [10], the SPARLS algorithm relies on the expectation-maximization technique to find estimates of the tap-weight vector output stream from its noisy observations. The lasso-Kalman smoother of [11] relies on the known dynamic model of the sparse signal to track it by regularizing the Kalman smoother cost function with the sparsity promoting l_1-norm of the signal vector. In [12], the time-varying sparse signal is estimated recursively in the scenario where the support changes very slowly. In particular, signal support at previous time instant is used as a known partial information for solving the problem at a given time instant. The optimization attempts to minimize the change in support which is expected to be much more sparse than the original signal. Under certain conditions, it was shown that much fewer measurements are needed than in the case when the optimization neglects the slowly changing support information. Another approach by the same author [13], referred to as LSCS-residual, imposes sparsity on the least squares residual computed using the support from the previous time instant. In [14], another Kalman filtering based method wherein the sparsity constraint is enforced via so-called pseudo-measurements was proposed. Two stages of Kalman filtering are employed—one for tracking the temporal changes and the other for enforcing the sparsity constraint at each stage. In [15], an unscented Kalman filter for the pseudo-measurement update stage was proposed.

None of these recursive compressed sensing techniques, however, consider quantization as a mean of further reduction of the required bandwidth and power resources. On the other hand, recently there has been considerable interest in developing algorithms for estimating sparse signals using quantized observations [16]. In [17], two methods for estimating sparse signals from quantized observations were proposed. The first one is a simple technique based on optimizing weighted least squares cost that relies on virtual measurements constructed from the centroids of the quantization bins. The other is a more sophisticated method which exploits the fact that when the noise is Gaussian (or has log-concave distribution), the negative log-likelihood function for x, given the measurements, is convex. The resulting convex optimization is solved after adding l_1-regularization to impose sparsity of the solutions. In [18], a generalized expectation-maximization algorithm for sparse signal reconstruction from quantized noisy measurements was proposed. Unlike [17], this work does not assume knowledge of the noise variance. In [19], matched sign pursuit for estimation of sparse signals on unit sphere from the signs of noiseless signal measurements was developed. Another work [20] proposes restricted step shrinkage to recover sparse signals from 1-bit measurements. This algorithm is similar in spirit to the trust region methods for non-convex optimization on the unit sphere. Moreover, [21, 22] propose message passing algorithms for estimating sparse signals from quantized measurements.

In this chapter, we study reconstruction of time-varying sparse signals in a sensor network with communication constraints. The network has a state-space representation, where the sparse state vector comprises many more zeros than non-zero components. In each time interval, the fusion center transmits the predicted signal estimate and its corresponding error covariance to a selected subset of sensors. The selected sensors compute quantized innovations and transmit them to the fusion center. We consider a general nonlinear dynamical system with linear observations. The fusion center employs a recursive signal estimation scheme based on the so-called Kalman-like particle filter [3], which we extend to nonlinear dynamical systems employing a bank of extended Kalman filter (EKF) [23] to track their dynamics. In the multiple measurement case, the scheme is implemented in a computationally efficient sequential form [23]. The proposed scheme imposes sparsity constraints on the estimates of the state vector at both particle and fused estimate levels, using either *projection* or *pseudo measurement* technique [14, 24]. We analyze the computational complexity of the proposed algorithms, demonstrating their practical feasibility. Simulation results show that the performance of the proposed algorithms is close to that of the filtering schemes with full (non-quantized) innovations, even in the extreme case where selected sensors transmit a single bit (i.e., the sign of innovation) to the fusion center. Moreover, the proposed algorithm is demonstrated to work well in the case of slowly varying sparsity pattern.

The chapter is organized as follows. In Sect. 12.2, we describe the system model. Recursive algorithms for tracking sparse signals with quantized innovations, and the analysis of their complexity, are presented in Sect. 12.3. Section 12.4 contains simulation results, and the chapter is concluded in Sect. 12.5.

12.1.1 Notation

Upper-case symbols are used to denote matrices, lower-case boldface symbols denote vectors. $\mathcal{N}(\mu, \sigma^2)$ denotes Gaussian distribution with mean μ and variance σ^2, $\mathcal{N}_t(s_1, s_2, \mu, \sigma^2)$ denotes truncated (to the interval $[s_1 \;\; s_2]$) Gaussian distribution with mean μ and variance σ^2, and $\phi(s_1, s_2, \mu, \sigma^2)$ denotes the probability that a random variable with distribution $\mathcal{N}(\mu, \sigma^2)$ belongs to $[s_1 \;\; s_2]$.

12.2 System Model and Problem Statement

Consider a network employing sensors which observe linear combinations of sparse time-varying signals. In each time interval, M sensors communicate their information to the fusion center. For a general non-linear dynamical system, the signals and measurement satisfy the following dynamical model

$$\mathbf{x}(n+1) = f(\mathbf{x}(n)) + \mathbf{w}(n) \tag{12.1}$$

$$\mathbf{y}(n) = H(n)\mathbf{x}(n) + \mathbf{v}(n) \tag{12.2}$$

In the case of linear dynamics $f(\mathbf{x}(n)) = A(n)x(n)$, in which case the dynamics and measurements are given by

$$\mathbf{x}(n+1) = A(n)\mathbf{x}(n) + \mathbf{w}(n) \tag{12.3}$$

$$\mathbf{y}(n) = H(n)\mathbf{x}(n) + \mathbf{v}(n). \tag{12.4}$$

Here $\mathbf{x}(n) \in R^N$ denotes the time-varying vector which is sparse in some transform domain, i.e., we can write $\mathbf{x}(n) = \Psi(n)\mathbf{x}_0(n)$, where the majority of components of $\mathbf{x}_0(n)$ are zero and where $\Psi(n)$ denotes an appropriate basis. Without a loss of generality, we assume that $\mathbf{x}(n)$ itself is sparse, having at most K non-zero components whose locations are unknown ($K \ll N$). The sensors, whose observations $y_1(n), \ldots, y_M(n)$ are collected in the M-dimensional real-valued vector $\mathbf{y}(n)$, are oblivious to this sparsity. Moreover, $\mathbf{w}(n) \in R^N$ and $\mathbf{v}(n) \in R^M$ denote uncorrelated Gaussian noise with zero mean and covariances $Q(n)$ and $R(n)$, respectively, and $n \geq 0$ is the time index. The initial state of the system, $\mathbf{x}(0)$, is uncorrelated with both $\mathbf{w}(n)$ and $\mathbf{v}(n)$. Furthermore, in (12.3)–(12.4) we introduced $A(n) \in R^{N \times N}$, and $H(n) = [\mathbf{h}_1(n)^T \; \mathbf{h}_2(n)^T \ldots \; \mathbf{h}_M(n)^T]^T \in R^{M \times N}$. The elements of $H(n)$ are drawn from a Gaussian distribution with zero mean and variance $1/M$. Note that this construction of $H(n)$ satisfies the so-called restricted isometry property (RIP) imposed in the design of compressed sensing schemes [4].

At each time step $n-1$, the fusion center uses past measurements collected from sensors to form a predicted estimate of $\mathbf{x}(n)$, $\hat{\mathbf{x}}(n|n-1)$, and computes the predicted observation for the lth sensor, $\hat{y}_l(n)$. We assume that the FC has sufficient power to transmit the predicted measurements and its corresponding error covariances to the sensors. However, sensors are limited in both power and allotted bandwidth, and hence transmit quantized innovations (i.e., the quantized difference between the sensor measurement and the estimate) to the FC. Quantizing innovations implies that, at the FC, there is an interval of uncertainty for the corresponding sensor measurements. Increasing the number of quantization levels reduces the uncertainty, but leads to higher bandwidth requirements and energy consumption.

12.3 The Algorithm

The proposed scheme is based on Kalman-like particle filter (KLPF) [3], which we generalize and apply to nonlinear systems. In particular, we employ an extended Kalman filter (EKF) in place of KF (first modifying the scheme so it can process multiple observations) and implement the resulting EKLPF algorithm in a computationally efficient sequential processing form [23]. To recover sparse signals, we impose sparseness constraints on either particle level, estimate level, or both. Details of the algorithm are described next.

At time n, the fusion center transmits predicted observation

$$\hat{y}_l(n) = \mathbf{h}_l(n)\hat{\mathbf{x}}(n|n-1), \quad l = 1, 2, \ldots, M, \tag{12.5}$$

and its corresponding error covariance $\sigma_l(n)$ to the lth sensor, where $\sigma_l(n)$ denotes the (l, l) entry of the matrix

$$R_{\hat{\mathbf{y}}}(n) = H(n)P(n|n-1)H(n)^T + Q(n). \tag{12.6}$$

The lth sensor computes the quantized innovation,

$$e_l(n) = \mathbf{Q}\left[\frac{y_l(n) - \hat{y}_l(n)}{\sigma_l(n)}\right]\sigma_l(n), \tag{12.7}$$

and transmits it to the fusion center ($\mathbf{Q}[\cdot]$ denotes the quantization operator). The EKLPF employs a bank of N_p parallel extended Kalman filters (in time-and-measurement-update form), where each extended Kalman filter performs the measurement update using an observation particle generated based on the received quantized innovation. In particular, the nth measurement update step of the ith extended Kalman filter uses an observation particle $\mathbf{y}^i(n) = [y_1^i(n) \ldots y_M^i(n)]^T$, where $y_l^i(n)$ $(1 \le l \le M)$ is generated from the truncated Gaussian distribution [3],

$$y_l^i(n) \sim \mathcal{N}_t(s_l^L(n), s_l^U(n), \mathbf{h}_l(n)\hat{\mathbf{x}}^i(n|n-1), \sigma_l(n)). \tag{12.8}$$

In (12.8), $\hat{\mathbf{x}}^i(n|n-1)$ denotes the state prediction computed by the ith extended Kalman filter in the previous time update step. The ith observation particle is assigned the weight $w^i(n) = \prod_{l=1}^{M} w_l^i(n)$, where

$$w_l^i(n) = \phi(s_l^L(n), s_l^U(n), \mathbf{h}_l(n)\hat{\mathbf{x}}^i(n|n-1), \sigma_l(n)). \tag{12.9}$$

In (12.8)–(12.9), $s_l^L(n) = \hat{y}_l(n|n-1) + L(n)$ and $s_l^U(n) = \hat{y}_l(n|n-1) + U(n)$, where $L(n)$ and $U(n)$ denote the lower and upper limits of the quantization interval containing $e_l(n)$, respectively (i.e., $L(n) < e_l(n) \le U(n)$). The measurement updates $\hat{\mathbf{x}}^i(n|n)$ $(1 \le i \le N_p)$ computed by the individual extended Kalman filters are then fused to obtain the overall filtered estimate,

$$\hat{\mathbf{x}}(n|n) = \sum_{i=1}^{N_p} w^i(n)\hat{\mathbf{x}}^i(n|n). \tag{12.10}$$

This is followed by the time update step, where the predicted estimate

$$\hat{\mathbf{x}}(n+1|n) = f\big(\mathbf{x}(n|n)\big) \tag{12.11}$$

and its error covariance matrix

$$P(n+1|n) = F(n+1)P(n|n)F(n+1)^T + R(n+1) \tag{12.12}$$

are computed. Here $F(n)$ is the Jacobian matrix of $f(x)$ evaluated at $\mathbf{x}(n|n)$. Note that when the dynamics are linear (i.e., $f(\mathbf{x}(n)) = A(n)x(n))$ $F(n) = A(n)$.

The KLPF in [3] is derived under the assumption that the system has access to only one measurement source at each time step. The extension to a multiple measurements scenario (e.g., multiple sensors) is straightforward but, in general, may lead to a computationally involved scheme. Note that the observations in our problem are mutually independent, which directly allows for a computationally efficient sequential processing implementation of the extended Kalman filters. In particular, we implement the measurement update step for each extended Kalman filter as

$$\begin{aligned} K_f^l &= \frac{P(n|n)\mathbf{h}_l(n)^T}{\mathbf{h}_l(n)P(n|n)\mathbf{h}_l(n)^T + R_{l,l}(n)} \\ P(n|n) &= P(n|n) - K_f^l\mathbf{h}_l(n)P(n|n) \\ \hat{\mathbf{x}}^i(n|n) &= \hat{\mathbf{x}}^i(n|n) + K_f^l(y_l^i(n) - \mathbf{h}_l(n)\hat{\mathbf{x}}^i(n|n)) \end{aligned} \tag{12.13}$$

where l runs from 1 to M. Note that the sequential processing avoids any matrix inversion and hence provides a computationally efficient implementation of the measurement update step.

To ensure that the proposed estimation scheme recovers sparsity pattern of the state vector, we impose sparsity constraints on either particle level (i.e., $\hat{\mathbf{x}}^i(n|n)$), fused estimate level (i.e., $\hat{\mathbf{x}}(n|n)$), or both. The sparseness is imposed either by introducing the so-called pseudo measurements, or via projections onto sparse domain.

Pseudo-measurements: The sparsity constraint can be imposed at each time step by bounding the l_1 norm of the estimate of the state vector, $\|\hat{\mathbf{x}}(n|n)\|_1 \le \epsilon$. This constraint is readily expressed as a fictitious measurement $0 = \|\hat{\mathbf{x}}(n|n)\|_1 - \epsilon$, where ϵ can be interpreted as a measurement noise [2, 14]. Now we construct an auxiliary state-space model of the form

$$\begin{aligned} \mathbf{z}(k+1) &= \mathbf{z}(k) \\ 0 &= \mathbf{h}_{pm}(k)\mathbf{z}(k) - \epsilon, \end{aligned} \tag{12.14}$$

where $\mathbf{z}(0) = \hat{\mathbf{x}}(n|n)$, $\mathbf{h}_{pm}(k+1) = [sign(\hat{z}_1(k|k)) \ \ldots \ sign(\hat{z}_N(k|k))]$, $k = 1, 2, \ldots, L$, $\hat{z}_j(k|k)$ denotes the jth component of the least-mean-square estimate of $\mathbf{z}(k)$ (obtained via Kalman filter), and $sign(\cdot)$ denotes the sign function. Finally, we reassign $\hat{\mathbf{x}}(n|n) = \hat{\mathbf{z}}(L|L)$, where the time-horizon of the auxiliary state-space model (12.14) L, is chosen such that $\|\hat{\mathbf{z}}(L|L) - \hat{\mathbf{z}}(L-1|L-1)\|^2$ is below some predetermined threshold. This iterative procedure is formalized below as Algorithm 1 (see [14] for more details).

The l_1-norm based pseudo-measurement which enforces sparsity leads to introducing a linear equation, allowing application of the Kalman filter directly. Since in general quasi-norms $||.||_p$ with $0 < p < 1$ are more accurate in approximating $||.||_0$ than the l_1-norm, [14] also proposes the use of pseudo-measurements technique with the quasi norms. In this case, the pseudo-measurement equation is given by

$$0 = \Big(\sum_{i=1}^{n} |z_k(i)|^p\Big)^{\frac{1}{p}} - \epsilon' \tag{12.15}$$

For implementation purposes, [14] linearizes the above equation around a nominal value, and then enforce the resulting constraint by employing an extended Kalman filter. Another alternative discussed in [14] is to replace the l_0 norm $||.||_0$ directly by

$$||z_k||_0 \approx n - \sum_{i=1}^{n} \exp(-\alpha|z_k(i)|) \tag{12.16}$$

and using this as a pseudo-measurement for imposing the sparsity constraint.

Algorithm 1 $[\hat{\mathbf{x}}(n|n), P(n|n)]$=PMKF$(\hat{\mathbf{x}}(n|n), P(n|n))$

Run Kalman filter updates for the system (Eq. 12.14) $||\hat{\mathbf{x}}(n|n)|| - \epsilon = 0$
$P_{pm}(1|1) = P(n|n)$, $\hat{\mathbf{z}}(1|1) = \hat{\mathbf{x}}(n|n)$
for $k = 1$ to L **do**
$\mathbf{h}_{pm}(k) = [sign(\hat{z}_1(k|k)), \ldots, sign(\hat{z}_N(k|k)))]$

$$K_{pm} = \frac{P_{pm}(k|k)\mathbf{h}_{pm}(k)}{\mathbf{h}_{pm}(k)P_{pm}(k|k)(\mathbf{h}_{pm}(k))' + R_\epsilon}$$

$$\hat{\mathbf{z}}(k+1|k+1) = (I - K_{pm}\mathbf{h}_{pm}(k))\hat{\mathbf{z}}(k|k)$$
$$P_{pm}(k+1|k+1) = (I - K_{pm}\mathbf{h}_{pm}(k))P_{pm}(k|k)$$

end for
$\hat{\mathbf{x}}(n|n) = \hat{\mathbf{z}}(L|L)$, $P(n|n) = P_{pm}(L+1|L+1)$

Projections onto sparse domain: This transformation finds the best K-sparse MMSE estimate of the signal by simply setting all but K largest magnitude components to zero. We formalize it as Algorithm 2.

Having introduced the two types of sparsity constraints, we define the following two algorithms:

Algorithm 1: The filtered particles are projected onto the constraint domain by retaining K components with largest magnitudes (i.e., we employ Algorithm 2). The constrained filtered particles are combined using the particle weights to obtain the filtered estimate. In general, this fused filtered estimate is not guaranteed to satisfy

Algorithm 2 $[\hat{\mathbf{x}}(n|n), P(n|n)]$=SPARSE($\hat{\mathbf{x}}(n|n)$,S,$P(n|n)$)

Zero out all but S components with largest magnitudes:
$\hat{\mathbf{z}} = sort(abs(\hat{\mathbf{x}}(n|n)))$ (in descending order), $\hat{\mathbf{z}}_{S+1:N} = \mathbf{0}$

$$P(n|n) = P(n|n) + (\hat{\mathbf{x}}(n|n) - \hat{\mathbf{z}})(\hat{\mathbf{x}}(n|n) - \hat{\mathbf{z}})', \hat{\mathbf{x}}(n|n) = \hat{\mathbf{z}}$$

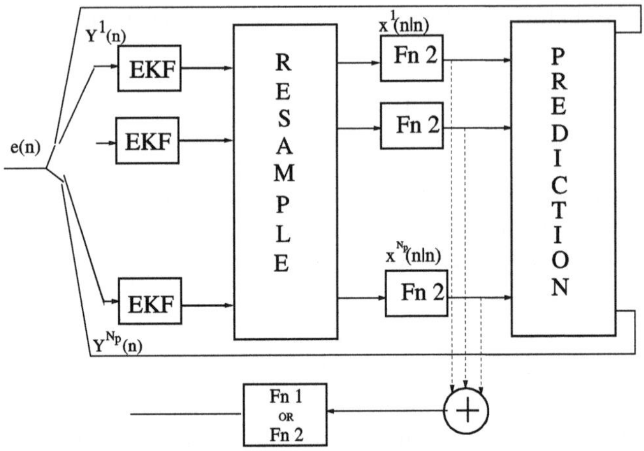

Fig. 12.1 Illustration of the proposed recursive signal processing scheme for time-varying sparse signals. Fn 1 and Fn 2 refer to Algorithms 1 and 2, respectively

the sparsity constraint and hence needs to be projected onto the constraint domain using Algorithm 2.

Algorithm 2: The unconstrained filtered particles are combined using the particle weights to obtain the filtered estimate. This fused filtered estimate is not guaranteed to satisfy the sparsity constraint and the pseudo measurement method is applied to impose the l_1 constraint (i.e., we employ Algorithm 1).

The general algorithm (illustrated in Fig. 12.1), of which Algorithm 1 and Algorithm 2 are special cases, is formalized below.

1. *Initialization:* $n = 0$,
   $\{\hat{\mathbf{x}}^i(0|-1), \hat{\mathbf{x}}(0|-1), P(0|-1)\}$.
2. *Fusion center transmits* $\sigma_l(n)$ *(Eq.* 12.6) *and* $\hat{y}_l(n)$ *(Eq.* 12.5) *to the* lth *sensor.*
3. *The* lth *sensor transmits the quantized innovation* $\mathbf{Q}\left[\frac{y_l(n)-\hat{y}_l(n)}{\sigma_l(n)}\right]$ *to the fusion center.*
4. *Using Eq.* (12.7), *the fusion center generates observation particles (Eq.* 12.8) *and determines corresponding weights (Eq.* 12.9).
5. *Run measurement updates in the sequential form (Eq.* 12.13) *using observations generated in step (4) above.*
6. *Specific to Algorithm 1*: Use Algorithm 2 to project $\hat{\mathbf{x}}^i(n|n)$ onto sparse domain.
7. *Resample the particles using the normalized weights.*

8. *Compute the fused filtered estimate* $\hat{\mathbf{x}}(n|n)$ *(Eq.* 12.10).
9. a. *Specific to Algorithm 1*: Project the estimate $\hat{\mathbf{x}}(n|n)$ onto sparse domain using Algorithm 2.
   b. *Specific to Algorithm 2*: Project the estimate $\hat{\mathbf{x}}(n|n)$ onto sparse domain using Algorithm 1.
10. *Determine time updates* $\hat{\mathbf{x}}^i(n+1|n)$, $\hat{\mathbf{x}}(n+1|n)$, $P(n+1|n)$, $\hat{y}_l(n+1)$, $R_{\hat{\mathbf{y}}}(n+1)$ *for the next time interval.*
    $\hat{\mathbf{x}}^i(n+1|n) = f(\hat{\mathbf{x}}^i(n|n))$
    $\hat{\mathbf{x}}(n+1|n) = f(\hat{\mathbf{x}}(n|n))$,
    $P(n+1|n) \;=\; A(n+1)P(n|n)A(n+1)^T + R(n+1)$
    $\hat{y}_l(n+1) = \mathbf{h}_l(n+1)\hat{\mathbf{x}}(n+1|n)$
    $R_{\hat{\mathbf{y}}}(n+1) = H(n+1)P(n+1|n)H(n+1)^T + Q(n+1)$.

Note that in the absence of steps (6) and (9), the above algorithm reduces to KLPF [3] as we extended it to the multiple measurements case.

12.3.1 Computational Complexity

The complexity of the sampling step (4) in the general algorithm is $O(N_p)$. The sequential processing step (5) is of the order $O(N^2M) + O(NN_p)$; the first term is the complexity of the first two steps in Eq. (12.13) which are common to all the particles, while the complexity of the last step in Eq. (12.13) is given by the second term. The step (6) which projects the filtered particles onto the sparse domain has a complexity $O(N_pNlogN)$. The resampling step (7) has a complexity $O(N_p)$. The step (9) has complexity either $O(NlogN)$ (Algorithm 1) or $O(N^2L)$ (Algorithm 2). The complexity of the time update step (10) is $O(N^2N_p) + O(N^2M)$. It depends on the specific problem which one of these terms dominate. We give an example of the complexity analysis in the next section.

12.4 Simulation Results

The system is simulated with the parameters set to $N = 200$, $M = 35$, $K = 4$, $N_p = 150$, $R_\epsilon = 200^2$, $L = 100$. Initially, there are 3 nonzero components, but we allow for slow change in the sparseness pattern. In particular, another component becomes nonzero at time index $n = 51$. The initial value of the nonzero components is distributed as $\mathcal{N}(0, 25)$. The nonzero components $x_i(n)$ then follow a Gaussian random walk independent of other components determined by $Q_{(i,i)}(n)$. In the simulations, we used $Q_{(i,i)}(n) = 4^2$ and $R_{(i,i)}(n) = 0.25^2$. We assume severely limited bandwidth resources, and transmit *1 bit quantized innovations*. We compare the performance of the proposed algorithms with the scheme considered in [14], which

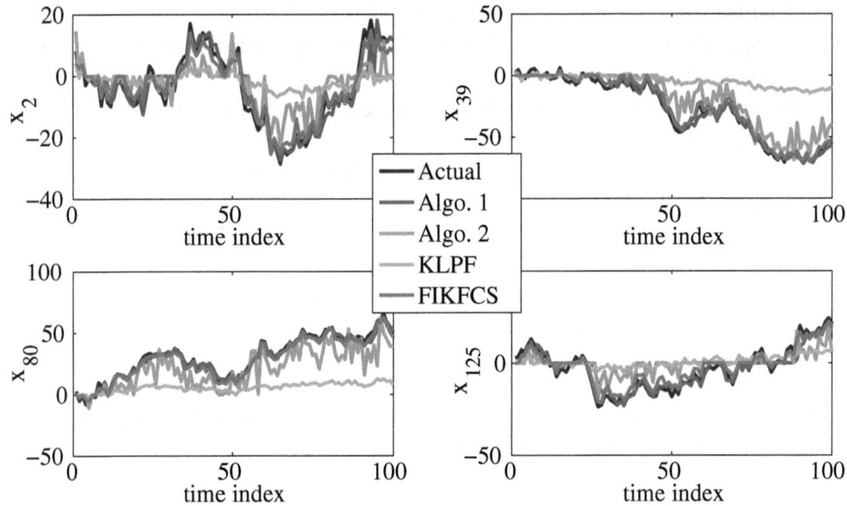

Fig. 12.2 Performance of tracking non-zero components of the sparse signal. $N = 200$, $M = 35$, $K = 4$, $N_p = 150$, $R_\epsilon = 200^2$, $L = 100$

investigates the scenario where the fusion center has full innovation (unquantized). For convenience, we refer to the scheme in [14] as FIKFCS.

Figure 12.2 shows how various algorithms track the non-zero components of the signal. The FIKFCS algorithm performs the best since it uses full innovations. Algorithm 1 performs almost as well as the FIKFCS algorithm. The KLPF clearly performs poorly, while Algorithm 2 does not fare much better. However, if the bandwidth constraint is relaxed (i.e., we allow more than 2 quantization levels), Algorithm 2 performs close to FIKFCS.

Figure 12.3 gives a comparison of the instantaneous values of the estimates at time index $n = 100$. Both Algorithm 1 and FIKFCS correctly identify the nonzero components, while Algorithm 2 erroneously implies signal content in the zero-signal region.

Finally, Fig. 12.4 shows the l_2 error performance of the algorithms. The top figure shows the error in the support set alone (nonzero components), while the bottom figure shows the error performance in the estimation of the zero components. From the figure, it is reaffirmed that Algorithm 1 performs very close to FIKFCS. Algorithm 2 and KLPF perform poorly in both the support set and zero-signal component. Hence, incorporating sparsity constraint at both the particle level and overall estimate level (as done by Algorithm 1) is preferable compared to imposing it only at the overall estimate level (as done by Algorithm 2). If only an approximate knowledge of K is known, then we can employ Algorithm 2 at step (6) using this approximate knowledge, and Algorithm 1 at step (9). This algorithm (Algorithm 3, omitted for brevity) results in a slightly inferior performance to Algorithm 1 but provides much better performance than Algorithm 2. The complexity of this algorithm is of the same order

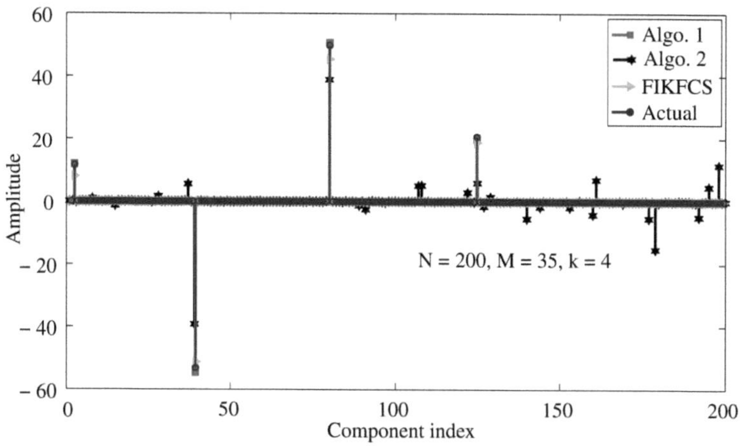

Fig. 12.3 Instantaneous values at $n = 100$. $N = 200$, $M = 35$, $K = 4$, $N_p = 150$, $R_\epsilon = 200^2$, $L = 100$

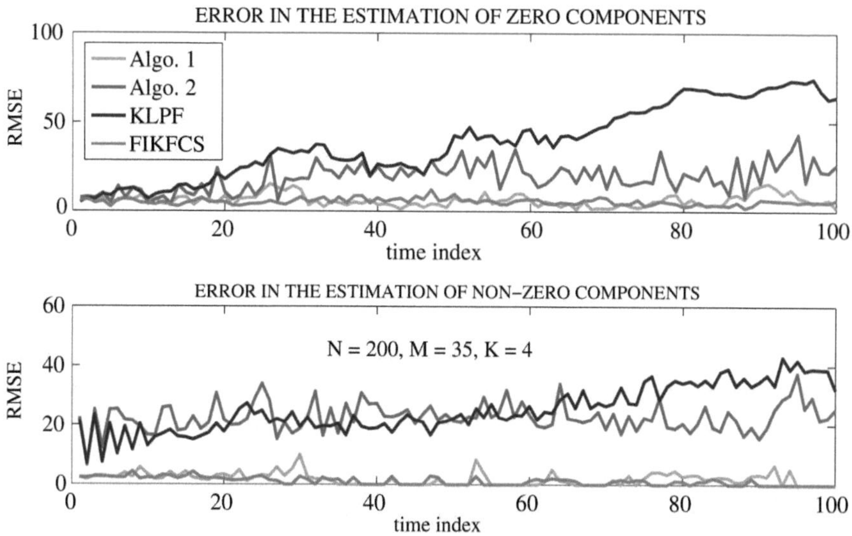

Fig. 12.4 Mean-square estimation error in the support part (non-zero components) and the zero components. $N = 200$, $M = 35$, $K = 4$, $N_p = 150$, $R_\epsilon = 200^2$, $L = 100$

as Algorithm 2. Hence, under bandwidth constraint, if the knowledge of the support is perfectly known, it is preferable to employ Algorithm 1; otherwise, Algorithm 3 is preferred. Algorithm 2 can be chosen in the case of relaxed bandwidth constraints (i.e., if more quantized levels are allowed). It is to be noted that these performances are achieved with significantly fewer measurements than unknowns (< 25 %). In this example, the complexity of Algorithm 1 is dominated by $O(N^2M) = 14 \times 10^5$,

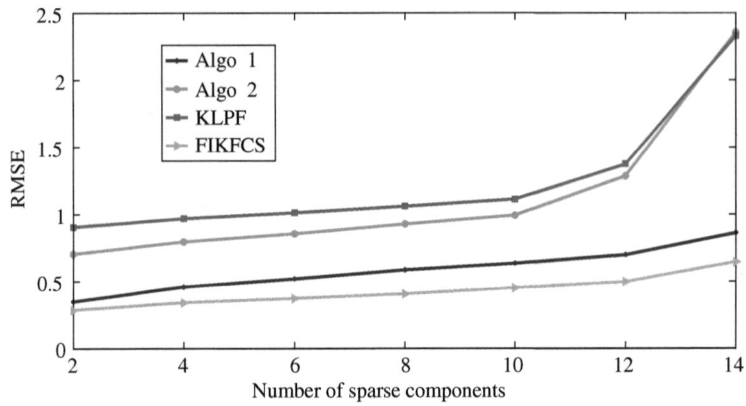

Fig. 12.5 RMSE versus Sparsity: $N = 140$, $M = 40$, $N_p = 150$, $R_\epsilon = 200^2$, $L = 100$

which is of the same order as that of KLPF, while the complexity of Algorithm 2 is dominated by step (9), $O(N^2 L) = 4 \times 10^6$.

Figure 12.5 shows the RMSE performance comparison of the proposed schemes as a function of non-zero components K (here $N = 140$, $M = 40$). For a fairness of the comparison, the error is normalized, i.e., we are plotting $\frac{\|x-\hat{x}\|_2}{\|x\|_2}$. While the error increases with the sparsity for all the estimators, Algorithm 1 and FIKFCS are more robust than the other two techniques.

12.5 Conclusions

We presented algorithms for estimating time-varying sparse signals under communication constraints. These algorithms incorporate the sparsity constraint at different levels (at the particle level, estimate level or both). For heavily bandwidth constrained (1 bit) scenarios, incorporating sparsity at both the particle level and fused estimate level is preferable. Effective tracking of the time variations is achieved with significantly smaller number of measurements than size of the state vector. These algorithms are demonstrated to work well in the case of slowly varying sparsity pattern.

References

1. Ribeiro A, Giannakis GB, Roumeliotis SI (2006) Soi-kf: distributed kalman filtering with low-cost communications using the sign of innovations. IEEE Trans Signal Process 54(12):4782–4795
2. Simon D, Chia TL (2002) Kalman filtering with state equality constraints. IEEE Trans Aerosp Electron Syst 38(1):128–136
3. Sukhavasi RT, Hassibi B (2009) The kalman like particle filter : optimal estimation with quantized innovations/measurements. In: IEEE CDC, pp 4446–4451

4. Candes EJ, Tao T (2008) Decoding by linear programming. IEEE Trans Inf Theory 51(12):4203–4215
5. Donoho DL (2002) Compressed sensing. IEEE Trans Inf Theory 52(4):1289–1306
6. Bajwa W, Haupt J, Sayeed A, Nowak R (2006) Compressive wireless sensing. In: IPSN, pp 134–142
7. Haupt J, Bajwa WU, Rabbat M, Nowak R (2008) Compressed sensing for networked data. IEEE Signal Process Mag 25(2):92–101
8. Angelosante D, Giannakis GB, Grossi E (2009a) Compressed sensing of time varying signals. In: IEEE DSP, pp 1–8.
9. Angelosante D, Giannakis GB (2009) RLS-weighted Lasso for adaptive estimation of sparse signals. In: IEEE ICASSP, pp 3245–3248.
10. Babadi B, Kalouptsidis N, Tarokh V (2010) SPARLS: the sparse RLS algorithm. IEEE Trans Signal Process 58(8):4013–4025
11. Angelosante D, Roumeliotis SI, Giannakis GB (2009b) Lasso-Kalman smoother for tracking sparse signals. In: IEEE ACSSC, pp 181–185
12. Vaswani N, Lu W (2010) Modified-CS: modifying compressive sensing for problems with partially known support. IEEE Trans Signal Process 58(9):4595–4607
13. Vaswani N (2010) LS-CS-residual (LS-CS): compressive sensing on least squares residual. IEEE Trans Signal Process 58(8):4108–4120
14. Carmi AY, Gurfil P, Kanevsky D (2010) Methods for sparse signal recovery using Kalman filtering with embedded pseudo-measurement norms and quasi-norms. IEEE Trans Signal Process Mag 58(4):2405–2409
15. Carmi AY, Mihaylova L, and Kanevsky D (2012) Unscented compressed sensing. In: IEEE ICASSP, pp 5249–5252
16. Dai W, Pham HV, and Milenkovic O (2009) A comparative study of quantized compressive sensing schemes. In: IEEE ISIT, pp 11–15
17. Zymnis A, Boyd S, Candes E (2010) Compressed sensing with quantized measurements. IEEE Signal Process Lett 17(2):149–152
18. Qui K, Dogandzic A (2012) Sparse signal reconstruction from quantized noisy measurements via GEM hard thresholding. IEEE Trans Signal Process 60(5):2628–2634
19. Boufounos PT (2009) Greedy sparse signal reconstruction from sign measurements. IEEE ACSSC, pp 1305–1309
20. Laska JN, Wen Z, Yin W, Baranuik RG (2011) Trust, but verify: fast and accurate signal recovery from 1-bit compressive measurements. IEEE Trans Signal Process 59(11):5289–5301, 1417–1420
21. Kamilov U, Goyal VK, Rangan S (2011) Message-passing estimation from quantized samples. eprint arXiv:1105.6368
22. Mezghani A, Nossek JA (2012) Efficient reconstruction of sparse vectors from quantized observations. IEEE ITG Workshop on smart antennas (WSA)
23. Kailath T, Sayed AH, Hassibi B (2000) Linear estimation. Prentice Hall, Upper Saddle River
24. Iltis RA (2006) A sparse Kalman filter with application to acoustic communications channel estimation. In: IEEE OCEANS, pp 1–5

Chapter 13
Sparsity and Compressed Sensing in Mono-Static and Multi-Static Radar Imaging

Ivana Stojanović, Müjdat Çetin and W. Clem Karl

Abstract This chapter is concerned with the application of sparsity and compressed sensing ideas in imaging radars, also known as synthetic aperture radars (SARs). We provide a brief overview of how sparsity-driven imaging has recently been used in various radar imaging scenarios. We then focus on the problem of imaging from undersampled data, and point to recent work on the exploitation of compressed sensing theory in the context of radar imaging. We consider and describe in detail the geometry and measurement model for multi-static radar imaging, where spatially distributed multiple transmitters and receivers are involved in data collection from the scene to be imaged. The mono-static case, where transmitters and receivers are collocated is treated as a special case. For both the mono-static and the multi-static scenarios we examine various ways and patterns of undersampling the data. These patterns reflect spectral and spatial diversity trade-offs. Characterization of the expected quality of the reconstructed images in these scenarios prior to actual data collection is a problem of central interest in task planning for multi-mode radars. Compressed sensing theory argues that the mutual coherence of the measurement probes is related to the reconstruction performance in imaging sparse scenes. With this motivation we propose a closely related, but more effective parameter we call the $t_{\%}$-average mutual coherence as a sensing configuration quality measure and examine its ability to predict reconstruction quality in various mono-static and ultra-narrow band multi-static configurations.

I. Stojanović (✉)
Scientific Systems Company, 500 West Cummings Park, Suite 3000, Woburn , MA 01801, USA
e-mail: istojanovic@ssci.com

M. Çetin
Sabancı University, Faculty of Engineering and Natural Sciences, Orhanlı Tuzla, Istanbul 34956, Turkey
e-mail: mcetin@sabanciuniv.edu

W. Clem Karl
Department of Electrical and Computer Engineering, Boston University, 8 Saint Mary's Street, Boston , MA 02215, USA
e-mail: wckarl@bu.edu

A. Y. Carmi et al. (eds.), *Compressed Sensing & Sparse Filtering*,
Signals and Communication Technology, DOI: 10.1007/978-3-642-38398-4_13,

13.1 Introduction

Synthetic aperture radar (SAR) is a microwave remote sensing system capable of producing high-resolution imagery of target scenes independent of time of day, distance, and weather. SAR constructs a large synthetic antenna by collecting data from multiple observation points and focusing the received information coherently to obtain a high-resolution description of the scene. Conventional SARs are mono-static, with collocated transmit and receive antenna elements. These SAR sensors coherently process multiple, sequential observations of a scene under the assumption that the scene is static. Imaging resolution is determined by the bandwidth of the transmitted signals and the size of the synthesized antenna. Greater resolution requires wider bandwidths and larger aspect angles obtained from a longer baseline observation interval. An alternative approach is based on multi-static configurations, wherein spatially dispersed transmitters and receivers sense the scene. Such configurations provide the opportunity for spatial as well as frequency diversity and offer potential advantages in flexible sensor planning, sensing time reduction, and jamming robustness.

Sparsity has been of interest for SAR imaging implicitly over many years, and more explicitly in the last decade or so [1]. Ideas based on sparse signal representation have recently led to advanced image formation methods offering a number of benefits for SAR, including increased resolvability of point scatterers, reduced speckle, and robustness to limitations in data quality and quantity [2, 3]. We provide an overview of how sparsity has been exploited in recently developed advanced SAR image formation methods in Sect. 13.4. Our primary focus in this chapter however is the use of sparsity in the context of SAR imaging from *undersampled data*, leading to consideration of ideas and analysis tools from compressed (or compressive) sensing (CS) [4, 5]. Compressed sensing seeks to acquire as few measurements as possible about an unknown signal, and given these measurements, reconstruct the signal either exactly or with provably small probability of error [6]. Reconstruction methods used in CS are related to sparsity-constrained, non-quadratic regularization. The compressed sensing literature has demonstrated accurate signal reconstructions from measurements involving extremely few, but randomly chosen Fourier samples of a signal [4, 7]. Since both mono-static and multi-static SAR sensing can be viewed as obtaining samples of the spatial Fourier transform of the scattering field [8], these results suggest interesting opportunities for SAR sensing. Compressed sensing is described in more detail in other chapters of this book, however for the sake of completeness, we provide a brief overview of pieces of compressed sensing that are particularly relevant for this chapter in Sect. 13.2. In the context of radar imaging, compressed sensing is primarily motivated by the fact that current radar sensing missions involve timeline constraints on data collection due to radar operation in multiple modes including searching, tracking, and imaging. Furthermore, the ability to use multi-platform sensor geometries and the possibility of passive sensing from transmitters of opportunity and quiet receivers such as unmanned aerial vehicles is increasingly important. These new mission requirements impose non-dense and irregular sampling patterns in the SAR measurement space, motivating the development of signal processing algorithms for such irregular and undersampled data scenarios. In Sect. 13.4 we mention how ideas from compressed sensing have recently been considered for radar imaging in such scenarios.

In the work presented in this chapter, we focus on radar imaging in undersampled data scenarios and on the development of tools to understand and evaluate the performance of various sensor operation and configuration choices in such scenarios in a straightforward, tractable manner. We consider both mono-static and multi-static sensing configurations. We describe the geometry and measurement model for general multi-static sensing in Sect. 13.3, which contains mono-static sensing as a special case. We examine various mono-static and ultra-narrowband multi-static configurations in Sect. 13.5. Different configurations lead to different Fourier sampling patterns, which trade off frequency and geometric diversity. In an effort to identify a tool for evaluating the imaging performance under various undersampled sensing configurations, we first observe that CS theory relates the accurate reconstruction of a signal to the mutual coherence of the corresponding measurement operator. We also note that mutual coherence can be a pessimistic measure of average performance. Accordingly, in Sect. 13.6, we propose a variant of mutual coherence, which we call the $t_{\%}$-mutual coherence as a more effective measure of expected sensing configuration quality. We then provide an experimental study of the impact of undersampled data collection in radar on the reconstruction quality of a scene of interest in Sect. 13.7. In particular, we consider various wide-band mono-static and narrow-band multi-static configurations, which trade off frequency and geometric diversity. We examine how the $t_{\%}$-mutual coherence of the corresponding measurement operator is affected by these distinct SAR sensing configurations and investigate how this easily computed metric is related to reconstruction quality. Portions of the analysis we present in this chapter can be found in a preliminary form in [9].

13.2 Compressed Sensing Overview

Compressed sensing enables reconstruction of sparse or compressible signals from a small set of linear, non-adaptive measurements, much smaller in size than that required by the Nyquist-Shannon theorem [4, 5]. A family of signals $\mathbf{s} \in \mathbf{S} \subset \mathcal{R}^{N\times 1}$ has a sparse representation in a dictionary $\mathbf{D} \in \mathcal{R}^{N\times K}$, if $\mathbf{s} = \mathbf{D}\boldsymbol{\alpha}$ and $\boldsymbol{\alpha} \in \mathcal{R}^{K\times 1}$ is a sparse vector. A vector α is considered sparse if the number of its non-zero components satisfies $\|\alpha\|_0 \leq T \ll K$, where the l_0 norm $\|\cdot\|_0$ counts the number of non-zero elements of the argument. Compressed sensing measures M projections of such a signal (where $T < M \ll K$), and then exploits its sparsity to obtain a reliable reconstruction. To represent this problem mathematically, let $\mathbf{r} \in \mathcal{R}^{M\times 1}$ represent the measured signal, $\mathbf{P} \in \mathcal{R}^{M\times N}$ the sensing (projection) matrix such that $M \ll K$ and

$$\mathbf{r} = \mathbf{P}\mathbf{s} = \mathbf{P}\mathbf{D}\alpha = \boldsymbol{\Phi}\alpha. \tag{13.1}$$

Since $M \ll K$ this set of equations is extremely under-determined, and many solutions are possible. To overcome this, a sparse solution with only a few non-zero elements is sought.

Optimal design of $\mathbf{P}$ and $\mathbf{D}$ is a topic of interest in compressed sensing. For now, let us assume that $\boldsymbol{\Phi} = \mathbf{PD}$ is given. For a given $\boldsymbol{\Phi}$, the problem is to find an optimally sparse solution. A direct formulation of this problem can be stated as:

$$\min_{\alpha} \|\alpha\|_0 \qquad \text{s.t.} \quad \mathbf{r} = \boldsymbol{\Phi}\alpha. \tag{13.2}$$

Unfortunately, this formulation is computationally difficult to solve, as it involves NP-hard enumerative search. The convex relaxation approach relies on the fact that besides the l_0 norm, the l_1 norm also promotes sparsity in a solution. The l_1 norm is defined as $\|\alpha\|_1 = \sum_{i=1}^{K} |\alpha_i|$, where α_i is the i-th element of α. This norm is a convex function of its arguments. The relaxed version of the problem then takes the form:

$$\min_{\alpha} \|\alpha\|_1 \qquad \text{s.t.} \quad \mathbf{r} = \boldsymbol{\Phi}\alpha, \tag{13.3}$$

which is essentially a linear optimization problem. The use of this formulation has recently been particularly motivated by the fact that under certain conditions on the matrix $\boldsymbol{\Phi}$, the original problem and the relaxed version can be shown to have the same solution [10–13].

When the signal $\mathbf{r}$ is noisy, the signal representation problem becomes a signal approximation problem. The convex relaxation formulation of the noisy signal approximation problem is given by:

$$\min_{\alpha} \|\alpha\|_1 \qquad \text{s.t.} \quad \|\mathbf{r} - \boldsymbol{\Phi}\alpha\|_2^2 \leq \sigma, \tag{13.4}$$

where σ represents a small noise allowance. Instead of enforcing perfect fidelity to the data, now the solution coefficient vector α is allowed to satisfy the relationship approximately. This problem is known in the literature as noisy basis pursuit [14]. Note that the problem (13.4) can also be cast in Lagrangian form as the following regularization problem for an appropriately chosen parameter λ:

$$\min_{\alpha} \|\mathbf{r} - \boldsymbol{\Phi}\alpha\|_2^2 + \lambda\|\alpha\|_1. \tag{13.5}$$

Computationally efficient algorithms for solution of the optimization problems given in (13.3), (13.4) and (13.5) have been developed [15–20].

Rather than relaxing the optimization problem in (13.2), an alternative approach has been to attempt to solve it using greedy algorithms, belonging to the family of matching pursuit algorithms [21]. Interestingly, such greedy algorithms have also been shown to solve (13.2) exactly under certain conditions [22].

Recent work in compressed sensing established that accurate reconstructions can be obtained with high probability even when only $\mathcal{O}(T)$ measurements are available [4, 5]. In particular it was shown that the number of measurements should satisfy

$$M \geq C_\sigma T \log(K), \tag{13.6}$$

with an appropriate coefficient C_σ, that depends on the desired accuracy of the reconstruction. These results require that the matrix $\boldsymbol{\Phi}$ satisfy the so-called restricted isometry property [4]. The restricted isometry property requires that all sub-matrices containing up to T columns of the matrix $\boldsymbol{\Phi}$ are near-isometries. Direct design of $\boldsymbol{\Phi}$ (and thus $\mathbf{P}$ and $\mathbf{D}$) based on this property is challenging, as it is combinatorial in nature. Thus, most of the work in CS simply assumes that the projections $\mathbf{P}$ are drawn at random, as such random projections can be shown to satisfy the required property with high probability.

An alternative approach is to focus on the so-called mutual coherence of the elements of the matrix $\boldsymbol{\Phi}$, seeking configurations with low mutual coherence. This measure is simple to compute, though less directly connected to performance. The mutual coherence $\mu(\boldsymbol{\Phi})$ of the matrix $\boldsymbol{\Phi}$ is formally defined as [23, 24]:

$$\mu(\boldsymbol{\Phi}) = \max_{i \neq j} \frac{|\phi_i^T \phi_j|}{\|\phi_i\|_2 \|\phi_j\|_2}, \tag{13.7}$$

where ϕ_i is the i-th column of the matrix $\boldsymbol{\Phi}$. Equivalently, the mutual coherence is the largest non-diagonal entry of the column normalized Gram matrix $\mathbf{G} = \boldsymbol{\Phi}^T \boldsymbol{\Phi}$, representing the worst case (i.e., the largest) similarity between the sensing columns. Orthonormal bases have zero mutual coherence and an overcomplete dictionary with a small mutual coherence is taken to be incoherent. Large mutual coherence indicates the presence of two closely related columns that may confuse reconstruction algorithms.

The mutual coherence provides a guarantee, although pessimistic, that the basis pursuit reconstruction algorithm solution employing l_1 relaxation of the l_0 norm yields the optimal solution of the original problem [10, 24]. It essentially provides a sufficient condition on the equivalence of the use of l_0 and l_1 norms, indicating that the NP hard l_0 problem in (13.2) can be solved by the tractable l_1 relaxation in (13.3). Namely, the signal $\mathbf{s}$ is perfectly recovered by both methods provided that the representation α satisfies the requirement [10, 24]:

$$\|\alpha\|_0 < \frac{1}{2}\left(1 + \frac{1}{\mu(\boldsymbol{\Phi})}\right).$$

This condition is pessimistic as it provides a worst-case guarantee, i.e., it guarantees zero-error recovery of *any* signal satisfying the above requirement. In general, successful compressed sensing recovery is possible for a significantly larger class of signals by introducing a small probability of error. However, one can still aim to optimize the projection probes such that mutual coherence is minimized in order to enlarge the class of signals with guaranteed successful compressed sensing recovery [25].

Finally, we note that it is known that Fourier measurements represent good projections for compressed sensing of sparse point-like signals [4], when $\boldsymbol{\Phi}$ represents random undersampling of the spatial frequency data. This suggests a natural application to the SAR sensing problem. In particular, we examine several mono-static and multi-static SAR sensing configurations with a number of undersampling schemes and study the relationship between their coherence (based on a variation of (13.7) introduced in Sect. 13.6) and reconstruction quality.

13.3 Multi-Static SAR Measurement Model

Most imaging radars use a transmitting and receiving antenna carried on a platform (aircraft or satellite) which travels along a path transmitting microwave pulses towards the ground. Through the motion of the antenna platform, the scene is observed from a diverse set of angles effectively creating a larger aperture than the real antenna aperture, leading to the concept of SAR. Having a moving sensor collect data at a certain number of observation points along its flight path is equivalent to collecting data from the same number of stationary sensors at those locations for a static scene. Some of the transmitted microwave energy is reflected back towards the sensor where it is received as a signal. This signal first undergoes some pre-processing, involving demodulation. The radar image formation problem is the problem of reconstruction of a spatial reflectivity distribution of the scene from the pre-processed radar returns. Under a single bounce (Born) scattering approximation, the observation model used in many practical radar imaging scenarios is linear. What we have described so far in this section assumes mono-static operation, i.e., collocation of the transmitter and the receiver on the same platform. Radar imaging can more generally be performed in a multi-static mode, where transmitters and receivers are on different, spatially dispersed platforms. In this section, we describe this general multi-static observation scenario.

We consider a general multi-static system with spatially distributed transmit and receive antenna elements within a cone positioned at the center of a scene of interest. The scene of interest is modeled by a set of point scatterers reflecting impinging electromagnetic waves isotropically to all receivers within the cone. We introduce a coordinate system with the origin in the center of the area of interest and, for simplicity, model the scene as two dimensional. Figure 13.1 illustrates this set up. The relative size of the scene is assumed to be small compared to distances from the origin of the coordinate system to all transmitters and receivers, such that transmit and receive angles would change negligibly if the coordinate origin moved to any point in the scene. Furthermore, we neglect signal propagation attenuation.

Let us present the received signal model for a pair of spatially separated transmit and receive antenna elements. The complex signal received by the l-th receiver, located at $\mathbf{x}_l = [x_l, y_l]^T$, for the narrow-band excitation from the k-th transmitter, located at $\mathbf{x}_k = [x_k, y_k]^T$, reflected from a point scatterer at the spatial location $\mathbf{x} = [x, y]^T$ is given by

$$g_{kl}(t) = s(\mathbf{x})\, \gamma_k\, (t - \tau_{kl}(\mathbf{x})),$$

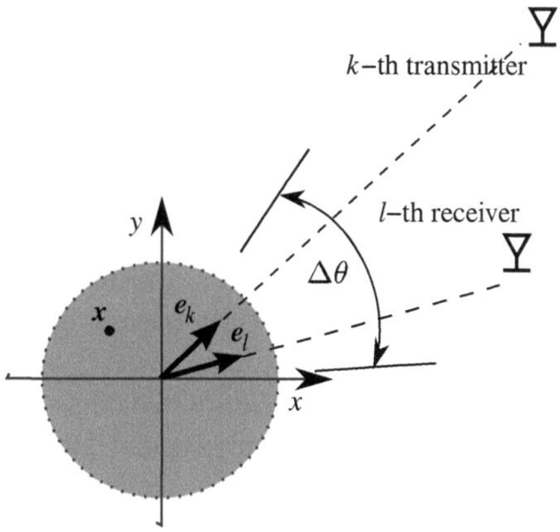

Fig. 13.1 Geometry of the kl-th transmit-receive pair with respect to the scene of interest. All transmit and receive pairs are restricted to lie within a cone of angular extent $\Delta\theta$

where $s(\mathbf{x})$ is the reflectivity of the scatterer, $\gamma_k(t)$ is the transmitted waveform from the k-th transmitter, and $\tau_{kl}(\mathbf{x})$ is the propagation delay from the transmitter to the scatterer and back from the scatterer to the receiver. The overall received signal from the entire ground patch of radius L is then modeled as a superposition of the returns from all the scattering centers and is given by:

$$g_{kl}(t) = \int_{\|\mathbf{x}\|\leq L} s(\mathbf{x})\gamma_k\left(t - \tau_{kl}(\mathbf{x})\right) d\mathbf{x}.$$

For narrow-band waveforms, defined by $\gamma_k(t) = \tilde{\gamma}_k(t)e^{j\omega_k t}$, where $\tilde{\gamma}_k(t)$ is a low-pass, slowly varying signal and ω_k is the carrier frequency, we can write:

$$g_{kl}(t) = \int_{\|\mathbf{x}\|\leq L} s(\mathbf{x})e^{j\omega_k(t-\tau_{kl}(\mathbf{x}))}\,\tilde{\gamma}_k\left(t - \tau_{kl}(\mathbf{x})\right)\, d\mathbf{x}. \tag{13.8}$$

In the far-field case, when $\|\mathbf{x}\| \ll \|\mathbf{x}_k\|$, $\|\mathbf{x}\| \ll \|\mathbf{x}_l\|$, and $\omega_k/c\|\mathbf{x}\|^2 \ll \|\mathbf{x}_k\|$, $\omega_k/c\|\mathbf{x}\|^2 \ll \|\mathbf{x}_l\|$, we can use the first order Taylor series expansion to approximate the propagation delay $\tau_{kl}(\mathbf{x})$ as:

$$\begin{aligned}\tau_{kl}(\mathbf{x}) &= \frac{1}{c}\left(\|\mathbf{x}_k - \mathbf{x}\| + \|\mathbf{x}_l - \mathbf{x}\|\right)\\ &\approx \tau_{kl}(\mathbf{0}) - \frac{1}{c}\mathbf{x}^T\mathbf{e}_{kl},\end{aligned}$$

where $\tau_{kl}(\mathbf{0}) \doteq (\|\mathbf{x}_k\|+\|\mathbf{x}_l\|)/c$ is the known transmitter-origin-receiver propagation delay, and $\mathbf{e}_{kl} \doteq \mathbf{e}_k + \mathbf{e}_l$ is the kl-th transmit-receiver pair's bistatic range vector. The vectors $\mathbf{e}_k \doteq [\cos\theta_k, \sin\theta_k]^T$ and $\mathbf{e}_l \doteq [\cos\theta_l, \sin\theta_l]^T$ are unit vectors in the direction of the k-th transmitter and l-th receiver respectively.

The chirp signal is one of the most commonly used pulses in SAR imaging [8], and is given by

$$\gamma_k(t) = \begin{cases} e^{j\beta_k t^2} \cdot e^{j\omega_k t}, & -\frac{\tau_c}{2} \le t \le \frac{\tau_c}{2} \\ 0 & \text{otherwise} \end{cases} \tag{13.9}$$

where ω_k is the center frequency and $2\beta_k$ is the so-called chirp rate of the k-th transmit element. The frequencies encoded by the chirp signal extend from $\omega_k - \beta_k \tau_c$ to $\omega_k + \beta_k \tau_c$, such that the bandwidth of this signal is given by $B_k = \frac{\beta_k \tau_c}{\pi}$. The narrow-band assumption is satisfied by choosing the chirp signal parameters such that $2\pi B_k / \omega_k \ll 1$. Ultra-narrow band waveforms are special cases of the chirp signal obtained by setting $\beta_k = 0$.

We use a general transmitted chirp signal in (13.8), along with the far-field delay approximation, and apply typical demodulation and baseband processing. In particular, the received signal $g_{kl}(t)$ is mixed with the transmitted signal referenced to the origin of the scene $e^{-j[\omega_k(t-\tau_{kl}(\mathbf{0})+\beta_k(t-\tau_{kl}(\mathbf{0}))^2]}$, and then low-pass filtered. After such preprocessing, and ignoring a quadratic phase term [8], we obtain the following observed signal model:

$$r_{kl}(t) \approx \int_{\|\mathbf{x}\|<L} s(\mathbf{x}) e^{j\Omega_{kl}(t)\mathbf{x}^T \mathbf{e}_{kl}} \, d\mathbf{x}, \tag{13.10}$$

where $\Omega_{kl}(t) = \frac{1}{c}[\omega_k - 2\beta_k(t - \tau_{kl}(\mathbf{0}))]$, depends on the frequency content of the transmitted waveform.

The observations of all receivers across all snapshots, i.e., pulses, are coherently processed. A discrete model can be obtained by discretizing the spatial variable $\mathbf{x}$, approximating the integral in (15) with a Riemann sum, and sampling in time. We stack the sampled data for all the receivers into a column vector. Likewise the reflectivity field is stacked into a column vector $\mathbf{s}$. Considering receiver noise $\mathbf{n}$ as well, we obtain the following noisy discrete observation model:

$$\mathbf{r} = \sum_{i=1}^{N} \mathbf{P}_i s_i + \mathbf{n} = \mathbf{P}\mathbf{s} + \mathbf{n}. \tag{13.11}$$

In this equation, $\mathbf{r} \in \mathcal{C}^{M \times 1}$ represents the observed, thus known, set of return signals at all receivers across time. Its elements are indexed by the tuple (k, l, t_s), with t_s being the sampling times associated with the kl-th transmit-receive pair. Thus, the discrete model implicitly assumes that the probes from different transmitters are separable at each receiver. This can be achieved by orthogonal waveform design or by ensuring sequential transmission. The reflectivity of the i-th spatial cell or pixel is denoted by $s_i \in \mathcal{C}^{1 \times 1}$ and $\mathbf{P}_i$ is the column vector capturing the contribution to the received signal of a reflector that is located in the i-th pixel.

In a particular undersampled data collection scenario, the specific matrix $\mathbf{P}$ to be used is derived from (13.10) with the spatial frequency $\Omega_{kl}(t)$ and aspect-vector samples $\mathbf{e}_{kl}$ determined by the specific sampling configuration. The received signal model for the mono-static configuration involves collocated transmit-receiver pairs, and is thus a special case of the multi-static model obtained by setting $\mathbf{x}_k = \mathbf{x}_l$.

13.4 Recent Use of Sparsity and Compressed Sensing in Radar Imaging

Radar reflectivity images can usually be approximated well through sparse representations either by using parametric models of physical scattering behaviors (see, e.g., [1] for a brief discussion) or by expressing the entire reflectivity field through appropriate spatial dictionaries. We focus on this latter perspective here. Ideas based on ℓ_1 formulations (see Sect. 13.2) and their variants have been successfully used in radar imaging in recent years. Here we first provide a brief overview of a subset of these developments in image formation. Our coverage is not comprehensive, but rather mostly highlights several lines of work in which the authors of this chapter were involved. We then mention how ideas from compressed sensing have recently been explored for analysis and design of radar imaging tasks with data limitations.

Using the same notation as in Sects. 13.2 and 13.3, let us start with the following noisy observation model for radar imaging:

$$\mathbf{r} = \mathbf{P}\mathbf{s} + \mathbf{n}, \tag{13.12}$$

where $\mathbf{r}$ denotes the observed radar data, $\mathbf{P}$ is the radar sensing matrix, $\mathbf{s}$ is the spatial reflectivity field to be imaged, and $\mathbf{n}$ denotes additive noise. All of these variables are complex-valued. The details of the sensing matrix $\mathbf{P}$ for general multi-static radar data collection can be found in Sect. 13.3. Given this observation model, let us first consider image formation of target scenes consisting of a sparse set of point reflectors. This is similar in nature to a special case of the sparse representation problem considered in Sect. 13.2, where the signal representation dictionary $\mathbf{D}$ is taken to be an identity matrix.[1] In this case, we can formulate the SAR reconstruction problem as the following optimization problem:

$$\begin{aligned} \widehat{\mathbf{s}} = \ & \arg\min_{\mathbf{s}} \ \|\mathbf{s}\|_1 \\ & \text{s.t.} \ \ \|\mathbf{r} - \mathbf{P}\mathbf{s}\|_2 \leq \sigma, \end{aligned} \tag{13.13}$$

where σ represents a parameter for noise allowance, $\|\mathbf{s}\|_1 = \sum_i \sqrt{(\Re s_i)^2 + (\Im s_i)^2}$ and $\|\mathbf{x}\|_2 = \sqrt{\sum_i \left((\Re x_i)^2 + (\Im x_i)^2\right)}$, where s_i and x_i are the i-th elements of $\mathbf{s}$ and $\mathbf{x}$, respectively.

As discussed in Sect. 13.2, such problems can also be expressed in Lagrangian form. Doing that, and considering more general ℓ_p quasi-norms ($0 < p \leq 1$), we reach an alternate formulation:

$$\widehat{\mathbf{s}} = \arg\min_{\mathbf{s}} \|\mathbf{r} - \mathbf{P}\mathbf{s}\|_2^2 + \lambda \|\mathbf{s}\|_p^p, \tag{13.14}$$

[1] Any desired spatial oversampling of the reflectivity field can be handled by appropriate modification of $\mathbf{P}$.

where $\|\mathbf{s}\|_p^p = \sum_i \left[(\Re s_i)^2 + (\Im s_i)^2\right]^{p/2}$. This is a special case of the feature-enhanced SAR imaging approach proposed in [2].

Next, let us consider more general signal representation dictionaries. In the complex-valued SAR imaging problem, what admits sparse representation is the magnitudes of the reflectivities, rather than the real and imaginary components. Phases of the complex-valued reflectivities are usually highly random and spatially uncorrelated. The scene should thus be encoded through the sparse representation of the reflectivity magnitudes. Hence it makes sense to use the representation $|\mathbf{s}| = \mathbf{D}\boldsymbol{\alpha}$, where $\mathbf{D}$ denotes the representation dictionary and $\boldsymbol{\alpha}$ denotes the representation coefficients as in Sect. 13.2. Now, introducing the notation $\mathbf{s} = \boldsymbol{\Psi}|\mathbf{s}|$, where $\boldsymbol{\Psi}$ is a diagonal matrix containing the reflectivity phases,[2] the sparsity-driven SAR imaging problem becomes [26]:

$$\hat{\boldsymbol{\alpha}}, \hat{\boldsymbol{\Psi}} = \arg\min_{\boldsymbol{\alpha},\boldsymbol{\Psi}} \|\mathbf{r} - \mathbf{P}\boldsymbol{\Psi}\mathbf{D}\boldsymbol{\alpha}\|_2^2 + \lambda\|\boldsymbol{\alpha}\|_p^p. \tag{13.15}$$

Note that due to the complex-valued nature of SAR, coupled with the observation that reflectivity magnitudes admit sparse representation, the optimization problem set up in (13.15) has a more complicated structure than that encountered in common real-valued sparse representation problems (cf. (13.5)). In particular, here one has to solve not only for the sparse representation coefficients $\boldsymbol{\alpha}$ but also for the reflectivity phases in $\boldsymbol{\Psi}$. Given the freedom in choosing the overcomplete dictionary $\mathbf{D}$ in (13.15), [26] demonstrates the use of a number of dictionaries in radar imaging including wavelets, combination of spikes and edges, as well as dictionaries of various geometric shapes matched to the expected scene structure. Note that after finding $\hat{\boldsymbol{\alpha}}$ and $\hat{\boldsymbol{\Psi}}$, one can compute the estimated complex-valued reflectivity field through $\hat{\mathbf{s}} = \hat{\boldsymbol{\Psi}}\mathbf{D}\hat{\boldsymbol{\alpha}}$.

The formulation in (13.15) is based on representing the signal of interest in terms of a dictionary and imposing sparsity on the dictionary coefficients. This is usually called a *synthesis* model. In contrast, in an *analysis* model, sparsity is imposed on some features of the signal of interest. In the context of SAR imaging, such an analysis-based formulation was proposed in [2]:

$$\widehat{\mathbf{s}} = \arg\min_{\mathbf{s}} \|\mathbf{r} - \mathbf{P}\mathbf{s}\|_2^2 + \lambda\|(\mathbf{L}|\mathbf{s}|)\|_p^p, \tag{13.16}$$

where $p \leq 1$. Here, the operator $\mathbf{L}$ is used to compute some *features* of the reflectivity *magnitudes* $|\mathbf{s}|$. This allows imposing sparsity on such features. For example, [2] considers the use of a discretized gradient operator for $\mathbf{L}$, leading to a sparsity constraint on the spatial derivatives of the reflectivities, and as a result indicating a preference for piecewise smooth fields. Such piecewise smoothness constraints have a long history in real-valued image restoration and reconstruction, under various names including edge-preserving regularization and total variation restoration.

The nonlinearity involved in the operation $\mathbf{L}|\mathbf{s}|$ makes the optimization problem for radar imaging more challenging than commonly used linear sparse representation

[2] In particular, $\boldsymbol{\Psi}$ is a diagonal matrix, the i-th diagonal element of which is $e^{j\varphi_i}$, with φ_i indicating the unknown phase of the i-th scene element s_i.

problems. Efficient algorithms matched to this problem structure have been developed [2, 3]. These algorithms are based on half-quadratic regularization [27], and can be viewed as quasi-Newton methods with a specific Hessian update scheme. Another interpretation is that the overall non-quadratic problem is turned into a series of iteratively reweighted quadratic problems, each of which is efficiently solved using conjugate gradients. These algorithms have initially been used on conventional mono-static SAR sensing scenarios involving narrow angular apertures, and observations over a contiguous band of frequencies. Existing experimental results of sparsity-driven radar imaging have demonstrated improvements in resolvability of dominant scatterers, as well as in terms of the potential for suppressing artifacts such as speckle. Furthermore, it has been shown that sparsity-driven radar imaging is capable of accurate reconstruction of point-like scatterers from data with significantly reduced spatial frequency coverage. Such improvements have partially been quantified in terms of feature extraction accuracy and object classification performance [28].

The benefits provided by sparsity-driven imaging extend to non-conventional sensing scenarios in which the sensing aperture or the data are sparse or limited in some sense. Examples include scenarios involving wide-angular apertures, multi-static active and passive sensing, data with frequency-band omissions, and circular apertures. Sparsity-driven imaging based on ℓ_p-norms has been extended to and applied in such scenarios [29–32]. When wide-angular apertures are considered, idealized isotropic point scattering assumptions need to be questioned, because the scattering response could exhibit angular dependence. Through multiple lines of work, sparse representation ideas have been used for joint wide-angle radar imaging and anisotropy characterization [30, 31, 33, 34]. Recently sparsity-driven imaging has also been applied to multiple-input multiple-output (MIMO) radar [35].

Two potential practical issues in the context of sparsity-driven radar imaging are worth mentioning. First, how do we choose the regularization parameter λ? While this is not a completely solved problem yet, preliminary results (see, e.g., [36]) offer some promise. Second, what if our sensing model $\mathbf{P}$ involves some uncertainties? For SAR imaging one of the most important model uncertainties is due to errors in the measurement of the time required for the transmitted signal to propagate from the SAR platform to the field and back. Such errors appear as phase errors in the SAR (spatial frequency domain) data. If uncompensated, such errors lead to various artifacts, including blurring, in the reconstructed imagery. Techniques to fix this problem are usually called *autofocus* methods. There exists some recent sparsity-driven work [37], that aims to address this problem by adding such phase errors as nuisance parameters to the optimization problem in (13.16), and performing joint imaging and model error correction. Results in [37] suggest that sparsity can be a valuable asset in the context of autofocusing as well.

Given all the prior work on ℓ_p-norm-based radar imaging, exploring the implications of existing compressed sensing theory on radar sensing design has recently been an emerging topic of interest [38–44]. A stylized compressed sensing radar was proposed in [38] in which the time-Doppler frequency plane was discretized into a grid and a small number of targets with unknown range-velocity are estimated via sparse recovery algorithms. The authors show that transmission of an Alltop sequence as the probing signal results in a sufficiently incoherent observation matrix **P** allowing for an accurate reconstruction of the sparse target scene. The authors in [40] propose the use of chirp pulses and pseudo-random sequences for compressed sensing with imaging radars. A compressed sensing technique for SAR is also discussed in [39], where the authors obtain measurements by random subsampling of a regular aspect-frequency grid in the **k**-space. Compressed sensing for multi-static SAR with reduced number of probes was first discussed in [45]. Mono-static compressed SAR sensing with a significantly reduced number of transmitted waveforms was also presented in [46]. Additionally, the application of compressed sensing to MIMO radar was studied with uniform linear array configurations for the transmit and receive antennas [47, 48] and for a network of randomly distributed antennas over a small area [49, 50]. Waveform design for distributed radar (a special case of which would be the MIMO radar) based on compressed sensing considerations has been considered in [43].

In the experimental work we describe in Sect. 13.7, we use sparsity-enforcing reconstruction for mono-static and multi-static SAR and investigate how different compressed sensing/sampling configurations affect (a variant of) the coherence of the SAR measurement operator as well as the implications of each sensing configuration choice on reconstruction quality. To provide the basis for that analysis, the following two sections describe the sensing/sampling configurations we consider, and the coherence-based measure we use.

13.5 Sampling Configurations for Compressed Sensing SAR

Based on (13.10), SAR data represent Fourier **k**-space measurements of the underlying spatial reflectivity field. Different mono-static and multi-static SAR measurement configurations produce different Fourier sampling patterns. These patterns reflect different spectral and spatial trade-offs that must be made during task planning. Compressed sensing theory argues that random Fourier measurements represent good projections for compressive sampling of point-like signals [4]. This suggests a natural application to the sparse aperture SAR sensing problem and opens a question of how different mono-static and multi-static SAR sensing configuration constraints influence reconstruction quality for a fixed number of measurements. In the following subsections, we describe various reduced data collection configurations with non-conventional SAR **k**-space sampling patterns.

13.5.1 Random Subsampling of the Conventional Mono-Static Grid

Using a chirp pulse in mono-static sensing, each radar return from the scene lies on a radial line at a particular angle in the spatial frequency domain [8]. With a linear flight path, the complete sampled data from a pre-specified diversity of angles (i.e., full-aperture) lie on a polar grid in an annular region in the spatial frequency domain. A simple way to achieve reduced data collection for mono-static SAR could be through random subsampling of this regular SAR polar grid. Such subsampling reduces the required on-board storage and data transmission to the fusion/processing center. However, the number of transmitted probes before and after subsampling remains the same with high probability. Random subsampling discards a random subset of frequencies at each synthetic aperture point and thus, it requires the transmission of all pulses used in the full-aperture case with high probability. Such random subsampling of the conventional SAR polar grid is illustrated in Fig. 13.2a.

13.5.2 Interrupted Aperture Mono-Static Data Collection

Another approach to reduced data collection is to directly reduce the number of transmitted probes with regular or random interrupts in the synthetic aperture. For the interrupted aperture collection scenarios, we directly reduce the number of transmitted probes without first collecting the data and then discarding a subset. We consider several cases of regular and random observation sampling patterns within a fixed aspect observation extent $\Delta\Theta$, coupled with regular or random frequency sampling within a desired chirp-signal bandwidth B. Here, we do not constrain random aspect/frequency samples to fall on the regular SAR polar grid points, corresponding to the full aperture case described in the previous subsection. Figure 13.2b illustrates the **k**-space sampling pattern when both aspect and frequency are sampled regularly, which we denote by (RegCS(θ), RegCS(f)). Figure 13.2c illustrates a realization of a **k**-space sampling pattern when aspect is sampled regularly, while frequency is sampled randomly, which we denote by (RegCS(θ), RandCS(f)). Finally, Fig. 13.2d illustrates a realization of a **k**-space sampling pattern when aspect is sampled randomly, and then frequency is sampled randomly at each of these aspects, denoted by (RandCS(θ), RandCS(f)).

13.5.3 Multistatic Data Collection

Multistatic SAR offers the possibility of different **k**-space sampling patterns with different trade-offs in temporal frequency and spatial transmit/receiver location diversity. In the mono-static case, the chirp signal bandwidth allowed for extended coverage of the **k**-space in the range direction. However, in the multi-static case, extended **k**-space coverage can also be achieved in theory using ultra-narrowband signals, provided that we exploit the spatial diversity of the transmitter and receiver locations. In this section, we consider circular multi-static SAR sensing based on transmission of a continuous wave, ultra-narrowband signal. For general multi-static SAR, the total number of measurements is calculated as $M = N_{tx}N_{rx}N_f$, where

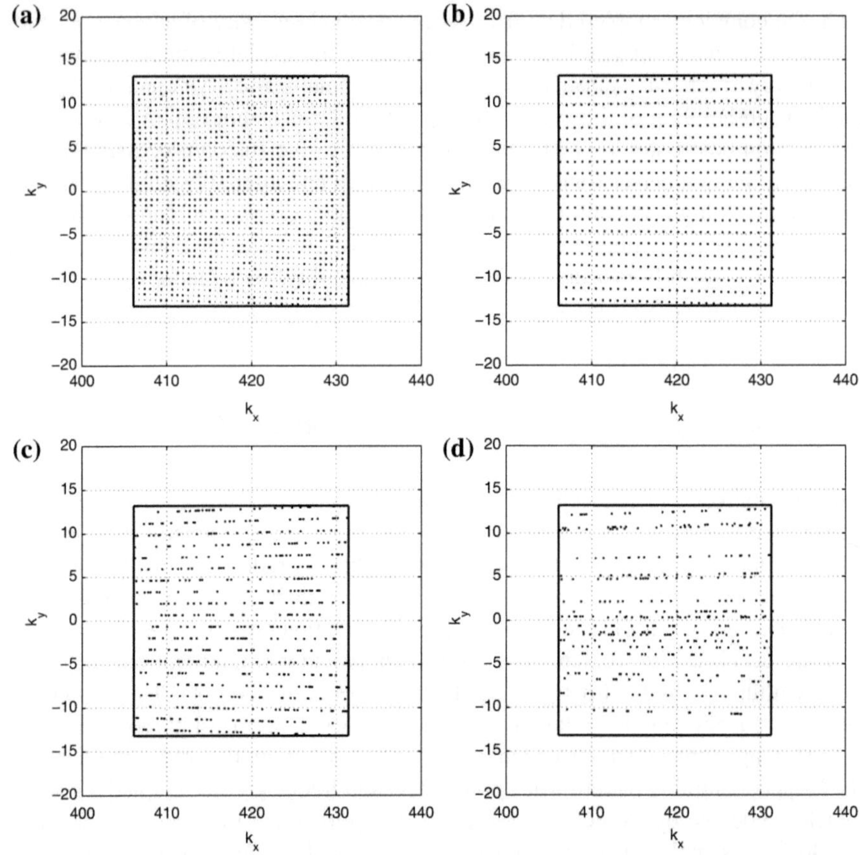

Fig. 13.2 Mono-static SAR **k**-space sampling patterns for a fixed **k**-space extent ($f_0 = 10$GHz, $B = 600$MHz, $\Delta\theta = 3.5$ deg). Each pattern is achieved with $M = 600$ measurements. The number of transmitted probes for patterns (**b–c**) is $N_{tx} = 20$. **a** Random subsampling of the conventional, SAR polar grid with (40, 40) aspect-frequency points, (NqGridRandCS). **b** Regular aspect-frequency sampling, (RegCS(θ), RegCS(f)). **c** Regular aspect, random frequency sampling, (RegCS(θ), RandCS(f)). **d** Random aspect, random frequency sampling, (RandCS(θ), RandCS(f))

N_{tx} is the number of transmitters/transmitted probes, N_{rx} is the number of receivers, and N_f is the number of frequency samples. In the case of ultra-narrowband transmission, we have $N_f = 1$ and different sampling patterns are achieved by varying transmitter and receiver angular locations. Figure 13.3a illustrates a **k**-space sampling pattern when both transmitter and receiver angular locations are sampled regularly, (RegCS(θ_{tx}), RegCS(θ_{rx})). Figure 13.3b illustrates a realization of a **k**-space sampling pattern when transmitters are positioned regularly, but receivers are dispersed randomly around the scene of interest, (RegCS(θ_{tx}), RandCS(θ_{rx})). Figure 13.3c illustrates a realization of a **k**-space sampling when both transmitter and receiver

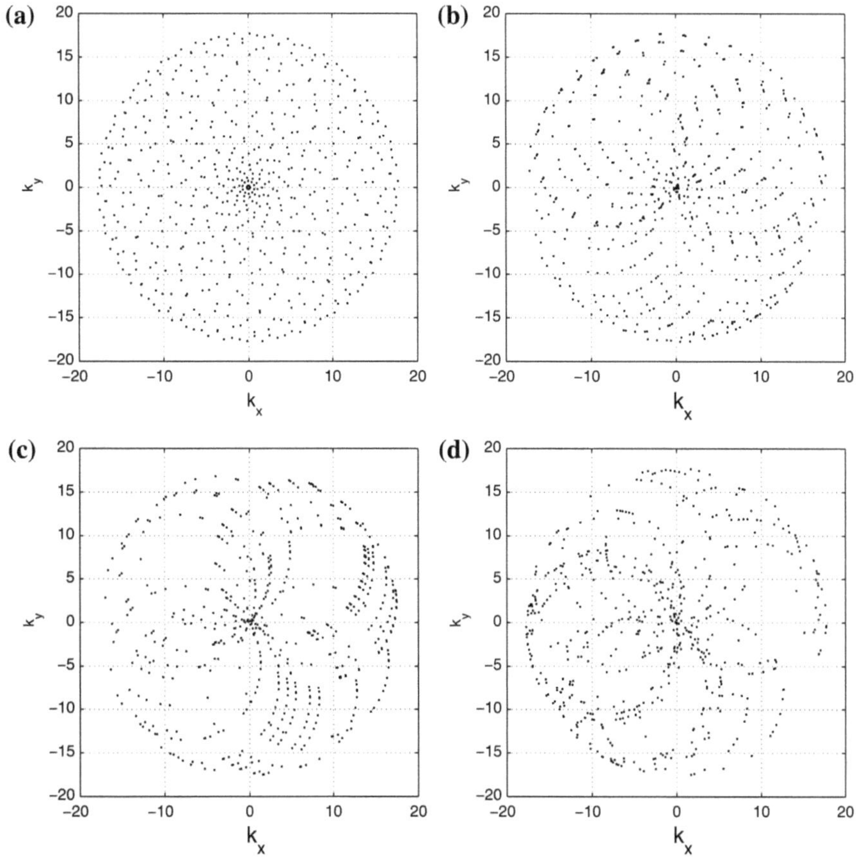

Fig. 13.3 Multi-static **k**-space sampling patterns for circular, ultra-narrowband SAR operator, with $N_f = 1$ and $N_{tx} = 20$ transmitters. For figures (**a**, **b**, **c**) the same set of $N_{rx} = 30$ receivers is used for each transmitted probe. **a** Regular transmitter-receiver aspect positioning, (RegCS(θ_{tx}), RegCS(θ_{rx})). **b** Regular transmitter, random receiver aspect positioning, (RegCS(θ_{tx}), RandCS(θ_{rx})). **c** Random transmitter, random receiver aspect positioning, (RandCS(θ_{tx}), RandCS(θ_{rx})). **d** Random transmitter, random receiver aspect positioning when a different set of $N_{rx} = 30$ receivers is used for each transmitted probe, (RandMulti)

locations are sampled randomly, (RandCS(θ_{tx}), RandCS(θ_{rx})). Finally, Fig. 13.3d illustrates a scenario in which each randomly located transmitter is associated with a different set of randomly located receivers, (RandMulti). This sampling pattern requires a prohibitively large number of transmitters and receivers, and is considered for theoretical comparison purposes only.

13.6 A Quality Measure: The t%-Average Mutual Coherence

We are interested in a simple quantitative measure that can predict reconstruction quality given a particular sensing configuration such as those above, even before sensing takes place. Such a quality predictor would allow a sensor management process to perform better task planning and resource utilization. From compressed sensing we know that the mutual coherence of the measurement probes is related to the reconstruction performance in sparse domains. With this motivation we examine the relationship of a closely related sensing geometry parameter to the reconstruction behavior of various mono-static and multi-static sensing geometries. In particular, in this section, we define a simple quantitative measure, which we call the $t_{\%}$-average mutual coherence, for *a priori* evaluation of sensing configurations, such as those described in Sect. 13.5.

The mutual coherence of a set of signals, described in Sect. 13.2, was proposed as a simple, but conservative measure of the quality of sparsity-enforcing reconstruction. The version of mutual coherence defined in Sect. 13.2 is based on the matrix $\boldsymbol{\Phi}$. In Sect. 13.4, we have discussed a variety of sparsity-driven SAR imaging methods. For the simplicity of exposition here, we will only consider image formation of target scenes consisting of a sparse set of point reflectors, i.e., sparsity will be directly imposed on the reflectivity field. In this case, image reconstruction can be achieved through the solution of (13.13) or (13.14). As discussed in Sect. 13.4, in this case we can take $\boldsymbol{\Phi} = \mathbf{P}$. Hence this is the matrix that will be involved in coherence computations in our study. The sparse reconstruction guarantees discussed in Sect. 13.2 were developed for real-valued signals. In the case of complex $\boldsymbol{\Phi}$, the mutual coherence of a sensing geometry can be similarly defined as:

$$\mu(\boldsymbol{\Phi}) = \max_{i \neq j} g_{ij}, \qquad g_{ij} = \frac{| < \phi_i, \phi_j > |}{\|\phi_i\|_2 \|\phi_j\|_2}, i \neq j \tag{13.17}$$

where ϕ_i is the i-th column of the matrix $\boldsymbol{\Phi}$, and the inner product is defined as $< \phi_i, \phi_j >= \phi_i^H \phi_j$. The i-th column vector ϕ_i can be viewed as a range-aspect "steering vector" of a SAR sensing geometry or the contribution of a scatterer at a specific spatial location to the received phase history signal. The mutual coherence measures the worst case correlation between responses of two distinct spatially distributed reflectors, and as such it is likely to be too conservatively connected to the average reconstruction quality of images reconstructed using (13.13) or (13.14).

A less conservative measure connected to the sparsity-enforcing reconstruction performance was proposed in [25] for compressed sensing projection optimization. In particular, the t-average mutual coherence was defined as the average value of the set $\{g_{ij} \mid g_{ij} > t\}$. Inspired by the t-average mutual coherence, we define and propose to use the $t_{\%}$-average mutual coherence as a measure that has the potential to be closely related to the average reconstruction performance of (13.13) or (13.14). We define the $t_{\%}$-average mutual coherence, $\mu_{t_{\%}}$, as follows. Let $\mathcal{E}_{t_{\%}}$ be the set containing the largest t percent column cross-correlations g_{ij}. The $t_{\%}$-average mutual coherence is defined as:

$$\mu_{t_{\%}}(\boldsymbol{\Phi}) = \frac{\sum_{i\neq j} g_{ij}\mathcal{I}_{ij}(t_{\%})}{\sum_{i\neq j} \mathcal{I}_{ij}(t_{\%})}, \quad \mathcal{I}_{ij}(t_{\%}) = \begin{cases} 1, & g_{ij} \in \mathcal{E}_{t_{\%}} \\ 0, & \text{otherwise.} \end{cases}$$

In other words, $\mu_{t_{\%}}(\boldsymbol{\Phi})$ measures the average cross-correlation value in the set of the $t_{\%}$ most similar column pairs. One should pick and use a small value for the parameter $t_{\%}$ in order to accurately represent the tail of the column cross-correlation distribution. This measure is more robust to outliers, which can unfairly dominate the mutual coherence. A large value of $\mu_{t_{\%}}(\boldsymbol{\Phi})$ indicates that there are many similar pairs of columns of $\boldsymbol{\Phi}$ that can potentially confuse the reconstruction algorithm.

13.7 Experimental Analysis

In this section we consider the problem of imaging random synthetic sparse scenes. From such scenes, we generate simulated radar returns under various reduced data scenarios within the mono-static and multi-static sensing configurations described in Sect. 13.5. From such data, we perform sparsity-driven image reconstruction. For each sensing configuration, we vary the number of transmitted probes and the number of measurements. For each such case, we compute the $t_{\%}$-average mutual coherence, and compute two metrics directly measuring image reconstruction quality. We then analyze the behavior of the $t_{\%}$-average mutual coherence and the two reconstruction metrics as we vary the configuration and various parameters. Such an analysis enables us to evaluate the power of $t_{\%}$-average mutual coherence as a predictor of reconstruction quality, as well as compare various sensing configurations.

The results we present are obtained by averaging over 100 Monte Carlo runs. For the regular aspect-frequency sampling of mono-static SAR, and the regular sampling of transmitter-receiver aspects of multi-static ultra-narrowband SAR, we average over different ground truth scene realizations. In cases involving random measurement sampling, we average over different SAR operators and different ground scene realizations. For each realization of $\boldsymbol{\Phi}$ we measure the $t_{\%}$-average mutual coherence $\mu_{t_{\%}}(\boldsymbol{\Phi})$, for $t_{\%} = 0.5\%$, and display its average over all Monte Carlo runs. To impose the sparsity constraints, we use l_1 norms. We solve the optimization problem (13.13) using the software described in [18, 51] and display reconstruction performance of different SAR sensing configurations by using two performance metrics directly measuring image quality. In particular, we compute and use the relative mean square error (RMSE), and the percentage of identified support, both of which we define next. The RMSE is defined as: $\text{RMSE} = E[\|\hat{\mathbf{s}} - \mathbf{s}_0\|_2/\|\mathbf{s}_0\|_2]$, where $\mathbf{s}_0$ is the ground truth signal, $\hat{\mathbf{s}}$ is its estimate from a reduced set of measurements, and $E[\cdot]$ stands for an empirical average over different Monte Carlo runs. The percentage of identified support measures the percentage of the correctly identified support of the T largest components of the estimated signal, where T is the number of point reflectors in the ground truth scene.

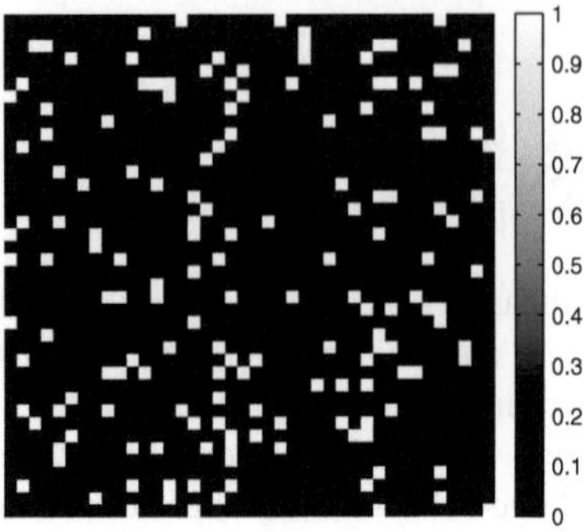

Fig. 13.4 The magnitude image of a random ground truth scene realization

13.7.1 Simulation Results for Mono-Static CS SAR

First, we consider mono-static SAR imaging of a small ground patch of size $(D_x, D_y) = (10, 10)$m, when observed over a narrow-angle aspect cone of $\Delta\theta = 3.5$ deg. The transmitted waveforms are chirp signals with $f_o = 10$GHz and $B = 600$MHz. The nominal range resolution is $\rho_x = \frac{c}{2B} = 0.25$m and the nominal cross-range resolution is $\rho_y = \frac{\lambda}{4\sin(\Delta\theta/2)} = 0.25$m. Assuming that the pixel spacing matches the nominal sensing resolution, we seek to reconstruct a 40×40 pixel reflectivity image. The ground truth scene consists of T randomly dispersed scatterers each with unit magnitude and random phase uniformly distributed in $[0, 2\pi]$. The magnitude image of a random ground truth scene realization is shown in Fig. 13.4. The nominal polar grid in the phase-history domain contains 40×40 elements. In the image reconstruction algorithm, we use the noise allowance parameter value of $\sigma = 0.1$, and set the maximum number of iterations to 1000 in all simulations.

Figure 13.5 shows the scatter plots of RMSE vs $\mu_{t\%}$ when random scenes with $T = 140$ scatters are sensed with $M = 600$ measurements using the (RandCS(θ), RandCS(f)) configuration. In order to obtain variability in $\mu_{t\%}$ over a wide range of values, we vary the number of aspect angles (i.e. the number of transmitted probes) such that $N_{tx} \in \{10, 20, 30, 40, 50, 60\}$. These scatter plots show that, in general, $\mu_{t\%}$ is indeed indicative of RMSE reconstruction quality. However, comparing Fig. 13.5a–c, we see that a desired level of RMSE reconstruction quality is better predicted with $\mu_{0.5\%}$ than the classical mutual coherence μ and $\mu_{0.1\%}$. In fact, RMSE vs μ and RMSE vs $\mu_{0.1\%}$ scatter plots are very close indicating that we do not properly capture full tails of the column cross-correlation distributions with $\mu_{0.1\%}$. On the other hand, if we pick $\mu_{t\%}$ too large the scatter plot behaves more like a step function. This is indicated with Fig. 13.5d when $\mu_{t\%} = \mu_{5\%}$. In the following, we present results for $\mu_{0.5\%}$.

In Fig. 13.6 we show the results for various mono-static sensing configurations as a function of the number of transmitted probes N_{tx} when the total number of measurements is fixed to $M = 600$ and the total number of randomly dispersed

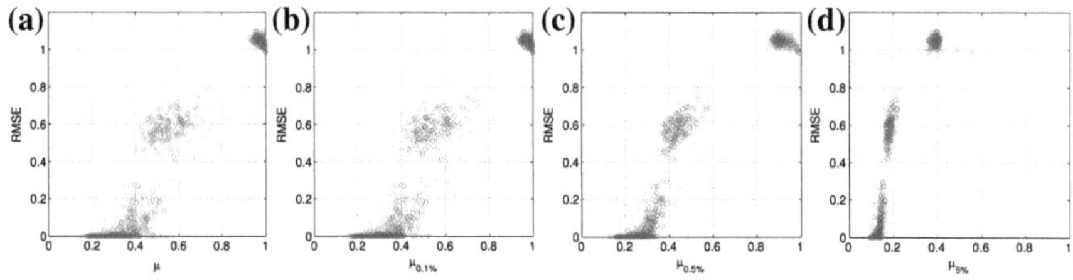

Fig. 13.5 Scatter plots of RMSE vs $\mu_{t\%}$ when scenes with $T = 140$ scatters are sensed with $M = 600$ measurements through the (RandCS(θ), RandCS(f)) configuration. **a** Mutual coherence μ. **b** $\mu_{0.1\,\%}$. **c** $\mu_{0.5\,\%}$. **d** $\mu_{5\,\%}$

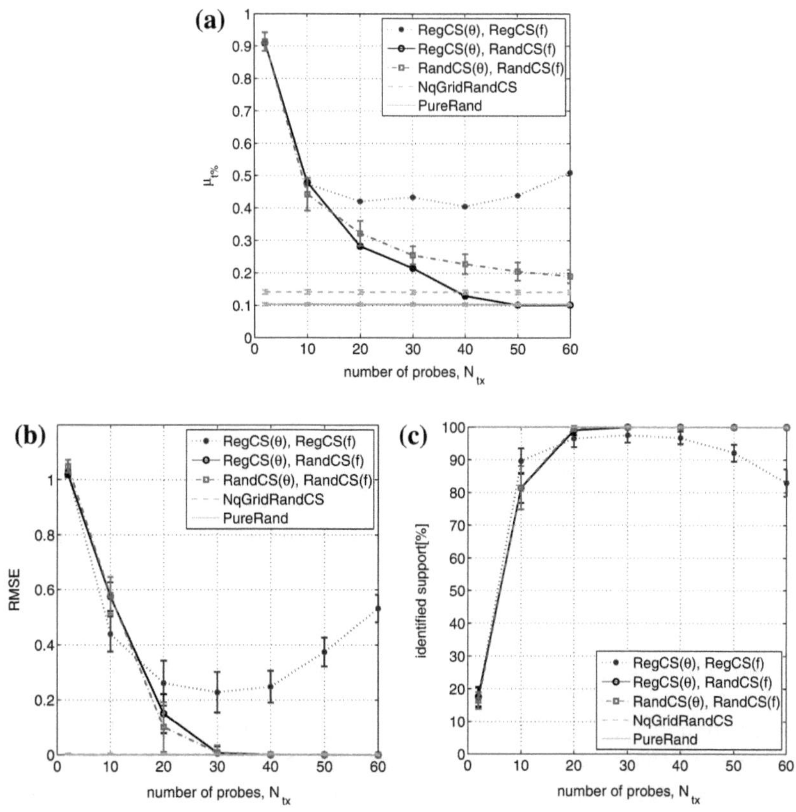

Fig. 13.6 Mono-static SAR performance versus the number of transmitted probes N_{tx} for various sensing configurations. In this experiment, the number of measurements, $M = 600$, and the signal support size, $T = 140$, are held fixed. **a** The $t_{\%}$-average mutual coherence, $\mu_{0.5\,\%}$. **b** RMSE. **c** Percentage of correctly identified support of T largest estimated signal peaks

scatterers in the scene is set to $T = 140$. In the mono-static case the number of transmitted probes N_{tx} corresponds to the number of aspect angles. The associated sampling patterns are illustrated in Fig. 13.2 when the number of transmitted probes is $N_{tx} = 20$. The number of aspect angles N_{tx} and the number of chirp frequency samples N_f are varied such that $M = N_{tx}N_f = 600$. As indicated earlier, regular sampling means that the aspect step and/or the frequency step are fixed. In random sampling, aspect angles and/or sampling frequencies are chosen independently and uniformly at random within their allowable ranges. In the NqGridRandCS case, the conventional SAR polar grid of $(N_{tx}, N_f) = (40, 40)$ points is down-sampled uniformly at random to 600 points.

In addition to the configurations discussed in Sect. 13.5, our plots also contain results for a scenario we call pure random sampling (PureRand). For PureRand, sampling frequencies in each direction are chosen uniformly at random. Such sampling is essentially impractical for SAR, as it requires a prohibitively large number of sampling probes and sensor locations. In other words, pure random sampling does not impose any structure on the sensing geometry and thus, a point sampled in the **k**-space may require a unique combination of the temporal frequency (pulse) and synthetic aperture point (sensor location). However, we consider this case for comparison purposes as it provides essentially maximal randomness in sampling and is expected to provide a lower bound on the coherence of the sensing operator.

The results in Fig. 13.6 indicate that the $t_{\%}$-average mutual coherence is the lowest when **k**-space sampling points cover the available **k**-space extent most uniformly. The most uniform coverage in the regular subsampling case with $M = N_{tx}N_f$ measurements is achieved when the ratio of the number of aspect angles to the number of frequency samples is approximately $N_{tx}/N_f = \Delta K_x/\Delta K_y$, where ΔK_x (ΔK_y) is the **k**-space extent in the cross-range (range) direction. On the other hand, when randomness is present in the aspect and/or frequency sampling and the number of transmitted probes increases, the uniformity of the **k**-space coverage approaches the SAR polar grid subsampling cases. This is reflected in the lower values of the $\mu_{t\%}$ curve as the number of transmitted probes increases. We expect that increasing the number of transmitted probes further after a certain critical value of $\mu_{t\%}$ is reached would have only a small impact on the reconstruction performance.

Comparing the $\mu_{t\%}$ curves to the corresponding reconstruction performance metrics shown in Fig. 13.6b–c, we see that as the mutual coherence is lowered, the reconstruction quality improves. Hence, $\mu_{t\%}$ can serve as a reasonable predictor of reconstruction quality. Regular aperture interrupts coupled with regular frequency sampling, the (RegCS(θ), RegCS(f)) case, introduces signal aliasing manifested as periodic and large column cross-correlation peaks that confuse the reconstruction algorithm. This case has consistently the worst reconstruction performance. However, if regular aperture interrupts are coupled with random frequency sampling, the (RegCS(θ), RandCS(f)) case, the reconstruction performance improves significantly. Similar observations hold for random sampling of both aspect and frequency, i.e., the (RandCS(θ), RandCS(f)) case. Overall, for the randomly interrupted aperture collection scenarios, the correct signal support is identified with a negligible error with $N_{tx} = 20$ probes, while the reconstruction error is negligible with $N_{tx} = 30$

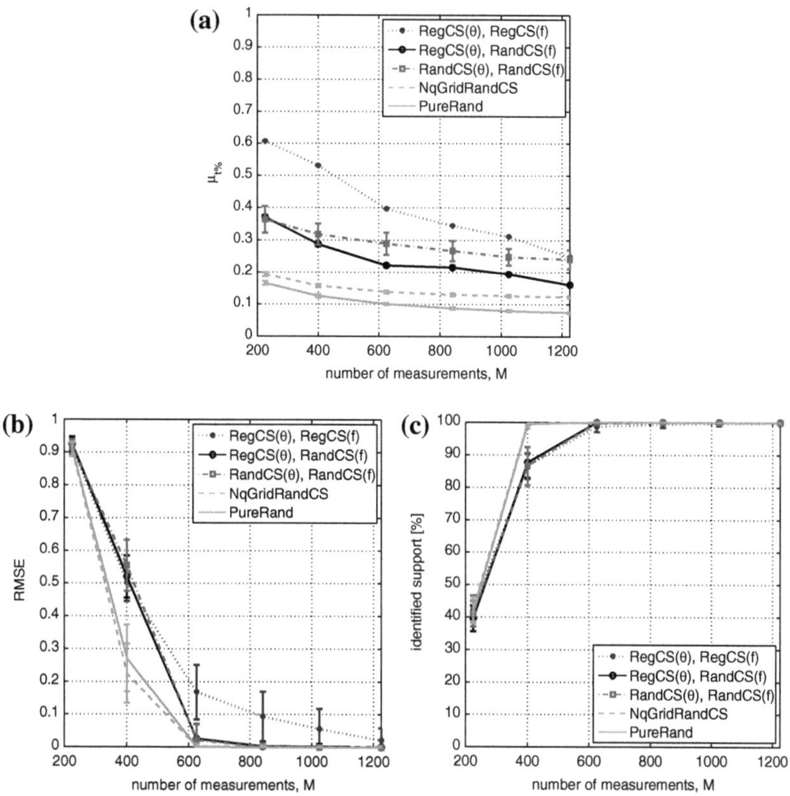

Fig. 13.7 Mono-static SAR performance versus the number of measurements M for various sensing configurations. In this experiment, the number of transmitted probes $N_{tx} = N_f$ and the signal support size $T = 140$ are held fixed. **a** The $t_{\%}$-average mutual coherence, $\mu_{0.5\%}$. **b** RMSE. **c** Percentage of correctly identified support of T largest estimated signal peaks

probes instead of the nominal $N_{tx} = 40$ probes. These results suggest that sparsity-driven imaging based on randomly interrupted aperture compressed sampling can produce very high-quality reconstructions with significantly lower number of transmitted probes than what would be used by conventional sensing and imaging. As expected, the impractical random SAR polar grid subsampling (NqGridRandCS) case offers high-quality reconstruction in terms of all reconstruction quality metrics.

In Fig. 13.7 we evaluate $\mu_{t\%}$ and the reconstruction quality as a function of the number of measurements. Signal sparsity is held fixed at $T = 140$. The number of measurements $M = N_{tx}N_f$ is varied such that $N_{tx} = N_f < 40$. We observe that (RegCS(θ), RegCS(f)) leads to the highest $\mu_{t\%}$ among all configurations, and as we add randomization $\mu_{t\%}$ decreases. As expected, increasing the number of measurements lowers the $t_{\%}$-average mutual coherence. It also reduces signal support estimate error and RMSE up to a point where a reconstruction with negligible error is achieved. Hence one can say that a lower value of $\mu_{t\%}$ implies better reconstruction quality in this experiment. The number of measurements needed for very accurate reconstruction in the random sampling configurations appears to be around 600 in this experiment, which is on the order of $4-5$ times the number of point reflectors in the scene, and which is significantly lower than the nominal number of measurements (i.e., 1600) that would be used by conventional sensing in this scenario.

13.7.2 Simulation Results for Multi-Static CS SAR

The primary benefit provided by compressed sensing in the mono-static, single-platform scenario is the reduction of data storage requirements and the reduction in the number of transmitted probes. Overall data collection time cannot be reduced, as the mono-static SAR platform covers the whole aspect range sequentially in time, although the time spent at the aperture positions where radar data collection is not performed could be used for a different task. On the other hand, multi-static SAR has the potential to further reduce the data acquisition time by using a multitude of spatially dispersed transmitters and receivers. Theoretically, there exist many multi-static geometries with similar **k**-space coverage as in the mono-static case, and thus, similar reconstruction results. As an extreme case, we consider a multi-static configuration with transmitters and receivers placed around the scene in a full circle [52]. In Fig. 13.3 we have illustrated several ultra-narrowband circular SAR **k**-space sampling patterns achievable with various regular and random transmitter-receiver angular location sampling strategies when $N_f = 1$, $N_{tx} = 20$, and the total number measurements is $M = N_{tx}N_{rx}N_f = 600$.

In order to carry out simulations comparable to the mono-static case presented earlier, the carrier wavelength is reduced, such that spatial resolutions of the two configurations are approximately the same. In our simulations, each transmitter sends out an ultra-narrow-band waveform with a frequency that satisfies $\rho_x = \rho_y = 0.25\text{m} = \sqrt{2}/4 \cdot c/f_o$. The scene size is the same as in the mono-static experiments. Similar to the mono-static case, we use a scene with isotropic scatterers. For a wide-angular observation scenario, a more realistic simulation could consider angular anisotropy of the scatterers, however we do not consider that additional complication here.

In Fig. 13.8, we show the $t_{\%}$-average mutual coherence as a function of the number of transmitted probes when the total number of measurements is held fixed to $M = 600 = N_{tx}N_{rx}N_f$. All sampling configurations result in **k**-space patterns that deviate significantly from a regular **k**-space grid. This translates into significantly reduced coherence of configurations with a few transmitted probes and higher-reconstruction quality as compared to the mono-static case with the same number of transmitted probes. While random sampling was the key to improved performance in the mono-static case, the circular multi-static configuration is robust to transmit/receive sensor aspects. Hence regular undersampling performs almost as well as random undersampling in the multi-static sensing scenario considered here. In the multi-static case with multiple transmitters and receivers, reduction in the

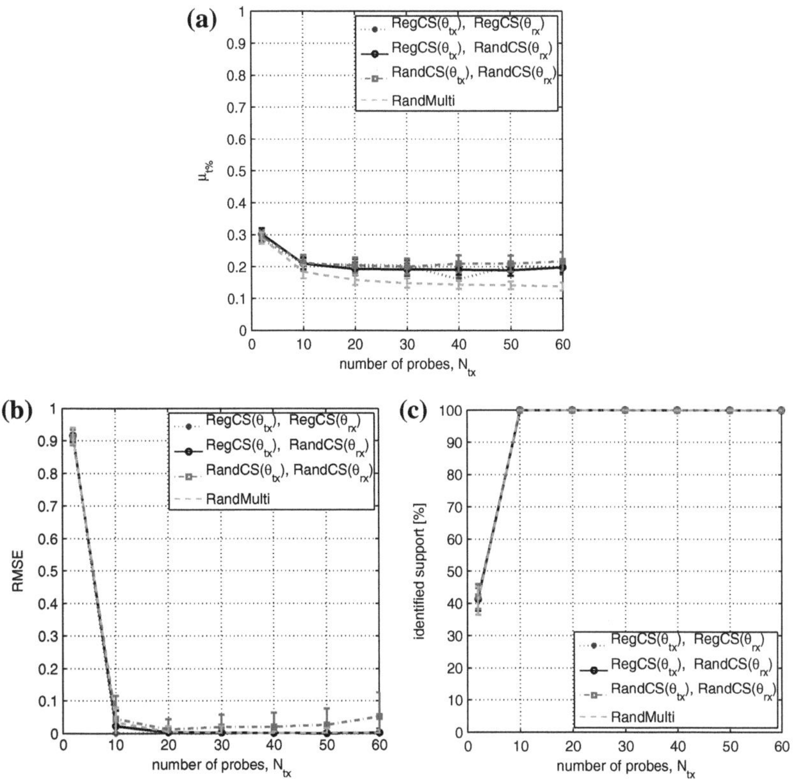

Fig. 13.8 Multi-static ultra-narrowband SAR performance versus the number of transmitted probes N_{tx} for various sensing configurations. In this experiment, the total number of measurements, $M = 600$, and the signal support size, $T = 140$, are held fixed. **a** The $t_{\%}$-average mutual coherence, $\mu_{0.5\,\%}$. **b** RMSE. **c** Percentage of correctly identified support of T largest estimated signal peaks

number of transmitted probes directly reduces the data acquisition time unlike the single-platform mono-static case, we do not have to wait for the SAR platform to fly through the eliminated aperture positions. Furthermore, when ultra-narrowband pulses used by different transmitters are non-overlapping in frequency, all transmitters can transmit simultaneously and the data can be acquired within the duration of data collection through a single transmitter receiver pair.

In Fig. 13.9, we show the $t_{\%}$-average mutual coherence and the image quality metrics as a function of the total number of measurements, when we set $N_{tx} = N_{rx}$ and use $T = 140$ point scatterers in the scene. We observe that different multi-static sampling patterns achieve similar performance. Similar to the mono-static case, the number of measurements required for very high-quality reconstructions is 4–5 times the number of scatterers in the scene, which in this case is around 600. This implies that sparsity-driven imaging can produce high-quality reconstructions with much smaller number of measurements than that would be used by conventional sensing and imaging.

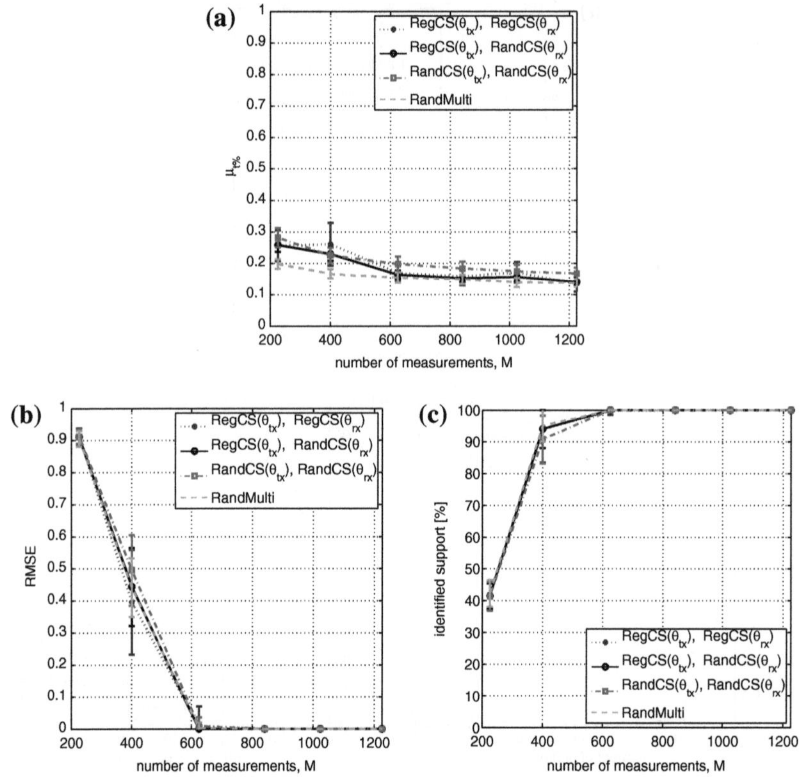

Fig. 13.9 Multistatic ultra-narrowband SAR performance versus number of measurements M for various sensing configurations. In this experiment, the number of transmitted probes $N_{tx} = N_f$ and the signal support size $T = 140$ are held fixed. **a** The $t_{\%}$-average mutual coherence, $\mu_{0.5\,\%}$. **b** RMSE. **c** Percentage of correctly identified support of T largest estimated signal peaks

13.8 Conclusion

In this chapter we have first provided a brief overview of a subset of the recent work applying sparsity and compressed sensing ideas in radar imaging. We have then focused on imaging from undersampled data, considering various mono-static and multi-static SAR measurement configurations for compressed sensing. Different regular and random data reduction approaches in each measurement configuration lead to different sampling patterns in the spatial frequency domain. In this context, we have presented the results of an experimental study analyzing the impact of such sampling patterns on the quality of reconstructed images of sparse scenes. We have shown that reconstructions of similar quality can be obtained using either wide-band mono-static or ultra-narrow-band multi-static configurations, effectively trading off frequency for geometric diversity. In the search for a quantitative measure that can potentially predict the expected reconstruction quality for a given sensing configuration prior to SAR data collection, we have proposed the $t_{\%}$-average mutual coherence. The $t_{\%}$-average mutual coherence is an easily computed parameter that can be used in real time design and evaluation of sensing configurations for, e.g., task planning of multi-mode radars. In both mono-static and multi-static cases, we have observed that configurations with sufficiently small values of the $t_{\%}$-average mutual coherence exhibit high-quality reconstruction performance. In the multi-static case, it is straightforward to obtain low coherence either by regular or random transmit/receive aspect sampling, whereas in the mono-static case randomness in the sampling pattern leads to lower coherence.

Our analysis shows that compressed sensing techniques when applied to SAR allow for reliable sparsity-driven imaging with dramatically reduced number of transmitted probes. In the mono-static case, compressed sensing and sparsity-driven reconstruction can enable reduced on-board data storage and sensing with a reduced number of transmitted probes relative to what is conventionally required. In the multi-static case, compressed sensing and sparsity-driven reconstruction can enable sensing not only with fewer transmitted probes, but also with reduced acquisition time as compared to conventional sensing and imaging.

References

1. Potter LC, Ertin E, Parker JT, Çetin M (2010) Sparsity and compressed sensing in radar imaging. Proc. IEEE 98:1006–1020
2. Çetin M, Karl WC (2001) Feature-enhanced synthetic aperture radar image formation based on nonquadratic regularization. IEEE Trans Image Process 10:623–631
3. Çetin M, Karl WC, Willsky AS (2006) Feature-preserving regularization method for complex-valued inverse problems with application to coherent imaging. Opt Eng 45(1):017003
4. Candes EJ, Romberg J, Tao T (2006) Robust uncertainty principles:exact signal reconstruction from highly incomplete frequency information. IEEE Trans Inf Theory 52(2):489–509
5. Donoho DL (2006) Compressed sensing. IEEE Trans Inf Theory 52(4):1289–1306
6. Donoho DL, Elad M, Temlyakov V (2006) Stable recovery of sparse overcomplete representations in the presence of noise. IEEE Trans Inf Theory 52(1):6–18
7. Lustig M, Donoho DL, Pauly JM (2007) Sparse MRI: the application of compressed sensing for rapid MR imaging. Magn Reson Med 58:1182–1195
8. Jakowatz CV, Wahl DE, Eichel PS, Ghiglian DC, Thompson PA (1996) Spotlight-mode synthetic aperture radar: a signal processing approach. Kluwer Academic Publishers, Norwell
9. Stojanovic I, Çetin M, Karl WC (2013) Compressed sensing of monostatic and multistatic SAR. IEEE Geosci Remote Sens Lett
10. Donoho DL, Elad M (2003) Optimally sparse representation in general (nonorthogonal) dictionaries via l1 minimization. Proc Nat Acad Sci 100(5):2197–2202
11. Gribonval R, Nielsen M (2003) Sparse representations in unions of bases. IEEE Trans Inf Theory 49(12):3320–3325
12. Fuchs J-J (2004) On sparse representations in arbitrary redundant bases. IEEE Trans Inf Theory 50(6):1341–1344
13. Malioutov DM, Çetin M, Willsky AS (2004) Optimal sparse representations in general overcomplete bases. Proc IEEE Int Conf Acoust Speech Signal Process 2: 793–796
14. Chen SS, Donoho DL, Saunders MA (1998) Atomic decomposition by basis pursuit. SIAM J Sci Comput 20:33–61

15. Daubechies I, Defrise M, De Mol C (2004) An iterative thresholding algorithm for linear inverse problems with a sparsity constraint. Commun Pure Appl Math 57(11):1413–1457
16. Figueiredo MAT, Nowak RD, Wright SJ (2007) Gradient projection for sparse reconstruction: application to compressed sensing and other inverse problems. IEEE J Sel Top Sign Proces 1(4):586–597
17. Kim S-J, Koh K, Lustig M, Boyd S, Gorinevsky D (2007) An interior-point method for large-scale l1-regularized least squares. IEEE J Sel Top Sig Process 1(4):606–617
18. Van den Berg E, Friedlander MP (2008) Probing the pareto frontier for basis pursuit solutions. SIAM J Sci Comput 31(2):890–912
19. Hale ET, Yin W, Zhang Y (2008) Fixed-point continuation for ℓ_1-minimization: methodology and convergence. SIAM J Optim 19:1107–1130
20. Wright SJ, Nowak RD, Figueiredo MAT (2009) Sparse reconstruction by separable approximation. IEEE Trans Sig Process 57(7):2479–2493
21. Mallat S, Zhang Z (1993) Matching pursuits with time-frequency dictionaries. IEEE Trans Sig Process 41(12):3397–3415
22. Tropp JA (2004) Greed is good: algorithmic results for sparse approximation. IEEE Trans Inf Theory 50(10):2231–2242
23. Candes EJ, Romberg J (2007) Sparsity and incoherence in compressive sampling. Inverse Prob 23(3):969–985
24. Donoho DL, Huo X (2001) Uncertainty principles and ideal atomic decomposition. IEEE Trans Inf Theory 47(7):2845–2862
25. Elad M (2007) Optimized projections for compressed sensing. IEEE Trans Sig Process 55(12):5695–5702
26. Samadi S, Çetin M, Masnadi-Shirazi MA (2009) Multiple feature-enhanced synthetic aperture radar imaging. In Zelnio EG, Garber FD (eds) Proceedings algorithms for synthetic aperture Radar imagery XVI. Proceedings SPIE, Orlando, FL, USA
27. Geman D, Yang C (1995) Nonlinear image recovery with half-quadratic regularization. IEEE Trans Image Process 4(7):932–946
28. Çetin M, Karl WC, Castañón DA (2003) Feature enhancement and ATR performance using non-quadratic optimization-based SAR imaging. IEEE Trans Aerosp Electron Syst 39(4):1375–1395
29. Çetin M, Lanterman A (2005) Region-enhanced passive radar imaging. IEE Proc Radar Sonar Navig 152(3):185–194
30. Moses RL, Potter LC, Çetin M (2004) Wide angle SAR imaging. In Zelnio EG, Garber FD (eds) Proceedings algorithms for synthetic aperture radar imagery XI. Proceedings SPIE, Orlando, FL, USA
31. Çetin M, Moses RL (2005) SAR imaging from partial-aperture data with frequency-band omissions. In Zelnio EG, Garber FD (eds) Proceedings algorithms for synthetic aperture radar imagery XII. Proceedings SPIE, Orlando, FL, USA, 2005
32. Ertin E, Austin CD, Sharma S, Moses RL, Potter LC (2007) GOTCHA experience report: three-dimensional SAR imaging with complete circular apertures. In Zelnio EG, Garber FD, (eds) Proceedings algorithms for synthetic aperture radar imagery XIV, volume 6568 of Proceedings SPIE, Orlando, FL, USA, April 2007
33. Stojanovic I, Çetin M, Karl WC (2008) Joint space aspect reconstruction of wide-angle SAR exploiting sparsity. In Zelnio EG, Garber FD (eds) Proceedings algorithms for synthetic aperture radar imagery XV volume 7337 of Proceedings SPIE, Orlando, FL, USA p 697005
34. Varshney KR, Çetin M, Fisher JW III, Willsky AS (2008) Sparse signal representation in structured dictionaries with application to synthetic aperture radar. IEEE Trans Sig Process 56(8):3548–3561
35. Tan X, Roberts W, Li J, Stoica P (2011) Sparse learning via iterative minimization with application to MIMO radar imaging. IEEE Trans Sig Process 59(3):1088–1101
36. Batu Ö, Çetin M (2008) Hyper-parameter selection in non-quadratic regularization-based radar image formation. In Zelnio EG, Garber FD (eds) Proceedings of algorithms for synthetic aperture radar imagery XV. Proceedings of SPIE, Orlando, FL, USA, March 2008

37. Önhon Ö, Çetin M (2012) A sparsity-driven approach for joint SAR imaging and phase error correction. IEEE Trans Image Process 21(4):2075–2088
38. Herman MA, Strohmer T (2009) High-resolution radar via compressed sensing. IEEE Trans Sig Process 57(6):2275–2284
39. Yoon YS, Amin MG (2008) Compressed sensing technique for high-resolution radar imaging. In Ivan Kadar (ed) Proceedings signal processing, sensor fusion, and target recognition XVII, volume 6968 SPIE Orlando, FL, USA
40. Baraniuk R, Steeghs P (2007) Compressive radar imaging. In: Proceedings IEEE radar conference, pp 128–133
41. Bhattacharya S, Blumensath T, Mulgrew B, Davies M (2007) Fast encoding of synthetic aperture radar raw data using compressed sensing. In: Proceedings IEEE 14th Workshop on Statistical signal processing, pp 448–452
42. Gürbüz C, McClellan J, Scott R Jr (2009) A compressive sensing data acquisition and imaging method for stepped frequency gprs. IEEE Trans Sig Process 57(7):2640–2650
43. Subotic NS, Thelen B, Cooper K, Buller W, Parker J, Browning J, Beyer H (2008) Distributed RADAR waveform design based on compressive sensing considerations. In Proceedings IEEE Radar Conference, p 1–6
44. Ender JHG (2010) On compressive sensing applied to radar. Sig Process 90(5):1402–1414
45. Stojanovic I, Karl WC, Çetin M (2009) Compressed sensing of mono-static and multi-static SAR. In Zelnio EG, Garber FD (eds) Proceedings algorithms for synthetic aperture radar imagery XVI volume 7337 of Proceedings SPIE, Orlando, FL, USA p 733705
46. Patel VM, Easley GR, Healy DM, Chellappa R (2010) Compressed synthetic aperture radar IEEE J Sel Top Sig Process 4(2): 244–254
47. Chen CY, Vaidyanathan PP (2008) Compressed sensing in MIMO radar. In: Proceedings 42nd asilomar conference on signals, systems and computers, pp 41–44 2008
48. Strohmer T, Friedlander B (2009) Compressed sensing for MIMO radar - algorithms and performance. In Proceedings asilomar conference on signals, systems and computers pp 464–468
49. Petropulu AP, Yu Y, Poor HV (2008) Distributed MIMO radar using compressive sampling. In: Proceedings asilomar conference on signals, systems and computers pp 203–207
50. Yu Y, Petropulu AP, Poor HV (2010) MIMO radar using compressive sampling. IEEE J Sel Top Sig Process 4(1):146–163
51. van den Berg E, Friedlander MP (2007) SPGL1: a solver for large-scale sparse reconstruction. http://www.cs.ubc.ca/labs/scl/spgl1
52. Himed B, Bascom H, Clancy J, Wicks MC (2001) Tomography of moving targets (TMT). In Fujisada H, Lurie JB, Weber K, (eds) Proceedings sensors, systems, and next-generation satellites V. vol 4540 of Proceedings SPIE, Toulouse, France pp 608–619

Chapter 14
Structured Sparse Bayesian Modelling for Audio Restoration

James Murphy and Simon Godsill

Abstract This chapter shows how sparse solutions can be obtained for a range of problems in a Bayesian setting by using prior models on sparsity structure. As an example, a model to remove impulse and background noise from audio signals via their representation in time-frequency space using Gabor wavelets is presented. A number of prior models for the sparse structure of the signal in this space are introduced, including simple Bernoulli priors on each coefficient, Markov chains linking neighbouring coefficients in time or frequency, and Markov random fields, imposing two dimensional coherence on the coefficients. The effect of each of these priors on the reconstruction of a corrupted audio signal is shown. Impulse removal is also covered, with similar sparsity priors being applied to the location of impulse noise in the audio signal. Inference is performed by sampling from the posterior distribution of the model variables using a Gibbs sampler.

14.1 Introduction

In many applications it is useful to represent a signal in terms of some set of basis functions. These might be useful to reveal certain structure in the signal or simply make the signal easier to store or process. In audio processing, for example, it is common to represent signals in terms of (local) frequency components. The transformation of the signal to the desired basis can be thought of as a regression problem, with the aim of determining the basis coefficients that best reconstruct the signal, for some meaning of 'best'. Some sets of basis functions will permit a range of different decompositions and in this case the problem can be thought of as an underdetermined regression problem with a range of possible solutions. The problem then becomes one of choosing a reconstruction that has desirable properties, for example sparsity of the decomposition, in which many basis coefficients are zero.

J. Murphy (✉) · S. Godsill
Department of Engineering, Cambridge University, Trumpington Street,Wuppertal, Cambridge
e-mail: jm362@cam.ac.uk

A. Y. Carmi et al. (eds.), *Compressed Sensing & Sparse Filtering*,
Signals and Communication Technology, DOI: 10.1007/978-3-642-38398-4_14,

Sparsity is a useful property for a number of reasons. Most simply, a sparse representation of a signal is efficient in terms of storage, which might allow more efficient signal processing. Some signals might be suspected of arising in a way that naturally leads to a sparse representation, with non-sparsity caused by the presence of corrupting noise. For example, the sound of a bell is likely to concentrate energy in relatively few frequency bands and only be present at certain times. In this case, if the sparse structure of the original source signal can be reconstructed, the signal can be accurately reconstructed without the corrupting noise. In other cases, having a signal represented by relatively few basis functions can be more revealing about the structure or source of that signal than having a representation composed of a wide spread of basis functions with small coefficients. In sparse representations, most basis coefficients will be zero, but the pattern of non-zero coefficients might be expected to have certain stucture, which, if incorporated into a model, can lead to more useful sparse representations. For example, in a signal arising from some occasional activity, non-zero coefficients might be expected to cluster together in times of activity and be zero at other times. By concentrating non-zero basis coefficients in these areas, a representation might be found that made activity clear and did not reconstruct random fluctuations in non-active periods. The idea of incorporating models of sparsity structure into signal representations (*structured sparsity*) will be the focus of this chapter.

The approach taken here focuses on explicitly modelling sparsity through the use of indicator variables ($\in \{0, 1\}$) that determine whether or not a particular regression component (basis function) is included in the signal representation. For the audio restoration example considered the aim is to reconstruct the true signal from a received input signal corrupted by noise. Signal reconstruction is tackled in probabilistic terms using a Bayesian approach, which requires a probabilistic model of the composition of the received signal in terms of the true signal to be reconstructed and the corrupting noise. This model allows the posterior probability distributions of the true signal and model parameters to be calculated and samples drawn from them, given the received signal and prior distributions for the model variables. If the model is a reasonably accurate reflection of reality this will yield a good estimate of the true signal. Such model based approaches make explicit any assumptions made about the structure of the signal and allow a prior structure to be imposed; whether or not this is an advantage depends largely on whether a model of the signal can sensibly be devised. In the methods described here, it is the prior structure on the indicator variables corresponding to the basis functions that express expectations of sparse structure in the signal representation.

This model-based approach is conceptually distinct from approaches that determine coefficients in such a way as to target sparsity directly, almost all of which attempt to limit or penalize the L_1 norm of the regression coefficients (i.e. the sum of the coefficient magnitudes). These include the basis-pursuit idea of [5] and the approximate greedy matching-pursuit algorithm of [15], which target a minimal L_1 representation, and the LASSO method of [22] which constrains the L_1 norm of the solution to be no greater than a certain value. A common mechanism to achieve these results is to treat the problem as one of optimization, introducing a *regularization* term consisting of a weighted penalty on the magnitude of the L_1 norm into the objective function. The minimal L_1 reconstruction of a signal has been shown to provide, with high probability, an exact reconstruction of a sparse signal under certain conditions; this is a key idea in compressed sensing, see, e.g. [3, 4, 6], amongst others. In a Bayesian setting, the targeting of sparsity in this way can be reproduced through the use of certain priors on regression coefficients, such as the Laplacian prior. Here, the use of indicator variables allows a different Bayesian approach to be taken, which, as Sect. 14.5 illustrates, allow further assumptions about the sparse structure of the coefficients to be explicitly modelled in a straightforward way.

The illustrative example used here is that of audio noise reduction accounting for two types of noise: background noise, which is assumed to be present at all times and have a Gaussian distribution, and impulse noise, caused, for example, by pops and clicks on a vinyl recording, and which is assumed to be present only occasionally and have a non-Gaussian distribution, with a wide range of scales. To cope with impulses, an indicator variable is associated with each audio sample and indicates whether or not an impulse is present during that sample. The reconstruction of the underlying signal uses an overcomplete set of Gabor basis functions, localized in the time and frequency, as the basis of a regression; structured sparsity is imposed on the coefficients of these through the use of a range of priors. Unlike previous techniques, the method here, based on that in [17], allows impulse and background noise to be removed *jointly* using a Bayesian, model-based approach.

In Sect. 14.2 the problem of audio restoration is introduced. Section 14.3 describes Gabor signal decomposition. Section 14.4 describes the Bayesian signal model used for restoration including many of the priors used in the model. In Sect. 14.5 priors are described for the modelling of sparse structure within the regression coefficients. Section 14.6 describes how the distributions of the variables in the model can be sampled via Gibbs sampling and derives the necessary conditional distributions for this. Section 14.7 presents some results from the model, particularly focussing on the sparsity structures obtained with various priors.

14.2 Audio Restoration

Noise reduction is an important component of audio restoration that aims to improve the perceived quality of corrupted audio signals. Early work in noise reduction can be found in [2], but the area continues to be active, e.g. [7, 12]. An overview of a range of methods can be found in [11] and the references therein, but alternative psychoacoustically-based approaches such as [13] are also popular. A technique common to many methods is the representation of the signal as a weighted sum of basis functions, with the aim being to reconstruct the true signal without reconstructing the noise. The basis functions used typically represent frequency components of the signal, localized in time. Such localized functions of varying frequencies are often called *wavelets* and a collection of such wavelets covering the full time span of the signal forms a *dictionary* of basis functions into which the original signal can be decomposed. There are many possible choices of wavelet dictionaries with various properties; the choice of one is likely to depend on the application in question. Such decompositions are useful for a range of audio processing tasks including noise reduction [21] and missing data interpolation [24]. In the audio application outlined here Gabor wavelets are used, though the method used to enforce sparsity can be used with any dictionary of basis functions.

Since the composition of audio signals varies with time, decomposition is performed in blocks on short sub-samples of the whole signal and, in order to reduce blocking effects, these sub-samples generally overlap. This *lapped* transform maps the signal from the time domain into a time-frequency plane, where time blocks overlap (see Fig. 14.2). This, combined with the fact that Gabor wavelets are not orthogonal, leads to multiple possible decompositions of the signal into the basis functions, a property known as *overcompleteness*. Which of these representations is best depends on the application. Sparse representations are frequently preferred as they give a parsimonious representation of the original signal, but even amongst these trade-offs can be made. For example, a representation that minimizes the number of non-zero coefficients might be best for compression, whilst one that has stronger temporal structure between components is likely to be better for missing data reconstruction. The method presented below allows a wide range of different modelling assumptions to be applied to the sparsity structure and is thus very flexible. It is based on the work of [17, 23, 24], in which the sparse reconstruction problem is formulated in a Bayesian setting, with the presence or absence of a particular basis function being treated as a model variable. The presence or absence of impulse noise in the signal is treated similarly. This formulation allows the straightforward incorporation of prior models of signal structure, meaning that expectations about likely sparsity structure can be embedded within the prior.

Observations of the data, combined with the model and the prior, allow a posterior density to be calculated for each of the model variables. In complicated Bayesian models involving many variables, the high dimensionality of the posterior and the fact that it can very often only be calculated up to proportionality means that its direct evaluation is usually infeasible. Sampling methods such as Markov chain Monte Carlo (MCMC) that allow samples to be drawn from the posterior distribution are thus commonly applied. MCMC methods set up a Markov chain with the target posterior distribution as its invariant distribution; by simulating from the chain for a sufficient number of samples, the invariant distribution can be reached (under very mild conditions) and a set of samples drawn from it that can be used to approximate the required posterior density. More details about MCMC methods can be found in, for example, [10].

The method described here allows the removal of both homogeneous background noise and impulse noise from an audio signal. Background noise is a common feature of many audio tracks and can arise from a number of sources, such as thermal noise in recording or processing equipment and is present, usually at the same scale, throughout the track. Here, following the work of [23] and [24], Gabor signal decomposition is used to remove homogeneous background noise. The idea behind this is that the Gabor coefficients can be found in such a way that they reconstruct the true signal features without reconstructing the noise.

Impulse noise, on the other hand, takes the form of large but brief deviations between the recorded value and the true signal. Such noise is often associated with old vinyl recordings and is usually perceived as audible pops or clicks, often caused by wear, dirt or scratches in vinyl tracks. Because it can derive from multiple sources and, in the case of vinyl recordings, involves uncontrolled deviation of the playback needle, impulse noise can vary across a very wide range of scales. Much previous impulse removal work has been carried out using autoregressive methods, described in [11], for example, though these can have a smoothing effect on the signal, acting as a form of low-pass filter and causing the loss of some high-frequency detail. Here, the work of [17] is followed in order to extend the noise removal method of [23] to allow the simultaneous removal of impulse and background noise.

14.3 Gabor Signal Decomposition

Gabor signal decomposition is the process of taking a signal and representing it as a weighted sum of *Gabor synthesis atoms* localized in time and frequency. A signal of length L can be decomposed into $M \times N$ Gabor synthesis atoms, representing M discrete frequency levels and N discrete time points, arranged as a grid. Such a transform maps a signal $x(t)$ onto an $M \times N$ time-frequency plane as shown in Fig. 14.2.

The Gabor synthesis atoms are defined in general by

$$\tilde{g}_{m,n}(t) = g\left(t - \frac{n}{N}L\right) \exp\left(2\pi i \frac{m}{M} t\right), \tag{14.1}$$

where $m \in \{0, 1, ..., M-1\}$, $n \in \{0, 1, ..., N-1\}$ and $t \in \{0, 1, ..., L-1\}$. The function g is the *Gabor window function*, typically a smooth bell-shaped window function with compact support that defines the temporal envelope of the corresponding Gabor atoms. Note that the Gabor synthesis atom in Eq. (14.1) has both real and imaginary components, allowing complex input signals to be reconstructed from these atoms. The method shown here uses a Hann window, defined as

$$g(t) = \begin{cases} 0.5 + 0.5\cos(2\pi t/\lambda) & |t| \leq \lambda/2 \\ 0 & |t| > \lambda/2 \end{cases} \tag{14.2}$$

where λ defines the window width, but many other choices are possible, including the Bartlett, Blackman, (truncated) Gaussian, Hamming, Kaiser, and Tukey windows, each centred at the parameter value. The width of the chosen window function must be such that it provides sufficient overlap between synthesis atoms (i.e. somewhat larger than L/N). The choice of window functions is discussed further in [8]. Figure 14.1 shows some examples of Gabor synthesis atoms.

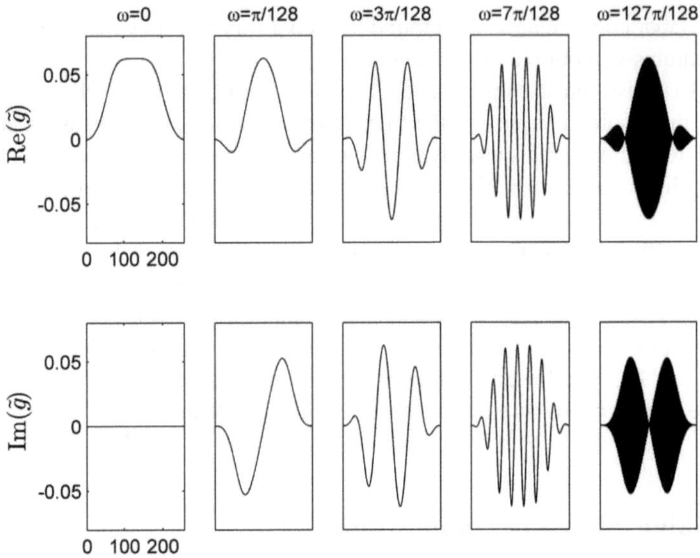

Fig. 14.1 A selection of Gabor synthesis atoms (real and complex parts) generated using a Hann window of width 256 (modified to generate basis functions forming a tight frame) with frequencies ω

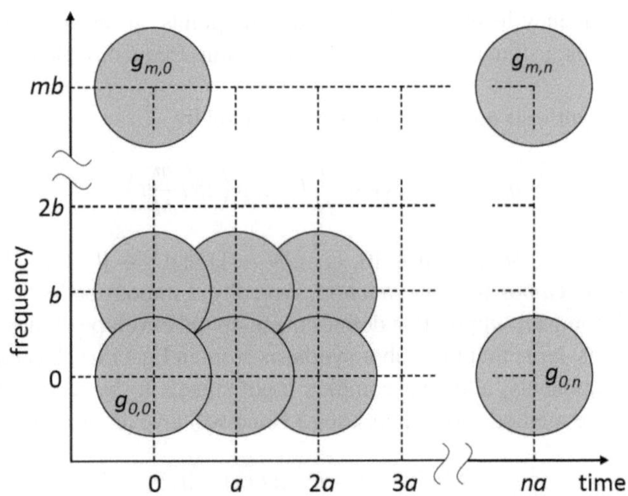

Fig. 14.2 A lapped transform, formed of overlapping atoms $g_{m,n}$ arranged in a regular grid in time-frequency space. These atoms form the basis for the representation of the signal in time-frequency space

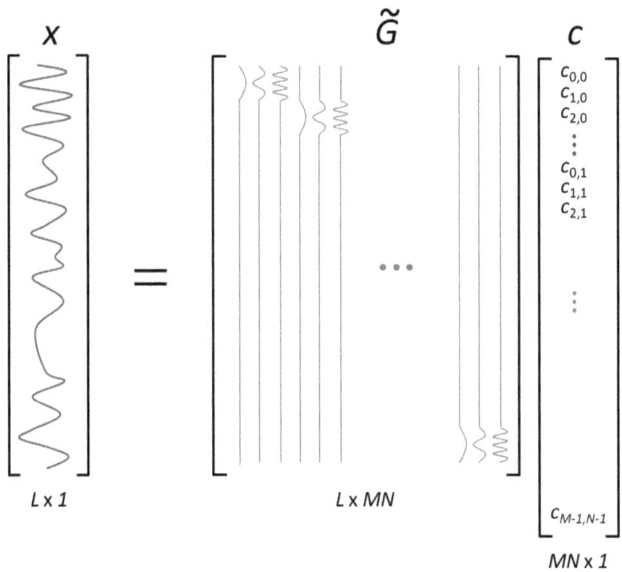

Fig. 14.3 Signal decomposition using a set of synthesis atoms can be thought of as regression aiming to reconstruct the signal x from the synthesis atoms. In matrix-vector form, each synthesis atom forms one column of the $\tilde{G}$ matrix. In the Gabor case, each atom has compact support and is a shifted version of the corresponding atom at the previous time location

Given a set of synthesis atoms $\tilde{g}_{m,n}(t)$, a given (complex) input signal $x(t)$ can be written as their weighted sum:

$$x(t) = \sum_{m=0}^{M-1} \sum_{n=0}^{N-1} \gamma_{m,n} c_{m,n} \tilde{g}_{m,n}(t), \tag{14.3}$$

where $c_{m,n} \in \mathbb{C}$ is the weighting coefficient for each atom and $\gamma_{m,n} \in \{0, 1\}$ are indicator variables that determine whether a particular atom is present in the decomposition. These are key to imposing sparse structure within the model and are discussed in detail in Sect. 14.5. Ignoring the $\gamma_{m,n}$ coefficients, this representation can be written in matrix-vector form as $x = \tilde{G}c$, where the input signal is represented as a column vector $x = [x(0)\ x(1)\ ...\ x(L-1)]^T$, $\tilde{G}$ is the $L \times MN$ *Gabor synthesis matrix*, consisting of the (m, n)th Gabor synthesis atom at each signal observation time as its $(m + nM)$th column, and the coefficient vector c is formed by stacking the individual coefficients $c_{m,n}$ in the appropriate order (see Fig. 14.3).

For decompositions in which the number of Gabor synthesis atoms is greater than the number of observations ($MN > L$), the system $x = \tilde{G}c$ is underdetermined with respect to the coefficients c. This will be the case in almost all real applications since redundancy in the Gabor dictionary is necessary in order to achieve good time-frequency localization. This is a consequence of the Balian-Low theorem [1, 14], which states that there is no well-concentrated Gabor basis in the critically sampled case where $MN = L$, discussed in more detail in [8] and [23]. The underdetermined system $x = \tilde{G}c$ can be solved via the *Gabor transform*, in which the coefficient of each atom is found by taking the inner product of that atom with the signal. Because atoms have compact support this can be performed efficiently using only the part of the signal that corresponds to the atom's region of support; [20] gives an algorithm for the discrete Gabor transform. Though this has the property that it recovers the coefficients that are minimal in an L_2 sense (i.e. they have minimal sum-of-squares), there is no guarantee of sparsity of coefficients. Indeed, this is unlikely in general as the L_2 norm will penalize the use of a few large coefficients as opposed to a larger number of smaller ones.

14.3.1 Real Input Signals

If the input signal is entirely real, as is the case with the audio signals considered here, the expansion on the right hand side of Eq. (14.3) must also be real. This can be arranged by setting $c_{m,n} = c^*_{M-m,n}$ and $\gamma_{M-m,n} = \gamma_{m,n}$ for all $m \in \{1, 2, ..., M/2\}$ (this relies on the assumption that M is even and that $\tilde{g}_{m,n} = \tilde{g}^*_{M-m,n}$, which can readily be shown from the definition of the Gabor synthesis atoms in Eq. (14.1)). In this case the decomposition in Eq. (14.3) can be written as

$$\begin{aligned} x(t) &= \sum_{m=0}^{M/2}\sum_{n=0}^{N-1} \gamma_{m,n}\alpha_m(c_{m,n}\tilde{g}_{m,n}(t) + c^*_{m,n}\tilde{g}^*_{m,n}(t)) \\ &= \sum_{m=0}^{M/2}\sum_{n=0}^{N-1} \gamma_{m,n}\left(\Re(\alpha_m c_{m,n})\Re(\tilde{g}_{m,n}(t)) - \Im(\alpha_m c_{m,n})\Im(\tilde{g}_{m,n}(t))\right), \end{aligned} \tag{14.4}$$

where α_m is 1 for all m except for $m = 0$ and $m = M/2$, when it is 1/2. This allows the decomposition to be reformulated in matrix-vector form using only real numbers by redefining $\tilde{G}$ and c as

$$\tilde{G} = \begin{bmatrix} \Re(\tilde{g}_{0,0}(0)) & \Im(\tilde{g}_{0,0}(0)) & \dots & \Re(\tilde{g}_{\frac{M}{2},N-1}(0)) & \Im(\tilde{g}_{\frac{M}{2},N-1}(0)) \\ \Re(\tilde{g}_{0,0}(1)) & \Im(\tilde{g}_{0,0}(1)) & \dots & \Re(\tilde{g}_{\frac{M}{2},N-1}(1)) & \Im(\tilde{g}_{\frac{M}{2},N-1}(1)) \\ \vdots & \vdots & \ddots & \vdots & \vdots \\ \Re(\tilde{g}_{0,0}(L-1)) & \Im(\tilde{g}_{0,0}(L-1)) & \dots & \Re(\tilde{g}_{\frac{M}{2},N-1}(L-1)) & \Im(\tilde{g}_{\frac{M}{2},N-1}(L-1)) \end{bmatrix} \tag{14.5}$$

$$c = \left[\Re(c'_{0,0}) - \Im(c'_{0,0})\Re(c'_{1,0}) - \Im(c'_{1,0})\dots\Re(c'_{\frac{M}{2},N-1}) - \Im(c'_{\frac{M}{2},N-1})\right]^T \tag{14.6}$$

where $c'_{m,n} = \alpha_m c_{m,n}$. Given these definitions, $\tilde{G}c$ will be the signal reconstruction in Eq. (14.4) ignoring the $\gamma_{m,n}$ indicators. The reconstruction including those indicators (as in Eqs. (14.3) and (14.4)) will be denoted R.

For practical purposes in what follows, the $c'_{m,n}$ coefficients will be treated as a two element vector of real numbers, representing the real and imaginary components of $c'_{m,n}$. This will be denoted $c_k \in \mathbb{R}^2$, with $k \in \{0, 1, ..., (M/2+1)N-1\}$ so that $c_k = c'_{m+nM}$ corresponds to $c'_{m,n}$. In the complex signal case, c_k will correspond to $c_{m,n}$ in the same way.

14.4 Bayesian Signal Model

The model of audio signals used for background noise reduction is as follows: at each sample time $t = 0, ..., L-1$ the received signal y_t is composed of the true signal $x(t)$ distorted by additive Gaussian noise v_t of scale σ_{v_t}, so that

$$y_t = x(t) + v_t, \tag{14.7}$$

with

$$v_t \sim \mathcal{N}(0, \sigma_{v_t}^2) \tag{14.8}$$

Homogenous background noise is modelled by having a constant noise scale across all samples, so that $\sigma_{v_t} = \sigma$ throughout, where σ is a parameter of the model that can be estimated. The noise model can also incorporate the presence of impulse noise in the received signal by allowing the scale of the noise process to increase in the presence of impulse noise. The noise scale is then given by

$$\sigma_{v_t}^2 = (1 + i_t \lambda_t)\sigma^2. \tag{14.9}$$

where $i_t \in \{0, 1\}$ is an indicator variable determining whether impulse noise is present at a particular sample time t and λ_t gives a scale for the impulse at that time if it exists. Thus the noise variance is σ^2 when no impulse is present and $(1 + \lambda_t)\sigma^2$ when it is.

A simple choice for λ_t is to set it to be constant, say λ. However, since impulsive noise can originate from a number of different physical sources, a single scale factor λ might not lead to a noise distribution sufficiently heavy-tailed to capture all impulses. Therefore the scale factor λ_t can be allowed to vary with time, giving an impulse scale at each sample time which can be estimated.

Although in principle many prior structures $p(\lambda_t)$ are possible for λ_t, a convenient one, as used in [12] in a different context, is a shifted inverse gamma model, the shape of which is shown in Fig. 14.4. This is a truncated and shifted version of the inverse gamma distribution and takes the following form:

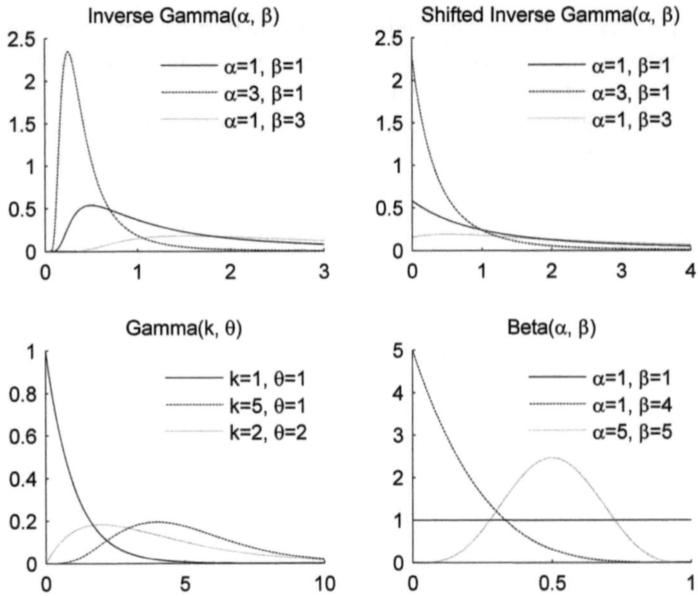

Fig. 14.4 Probability density functions for a number of priors used in the model with a selection of parameter values

$$p(\lambda_t) = \frac{\beta_\lambda^{\alpha_\lambda}(1+\lambda_t)^{-(\alpha_\lambda+1)}\exp(-\beta_\lambda/(1+\lambda_t))}{\gamma(\alpha_\lambda, \beta_\lambda)}, \quad \lambda \geq 0,$$
$$\propto \mathcal{IG}(1+\lambda_t; \alpha_\lambda, \beta_\lambda) \tag{14.10}$$

where $\mathcal{IG}(1+\lambda_t; \alpha_\lambda, \beta_\lambda)$ is the inverse gamma pdf with parameters α_λ and β_λ, evaluated at $1+\lambda_t$ and $\gamma(\alpha_\lambda, \beta_\lambda)$ is the lower incomplete gamma function defined as

$$\gamma(\alpha_\lambda, \beta_\lambda) = \int_0^\beta t^{\alpha-1}e^{-t}dt. \tag{14.11}$$

The Gabor-based inference mechanism described in [23] for noise removal is based on the model of homogenous background noise, but the introduction of impulse noise at different scales for some samples invalidates this assumption. In [17] this is dealt with by introducing an artificial latent process z_t with the required homogenous noise distribution such that

$$z_t = x(t) + w_t. \tag{14.12}$$

with $w_t \sim \mathcal{N}(0, \sigma^2)$. The original Gabor decomposition algorithm can then be used as a sampling step to sample the x_t variables corresponding to the true signal, conditioned on the z_t variables in Eq. (14.12) (rather than conditioning on the observations as in [23]). The observed process y_t is then given by

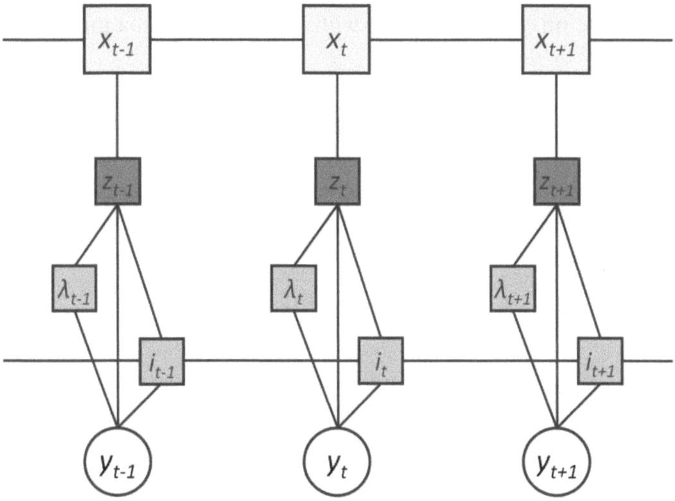

Fig. 14.5 Logical structure of model variables with the artificial latent z process introduced along with impulse indicators i and scale factors λ. This process is the true signal distorted by homogenous Gaussian noise (whereas y observations may be subject to noise at multiple scales)

$$y_t = z_t + i_t u_t, \tag{14.13}$$

with

$$u_t \sim \mathcal{N}(0, \lambda_t \sigma^2). \tag{14.14}$$

This structure is shown in Fig. 14.5 and has the property that the underlying true signal $x(t)$ is conditionally independent of the observations y and impulse indicators i, given the z process, so that

$$p(x \mid y, z, i, \lambda) \propto p(x \mid z), \tag{14.15}$$

where here un-subscripted variables have been used to indicate the full set of such variables (e.g. $i = \{i_t \mid t \in 0, \ldots, L-1\}$). This means that samples from the posterior distribution $p(x \mid z)$ can be drawn in the same way as in [23] (given by the weighted sum of synthesis atoms) but applying the model in [23] to the latent process z rather than directly to the input samples y. In the modified algorithm a sampling iteration consists of sampling both the z and x processes along with the other model variables and parameters. In fact, as shown in [18] it is possible to derive distributions for many of the model variables, including the x process *without* directly inferring the z process (i.e. by marginalizing it out of the inference). This marginalized approach can lead to faster convergence of the MCMC method used for sampling.

The Gabor synthesis coefficients $c_k \in \mathbb{R}^2$ can be expected to take a wide range of values, with very large values being comparatively common. The prior chosen for these variables is, therefore, a heavy-tailed Student t distribution, which can be realized as a scale mixture of normals with an inverse gamma mixing distribution, so that

$$p(c_k \mid \sigma_{c_k}, \gamma_k) = (1 - \gamma_k)\delta_0(c_k) + \gamma_k \mathcal{N}\left(c_k; 0, \sigma_{c_k}^2 I_2\right), \tag{14.16}$$

where ε_0 is a Dirac delta function centered at 0, which ensures that c_k is set to 0 when γ_k is 0, I_2 is the 2×2 identity matrix (for the case when $c_k \in \mathbb{R}^2$) and $\sigma_{c_k}^2$ is distributed according to the inverse gamma mixing distribution

$$p(\sigma_{c_k}^2 \mid \gamma_k = 1) = \mathcal{IG}(\sigma_{c_k}^2; \kappa, \nu_k). \tag{14.17}$$

Here κ is a shape parameter that determines the heaviness of the tails of the prior distribution. ν_k is a scale parameter that is itself assigned a gamma prior (see Fig. 14.4) so that

$$\nu_k = f(k)\nu, \tag{14.18}$$

with

$$\nu \sim \mathcal{G}(\alpha_\nu, \beta_\nu), \tag{14.19}$$

where $f(k)$ is a fixed weighting function that can be used to express a prior belief about the expected degree of smoothness in the reconstructed signal. The choice of $f(k)$ is discussed in more detail in [23], where the authors suggest using the reciprocal of the frequency modulation number m corresponding to the coefficient k.

The effect of the heavy-tailed prior on the coefficients c_k is to allow them to range very widely compared to a Gaussian prior with constant scale and means that the prior does not induce excessive smoothing across these coefficients. This is particularly important when the coefficients are sparse because then the signal will be represented by a relatively small number of coefficients and consequently it can be necessary for these to take large values.

14.5 Structured Sparsity

Prior distributions for the indicator variables (for both impulses and Gabor coefficients) are important components of the model. It is through these priors that a preference for sparsity can be incorporated, since they can encode a belief that sparse solutions are more likely than dense ones. Unlike methods that specifically seek a minimal solution in some norm, Bayesian inference does not inherently favour any particular solution unless that solution is more probable according to the modelling and prior assumptions and in light of the observations. The overcompleteness of the Gabor dictionary and the flexibility that this introduces means that without some sort of regularization there is a strong risk of overfitting the Gabor coefficients to the noisy signal; the modelling and prior assumptions are what prevent this.

The priors on the sets of indicator variables $\gamma = \{\gamma_{m,n} \mid \forall m, n\}$ and $i = \{i_t \mid t = 0, ..., L-1\}$ can be used to encode a prior belief that solutions will be sparse in terms of Gabor coefficients or impulses. However, in many cases further prior information about the structure of the non-zero indicators is available and it is desirable to incorporate this in the model via the indicator priors, leading to the idea of structured sparsity.

Consider the impulse process represented by the i variables, indicating the presence or absence of an impulse at a particular sample time. It is likely that impulses will be present in relatively few samples (i will be sparse) and this simple expectation can be incorporated into the prior in a straightforward way, through a prior belief that an indicator value of 0 (no impulse) is more likely than 1. A more sophisticated prior model could incorporate the belief that impulses will be relatively rare but, when they do occur, are likely to last for a number of samples, since the time taken to traverse a damaged section of record surface is likely to be longer than a single sample. In this case, the prior encodes a belief about the likely structure of the i process.

The simplest prior for i is to treat each i_t as a Bernoulli random variable with some prior probability p of a sample being subject to an impulse. This alone is sufficient to favour sparse solutions, since if p is small, a sparse solution is, all other things being equal, more likely than a dense one. Under these assumptions, the prior probability p indicates the proportion of samples that might be expected to be affected by impulse noise. The prior on the full set of indicators i in this case is given by

$$p(i \mid \phi_i) = \prod_{t=0}^{L-1} p(i_t \mid \phi_i), \tag{14.20}$$

where ϕ_i is the set of parameters for the prior on i. In the Bernoulli case this set contains only the prior probability $p \in [0, 1]$ of an indicator being 1, so that

$$p(i_t = 1) = p, \tag{14.21}$$
$$p(i_t = 0) = 1 - p. \tag{14.22}$$

Here a link can be made to penalized likelihood estimation, a common alternative method for finding sparse solutions. In such methods the sparse estimator is one that maximizes a version of the (log) likelihood function penalized according to the number of non-zero coefficients, with the strength of the penalty being determined by a penalty coefficient λ, chosen by the user. For the impulse indicator variable this can be expressed as

$$\hat{i}_{\text{PLE}} = \arg\max_{i} \log p(y \mid i) - \lambda \|i\|_0. \tag{14.23}$$

where $\|i\|_0$ is the number of non-zero elements of i.

The Bayesian posterior distribution of the indicator variables i given the observations is

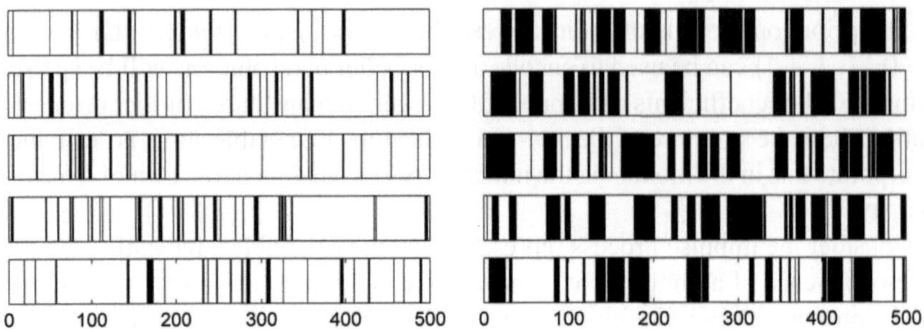

Fig. 14.6 Sample draws of length 500 from the Markov chain prior with $p_{00} = 0.95$, $p_{11} = 0.5$ for the left group of five draws and $p_{00} = 0.9$, $p_{11} = 0.9$ for the right group (*black* indicates a value of 1)

$$\log p(i \mid y) = \log p(y|i) + \log p(i) + C, \tag{14.24}$$

where C is constant with respect to i. For the Bernoulli prior above, this becomes

$$\log p(i \mid y) = \log p(y|i) + \log\left(\frac{p}{1-p}\right) \|i\|_0 + C', \tag{14.25}$$

and thus the penalized likelihood estimate in Eq. (14.23) is equivalent to a *maximum a posteriori* (MAP) estimate from the Bayesian model (that is, the estimate that maximizes the posterior density) with the Bernoulli prior, where $\lambda = \log(p/1-p)$. When $p < 0.5$ this is negative, resulting in a penalty term for additional non-zero coefficients. This gives an intuitive way of interpreting the penalty coefficient λ in the penalized likelihood estimator in terms of a Bayesian prior probability p that an impulse is present in any given sample.

The Bayesian formulation also allows further complexity to be built into the prior assumption in a simple and explicit way. In order to incorporate a belief that impulses, when they do occur, are likely to last for several samples, the prior for the impulse indicator can be modelled as a two-state Markov chain. The idea behind this is that in the 'no impulse' state the next state of the indicator process is very likely to also be 'no impulse', with only a small probability of a transition to the 'impulse' state. However, in the 'impulse' state, the next state is reasonably likely to also be 'impulse', with some probability of a transition back to 'no impulse'. Figure 14.6 shows some draws from such a Markov chain prior with different transition probabilities. In this case,

$$p(i \mid \phi_i) = p(i_0 \mid \phi_i) \prod_{t=1}^{L-1} p(i_t \mid i_{t-1}, \phi_i), \tag{14.26}$$

and the conditional distribution of a particular indicator i_t given the rest of the indicator process is given by

$$p(i_t \mid i_{-t}, \phi_i) \propto p(i_{t+1} \mid i_t, \phi_i) p(i_t \mid i_{t-1}, \phi_i), \tag{14.27}$$

where $p(i_{t+1} \mid i_t, \phi_i)$ is determined by the transition probabilities of the Markov chain (the notation i_{-t} refers to the set of all i indicators, excluding that at sample time t, i.e. $i_{-t} = i \backslash i_t$). The transition probabilities can be parameters of the model or can themselves be learnt. Two parameters are necessary to define the Markov chain transition matrix, the probability of remaining in state 0, and that of remaining in state 1 (the other entries in the transition matrix being calculable from these). By treating these as the probabilities of independent Bernoulli variables, they can be learnt from the data; further details are given in Sect. 14.6.

In general the inference methods given in Sect. 14.6 can use any conditional prior $p(i_t \mid i_{-t}, \phi)$ for the indicator variables. This is a very flexible class of possible prior functions and means that very many different forms of prior knowledge can be incorporated in this framework. Incorporating more structure in the prior can lead to less sparse results, since structural priors impose additional restrictions on the solution compared to simple Bernoulli priors. However, the ability to incorporate a more realistic prior model of the process is likely to lead to better results in many cases.

Similar prior structures can also be used for the Gabor coefficient indicators γ. A simple Bernoulli prior giving a prior probability for each atom being zero will lead to sparse solutions in time-frequency space though possibly with little structure between atoms, especially if the prior on the probability of a non-zero coefficient is small. This might be most suitable for compression, where minimizing the number of non-zero coefficients is paramount.

As with the impulse process i, Markov chain priors can be imposed in time, implying that frequency components have some tendency to remain consistent from one sample block to the next. Such a prior structure might be appropriate for signals expected to consist of slowly time-varying oscillations. In this case, a Markov chain prior is applied to the indicators γ_m for each frequency scale m. As with the impulse indicator prior, the transition probabilities can be estimated from the data as shown in Sect. 14.6. Similarly, a Markov chain structure can be imposed in the frequency direction, implying a prior expectation of local frequency clustering in each of the N sample blocks.

Another easy to implement prior for the Gabor coefficients is a Markov random field (MRF) prior. This can be used to impose two dimensional structure on the coefficients and such priors will tend to favour signals in which activity occurs in patches on the time-frequency plane. For the lattice of time-frequency indicator variables the MRF is arranged so that each indicator is connected to its four nearest neighbours corresponding to previous and subsequent times at the same frequency level and immediately lower and higher frequencies in the same time period (for $\gamma_{m,n}$ these are $\gamma_{m,n-1}$, $\gamma_{m,n+1}$, $\gamma_{m-1,n}$ and $\gamma_{m+1,n}$, respectively). Noting that $(2\gamma_k - 1)$ is 1 if $\gamma_k = 1$ and -1 if $\gamma_k = 0$, the conditional prior for the indicators is

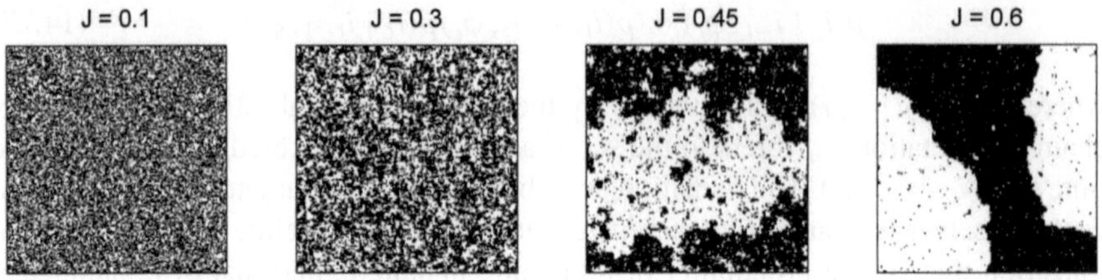

Fig. 14.7 Draws from the Ising model prior on a 200×200 grid with different values of the J parameter leading to structure at different scales. $K = 0$ for all these examples, since otherwise the favoured value predominates

$$p(\gamma_k \mid \gamma_{-k}, \phi_\gamma) \propto \exp\left(J \sum_{j \in \mathcal{N}(k)} (2\gamma_k - 1)(2\gamma_j - 1) + K(2\gamma_k - 1) \right), \quad (14.28)$$

where $\mathcal{N}(k)$ is the neighbourhood of γ_k. With $K = 0$, this prior is the Ising model and reflects the belief that two neighbouring indicators are more likely to be the same than they are to be different. This model was originally proposed in the physics literature as a model of ferromagnetism [19], but later (and then commonly) applied in statistics [9]. The J parameter can be thought of as an 'inverse temperature', which, at low values leads to disordered 'random'-looking states and at high values leads to more emphasis being placed on consistency between neighbours and hence favours stronger patterns with clearer 'patches'. The K parameter can be used to favour patterns in which one value (0 or 1) is expected to be more likely for γ_k than the other. Negative values of K act as a penalty on non-zero indicators and so help to induce sparsity in the solutions. Some draws from this prior model with various parameter values are shown in Fig. 14.7.

That the prior is only known to proportionality does not matter, since it will be the ratio of this prior for $\gamma_k = 1$ and $\gamma_k = 0$ that is of interest. This can be calculated as

$$\frac{p(\gamma_k = 0 \mid \gamma_{-k}, \phi_\gamma)}{p(\gamma_k = 1 \mid \gamma_{-k}, \phi_\gamma)} = \exp\left(2J \left(|\mathcal{N}(k)| - 2 \sum_{i \in \mathcal{N}(k)} \gamma_i \right) - 2K \right), \quad (14.29)$$

where $|\mathcal{N}(k)|$ is the size of the neighbourhood of γ_k (which will be 4, except at the edges of the lattice). More complicated MRF models, for example with more or different connectivity between lattice elements, can also be used.

Each of the priors proposed in this section has a conditionally Markovian structure so that

$$p(\gamma_k \mid \gamma_{-k}, \phi_\gamma) = p(\gamma_k \mid \gamma_{\mathcal{N}(k)}, \phi_\gamma). \quad (14.30)$$

These models are easy to use and flexible, although, as with the impulse indicators, the inference method can make use of any other conditional structure prior $p(\gamma_k \mid \gamma_{-k}, \phi_\gamma)$.

14.6 Inference

The joint posterior distribution of the Gabor reconstruction variables (c, σ_c, ν, γ and ϕ_γ), latent process variables (z), impulse process variables (i, ϕ_i and λ) and noise scale parameter (σ) can be sampled using a Markov chain Monte Carlo sampling procedure. In particular, a Gibbs sampler [9, 10] is used so that blocks of variables can be sampled from their conditional distributions given all other model variables. By iterating over all variables in the system, samples can be drawn from a Markov chain having the posterior distribution of the variables given the observations as its invariant distribution. As with all MCMC methods, convergence to the invariant distribution from the initial values will take a number of steps (which is, in general, difficult to determine), so an initial *burn-in* period during which samples are discarded is necessary. The flexibility of Gibbs sampling means that these variables can be sampled from the conditional distributions given below in any order.

In general, the conditional distributions given in this section are derived by noting that the variable under consideration v is only conditionally dependent on a subset of the other variables $V \subset \Omega$, where Ω is the set of all variables in the model. Conditioned on these, an application of Bayes' rule leads to a formulation such as

$$p(v \mid \Omega_{-v}) = p(v \mid V) \propto p(V_1 \mid v, V_2)p(v \mid V_2), \tag{14.31}$$

where $V = V_1 \cup V_2$. In this formulation, $p(V_1 \mid v, V_2)$ can be thought of as a likelihood of V_1 given v (under parameters V_2) and $p(v \mid V_2)$ can be thought of as a (conditional) prior on v (also under V_2). In the noise reduction model, most cases permit the use of priors on v that are conjugate with the corresponding 'likelihood', meaning that the conditional posterior has the same form as the prior with parameters that can be found in closed form. This is advantageous because the resulting conditional distributions can be efficiently sampled.

14.6.1 Background Noise Reduction

The variables involved in estimating the Gabor reconstruction of the signal, σ_c, c, γ, ν, σ and ϕ_γ, can each be sampled as steps in a Gibbs sampler using the conditional distributions given in this section.

Sampling σ_c^2: Conditioned on the other model variables, the distribution of $\sigma_{c_k}^2$ is given by

$$\begin{aligned} p(\sigma_{c_k}^2 \mid c, \sigma_{-c}, \sigma, \nu, \gamma, i, z, \lambda, y, \phi) &\propto p(c_k \mid \sigma_{c_k}, \gamma_k)p(\sigma_{c_k} \mid \nu_k) \\ &= \mathcal{IG}\left(\gamma_k + \kappa, \gamma_k \frac{\|c_k\|^2}{2} + \nu_k\right) \end{aligned} \tag{14.32}$$

This can be seen by considering the prior for c_k from Eq. (14.16) and the prior for $\sigma_{c_k}^2$ from Eq. (14.17) and noting that if $\gamma_k = 0$, the c_k prior is a delta function centred on the current value of c_k. This means that when $\gamma_k = 0$, $\sigma_{c_k}^2$ can be drawn from its prior distribution, $\mathcal{IG}(\kappa, \nu_k)$. If $\gamma_k = 1$ then the prior for $\sigma_{c_k}^2$ is the conjugate prior for the distribution $p(c_k \mid \sigma_{c_k}, \gamma_k = 1) = \mathcal{N}\left(c_k; 0, \sigma_{c_k}^2 I_2\right)$ from Eq. (14.16) and the resulting distribution is therefore also inverse gamma with the parameters shown in Eq. (14.32), owing to the fact that c_k consists of two elements.

Sampling c_k and γ_k: The conditional distribution of the Gabor coefficients c is a multivariate Gaussian, due to the assumption of Gaussian noise with variance σ^2 in the z process. For a single c_k coefficient a sample can be drawn jointly with the corresponding indicator variable γ_k. The joint conditional distribution can be decomposed as

$$p(c_k, \gamma_k \mid c_{-k}, \gamma_{-k}, z) = p(c_k \mid \gamma, c_{-k}, z)p(\gamma_k \mid \gamma_{-k}, c_{-k}, z), \tag{14.33}$$

where here (and throughout this section) dependence of all terms on σ and σ_c has been dropped from the notation for brevity. In the case when $\gamma_k = 0$, the first term on the right hand side is simply a delta function at c_k, owing to the prior on c_k in Eq. (14.16) and therefore

$$p(c_k, \gamma_k = 0 \mid c_{-k}, \gamma_{-k}, z) = p(\gamma_k = 0 \mid \gamma_{-k}, c_{-k}, z). \tag{14.34}$$

On the other hand, when $\gamma_k = 1$,

$$\begin{aligned} p(c_k \mid \gamma_k = 1, \gamma_{-k}, c_{-k}, z) &\propto p(z \mid c, \gamma_k = 1, \gamma_{-k})p(c_k \mid \gamma_k = 1) \\ &\propto \mathcal{N}\left(z; R, \sigma^2 I_L\right) \mathcal{N}\left(c_k; 0, \sigma_{c_k}^2 I_2\right), \end{aligned} \tag{14.35}$$

with the first term on the right coming from the fact that z is the reconstructed signal R (given by the reconstruction in Eq. (14.3)) corrupted at every sample by independent additive Gaussian noise with zero mean and variance σ^2, as described in Eq. (14.12). Through algebraic manipulation of the expression for the density of a multivariate Gaussian distribution, this first normal distribution can be expressed instead as a bivariate Gaussian in terms of c_k, which can similarly be combined with the second Gaussian distribution to give a Gaussian distribution in terms of c_k, giving

$$p(c_k \mid \gamma_k = 1, \gamma_{-k}, c_{-k}, z) = \mathcal{N}\left(c_k; \mu_k, \sigma^2 \Sigma_k\right) \tag{14.36}$$

with

$$\Sigma_k = \left(\tilde{G}_k^T \tilde{G}_k + \frac{\sigma^2}{\sigma_{c_k}^2} I_2\right)^{-1}, \tag{14.37}$$

$$\mu_k = \Sigma_k \tilde{G}_k^T (z - R_{-k}), \tag{14.38}$$

where R_{-k} is the reconstruction of the true signal given in Eq. (14.3) with the kth atom excluded. The proportionality in Eq. (14.35) can be replaced with an equality in Eq. (14.36) because both sides in that equation are probability distributions with respect to c_k, so must be normalized. The expression for $p(\gamma_k \mid \gamma_{-k}, c_{-k}, z)$ in Eq. (14.33) cannot be directly evaluated, since the dependency between z and γ_k depends on the value of c_k, and so it is necessary to consider the joint distribution of γ_k and c_k, integrated over all c_k, i.e.

$$\begin{aligned} p(\gamma_k \mid \gamma_{-k}, c_{-k}, z) &= \int p(\gamma_k, c_k \mid \gamma_{-k}, c_{-k}, z) \mathrm{d}c_k \\ &\propto p(\gamma_k \mid \gamma_{-k}) \int p(z \mid c, \gamma) p(c_k \mid \gamma_k) \mathrm{d}c_k. \end{aligned} \tag{14.39}$$

The constant of proportionality here is $p(z \mid \gamma_{-k}, c_{-k})$ and, as this does not depend on γ_k is the same for both its possible values. Therefore, it suffices to determine the ratio τ_k between these terms when $\gamma_k = 0$ and $\gamma_k = 1$, i.e.

$$\tau_k = \frac{p(\gamma_k = 1 \mid \gamma_{-k}, c_{-k}, z)}{p(\gamma_k = 0 \mid \gamma_{-k}, c_{-k}, z)}, \tag{14.40}$$

and use the fact that the numerator and denominator in Eq. (14.44) sum to 1 to give

$$p(\gamma_k = 0 \mid \gamma_{-k}, c_{-k}, z) = \frac{1}{1 + \tau_k}, \tag{14.41}$$

$$p(\gamma_k = 1 \mid \gamma_{-k}, c_{-k}, z) = \frac{\tau_k}{1 + \tau_k}. \tag{14.42}$$

The ratio τ_k is given by

$$\tau_k = \frac{p(\gamma_{k=1} \mid \gamma_{-k}) \int p(z \mid c, \gamma_{k=1}, \gamma_{-k}) p(c_k \mid \gamma_{k=1}) \mathrm{d}c_k}{p(\gamma_{k=0} \mid \gamma_{-k}) \int p(z \mid c, \gamma_{k=0}, \gamma_{-k}) p(c_k \mid \gamma_{k=0}) \mathrm{d}c_k}. \tag{14.43}$$

The expressions inside the integrals are given in the same way as Eqs. (14.36) and (14.34) above, but here attention must be paid to the normalizing constants of these distributions since the numerator and denominator of this ratio are not probability distributions for c_k and so do not normalize to 1 with respect to c_k. Further algebraic manipulation of these probability distributions leads to the expression for τ_k

$$\tau_k = \frac{p(\gamma_k = 1 \mid \gamma_{-k})}{p(\gamma_k = 0 \mid \gamma_{-k})} \frac{\sigma^2}{\sigma_{c_k}^2} |\Sigma_k|^{\frac{1}{2}} \exp\left(\frac{\mu_k^T \Sigma_k^{-1} \mu_k}{2\sigma^2} \right), \tag{14.44}$$

which allows the γ_k and c_k to be sampled by first sampling γ_k as a Bernoulli sample with probabilities given by Eqs. (14.41) and (14.42) and then, if this sample for γ_k is 1, sampling c_k from the Gaussian distribution in Eq. (14.36), but otherwise setting it to zero. Note that these distributions apply to the case where the signal is constrained to be real valued and hence $\Sigma_k \in \mathbb{R}^{2\times 2}$, $\mu_k, c_k \in \mathbb{R}^2$, and $\tilde{G}_k \in \mathbb{R}^{L\times 2}$ (given by the corresponding columns of the $\tilde{G}$ matrix in Eq. (14.5)).

It is also possible to sample from the complete joint conditional distribution for all c at once. This is derived in Appendix A.2 of [23], although sampling from the resulting multivariate Gaussian distribution can be computationally prohibitive in the case of long time series.

Sampling ν: The parameter ν that controls the scale of the prior for the σ_k can be updated by noting the conditional distribution of $\sigma_k^2/f(k)$ given ν and the parameter κ (which is taken to be a fixed model parameter) can be obtained from the distribution of $\sigma_{c_k}^2$ in Eq. (14.17) as

$$p\left(\frac{\sigma_k^2}{f(k)} \mid \kappa, \nu\right) = \mathcal{IG}\left(\frac{\sigma_k^2}{f(k)}; \kappa, \nu\right). \tag{14.45}$$

Since the gamma distributed prior for ν in Eq. (14.19) is a conjugate prior for this inverse gamma distribution with unknown scale parameter ν, the conditional distribution of ν given the other model variables can be expressed as a gamma distribution, given by standard results. However, sampler convergence is improved for sparse signals (where many γ_k are 0) if only $\sigma_{c_k}^2$ where the corresponding γ_k are non-zero are considered (since otherwise the scale of the c_k prior conveys little information from the data). This can be arranged in the Gibbs sampler by drawing a block sample from the joint distribution of ν and the $\sigma_{c_k}^2$ such that $\gamma_k = 0$ (written as $\sigma_{c_k}^{(r)} = \{\sigma_{c_k} : \gamma_k = r\}$, with $r \in 0, 1)$, i.e.

$$p(\nu, \sigma_k^{(0)} \mid \sigma_{c_k}^{(1)}, \kappa) = p(\sigma_{c_k}^{(0)} \mid \nu, \kappa) p(\nu \mid \sigma_{c_k}^{(1)}, \kappa), \tag{14.46}$$

with $p(\nu \mid \sigma_{c_k}^{(1)}, \kappa)$ being a gamma distribution due to prior conjugacy as described above, so that

$$p(\nu \mid \sigma_k^{(1)}, \kappa) = \mathcal{G}\left(\kappa|\gamma| + \alpha_\nu, \sum_{k:\gamma_k=1} \frac{f(k)}{\sigma_{c_k}^2} + \beta_\nu\right), \tag{14.47}$$

where $|\gamma|$ is the number of non-zero γ_k. The joint distribution can be sampled by drawing a sample of ν from the distribution in Eq. (14.47), followed by drawing the $\sigma_{c_k} \in \sigma_{c_k}^{(0)}$ from their prior in Eq. (14.18), given this new value of ν.

Sampling σ^2: The conditional distribution of the noise variance σ^2 can be found by observing that, conditional on the signal reconstruction given by the Gabor coefficients, the z process (with homogenous Gaussian noise of variance σ^2) can be treated as a series of observations of a Gaussian distributed random variable with unknown variance. In this situation, the prior on σ^2 can be chosen to be the inverse-Gamma conjugate prior with distribution $\mathcal{IG}(\frac{\alpha}{2}, \frac{\beta}{2})$. Considering the L samples in the z series, the conditional distribution for σ^2 is

$$p(\sigma^2|z, c, \gamma) = \mathcal{IG}\left(\sigma^2; \frac{L+\alpha}{2}, \frac{\|z-R\|^2+\beta}{2}\right). \tag{14.48}$$

where R is the reconstruction of the true signal given in Eq. (14.3). Unless something is known about the scale of the noise in advance, the parameters α and β should be chosen to give a vague prior on σ^2 (see Fig. 14.4).

Sampling ϕ_γ: The parameters of the indicator prior ϕ_γ depend on the prior structure chosen for γ. In Sect. 14.5, three possible prior structures were discussed: Bernoulli priors, Markov chain priors and Markov random field priors. The parameters for these priors are, respectively, the prior probability of a non-zero indicator p; two transition probabilities for the Markov chain, p_{00} and p_{11}; and the distribution temperature, J and value preference K from Eq. (14.28). In each of these Markovian cases, the distribution of the parameter(s) is given by

$$p(\phi_\gamma \mid \gamma) \propto p(\gamma \mid \phi_\gamma)p(\phi_\gamma) = p(\phi_\gamma)\prod_k p(\gamma_k \mid \gamma_{\mathcal{N}(k)}, \phi_\gamma). \tag{14.49}$$

- *Bernoulli*: In the Bernoulli case, the neighbourhood of each γ_k is empty ($\mathcal{N}(k) = \emptyset$) for all k and the 'likelihood' term inside the product is given simply by $p^{|\gamma|}(1-p)^{L-|\gamma|}$, where $|\gamma|$ is the number of non-zero elements of γ. The conjugate prior for this Bernoulli likelihood is the beta distribution $\mathcal{B}(\alpha_\gamma, \beta_\gamma)$, which, for $\alpha_\gamma = \beta_\gamma = 1$, gives a uniform prior (see Fig. 14.4). Using this prior it is possible to marginalize out the Bernoulli parameter p when calculating the ratio $\frac{p(\gamma_k=1|\gamma_{-k})}{p(\gamma_k=0|\gamma_{-k})}$ in the expression for τ_k in Eq. (14.44) (see [23], Appendix A.3). In this case, that ratio is given by

$$\frac{p(\gamma_k = 1 \mid \gamma_{-k})}{p(\gamma_k = 0 \mid \gamma_{-k})} = \frac{|\gamma_{-k}| + \alpha_\gamma}{K - |\gamma_{-k}| - 1 + \beta_\gamma}, \tag{14.50}$$

  where $|\gamma_{-k}|$ is the number of non-zero indicators, excluding γ_k and K is the total number of indicator variables in γ.
- *Markov chain*: For the Markov chain prior, the transition matrix for any particular chain can be fully determined by the probability of remaining in state 0, p_{00}, and the probability of remaining in state 1, p_{11} (since $p_{01} = 1 - p_{00}$ and similarly for p_{10}). For a given chain (e.g. linking indicators in the time direction at a particular frequency scale m), these can be estimated by treating them as independent Bernoulli variables with beta prior distributions. The initial distribution of the chain $p(\gamma_{m,0} \mid p_{00}^m)$ is taken to be the chain's stationary distribution. Then, for example for p_{00}^m,

$$\begin{aligned}p(p_{00}^m \mid \gamma_{m,\cdot}) &\propto p(\gamma_{m,\cdot} \mid p_{00}^m)p(p_{00}^m) \\ &\propto p(p_{00}^m)p(\gamma_{m,0} \mid p_{00}^m) \prod_{t:\gamma_{m,t-1}=0} p(\gamma_{m,t} \mid \gamma_{m,t-1}, p_{00}^m), \quad (14.51)\end{aligned}$$

where $\gamma_{m,\cdot}$ is the set of indicators for all time blocks at frequency scale m. Note that since the transition probabilities from state 0 are being considered as independent of those from state 1, only indicators whose predecessor is 0 need be considered. A similar expression can be derived for p_{11}^m.

Sampling can be performed using a Metropolis-within-Gibbs step. This is particularly convenient if a beta prior $\mathcal{B}(\alpha_{p_{00}^m}, \beta_{p_{00}^m})$ is applied to p_{00}^m and proposals p_{00}^{m*} are drawn from the full conditional distribution in Eq. (14.51) where the initial state instead has a fixed, uniform distribution (i.e. $p(\gamma_{m,0} \mid p_{00}^m) = p$), leading to a tractable proposal distribution defined as

$$\begin{aligned}q(p_{00}^{m*} \mid p_{00}^{m(i)}) &= p(p_{00}^{m*}) \prod_{t:\gamma_{m,t-1}=0} p(\gamma_{m,t} \mid \gamma_{m,t-1}, p_{00}^{m*}) \\ &= \mathcal{B}\left(|A_{00}^m| + \alpha_{p_{00}^{m*}}, |A_{01}^m| + \beta_{p_{00}^{m*}}\right), \quad (14.52)\end{aligned}$$

where $p_{00}^{m(i)}$ is the current value of p_{00}^m, and A_{00}^m is the set of times of transitions from 0 to 0, i.e.

$$A_{00}^m = \{t \mid \gamma_{m,t-1} = 0, \gamma_{m,t} = 0\}, \quad (14.53)$$
$$A_{01}^m = \{t \mid \gamma_{m,t-1} = 0, \gamma_{m,t} = 1\}. \quad (14.54)$$

Thus $|A_{00}^m|$ is the number of transitions from 0 to 0 and similarly $|A_{01}^m|$ is the number of transitions from 0 to 1. The acceptance ratio for the Metropolis-Hastings step is given by

$$\begin{aligned}p_{\text{accept}} &= \min\left(\frac{p(p_{00}^{m*} \mid \gamma_{m,\cdot})q(p_{00}^{m(i)} \mid p_{00}^{m*})}{p(p_{00}^{m(i)} \mid \gamma_{m,\cdot})q(p_{00}^{m*} \mid p_{00}^{m(i)})}, 1\right) \\ &= \min\left(\frac{p(\gamma_{m,0} \mid p_{00}^{m*})}{p(\gamma_{m,0} \mid p_{00}^{m(i)})}, 1\right) \quad (14.55)\end{aligned}$$

where the simplification here is due to the specific form of the proposal in Eq. (14.52). Finally, the initial state is assumed to be distributed according to the stationary distribution of the chain, which is given using the standard result from the theory of Markov chains,

$$p(\gamma_{m,0} \mid p_{00}^m) = \frac{1 - \gamma_{m,0}p_{00}^m - (1 - \gamma_{m,0})p_{11}^m}{2 - p_{00}^m - p_{11}^m}. \quad (14.56)$$

This allows the parameters of each Markov chain to be sampled efficiently.

- *Markov random field*: Estimation of the parameters J and K in Eq. (14.28) is greatly complicated by the fact that the constant of proportionality in that equation depends on their values and so cannot be ignored when sampling them. This means that straightforward Metropolis-within-Gibbs sampling cannot be performed without evaluating this proportionality constant for the current and proposed value, which is difficult. In this work fixed value have been used for these parameters (as in e.g. [9]), although some approximate Bayesian estimation methods for them are available, for example [16].

14.6.2 Impulse Noise Removal

In addition the removal of background noise as in [23], the method presented in this chapter allows the removal of impulse noise from a signal, following the method set out in [17]. The variables corresponding to the impulse process, i_t, z_t and λ_t, can be sampled within the Gibbs sampler as a block from their joint conditional distribution

$$p(i_t, z_t, \lambda_t \mid x, y, i_{-t}, z_{-t}, \lambda_{-t}, \sigma^2, \phi_i) = \\ p(z_t \mid i_t, \lambda_t, x_t, y_t, \sigma^2) p(\lambda_t \mid i_t, x_t, y_t, \sigma^2) p(i_t \mid i_{-t}, x_t, y_t, \sigma^2, \phi_i), \quad (14.57)$$

where x denotes the signal reconstruction from the Gabor synthesis atoms, with x_t denoting its value at the time of input sample t. A joint sample can be drawn by by sampling i_t, λ_t and z_t sequentially (in that order) from the distributions on the right of Eq. (14.57).

The distribution from which to sample i_t is given by

$$p(i_t \mid i_{-t}, x_t, y_t, \sigma^2, \phi_i) \propto p(i_t \mid i_{-t}, \phi_i) p(y_t \mid x_t, i_t, \sigma^2), \quad (14.58)$$

In the simple case where $\lambda_t = \lambda_{\text{fixed}}$ for all t, the observation likelihood is given by

$$p(y_t \mid x_t, i_t, \sigma^2) = \mathcal{N}\left(y_t; x_t, (1 + i_t \lambda_{\text{fixed}}) \sigma^2\right). \quad (14.59)$$

As the impulse indicator can only take one of two values, the distribution in Eq. (14.58) can be sampled directly by evaluating the expression for both $i_t = 0$ and $i_t = 1$ and normalizing to give the probabilities for a sample from a Bernoulli distribution as with the Gabor component indicators γ_k in Eqs. (14.39) and (14.40).

For non-constant impulse noise scale, the likelihood is given by

$$p(y_t \mid x_t, i_t, \sigma^2) = \begin{cases} \mathcal{N}\left(y_t \mid x_t, \sigma^2\right), & i_t = 0 \\ p(y_t \mid x_t, i_t = 1, \sigma^2), & i_t = 1. \end{cases} \quad (14.60)$$

Using the inverse gamma prior $p(\lambda_t)$ given in Eq. (14.10), it is possible to find $p(y_t \mid x_t, i_t = 1, \sigma^2)$ in closed form, as described in [12]:

$$
\begin{aligned}
p(y_t \mid x_t, i_t = 1, \sigma^2) &= \int_0^\infty p(y_t \mid \lambda_t, x_t, i_t = 1, \sigma^2) p(\lambda_t) d\lambda_t \\
&= \frac{1}{\sqrt{2\pi\sigma^2}} \frac{\gamma(\alpha_p, \beta_p)}{\gamma(\alpha_\lambda, \beta_\lambda)} \frac{\beta_\lambda^{\alpha_\lambda}}{\beta_p^{\alpha_p}},
\end{aligned} \tag{14.61}
$$

where

$$
\alpha_p = \alpha_\lambda + 1/2 \tag{14.62}
$$

$$
\beta_p = \beta_\lambda + \frac{(y_t - x_t)^2}{2\sigma^2}. \tag{14.63}
$$

Again, i_t can be sampled by evaluating the the distribution in Eq. (14.58) for the cases $i_t = 0$ and $i_t = 1$ and normalizing to get the probability for a Bernoulli sample.

With a sample drawn for i_t, λ_t can be drawn from the conditional distribution

$$
\begin{aligned}
p(\lambda_t \mid i_t, x_t, y_t, \sigma^2) &\propto p(y_t \mid x_t, \lambda_t, i_t, \sigma^2) p(\lambda_t) \\
&= \mathcal{N}\left(y_t \mid x_t, (1 + i_t\lambda_t)\sigma^2\right) p(\lambda_t) \\
&\propto \begin{cases} p(\lambda_t), & i_t = 0 \\ \mathcal{IG}\left(1 + \lambda_t; \alpha_p, \beta_p\right), & i_t = 1. \end{cases}
\end{aligned} \tag{14.64}
$$

For the inverse gamma prior $p(\lambda_t)$ given in Eq. (14.10), both distributions in the last line of Eq. (14.64) are shifted inverse gamma distributions (see Fig. 14.4) and can be sampled using a rejection sampling trick. First, a variable $l_t = 1 + \lambda_t$ is defined. This is then sampled from the inverse gamma distribution with the appropriate parameters. If the sampled value is less than 1, it is rejected and the variable resampled, otherwise it is accepted and 1 is subtracted from it to give a sample for λ_t. This can be shown to result in a sample from the required distribution.

Finally, once i_t and λ_t have been sampled, z_t can be sampled from the conditional distribution

$$
\begin{aligned}
p(z_t \mid i_t, \lambda_t, x, y) &\propto p(y_t \mid z_t, i_t) p(z_t \mid x_t) \\
&= \mathcal{N}(y_t \mid z_t, i_t\lambda_t\sigma^2) \mathcal{N}(z_t \mid x_t, \sigma^2) \\
&\propto \mathcal{N}\left(z_t \mid \frac{y_t + i_t\lambda_t x_t}{1 + i_t\lambda_t}, \frac{i_t\lambda_t\sigma^2}{1 + i_t\lambda_t}\right).
\end{aligned} \tag{14.65}
$$

Note that if $i_t = 0$ then $z_t = y_t$. A scheme that avoids sampling of the z variables can also be developed, as described in [18]. Since all the distributions of interest can be sampled directly, the Gibbs sampling scheme here is computationally efficient. The indicator variable prior parameters ϕ_i can be sampled in a similar way as those for the γ_k indicators described above (with Bernoulli or Markov chain priors).

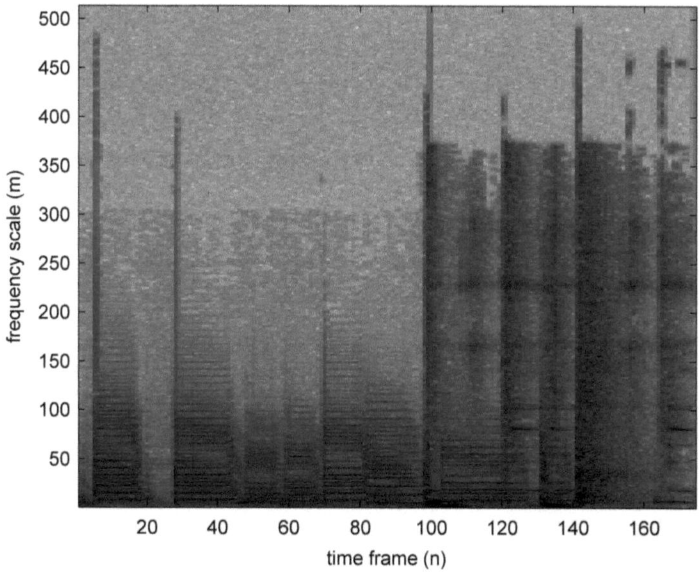

Fig. 14.8 The Gabor transform of the clean audio sample, showing log of coefficient magnitudes in time-frequency space (*dark areas* indicate high coefficient values)

14.7 Results

In this section the effects of the various priors are shown on the reconstruction of the signal, paying particular attention to the structure of the Gabor coefficients in each case. All results are shown for the same piece of audio data, an approximately 3s sample of music (from the song *Kalimba* by Mr. Scruff) sampled at 44.1 kHz, corrupted with artificially generated noise, generated using the noise model described in Sect. 14.4. This allows the reconstruction to be assessed against a clean 'ground truth' signal. To start the algorithm, both the final and noisy signal reconstruction (x_t and z_t, respectively) were initialized to the observed signal y_t.

Figure 14.8 shows the Gabor transform of the clean signal. Note that in this representation, every coefficient has a non-zero value, giving a completely dense representation of the signal in time-frequency space. Figures 14.10, 14.11, 14.12, 14.13 and 14.14 show sparse reconstructions of the signal derived using the range of priors described in Sect. 14.5. All other parameters and hyperparameters were kept the same between runs. The figures show the mean reconstruction of the signal over the MCMC samples drawn after the burn in period, alongside a single representative draw of the γ indicator variables. The intensity of the shading indicates the logarithm of the magnitude of the average signal reconstruction in terms of Gabor coefficients over that period. The burn in period was taken to be 100 samples, and a further 100 samples were used for reconstruction; the convergence results shown in Fig. 14.9

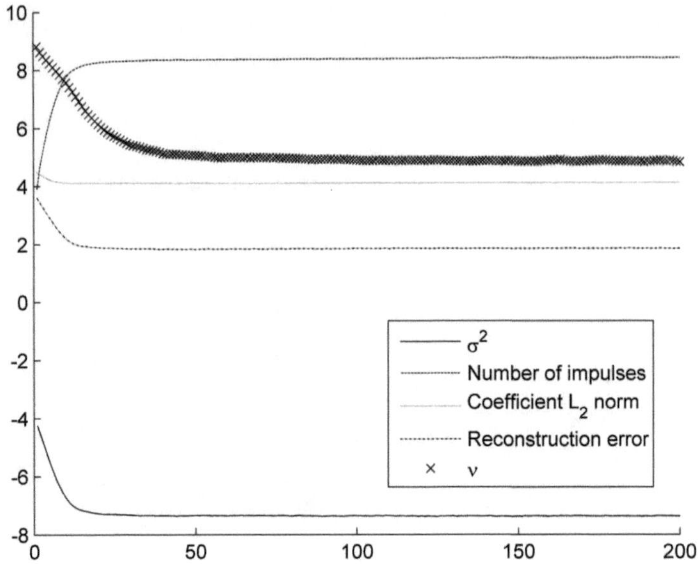

Fig. 14.9 Convergence of the log of a selection of variables with the iteration number for a typical run of the sampler (here run using Ising prior)

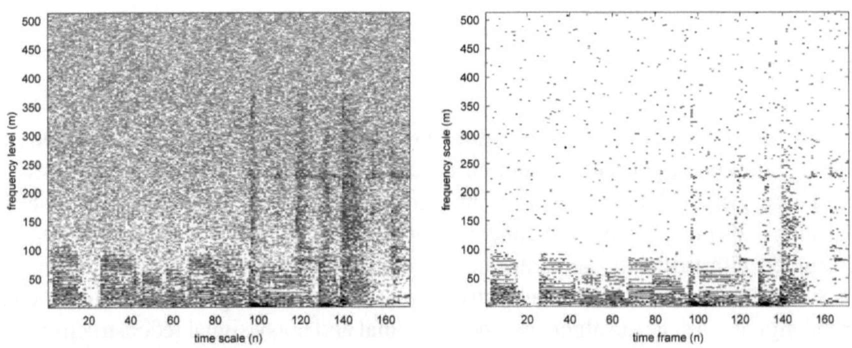

Fig. 14.10 Sparse reconstruction of the signal using a Bernoulli prior on γ with fixed $p(\gamma_k = 1) = 0.01$, showing the log mean reconstruction (*left*) and a single sample of the γ indicators (*right*)

indicate that this is a reasonable choice. Table 14.1 gives some statistics for each reconstruction.

Figures 14.10, 14.11, 14.12, 14.13 and 14.14 illustrate the key properties of the reconstructions derived using each of the priors. In Figs. 14.10 and 14.11, based on Bernoulli priors applied to each individual indicator variable without reference to their neighbours, the reconstruction is sparse (although less so in the case where the Bernoulli parameter is estimated), but with fairly randomly distributed non-zero coefficients in the reconstruction. The Markov chain priors used in the reconstructions in Figs. 14.12 and 14.13 look very different, with each showing strong patterns in

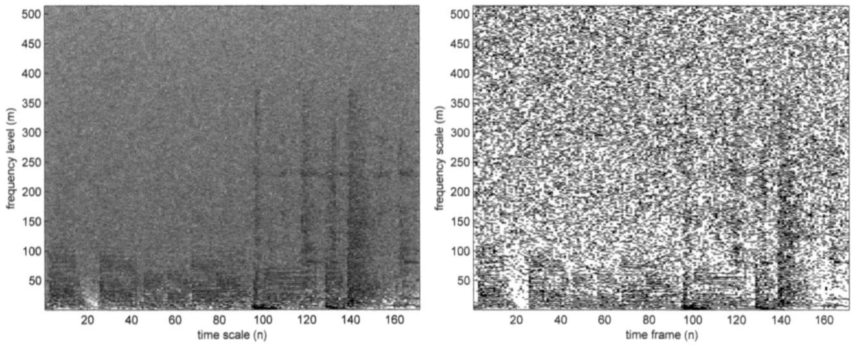

Fig. 14.11 Sparse reconstruction of the signal using a Bernoulli prior on γ with estimated prior probability of non-zero indicators (with beta prior, parameters $\alpha = 1$, $\beta = 40$, favouring low probabilities), showing the log mean reconstruction (*left*) and a single sample of the γ indicators (*right*)

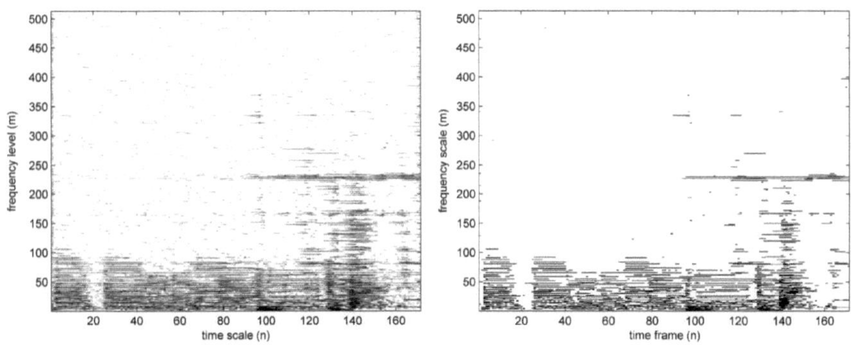

Fig. 14.12 Sparse reconstruction of the signal using a Markov chain prior in time on γ, with transition probabilities estimated from data (with uniform prior), showing the log mean reconstruction (*left*) and a single sample of the γ indicators (*right*)

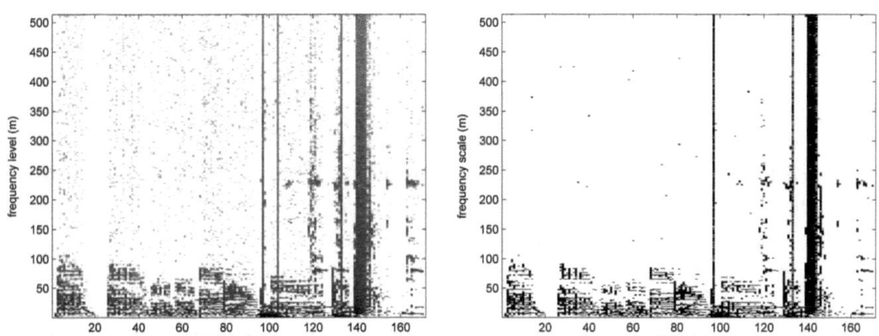

Fig. 14.13 Sparse reconstruction of the signal using a Markov chain prior in frequency on γ, with transition probabilities estimated from data (with uniform prior), showing the log mean reconstruction (*left*) and a single sample of the γ indicators (*right*)

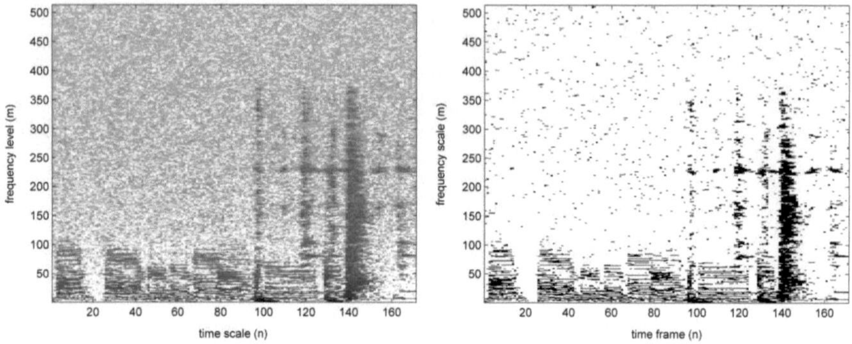

Fig. 14.14 Sparse reconstruction of the signal using an Ising model prior on γ, with $J = 0.5$ and $K = -0.1$, encouraging sparse, spatially cohesive solutions, showing the log mean reconstruction (*left*) and a single sample of the γ indicators (*right*)

Table 14.1 Statistics for the restorations produced by each prior type

Method	SNR(dB)	Non-zero γ	L_1 norm	L_2 norm
Gabor transform (true signal)	–	100%	3,061	39.5
Gabor transform (noisy signal)	–	100%	7,983	43.5
Bernoulli prior (fixed p)	15.3	4 %	1,765	60.7
Bernoulli prior (estimated p)	17.7	30 %	2,711	63.0
Markov chain (in time)	15.7	5 %	1,826	59.3
Markov chain (in frequency)	16.6	8 %	2,176	63.4
Ising prior	17.9	8 %	2,137	61.1

The signal to noise ratio (SNR) of the corrupted signal was 6.7 dB. The L_1 and L_2 norms quoted here refer to the mean norms of the reconstruction over all non-burn in samples

the expected direction (horizontal with the prior applied to time and vertical with it applied to frequency) and much less structure in the other direction. Finally, the Ising prior used to generate the reconstruction in Fig. 14.14 gives a reconstruction in which non-zero coefficients tend to cluster together, although without a directional bias, and with relatively few non-zero elements outside those clusters. These results are typical of the structures imposed by the respective priors and show how structure embedded in the prior has a substantial effect on the final reconstruction and its sparsity structure. Table 14.1 shows the expected relationship between the norms of the calculated Gabor transform and the sparse reconstructions: the transform has a smaller L_2 norm but larger L_1 norm that the sparse reconstructions. It also suggests that reconstruction using a structured prior such as a Markov chain or Ising model, can improve reconstruction performance as measured by the signal to noise ratio, whilst keeping the proportion of non-zero coefficients used in the reconstruction low.

Table 14.2 Signal to noise ratios of noisy and restored signals (in dB) before and after restoration using different impulse models

Impulse model	Impulse type	Noisy SNR	Final SNR
No impulse	Fixed λ	6.97	12.32
No impulse	Variable λ	6.96	9.65
No impulse	Real	6.61	8.83
Fixed λ	Fixed λ	6.97	13.83
Fixed λ	Variable λ	6.96	13.36
Fixed $\lambda = 15$	Real	6.61	11.54
Fixed $\lambda = 100$	Real	6.61	12.79
Variable λ	Fixed λ	6.97	13.36
Variable λ	Variable λ	6.96	13.47
Variable λ	Real	6.61	12.81

14.7.1 Impulse Removal

For signals containing impulses, their removal is crucial to good signal reconstrution. This can be seen from the results in Table 14.2, in which the results of restoration using different impulse models with a range of different types of impulse present are compared. For this evaluation, the same clean audio signal was taken and again corrupted with various types of additive noise at a signal to noise ratio (SNR) of around 7 dB. The tests shown used a selection of artificially generated and real impulse noise. For the artifical impulses, homogenous Gaussian noise was first added to a SNR of 15 dB followed by impulse noise with either a constant ('Fixed λ') or variable ('Variable λ') impulse variance. The former used a scale factor of $\lambda = 100$, while the latter had a scale factor with an inverse gamma distribution with parameters $\alpha = 1$ and $\beta = 20$, giving roughly the same SNR. For the 'Real' impulse noise, a suitable multiple of noise from the run-in track of an old vinyl recording was added to the signal, high-pass filtered to remove low-frequency distortions.

For each impulse model the SNR shown is that of the restored signal to the original signal, taken as the mean signal over the last 100 MCMC samples, after a 100 sample burn-in. Parameter and impulse estimates were generally observed to converge within this time (see Fig. 14.9). When using the fixed variance algorithm with variable variance impulses, the fixed impulse variance parameter λ was set to the mean impulse variance. Figure 14.15 shows the impulse detection and removal results for a small section of audio using the variable impulse size algorithm.

The results in Table 14.2 show that without an impulse model, the restoration performs poorly in the presence of impulses. For variable size and real impulses the variable impulse model performs well, with no need for tuning the value of λ, which has a considerable impact on performance of the fixed size impulse model.

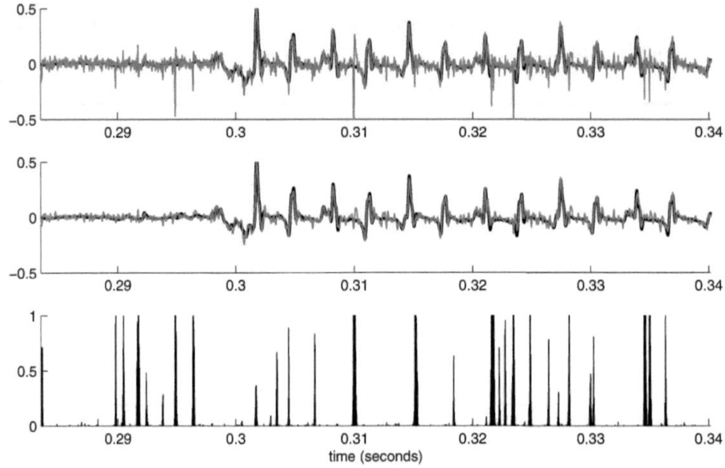

Fig. 14.15 A short excerpt showing the removal of impulses from a track using the variable impulse size algorithm. The top chart shows the noisy waveform (*grey*) superimposed on the clean signal (*heavy black*). The second chart shows the reconstructed waveform superimposed on the clean signal. The final chart shows the estimated posterior probability of an impulse being present at each location. (Reproduced from [17])

14.8 Conclusion

This chapter has shown how the idea of structured sparsity can be used in a Bayesian modelling context. The technique has been demonstrated with a number of different prior structures and, as an example of its use, has been applied to the problem of audio signal restoration. The effect on signal reconstruction of different prior models for the sparse signal structure has been demonstrated in this setting and seen to make a considerable difference to the results obtained.

Sparsity is often a useful property of a signal reconstruction, whether for compression or because the signal is thought to derive from some source that naturally leads to a sparse representation. Particularly in this latter case, prior knowledge about the expected structure of the sparsity pattern in the signal can aid in efficient representation of the signal and so it is desirable to incorporate this into any algorithm aiming to find such a sparse representation. Bayesian modelling of the type described in this chapter offers a principled way to do this and so can provide a powerful tool when designing methods that incorporate sparsity.

References

1. Balian R (1981) Un principe dincertitude fort en théorie du signal ou en mécanique quantique. CR Acad Sci Paris 292(2):1357–1361

2. Boll S (1979) Suppression of acoustic noise in speech using spectral subtraction. IEEE Trans Acoust Speech Signal Process 27(2):113–120
3. Candès EJ, Romberg J, Tao T (2006) Robust uncertainty principles: exact signal reconstruction from highly incomplete frequency information. IEEE Trans Inf Theory 52(2):489–509
4. Candès EJ, Tao T (2006) Near-optimal signal recovery from random projections: universal encoding strategies? IEEE Trans Inf Theory 52(12):5406–5425
5. Chen SS, Donoho DL, Saunders MA (2001) Atomic decomposition by basis pursuit. SIAM Rev 43(1):129–159
6. Donoho DL (2006) Compressed sensing. IEEE Trans Inf Theory 52(4):1289–1306
7. Erkelens JS, Heusdens R (2008) Tracking of nonstationary noise based on data-driven recursive noise power estimation. IEEE Trans Audio Speech Lang Process 16(6):1112–1123
8. Feichtinger HG, Strohmer T (1998) Gabor analysis algorithms: theory and applications. Birkhäuser, Boston
9. Geman S, Geman D (1984) Stochastic relaxation, Gibbs distributions and the Bayesian restoration of images. IEEE Trans Pattern Anal Mach Intell 6:721–741
10. Gilks WR, Gilks WR, Richardson S, Spiegelhalter DJ (1996) Markov chain Monte Carlo in practice. Chapman & Hall/CRC, London
11. Godsill SJ , Rayner PJW (1998) Digital audio restoration:a statistical model-based approach. Springer, Berlin (ISBN 3 540 76222 1, Sept 1998)
12. Godsill SJ (2010) The shifted inverse-gamma model for noise floor estimation in archived audio recordings. Appl Signal Process 90.991-999(Special Issue on Preservation of Ethnological Recordings)
13. Gustafsson S, Martin R, Jax P, Vary P (2002) A psychoacoustic approach to combined acoustic echo cancellation and noise reduction. IEEE Trans Speech Audio Process 10(5):245–256
14. Low F (1985) Complete sets of wave packets. A passion for physics-essays in honor of Geoofrey Chew. World Scientific, Singapore, pp 17–22
15. Mallat SG, Zhang Z (1993) Matching pursuits with time-frequency dictionaries. IEEE Trans sig process 41(12):3397–3415
16. McGrory CA, Titterington DM, Reeves R et al (2009) DM Titterington, R. Reeves, and A.N. Pettitt. Variational Bayes for estimating the parameters of a hidden Potts model. Stat Comput 19(3):329–340
17. Murphy J, Godsill S (2011) Joint Bayesian removal of impulse and background noise. In: Proceedings of the IEEE international conference on acoustics, speech and signal processing (ICASSP), pp 261–264
18. Murphy J, (2013) Sparse audio restoration in hidden states, hidden structures: bayesian learning in time series models, PhD Thesis, Cambridge University
19. Niss M (2005) History of the Lenz-Ising model 1920–1950: from ferromagnetic to cooperative phenomena. Arch Hist Exact Sci 59(3):267–318
20. Qian S, Chen D (1993) Discrete Gabor transform. IEEE Trans Sig Process 41(7):2429–2438
21. Soon IY, Koh SN, Yeo CK (1998) Noisy speech enhancement using discrete cosine transform. Speech Commun 24(3):249–257
22. Tibshirani R (1996) Regression shrinkage and selection via the lasso. J R Stat Soc Ser B (Methodol) 58: 267–288
23. Wolfe PJ, Godsill SJ, Ng WJ (2004) Bayesian variable selection and regularisation for time-frequency surface estimation. J R Stat Soc Ser B 66(3):575–589 Read paper (with discussion)
24. Wolfe PJ, Godsill SJ (2005) Interpolation of missing data values for audio signal restoration using a Gabor regression model. In: Proceedings of the IEEE international conference on acoustics, speech and signal processing (ICASSP), pp. 517–520

Chapter 15
Sparse Representations for Speech Recognition

Tara N. Sainath, Dimitri Kanevsky, David Nahamoo, Bhuvana Ramabhadran and Stephen Wright

Abstract This chapter presents the methods that are currently exploited for sparse optimization in speech. It also demonstrates how sparse representations can be constructed for classification and recognition tasks, and gives an overview of recent results that were obtained with sparse representations.

15.1 Introduction

Sparse representation techniques for machine learning applications have become increasing popular in recent years [1, 2]. Since it is not obvious how to represent speech as a sparse signal, sparse representations have received attention only recently from the speech community [3], where they were proposed originally as a way to enforce exemplar-based representations. Exemplar-based approaches have also found a place in modern speech recognition [4] as an alternative way of modeling observed data. Recent advances in computing power and improvements in machine learning algorithms have made such techniques successful on increasingly complex speech tasks. The goal of exemplar-based modeling is to establish a generalization

T. N. Sainath (✉) · D. Kanevsky · D. Nahamoo · B. Ramabhadran
IBM T. J. Watson Research Center, Yorktown Heights, NY, USA
e-mail: tsainath@us.lbm.com

D. Kanevsky
e-mail: kanevsky@us.lbm.com

D. Nahamoo
e-mail: nahamoo@us.lbm.com

B. Ramabhadran
e-mail: bhuvana@us.lbm.com

S. Wright
University of Wisconsin, Madison, WI, USA
e-mail: swright@us.lbm.com

A. Y. Carmi et al. (eds.), *Compressed Sensing & Sparse Filtering*,
Signals and Communication Technology, DOI: 10.1007/978-3-642-38398-4_15,

from the set of observed data such that accurate inference (classification, decision, recognition) can be made about the data yet to be observed the " unseen" data. This approach selects a subset of exemplars from the training data to build a local model for every test sample, in contrast with the standard approach, which uses all available training data to build a model before the test sample is seen.

Exemplar-based methods, including k-nearest neighbors (kNN) [1], support vector machines (SVMs) and sparse representations (SRs) [3], utilize the details of actual training examples when making a classification decision. Since the number of training examples in speech tasks can be very large, such methods commonly use a small number of training examples to characterize a test vector, that is, a *sparse representation*. This approach stands in contrast to such standard regression methods as ridge regression [5], nearest subspace [6], and nearest line [6] techniques, which utilize information about *all* training examples when characterizing a test vector.

An SR classifier can be defined as follows. A dictionary $H = [h_1; h_2 \ldots; h_N]$ is constructed using individual examples of training data, where each $h_i \in Re^m$ is a feature vector belonging to a specific class. H is an over-complete dictionary, in that the number of examples n is much greater than the dimension of each h_i (that is, $m \ll N$). To reconstruct a signal y from H, SR requires that equation $y \approx H\beta$, but imposes a sparseness condition on β, meaning that it requires only small number of examples from H to describe y. A classification decision can be made by looking at the values of β coefficients for columns in H belonging to the same class.

The goal of this chapter is to explain how sparse optimization methods can be exploited in speech, how sparse representation can be constructed for classification and recognition tasks, and to give an overview of results obtained using sparse representation.

15.1.1 Chapter Organization

The remainder of the chapter is organized as follows. The second section deals with mathematical aspects of sparse optimization. We describe two SR methods: approximate Bayesian compressive sensing (ABCS) [7] and convex hull extended Baum-Welch (CHEBW) [8]. We discuss too their relation with the Extended Baum-Welch (EBW) optimization framework [9].

The third section is concerned with a variety of different sparseness techniques employing different types of regularization [2, 3]. Following [10] we explore what type of sparseness regularization should be employed. Typically sparseness methods such as LASSO [11] and Bayesian compressive sensing (BCS) [12] use an l_1 sparseness constraint. Other possibilities include the Elastic Net [13], which uses a combination of an l_1 and l_2 (Gaussian prior) constraint, and ABCS [3], which uses an l_1^2 constraint, known as a Semi-Gaussian prior. We analyze the difference in the spareness objectives for the above methods and we compare the performance of these methods for phonetic classification in TIMIT.

In the fourth section, we explore the application of ABCS to phoneme classification task in TIMIT. The benefit of this Bayesian approach is that it allows us to build compressive sensing (CS) on top of other Bayesian classifiers, for example a Gaussian mixture model (GMM). It was shown, following [3], that the CS technique allows attaining an accuracy of 80.01 %, outperforming the GMM, kNN, and SVM methods.

In the fifth section, we describe a novel exemplar-based technique for classification problems, in which for every new test sample the classification model is re-estimated from a subset of relevant samples of the training data. We formulate the exemplar-based classification paradigm as a SR problem and explore the use of convex hull constraints to enforce both regularization and sparsity. Finally, we utilize the EBW optimization technique to solve the SR problem, and apply our proposed methodology for the TIMIT phonetic classification task, showing statistically significant improvements over common classification methods.

In the sixth section, following [14], we explore the use of exemplar-based SR to map test features into the linear span of training examples. Given these new SR features, we train a Hidden Markov Model (HMM) and perform recognition. On the TIMIT corpus, we show that applying the SR features on top of our best discriminatively trained system yields a reduction in phonetic error rate (PER) from 19.9 % to 19.2 %. In fact, after applying model adaptation we reduce the PER further to **19.0** %, which was the best result on TIMIT reported in 2011. Furthermore, on a large vocabulary 50-h broadcast news task, we achieve a reduction in word error rate (WER) of **0.3** %.

In the seventh section, following [15], we discuss using SRs to create a new set of sparse representation phone identification features (S_{pif}). We describe the S_{pif} features for both small and large vocabulary tasks. On the TIMIT corpus [16], we show that the use of SR in conjunction with our best context-dependent (CD) HMM system allows for a 0.7 % absolute reduction in phonetic error rate (PER), to 23.8 %. Furthermore, on a 50-h Broadcast News task [17], we achieve a reduction in word error rate (WER) of $0.9 - 17.8$ %, using the SR features on top of our best discriminatively trained HMM system.

In the eighth section we describe how one can improve sparse exemplar modeling for speech tasks via enhancing exemplar-based posteriors.

15.2 Sparse Optimization

Recent studies have shown that sparse signals can be recovered accurately using fewer observations than the Nyquist/Shannon sampling principle would imply. The emergent theory that brought this insight to light is known as compressive sensing (CS) [22, 23]. Problems of reconstructing signals from compressive sensing data can be represented in several equivalent ways. One such formulation is the following optimization problem:

$$\min_{\beta} \parallel y - H\beta \parallel_2 \quad \text{subject to} \quad \parallel \beta \parallel_1 \leq \epsilon, \tag{15.1}$$

where y is an m-dimensional vector, x is an N-dimensional vector, H is an $m \times N$ matrix. The parameter ϵ controls the sparsity of the recovered solution. Provided H satisfies certain properties, the signal β can be reconstructed even when the number of observations m is much less than the dimension N of the ambient space in which β resides. In fact, the required number of observations m is related more strongly to the number of nonzeros in β.

This formulation can be generalized to handle other types of sparse and regularized optimization. We can write

$$\min_{\beta} f(\beta) \quad \text{subject to} \quad \phi(\beta) \leq \epsilon, \tag{15.2}$$

where f and ϕ are typically convex functions mapping $\mathbb{R}^n$ to $\mathbb{R}$. Typically, f is a loss function or maximum likelihood function, while the regularization function ϕ is typically nonsmooth, and chosen so as to induce the desired type of structure in β. As noted above, the popular choice $\phi(\beta) = \|\beta\|_1$ induces sparsity into β. An alternative to (15.2) is the following weighted formulation:

$$\min_{\beta} f(\beta) + \lambda\phi(\beta), \tag{15.3}$$

for some parameter $\lambda \geq 0$. It can be shown that (15.2) and (15.3) are equivalent: Under certain assumptions on f and ϕ, the solution of (15.2) for some value of $\epsilon > 0$ is identical to the solution of (15.3) for some value of $\lambda \geq 0$, and vice versa.

We can generalize the formulations (15.2) and (15.3) further by considering nonconvex loss functions f and regularization functions ϕ, and adding an explicit constraint on the values of β. Nonconvex f arise in, for example, deep belief networks, in which the outputs are highly nonconvex functions of the parameters in the network. Nonconvex regularizers ϕ such as SCAP and MCP are sometimes used to avoid biasing effects associated with the use of convex penalties. Explicit constraints such as nonnegativity ($\beta \geq 0$) and simplex ($\beta \geq 0$ and $\sum_{i=1}^{n} \beta_i = 1$) are common in many settings.

Many algorithms have been proposed to solve (15.2) and (15.3), many of which exploit the particular structure of f and ϕ in various applications. One general approach that has been applied successfully in several settings is the *prox-linear* approach in which f in (15.3) is replaced by a linear approximation and a prox-term that discourages the new iterate β^{k+1} from being moved too far from the current iterate β^k. The subproblem to be solved at each iteration is:

$$\beta^{k+1} = \arg\min_{\beta} \nabla f(\beta^k)^T(\beta - \beta^k) + \frac{1}{2\alpha_k}\|\beta - \beta^k\|_2^2 + \lambda\phi(\beta), \tag{15.4}$$

where α_k is a positive parameter that plays the role of a line-search parameter. If the new iterate does not give satisfactory descent in the objective function of (15.3), we can decrease α_k and recompute a more conservative alternative value of β^{k+1}, repeating as necessary.

The approach based on (15.4) is potentially useful when (a) the gradient $\nabla f(\cdot)$ can be computed at reasonable cost and (b) the subproblem (15.4) can be solved efficiently. Both situations typically hold in compressed sensing, under the formulation (15.3) with $f(\beta) = \|H\beta - y\|_2^2$ and $\phi(\cdot) = \|\cdot\|_1$. In this situation, the solution of (15.4) can be computed in $O(n)$ operations.

In the remainder of this chapter, we consider two fundamental methods for sparse optimization: an extended Baum-Welch (EBW) method (which can be expressed via a line-search $\mathcal{A}$-function (LSAF)) and an Approximate Bayesian Compressive Sensing (ABCS) algorithm, which is also closely related to EBW. The LSAF derivation is closely related to the prox-linear approach described above; in fact, the $\mathcal{A}$-function can be thought of as a generalization of the simple quadratic approximation to f that is used in (15.4).

Both EBW and ABCS have been applied to speech classification and recognition problems, as we discuss in subsequent sections.

15.2.1 An EBW Compressed Sensing Algorithm

The Extended Baum-Welch (EBW) technique was introduced initially for estimating the discrete probability parameters of multinomial distribution functions of HMM speech recognition problems under the Maximum Mutual Information discriminative objective function [24]. Later, in [25], EBW was extended to estimating parameters of Gaussian Mixture Models (GMMs) of HMMs under the MMI discriminative function for speech recognition problems. In [9] the EBW technique was generalized to the novel Line Search A-functions (LSAF) optimization technique. A simple geometric proof was provided to show that LSAF recursions result in a growth transformation (that is, the value of the original function increases for the new parameters values). In [26] it was shown that a discrete version of EBW invented in more than 24 years ago can be also represented using $\mathcal{A}$-functions. This connection allowed a convergence proof for a discrete EBW to be developed [26].

15.2.2 Line Search A-Functions

Let $f(x) : \mathcal{U} \subset \mathbb{R}^n \to \mathbb{R}$ be a real valued differentiable function in an open subset $\mathcal{U}$. Let $\mathbf{A}_f = \mathbf{A}_f(x, y) : \mathbb{R}^n \times \mathbb{R}^n \to \mathbb{R}$ be twice differentiable in $x \in \mathcal{U}$ for each $y \in \mathcal{U}$. We define $\mathbf{A}_f$ as an $\mathcal{A}$-function for f if the following properties hold.

1. $\mathbf{A}_f(x, y)$ is a strictly convex or strictly concave function of x for any $y \in \mathcal{U}$. (Recall that twice differentiable function is strictly concave or convex over some domain if its Hessian function is positive or negative definite in the domain, respectively.)
2. Hyperplanes tangent to manifolds defined by $z = g_y(x) = \mathbf{A}_f(x, y)$ and $z = f(x)$ at any $x = y \in \mathcal{U}$ are parallel to each other, that is,

$$\nabla_x \mathbf{A}_f(x, y)|_{x=y} = \nabla_x f(x) \tag{15.5}$$

It was shown in [9] that a general optimization technique can be constructed based on $\mathcal{A}$-function. We formulated a growth transformation such that the next step in the parameter update that increases $f(x)$ is obtained as a linear combination of the current parameter values and the value $\tilde{x}$ that optimizes the $\mathcal{A}$-function, for which $\nabla_x \mathbf{A}_f(x, y)|_{x=\tilde{x}} = 0$. More precisely, we stated that $\mathcal{A}$-function gives a set of iterative update rules with the following "growth" property: let x_0 be some point in $\mathcal{U}$ and $\mathcal{U} \ni \tilde{x}_0 \neq x_0$ be a solution of $\nabla_x A(x, x_0)|_{x=\tilde{x}_0} = 0$. Defining

$$x_1 = x(\alpha) = \alpha \tilde{x}_0 + (1 - \alpha) x_0, \tag{15.6}$$

we have for sufficiently small $|\alpha| \neq 0$ that $f(x(\alpha)) > f(x_0)$, where $\alpha > 0$ if $A(x, x_0)$ concave and $\alpha < 0$ if $A(x, x_0)$ convex. The technique of generating $\tilde{x}$ in this way and performing the line search is termed "Line Search A-Function" (LSAF).

15.2.3 Discrete EBW

Here we show that discrete EBW can be described using the LSAF framework. Our descriptiong ins limited to the case of a single distribution, but the technique generalizes readily to several distributions.

Let the simplex $\mathcal{S}$ be defined as

$$\mathcal{S} := \{\beta : \beta \in \mathbb{R}^n, \beta_i \geq 0, i = 1, \ldots n, \sum \beta_i = 1\},$$

and suppose that $f : \mathbb{R}^n \to \mathbb{R}$ is a differentiable function on some subset $X \subset \mathcal{S}$. We wish to solve the following maximization problem for a function $f(\beta)$:

$$\max f(\beta) \text{ subject to } \beta \in \mathcal{S}. \tag{15.7}$$

Let $\beta \in X$ and define $a_i^k := \frac{\partial f(\beta^k)}{\partial \beta_i^k}, i = 1, \ldots n$. For any $D \in \mathbb{R}$ and $\beta^k \in \mathbb{R}^n$ such that $\sum_{j=1}^n a_j^k \beta_j^k + D \neq 0$, we define a recursion $T_D : \mathbb{R}^n \to \mathbb{R}^n$ as follows:

$$\beta_i^{k+1} = T_D(\beta^k) = \frac{a_i^k \beta_i^k + D\beta_i^k}{\sum_{j=1}^n a_j^k \beta_j^k + D}. \tag{15.8}$$

It was shown in [27] that for sufficiently large D, we have $f(\beta^{k+1}) > f(\beta^k)$, unless $\beta^{k+1} = \beta^k$.

An $\mathcal{A}$-function $\mathbb{A}_f$ for the function f in (15.7) that is differentiable in some compact neighborhood $\mathcal{U} \subset X$ of a point $\beta_0 \in \mathcal{S}$ is given as:

$$\mathbb{A}_f(\beta_0, \beta) = \sum (c_i + \beta_{0i} D) \log \beta_i, \tag{15.9}$$

where $c_i = c_i(\beta_0) = \beta_{0i} \frac{\partial f(\beta)}{\partial \beta_i}|_{\beta=\beta_0} = \beta_{0i} a_i(\beta_0)$ and D is any number such that $a_i(\beta) + D > 0$ for all i and any $\beta \in \mathcal{U}$. (Existence of D is guaranteed by differentiability of f in $\mathcal{U}$ and compactness of $\mathcal{U}$.) To show that the function $\mathbb{A}_f(\beta_0, \beta)$ in (15.9) is an $\mathcal{A}$-function, one needs to check (15.5) as follows. Replace $\beta_n = 1 - \sum \beta_i$ in (15.7), (15.9), that is, consider the functions $g(\beta') = f(\beta_1, ..., \beta_{n-1}, 1 - \sum_1^{n-1} \beta_i)$, $\mathbb{A}_g(\beta_0; \beta') = \mathbb{A}_f(\beta_0, \{\beta_1, ..., \beta_{n-1}\}, 1 - \sum_1^{n-1} \beta_j)$ where $\beta' = \{\beta_1, ...\beta_{n-1}\}$. We have

$$\frac{\partial \mathbb{A}_g(\beta_0, \beta')}{\partial \beta_i}|_{\beta_i=\beta_{0i}} = a_i(\beta_0) \frac{\partial f(\beta)}{\partial \beta_i}|_{\beta_i=\beta_{0i}} + D\beta_{0i} \frac{\partial \log \beta_i}{\partial \beta_i}|_{\beta_i=\beta_{0i}} +$$
$$D(1 - \sum_1^{n-1} \beta_{0i}) \frac{\partial \log(1 - \sum_1^{n-1} \beta_i)}{\partial \beta_i}|_{\beta=\beta_0} = \frac{\partial g(\beta')}{\partial \beta'}|_{\beta_i'=\beta_{0i}'}.$$

It can be shown that adding a quadratic penalty $C\beta^T\beta$ to the objective function $f(\beta)$ is equivalent to substituting the term D with $D + 2C$ in the discrete EBW recursion (15.8). Moreover, for sufficiently large C, the function $f(\beta) + C\beta^T\beta$ is concave in a simplex $\mathcal{S}$. Therefore, it achieves its maximum on the boundary of the of the simplex $\mathcal{S}$. This fact implies that for sufficiently large D, the EBW recursion enforces a sparse solution.

Discrete EBW methods can be applied to optimization of objective functions with fractional norm constraints, as suggested in [28]. We have

$$\max\ f(\{\beta_i\}) \ \text{ subejct to } \ \|\beta\|_q = 1 \ \text{ and } \ \beta_i \geq 0, \ \mathrm{i} = 1, 2, \ldots, \mathrm{n}, \tag{15.10}$$

where $\| \beta \|_q := (\sum \beta_i^q)^{1/q}$. Setting

$$\gamma_i = \beta_i^{1/q}, \quad g(\{\gamma_i\}) = f(\{\beta_i\}), \tag{15.11}$$

transforms the problem (15.10) into a discrete EBW problem for which the recursion (15.8) could be applied. In [26], this optimization method with fractional norm constraints was applied to TIMIT classification tasks.

15.2.4 An ABCS Compressed Sensing Algorithm

Following [29], we describe the approximate Bayesian CS (ABCS) method. The key idea behind this algorithm is based on an approximate sparseness promoting prior which is a sort of mixture of Gaussian and Laplace distributions. ABCS is a variant of the algorithm in [30] and [31]. In what follows we gradually develop this underlying concept and a few others which form the core of the new method.

15.2.4.1 Bayesian Estimation

The Bayesian estimation methodology provides a convenient representation for dealing with complex observation models. In this work, however, we restrict ourselves to the conventional linear model used in CS theory

$$y_k = H\beta + n_k \tag{15.12}$$

where y_k, $H \in \mathbb{R}^{m\times N}$, and n_k denote the kth $\mathbb{R}^m$-valued observation, a fixed sensing matrix, and the observation noise of which the pdf $p(n_k)$ is known, respectively. The sought-after random parameter (the signal) β is a $\mathbb{R}^N$-valued vector for which the prior pdf $p(\beta)$ is given. Following this, the complete statistics of β conditioned on the entire observation set consisting of k elements, $\mathcal{Y}_k = [y_1, \ldots, y_k]$ can be sequentially computed via the Bayesian recursion

$$p(\beta \mid \mathcal{Y}_k) = \frac{p(y_k \mid \beta)p(\beta \mid \mathcal{Y}_{k-1})}{\int p(y_k \mid \beta)p(\beta \mid \mathcal{Y}_{k-1})d\beta} \tag{15.13}$$

where the likelihood $p(y_k \mid \beta) = p_{n_k}(y_k - H\beta)$. One can rarely obtain a closed-form analytic expression of the posterior pdf (15.13), so approximation techniques are often used. One well-known example in which (15.13) does admit a closed-form solution is given by the following theorem, which plays a fundamental role in this work. (This is a well known result in estimation theory which is revisited here for completeness.)

Theorem 1 (Gaussian pdf Update). *Assume that $p(\beta \mid \mathcal{Y}_{k-1})$ is a Gaussian pdf of which the first two statistical moments are given by $\hat{\beta}_{k-1} \in \mathbb{R}^n$ and $P_{k-1} \in \mathbb{R}^{n\times n}$, that is $p(\beta \mid \mathcal{Y}_{k-1}) = \mathcal{N}(\beta \mid \hat{\beta}_{k-1}, P_{k-1})$. Assume also that the observation y_k satisfies the linear model* (15.12) *where n_k is a $\mathbb{R}^m$-valued zero-mean Gaussian random variable $n_k \sim \mathcal{N}(0, R)$ that is statistically independent of β. Then the Bayesian recursion* (15.13) *yields $p(\beta \mid \mathcal{Y}_k) = \mathcal{N}(\beta \mid \hat{\beta}_k, P_k)$ where*

$$\hat{\beta}_k = \hat{\beta}_{k-1} + P_{k-1}H^T\left(HP_{k-1}H^T + R\right)^{-1}\left[y_k - H\hat{\beta}_{k-1}\right] \tag{15.14a}$$

$$P_k = \left[I - P_{k-1} H^T \left(H P_{k-1} H^T + R \right)^{-1} H \right] P_{k-1} \tag{15.14b}$$

The initial values of the above quantities are set according to the Gaussian prior $p(\beta) = \mathcal{N}(\beta \mid \hat{\beta}_0, P_0)$.

The proof of this statement can be found in [29]. Note that the quantity P_k in Theorem 1 is the estimation error covariance, i.e.,

$$P_k := E\left[(\beta - \hat{\beta}_k)(\beta - \hat{\beta}_k)^T \mid \mathcal{Y}_k \right]$$

where $\beta - \hat{\beta}_k$ is the estimation error of the unbiased estimator $\hat{\beta}_k$.

15.2.5 Sparseness-Promoting Semi-Gaussian Priors

Compressed sensing was embedded in the framework of Bayesian estimation by utilizing sparseness promoting priors such as Laplace and Cauchy [32]. Here we consider a different type of prior that facilitates the application of the closed-form recursion of Theorem 1. The sparseness-promoting prior used here is termed "semi-Gaussian" (SG) owing to its form

$$p(\beta) = c \exp\left(-\frac{1}{2} \frac{\| \beta \|_1^2}{\sigma^2} \right). \tag{15.15}$$

The motivation for using a SG prior can be motivated by analyzing the characteristics of the SG constraint $\| \beta \|_1^2 = (\sum_i |\beta_i|)^2$ and the Laplacian constraint $\| \beta \|_1 = (\sum_i |\beta_i|)$. We can denote the SG density function as proportional to $p_{semi-gauss} \propto exp(- \| \beta \|_1^2)$ and the Laplacian density function proportional to $p_{laplace} \propto exp(- \| \beta \|_1)$. When $\| \beta \|_1 < 1$, it is straightforward to see that $p_{semi-gauss} > p_{laplace}$. When $\| \beta \|_1 = 1$, the density functions are the same, and when $\| \beta \|_1 > 1$ then $p_{semi-gauss} < p_{laplace}$. Therefore the semi-Gaussian density is more concentrated than the Laplacian density in the convex area inside $\| \beta \|_1 < 1$. Given the sparseness constraint $\| \beta \|_q$, as the fractional norm q goes to 0, the density becomes concentrated at the coordinate axes and the problem of solving for β becomes a non-convex optimization problem where the reconstructed signal has the least mean-squared-error (MSE). Intuitively, we expect the solution using the semi-Gaussian prior to behave closer to the non-convex solution.

This observation is further illustrated in Fig. 15.1, in which the level maps are shown for Laplace, semi-Gaussian, and Gaussian pdfs in the 2-dimensional case. The embedding of the prior (15.15) within the Gaussian variant of the Bayesian recursion in Theorem 1 is not straightforward. This follows from the fact that the restrictions under which Theorem 1 is derived involve a purely Gaussian prior and a

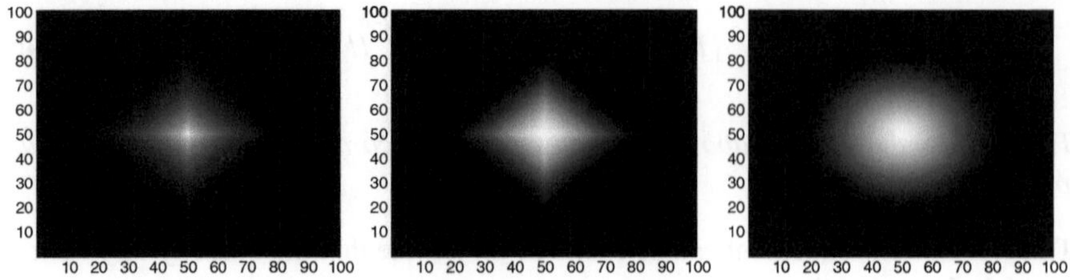

Fig. 15.1 Laplace, semi-Gaussian, and Gaussian pdfs in $\mathbb{R}^2$

likelihood pdf that is based on a deterministic sensing matrix H,

$$p(y_k \mid \beta) \propto \exp\left(-\frac{1}{2}(y_k - H\beta)^T R^{-1}(y_k - H\beta)\right). \tag{15.16}$$

Theorem 1 provides an exact recursion for computing the Gaussian posterior based exclusively on the factors composing the above likelihood: the observation y_k, the sensing matrix H and the observation noise covariance R. This fact has motivated the following approach which allows enforcing an approximate semi-Gaussian prior without changing the fundamental structure of the underlying update equations as obtained in Theorem 1.

15.2.6 Approximate Semi-Gaussian Prior

We introduce a state-dependent matrix $\hat{H} \in \mathbb{R}^{1\times N}$ whose entries are $\hat{H}^i = \text{sign}(\beta^i)$, $i = 1, 2, \ldots, N$ (that is, $\hat{H}^i = +1$ and $\hat{H}^i = -1$ for $\beta^i > 0$ and $\beta^i < 0$, respectively). The semi-Gaussian prior can be expressed based on (15.16) while replacing H and R with $\hat{H}$ and σ, respectively, and assuming a fictitious observation $y = 0$, that is

$$p(\beta) = p(y = 0 \mid \beta, \hat{H}, \sigma) \propto \exp\left(-\frac{1}{2}\frac{(0 - \hat{H}\beta)^2}{\sigma^2}\right) \tag{15.17}$$

The only difficulty in using (15.14a) for enforcing the semi-Gaussian prior (15.17) is the dependency of $\hat{H}$ on β. We recall that Theorem 1 relies on possibly varying a deterministic H as opposed to the formulation in (15.17). This problem can be alleviated by letting

$$\hat{H}^i = \text{sign}(\hat{\beta}^i_k), \quad i = 1, 2, \ldots, N, \tag{15.18}$$

that is, by substituting the conditional mean instead of the actual β. This modification renders $\hat{H}$ a $\mathcal{Y}_k$-measurable quantity, as it depends on $\hat{\beta}_k$ which is a function of the entire observation set. This fact clearly does not affect the expressions in Theorem 1 as the derivations are conditioned on $\mathcal{Y}_k$ (see [29]). Applying this approximation facilitates the implementation of Theorem 1 based on the likelihood (15.17). Hence, an additional processing stage is needed to apply the approximate sparseness-promoting prior:

$$\hat{\beta}_{k+1} = \left[I - \frac{P_k \hat{H}^T \hat{H}}{\hat{H} P_k \hat{H}^T + \sigma^2} \right] \hat{\beta}_k \tag{15.19a}$$

$$P_{k+1} = \left[I - \frac{P_k \hat{H}^T \hat{H}}{\hat{H} P_k \hat{H}^T + \sigma^2} \right] P_k. \tag{15.19b}$$

This stage is implemented after the usual processing of the observations set $\mathscr{Y}_k$ (see (15.14)), where the initial covariance is taken as $P_0 \to \infty$.

At this point, a natural question is raised concerning the validity of the approximation suggested above. The following theorem, proved in [29], bounds the discrepancy between the exact posterior which uses the semi-Gaussian prior (15.15) and the approximate posterior in terms of the estimation error covariance $\hat{P}_k$.

Theorem 2 *Denote $\hat{p}(\beta \mid \mathscr{Y}_k)$ the Gaussian posterior pdf obtained by using the approximate semi-Gaussian prior technique, and let $p(\beta \mid \mathscr{Y}_k)$ be the posterior pdf obtained by using the exact semi-Gaussian prior (15.15). Then*

$$\mathrm{KL}\left(\hat{p}(\beta \mid \mathscr{Y}_k) \parallel p(\beta \mid \mathscr{Y}_k)\right) = \mathcal{O}\left(\sigma^{-2} \max\left\{\mathrm{Tr}(\hat{P}_k), \mathrm{Tr}(\hat{P}_k)^{1/2}\right\}\right), \tag{15.20}$$

where KL *and* Tr *denote the Kullback-Leibler divergence and the matrix trace operator, respectively.*

In practical applications for speech classification and recognition tasks, it was observed that the classification and recognition accuracy is not affected if ones computes a term P_k in (15.19b) only once, then fixes this term for all subsequent iterations. This trick provides a significant speed up without significant degradation of accuracy.

15.2.7 ABCS Representations via LSAF

We recall the ℓ_1-constrained problem (15.1), modified slightly by the use of a weighted data-fitting term

$$\min \parallel y - H\beta \parallel_R^2 \quad \text{subject to} \quad \parallel \beta \parallel_1 \leq \epsilon.$$

In many practical application it is useful to add an l_2 regularization term to this formulation, to yield

$$\min \parallel y - H\beta \parallel_R^2 + \parallel \beta - \beta_0 \parallel_{P_0}^2 \quad \text{subject to} \quad \parallel \beta \parallel_1 \leq \epsilon.$$

Using $\| y - H\beta \|_R^2 + \| \beta - \beta_0 \|_{P_0}^2 = \| \beta - \beta_1 \|_{P_1}^2$ we can represent this problem as

$$\min \| \beta - \beta_1 \|_{P_1}^2 \quad \text{subject to} \quad \| \beta \|_1 \leq \epsilon,$$

where P_1 is assumed to be positive-definite. We can now represent (15.1) by

$$\min F(\beta) := \| \beta - \beta_1 \|_{P_1}^2 + \| \beta \|_1^i / \sigma^2, \tag{15.21}$$

and define the $\mathcal{A}$-function as:

$$\mathbb{A}(\beta, \beta^*) = \| \beta - \beta^* \|_{P_1}^2 + \{\text{sign}(\beta^*)\beta\}^i / \sigma^2, \tag{15.22}$$

where $i = 1$ (Laplacian) or $i = 2$ (squared l_1 norm). In [26] we show that $\mathbb{A}(\beta, \beta^*)$ is $\mathcal{A}$-function of $F(\beta)$. According to the definition of the $\mathcal{A}$-function, we consider $\mathbb{A}(\beta, \beta^*)$ and $F(\beta)$ in an open domain where they are both differentiable and construct an update of parameters when the extremum of $\mathbb{A}(\beta, \beta^*)$ belongs to this domain. Our open domain excludes the origin $\beta = 0$. If some coordinates of β approach 0 we can remove them by reducing the dimension of the problem. Using LSAF, we have the recursion:

$$\beta_k = \alpha \tilde{\beta}_{k-1} + (1 - \alpha)\beta_{k-1}.$$

The ABCS algorithm corresponds to a squared l_1-norm. Analysis of various regularization penalties for speech classification problems is given in Sect. 15.3. The ABCS method gives a solution of (15.21) via the recursion:$\tilde{\beta}_{k-1} = \arg\max_\beta A(\beta, \beta_{k-1})$. Numerical experiments show that for a suitable choice of α, the parameter β_k converges to a solution of (15.21) more rapidly than the one obtained through the ABCS recursion. One can expect that LSAF with appropriate choices of α is more efficient than the ABCS.

15.3 An Analysis of Sparseness and Regularization in Exemplar-Based Methods for Speech Classification

Following [10] we describe and compare a variety of different sparseness techniques, which employ different types of regularization, and that have been explored for speech tasks [2, 3]. Firstly, we describe the main framework behind exemplar-based classification. Then we give a brief description of the TIMIT corpus. Next we discuss how sparseness can be useful in classification tasks. Finally, we compare the performance of different sparseness methods for classification.

15.3.1 Classification Based on Exemplars

The goal of classification is to use training data from k different classes to determine the best class to assign to test vector y. First, let us consider taking all training examples n_i from class i and concatenate them into a matrix H_i as columns, in other words $H_i = [x_{i,1}, x_{i,2}, \ldots, x_{i,n_i}] \in \mathbb{R}^{m \times n_i}$, where $x \in \mathbb{R}^m$ represents a feature vector from the training set of class i with dimension m. Given sufficient training examples from class i, [6] shows that a test sample y from the same class can be represented as a linear combination of the entries in H_i weighted by β, that is:

$$y = \beta_{i,1}x_{i,1} + \beta_{i,2}x_{i,2} + \ldots + \beta_{i,n_i}x_{i,n_i} \tag{15.23}$$

However, since the class membership of y is unknown, we define a matrix H to include training examples from all k classes in the training set, in other words the columns of H are defined as $H = [H_1, H_2, \ldots, H_k] = [x_{1,1}, x_{1,2}, \ldots, x_{k,n_k}] \in \mathbb{R}^{m \times N}$. Here N is the total number of all training examples from all classes. We can then write test vector y as a linear combination of all training examples, in other words $y = H\beta$. We can solve this linear system for β and use information about β to make a classification decision. Specifically, large entries of β should correspond to the entries in H with the same class as y. Thus, one proposed classification decision approach [3] is to compute the l_2 norm for all β entries within a specific class, and choose the class with the largest l_2 norm support.

15.3.2 Exemplar-Based Methods

Various types of exemplar-based classifiers can be cast in the framework of representing the test vector y as a linear combination of training examples H, subject to a constraint on β. Below, we review a few popular techniques that are based on the following optimization problem for various values of q and α

$$\min_{\beta} \parallel y - H\beta \parallel_2 \quad \texttt{s.t.} \quad \parallel \beta \parallel_q^{\alpha} \leq \epsilon \tag{15.24}$$

1. Ridge regression (RR) methods [5] use information about all training examples in H to make a classification decision about y, in contrast to a nearest-neighbor (NN) approach to exemplar-based classification, which uses information about just 1 training example. Specifically, the RR method looks to project y into the linear space of all training examples and solves for the β which minimizes (15.24) for $q = 2, \alpha = 2$. The term $\parallel \beta \parallel_2^2 \leq \epsilon$ is an l_2 norm on β (i.e. a Gaussian constraint) but does not enforce any sparseness.
2. Sparse representations: like RR methods, sparse representation (SR) techniques (i.e., [3, 6], project y into the linear span of examples in H, but constrain β to be sparse. Specifically, SR methods solve for β by minimizing (15.24), given

various settings for α and q. For example, in a probabilistic setting $q = 1, \alpha = 1$ leads to a Laplacian constraint, whereas $q = 1, \alpha = 2$ leads to a Semi-Gaussian constraint. The remainder of this section is focused on comparing the RR method to various SR methods with different types of regularizations.

15.3.3 Description of TIMIT

We analyze the behavior of various exemplar-based methods on the TIMIT [16] corpus. The corpus contains over 6,300 phonetically rich utterances divided into three sets, namely the training, development, and core test set. For testing purposes, the standard practice is to collapse the 48 trained labels into a smaller set of 39 labels. All methods are tuned on the development set and all experiments are reported on the core test set.

The complete experimental setup, as well as the features used for classification, are similar to [3]. First, we represent each frame in our signal by a 40 dimensional discriminatively trained Space Boosted Maximum Mutual Information (fBMMI) feature. We split each phonetic segment into thirds, taking the average of these frame-level features around 3rds, and splice them together to form a 120 dimensional vector. This allows us to capture time dynamics into each segment. Then, at each segment, segmental feature vectors to the left and right of this segment are joined together and a Linear Discriminative Analysis (LDA) transform is applied to project 200 dimensional feature vector down to 40 dimensions.

Similar to [3], we find a neighborhood of closest points to y in the training set using a kd-tree. These k neighbors become the entries of H. We explore classification performance for different sizes of H. In what follows, we explore the following two questions, using TIMIT to provide experimental results to support our framework.

- Why and when is sparseness important for exemplar-based methods?
- If sparseness is used, what type of regularization constraint should be utilized?

15.3.4 Why Sparse Representations?

We will motivate the difference between the RR and SR methods further with the following example. Let us consider a 2×7 matrix

$$H = [h_1, h_2, h_3, h_4, h_5, h_6, h_7] = \begin{bmatrix} 0.2 & 0.1 & 0.4 & 0.3 & -0.6 & 0.6 & -0.6 \\ 0.2 & 0.3 & 0.35 & 0.3 & 0.1 & 0.3 & 0.4 \end{bmatrix},$$

where first three columns h_1, h_2, h_3 are "training" utterances that belong to a class C_1 and last four columns are "training" utterances that belong C_2. Assume also that a vector $y = [0.29; 0.29]$ is "test" data that belong to a class C_1. thus will

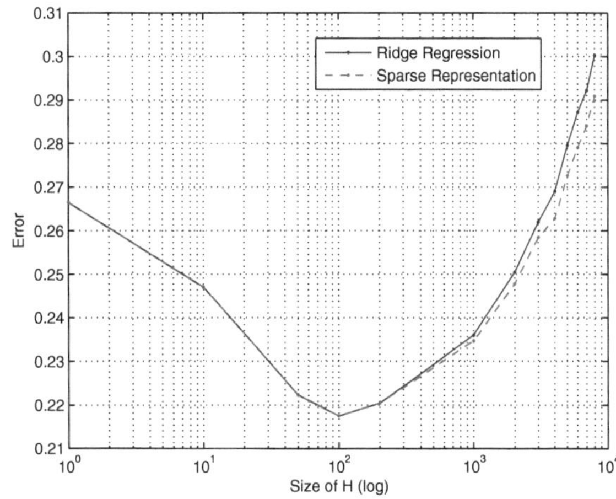

Fig. 15.2 Error for RR and SR methods for varied H

include the outlier points of C_2. Solving (15.24) with $q = 2, \alpha = 2$ (i.e., the RR method) produces the vector $\beta \approx [0.12; 0.15; 0.21; 0.18; -0.05; 0.1220.08]$ and the best class is C_2. However using the SR method in (15.24) (for example, using ABCS method with a SG constraint as explained in Sect. 15.2) produces a vector $\beta \approx [0.00; 0.01; 0.77; 0.00; 0.00; 0.00; 0.03]$ with the support located at the third entry in H. In this case, the C_1 is identified as the correct class. Thus, by using a subset of examples in H, the classification decision for SR and RR can be vastly different, particularly in the case of outliers.

To analyze the behavior of the SR and RR methods in a practical speech example, we explore phonetic classification on TIMIT as the size of H is varied from 1 to 10,000. A plot of the error rate for the two methods for varied H is shown Fig. 15.2. For this figure, we again used the ABCS SR method. First notice that as the size of H increases up to 1,000 the error rates of the RR and SR both decrease, showing the benefit of including multiple training examples when making a classification decision. Also notice that there is no difference in error between the RR and SR techniques, suggesting that regularization does not provide any extra benefit. However, as the size of H increases past 1,000 and there are more number of training examples for each class, the SR method performs better than the RR method, demonstrating the advantage of using sparseness to select only a few examples in H to explain y rather than all examples in H.

15.3.5 What Type of Regularization?

Now that we have motivated the use of regularization, in this section we analyze different forms of regularization. As illustrated by (15.24), with $q = 1$, a sparse representation solution can be formulated by finding the β which minimizes the residual error $\| y - H\beta \|_2$, subject to a regularization $\| \beta \|_q \leq \epsilon$ on β. There are four common types of regularizations on β.

1. If $q = 2$ and $\alpha = 2$, then the regularization becomes $\| \beta \|_2 \leq \epsilon$. This constraint can be modeled as a Gaussian prior. Common techniques which impose an l_2 constraint on β include Ridge Regression [5]. The effect of the l_2 norm is to spread values of entries in β equally. Therefore the optimization problem (15.24) for $q = 2$ tries to find a balance between keeping the residual $\| y - \beta \|_2$ small and trying to keep all the entries in the vector β to be non-zero.
2. If $q = 1$ and $\alpha = 1$, then the regularization becomes $\| \beta \|_1 \leq \epsilon$. This constraint can be modeled as a Laplacian prior. Common techniques which impose an l_1 constraint on β include LASSO [11] and Bayesian Compressive Sensing (BCS) [12]. The Lasso problem can be formulated as follows:

$$\min_{\beta} \| y - H\beta \|_2 + \lambda \| \beta \|_1, \tag{15.25}$$

as in (15.3), where λ controls the weight of the l_1 norm. The Least Angle Regression (LARS) ([33]) solves LASSO through a forward stepwise regression, computing point estimates of β at each step. The effect of the l_2 norm is to spread values of entries in β equally. Therefore the optimization problem (15.24) for $q = 2$ tries to find a balance between keeping the residual $\| y - \beta \|_2$ small while at the same time preventing all the entries in β from vanish. In contrast, the norm l_1 tries to enforce sparsity in β while keeping the residual $\| y - H\beta \|_2$ small.
Bayesian Compressive sensing [12] can be formulated in a fashion similar to (15.25). BCS introduces a probabilistic framework to estimate the spareness parameters required for signal recovery. This technique limits the effort required to tune the sparseness constraint and also provides complete statistics for the estimate of β.
3. Many techniques also impose a combination of an l_1 and l_2 constraint on β. These methods include the popular Elastic Net [13]. The Elastic Net [13] method imposes a mixture of an l_1 and l_2 constraints, i.e.,

$$\min_{\beta} \| y - H\beta \|_2 + \lambda_1 \| \beta \|_1 + \lambda_2 \| \beta \|_2^2 . \tag{15.26}$$

Here λ_1 and λ_2 are weights controlling the l_1 and l_2 constraint. In the elastic net formulation the l_1 term enforces the sparsity of solution, whereas the l_2 penalty ensures democracy among groups of correlated variables. The second term has also a smoothing effect that stabilizes the obtained solution.
4. The previously described ABCS explores the use of a semi-Gaussian prior and solves for β in a Bayesian framework. The ABCS essentially solves

$$\min_{\beta} \| y - H\beta \|_2 + \lambda_1 (\beta - \beta_0)^T P_0^{-1} (\beta - \beta_0) + \lambda_2 \| \beta \|_1^2 . \tag{15.27}$$

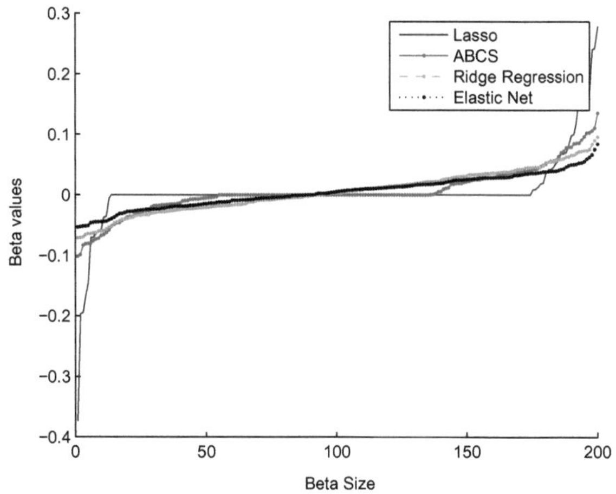

Fig. 15.3 Plot of β for different regularization constraints

Visualization of Sparsity

We analyze the difference in β coefficients for different sparseness methods. For a randomly selected classification frame y in TIMIT and an H of size 200, we solve (15.24) for β. Figure 15.3 plots the sorted 200 β coefficients for four different techniques employing different reguliarizations, namely Ridge Regression, Lasso, Elastic Net and ABCS. The plot shows that the β coefficients for the RR method are the least sparse, as we would expect. In addition, the LASSO technique has the sparsest β values. The sparsity of the Elastic Net and ABCS techniques methods are in between RR and LASSO, with ABCS being more sparse than Elastic Net due to the Semi-Gaussian constraint in ABCS, which is more sparse than the l_1 constraint in the Elastic Net.

TIMIT Results

Table 15.1 shows the results comparing various sparseness methods on TIMIT for a size of $H = 200$. As one can see from the table, the three methods which combine a sparseness constraint with and l_2 norm, namely ABCS, Elastic Net and CSP, all achieve statistically the same accuracy. The two methods which use the l_1 norm, namely BCS and LASSO, have slightly lower accuracy, showing the decrease in accuracy when a high degree of sparseness is enforced. Thus, it appears that using a combination of a sparsity constraint on β, coupled with an l_2 norm, does not force unnecessary sparseness and offers the best performance.

Table 15.1 Accuracies for different sparseness methods

Method	PER
LASSO	74.40
BCS	73.58
Elastic net	**77.89**
ABCS	**77.80**
CSP	**77.55**

15.4 ABCS for Classification

In this section we follow [3] and describe application of ABCS for Timit classification tasks. We perform classification as described in Sect. 15.3.1 solving (15.23) for $\alpha = 2$ and $q = 1$ via (15.14a), (15.14b), (15.19a), and (15.19b). We compute the l_2 norm for all β entries within a specific class and choose the class with the largest l_2 norm support. Pooling together all training data from all classes into H will make the columns of H large (i.e., can be greater than 100,000 for TIMIT), and will make solving for β intractable. Therefore, to reduce the size of N and make ABCS problem more solvable, for each y, we find a neighborhood of closest points to y in the training set using a kd-tree [35]. These k neighbors become the entries of H. k is chosen to be in the large to ensure that β is sparse and all training examples are not chosen from the same class.

Constants P_0 and β_0 must be chosen to initialize the ABCS algorithm. Recall that β_0 and the diagonal elements of P_0 all correspond to a specific class. We choose β_0 to be 0 since we do not have a very confident estimate of β and we assume its sparse around 0 anyways. We choose to initialize a diagonal P_0 where the entries corresponding to a particular class are proportional to the GMM posterior for that class. The intuition behind this is that the larger the initial P_0, the more weight is given to examples in H belonging to this class in ABCS. Therefore, the GMM posterior picks out the most likely supports, and ABCS provies an addition step by using the actual training data to refine these supports.

15.4.1 Nonlinear Compressive Sensing

The traditional CS implementation represents y as a linear combination of samples in H. Many pattern recognition algorithms, such as SVMs [36] have shown better performance can be achieved by a nonlinear mapping of the feature set to a higher dimensional space. After this mapping, a weight vector w is found which projects all dimensions within a particular feature vector to a single dimension where different classes are linearly separable. We can think of this weight vector w as selecting some linear combination of dimensions within a feature vector to make it linearly separable. The goal of CS is to find a linear combination of actual features, not dimensions within a feature vector. Therefore, we introduce nonlinearity into CS, by constructing H such that the entries of H themselves are nonlinear. For example, one such nonlinearity is to square all the elements within H. That is if we define $H_{lin} = [x_{1,1}, x_{1,2}, \ldots, x_{k,n_k}]$, then H^2 is defined as $H^2 = [x_{1,1}^2, x_{1,2}^2, \ldots, x_{k,n_k}^2]$ and similarly H^3 would take cubed products of each of the x entries. We could also take products between different x_i as $H_{inner} = [x_{1,1}x_{1,2}, x_{1,1}x_{1,3} \ldots, x_{k,8}x_{k,n_k}]$. We then take a specific nonlinear H_{nonlin} and combine it with the linear H_{lin} to form a new $H_{tot} = [H_{lin}, H_{nonlin}]$ and use ABCS to solve for β. In Sect. 5.1 , we discuss the performance of the ABCS algorithm for different choices of nonlinear H.

15.4.2 Experiments

Classification experiments are conducted on TIMIT [16] acoustic phonetic corpus as described in Sect. 15.3.3. First, we analyze the performance of the CS classifier for different choices of linear and nonlinear H as described in Sect. 3.4. Next, we compare the performance of CS with three other standard classifiers used on this task, namely a Gaussian Mixture Model (GMM), Support Vector Machine (SVM) [36] and k-nearest Neighbors (kNN) classifier [35]. The parameters of each classifier were optimized for each feature set on the development set. Specifically, we found that modeling each phone as a 16-component GMM was appropriate. The kernel type and parameters within this kernel were optimized for the SVM. In addition, the number of k closest neighbors for kNN was also learned. And finally, for CS the size of H_{lin} was optimized to be 200 examples from the kd-tree. In addition to compute H_{nonlin}, 100 columns were randomly chosen from H_{lin} to compute each type of nonlinear H.

Performance for Different H

Table 15.2 shows the accuracy on the development set for different choices of H using Mel-frequency cepstral coefficients (MFCC) features. Notice that the nonlinear CS-$H_{lin}H^2$ method offers improvements over the linear CS-H_{lin} method. Taking $H_{lin}H^2H^3$ offers addition improvements, though overtraining occurs when higher order features past H^3 are used. Furthermore, there is very little difference between squaring individual entries of H (i.e. $H_{lin}H^2$) or taking products between differ-

Table 15.2 Accuracy for different H using MFCC features

Method	Dev-MFCC
CS-H_{lin}	76.64
CS-$H_{lin}H^2$	76.84
CS-$H_{lin}H^2H_{inner}$	76.53
CS-$H_{lin}H^2H^3$	**76.89**
CS-$H_{lin}H^2H^3H^4$	76.86

Table 15.3 Accuracy for different classifiers on TIMIT testcore set

Method	MFCC	fBMMI
GMM	74.19	78.31
kNN	73.69	79.58 (=)
SVM	76.20 (=)	78.38
CS-$H_{lin}H^2H^3$	**76.44**	**80.01**

ent entries of H (i.e., $H_{lin}H_{inner}$). While not shown here, similar trends were also observed for fBMMI features. Since the CS-$H_{lin}H^2H^3$ method offers the best performance of the CS methods, we will report the results for this classifier in subsequent sections.

Comparing Different Classifiers

Table 15.3 compares the performance of the CS classifier with the GMM, kNN and SVM methods for both MFCC and fBMMI features. Classifiers which are not statistically significant from the CS classifier, as confirmed by McNemar's Test, are also indicated by '='. First, notice that when MFCC features are used, CS outpeforms both then kNN and GMM methods, and offers similar performance to the SVM. When discriminative features are used, the GMM technique is closely matched to the SVM though CS is able provide further gains over these two methods. This is one of the benefits of CS—a discriminative non-parametric classifier built on top of the GMM.

Analysis of Results

To better understand the gains achieved by the CS classifier compared to the other three techniques, Fig. 15.4 plots the relative difference in error rates within 6 broad phonetic classes (BPCs) for CS compared to the three other methods. First, notice that CS offers improvements over the GMM in all BPCs, again confirming its benefit of a non-parametric discriminative classifier on top of the GMM. Secondly, while the SVM technique offers improvements over the CS method in the vowel/semi-vowel class, the CS method significantly outperforms the SVM in the weak fricative, stop and closure classes. Finally, the CS method offers slight improvements over the kNN method in the nasal, strong fricative and stop classes, while kNN offers slight improvements in the vowel, weak fricative and closure classes. Thus, we can see that with the exception of the GMM, the gains from CS do not come from it outperforming the kNN and SVM techniques within all BPCs, but only within certain BPCs.

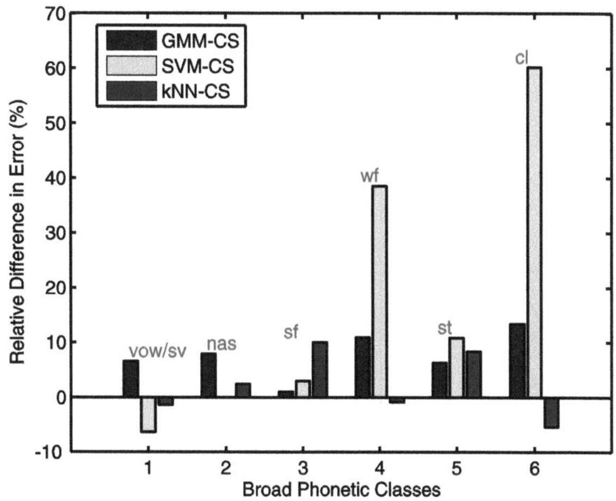

Fig. 15.4 Relative difference in error rates between CS and other methods

15.5 A Convex Hull Approach to Sparse Representations

A typical SR formulation in (15.24) does not constrain β to be positive and normalized, which can result in the projected training points $h_i\beta_i \in H\beta$ being reflected and scaled. While this can be desirable when data variability exists, allowing for too much flexibility when data variability is minimized can reduce the discrimination between classes. Driven by this intuition, below we present two examples where data variability is minimized, and demonstrate how SRs manipulate the feature space, thus leading to classification errors.

First, consider two clusters in a 2-dim space as shown in Fig. 15.5a with sample points $\{a_1, a_2, \dots, a_6\}$ belonging to Class 1 and $\{b_1, b_2, \dots, b_6\}$ belonging to Class 2. Assume that points a_i and b_i are concatenated into a matrix $H = [h_1, h_2, \dots, h_{12}] = [a_1, \dots, a_6, b_1, \dots b_6]$, with a specific entry being denoted by $h_i \in H$. In a typical SR problem, given a new point y indicted in Fig. 15.5b, we project y into the linear span of training examples in H by trying to solve:

$$\arg\min \parallel \beta \parallel_0 \quad \texttt{s.t.} \quad y = H\beta = \sum_{i=1}^{12} h_i\beta_i \tag{15.28}$$

As shown in Fig. 15.5a, the best solution will be obtained by setting all $\beta_i = 0$ except for $\beta_8 = -1$, corresponding to the weight on point b_2. At this point $|\beta|_0$ takes the lowest value of 1 and $y = -b_2$, meaning it is assigned to Class 2. The SR method misclassifies point y, as it is clearly in Class 1, because it puts no constraints on the β values. Specifically, in this case, the issue arises from the possibility of β entries taking negative values.

Second, consider two clusters in a 2-dimensional space as shown in Fig. 15.5b with sample points belonging to Class 1 and 2. Again, we try to find the best representation

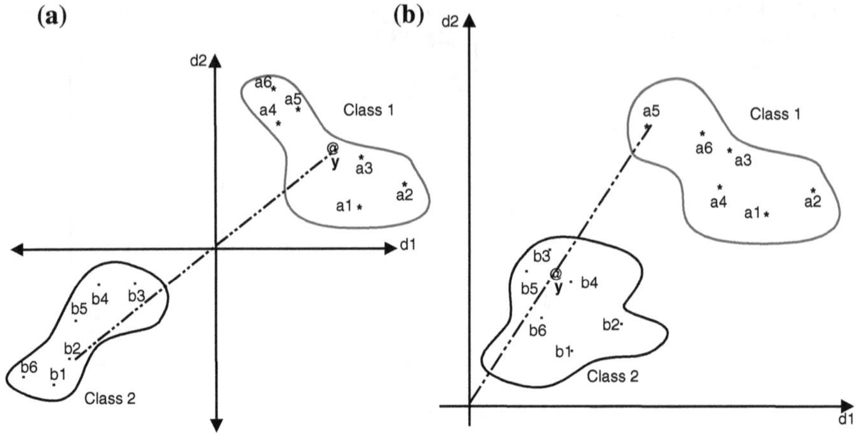

Fig. 15.5 **a** Reflective issue with negative β, **b** Scaling issue with unnormalized β

for test point y by solving (15.28). The best solution will be obtained by setting all $\beta_i = 0$ except for $\beta_5 = 0.5$. At this value, $|\beta|_0$ will take the lowest possible value of 1 and $y = 0.5 \times a_5$. This leads to a wrong classification decision as y clearly is a point in Class 2. The misclassification is due to having no constraint on the β elements. Specifically, in this case, the issue arises from total independence between the β values and no normalization criteria as a way to enforce dependency between the β elements. If we enforce β to be positive and normalized, then training points $h_i \in H$ form a convex hull. Mathematically speaking, a convex hull of training points H is defined by the set of all convex combinations of finite subsets of points from H, in other words a set of points that satisfy the following: $\sum_{i=1}^{n} h_i \beta_i$. Here n is any arbitrary number and the β_i components are positive and sum to 1.

Since many classification techniques can be sensitive to outliers, we examine the sensitivity of our convex hull SR method. Consider two clusters shown in Fig. 15.6 with sample points in Classes 1 and 2. Again, given point y, we try to find the best representation for y by solving (15.28), where now we will use a convex hull approach to solve, putting extra positivity and normalization constraints on β.

As shown in Fig. 15.6, if we project y onto the convex hulls of Class 1 and Class 2, the distance from y to the convex hull of Class 1 (indicated by r_1) is less than the distance from y to the convex hull of Class 2 (i.e. r_2). This leads to a wrong classification decision as y clearly is a point in Class 2. The misclassification is due to the effect of outliers a_1 and a_4, which create an inappropriate convex hull for Class 1.

However, all-data methods, such as GMMs, are much less susceptible to outliers, as a model for a class is built by estimating the mean and variance of training examples belonging to this class. Thus, if we include the the distance between the projection of y onto the two convex hulls of Class 1 and Class 2, as well as the distance between this projection and the means m_i of Class 1 and 2 (distance indicated by q_1 and q_2)

Fig. 15.6 Outliers effect

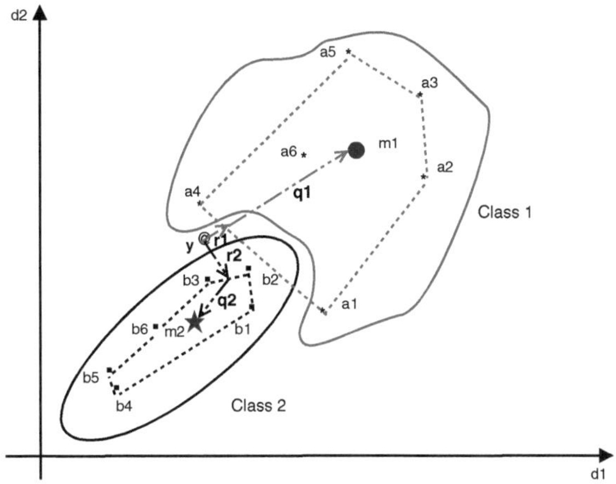

respectively, then test point y is classified correctly. Thus combining purely exemplar-based distances (r_i) with GMM-based distances (q_i), which are less susceptible to outliers, provides a more robust measure.

15.5.1 Convex Hull Formulation

In our sparse representations convex hull (SR-CH) formulation, first we seek to project test point y into the convex hull of H. After y is projected into the convex hull of H, we compute how far this projection (which we call $H\beta$) is from the Gaussian means[1] of all classes in H. The full convex hull formulation, which tries to find the optimal β to minimize both the exemplar and GMM-based distances [8]. Here $N_{classes}$ represents the number of unique classes in H, and $\| H\beta - \mu_t \|_2^2$ is the distance from $H\beta$ to the mean μ_t of class t,

$$\arg\min_{\beta} \| y - H\beta \|_2^2 + \sum_{t=1}^{N_{classes}} \| H\beta - \mu_t \|_2^2 \quad \texttt{s.t.} \quad \sum_i \beta_i = 1 \texttt{ and } \beta_i \geq 0 \tag{15.29}$$

In our work, we associate these distance measures with probabilities. Specifically, we assume that y satisfies a linear model as $y = H\beta + \zeta$ with observation noise $\zeta \sim N(0, R)$. This allows us to represent the distance between y and $H\beta$ using the term $p(y|\beta)$

$$p(y|\beta) \propto \exp(-1/2(y - H\beta)^T R^{-1}(y - H\beta)) \tag{15.30}$$

[1] Note that the Gaussian means we refer to in this work are built from the original training data, not the projected $H\beta$ features.

which we will refer to as the exemplar-based term.

We also explore a probabilistic representation for the $\sum_{t=1}^{N_{classes}} \parallel H\beta - \mu_t \parallel_2^2$ term. Specifically, we define the GMM-based term $p_M(\beta)$, by seeing how well our projection of y onto the convex hull of H, as represented by $H\beta$, is explained by each of the $N_{classes}$ GMM models. We score $H\beta$ against the GMM from each of the classes and sum the scores (in log-space) from all classes. This is given more formally as (log-space)

$$\log p_M(\beta) = \sum_{t=1}^{N_{classes}} \log p(H\beta|GMM_t) \tag{15.31}$$

where $p(H\beta|GMM_t)$ indicates the score from GMM t. Given the exemplar-based term $p(y|\beta)$ and GMM-based term $p_M(\beta)$, the total objective function we would like to maximize is given in the log-space by

$$\max_{\beta} F(\beta) = \{\log p(y|\beta) + \log p_M(\beta)\} \quad \texttt{s.t.} \quad \sum_i \beta_i = 1 \quad \texttt{and} \quad \beta_i \geq 0 \tag{15.32}$$

Equation (15.32) can be solved using a variety of optimization methods. We use a technique widely employed in speech recognition, namely the Extended Baum-Welch transformations (EBW) [24], to solve this problem. In [37], it is shown that EBW optimization technique can be used to maximize objective functions which are differentiable and satisfy constraints given in (15.32) (see also Sect. 15.2.3 and the recursion (15.8)). In [8], we provide a closed-form solution for β_k^i given the exemplar-based term (15.30) and a GMM-based term (15.31).

The parameter D in (15.8) controls the growth of the objective function. We explore setting D to a small value to ensure a large jump in the objective function. However, for a specific choice of D if we see that the objective function value has decreased when estimating β^k, i.e. $F(\beta^k) < F(\beta^{k-1})$, or one of the β_i^k components is negative, then we double the value of D and use this to estimate a new value of β^k in (15.8). We continue to increase the value of D until we guarantee a growth in the objective function, and all β_i components are positive. This strategy of setting D is similar to other applications in speech where the EBW transformations are used [38]. The process of iteratively estimating β continues until there is very little change in the objective function value.

15.5.2 Convex Hull Classification Rule

Because we are trying to solve for β which maximizes the objective function (15.32), it seems natural to also explore a classification rule which defines the best class as that which maximizes this objective function. Using (15.32) with the exemplar-based term (15.30) and the GMM-based term (15.31), the objective-function linked classification rule for the best class t^* is given by

$$t^* = \max_t\{\log p(y|\delta_t(\beta)) + \log p(H\delta_t(\beta)|GMM_t)\} \tag{15.33}$$

where $\delta_t(\beta)$ is a vector which is only non-zero for entries of β corresponding to class t.

15.5.3 Experiments

We compare the performance of our SR-CH method to other standard classifiers used on the TIMIT task, including the GMM, SVM, kNN and ABCS sparse representation methods. For the GMM, we explored training it via a maximum likelihood objective function, and a discriminative BMMI objective function [38]. The parameters of each classifier were optimized for each feature set on the development set. We compare SR-CH to this method. Note that for the ABCS classification rule, the best class is defined as that which has the maximum l_2 norm of β entries.

Algorithmic Behavior

As discussed in Sect. 15.5.1, for an appropriate choice of D, the objective function of the SR-CH method is guaranteed to increase on each iteration. To observe this behavior experimentally on TIMIT, we chose a random test phone segment y, and solve $y = H\beta$ using the SR-CH algorithm. Figure 15.7 plots the value of the objective function at each iteration. Notice that the objective function increases rapidly until about iteration 30 and then increases slower, experimentally confirming growth.

We also analyze the sparsity behavior for the SR-CH method. For a randomly chosen test segment y, Fig. 15.7 plots the sparsity level (defined as the number of

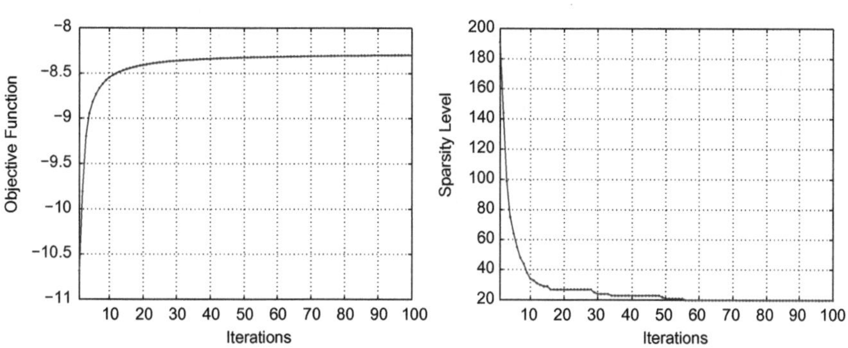

Fig. 15.7 *Left* Iterations versus objective function. *Right* Iterations versus sparsity

Table 15.4 Accuracy of sparse representation methods

Method	Accuracy
SR-CH (exemplar-only)	**83.86**
ABCS (exemplar-only)	78.16

non-zero β coefficients), for each iteration of the SR-CH algorithm. Notice that as the number of iterations increases, the sparsity level continues to decrease and eventually approaches 20. Our intuitive feeling is that the normalization and positive constraints on β in the convex hull formulation allow for this sparse solution. Recall that all β coefficients are positive and the sum of the β coefficients is small (i.e., $\sum_i \beta_i = 1$). Given that the initial β values are chosen to be uniform, and the fact we seek to find a β to maximize (15.32), then naturally only a few β elements will dominate and most β values would evolve to be close to zero.

Comparison with ABCS

To explore the constraints on β in the CH framework, we compare SR-CH to ABCS, an SR method which puts no positive and normalization constraints on β. To fairly analyze the different β constraints in the SR-CH and ABCS methods, we compare both methods only using the exemplar terms, since the GMM-based terms for the two are different. Table 15.4 shows that SR-CH method offers improvements over ABCS on the fBMMI feature set, experimentally demonstrating that constraining β values to be positive and normalized, and not allowing data in H to be reflected and shifted, allows for improved classification accuracy.

GMM-Based Term

In this section we analyze the behavior of SR-CH when using the exemplar-term only versus including the additional model-based term given in (15.31). Table 15.5 shows the classification accuracy on the development set with the fBMMI features. Notice that including the additional $H\beta$ GMM modeling term over the exemplar-based term offers a slight improvement in classification accuracy, demonstrating that including the GMM term allows for a slightly better classifier.

Table 15.5 SR-CH accuracy, TIMIT development set

SR-CH GMM-based term	Accuracy
Exemplar term only	83.86
Exemplar term+ $H\beta$ GMM term	**84.00**

Table 15.6 Classification accuracy, TIMIT core test set

Method	Accuracy fBMMI	Accuracy SA+fBMMI
SR-CH (Ex. + GMM)	**82.87**	**85.14**
ABCS (Ex. + GMM)	81.37	83.22
kNN	81.30	83.56
GMM—BMMI trained	80.82	82.84
SVM	80.79	82.62
GMM—ML trained	79.75	82.02

Comparison with Other Techniques

Table 15.6 compares the classification accuracy of the SR-CH method on the TIMIT core test set to other common classification methods. Note that for ABCS, the best numbers for this method, which include the exemplar and GMM-based terms, are reported. Results are provided for the fBMMI and SA+fBMMI feature sets. Notice that SR-CH outperforms the GMM, kNN and SVM classifiers. In addition, enforcing β to be positive allows for improvements over ABCS. A McNemar's Significance Test indicates that the SR-CH result is statistically significant from other classifiers with a 95 % confidence level. The classification accuracy of 82.87 % achieved in [8] was in 2011 the best number on the TIMIT phone classification task reported when discriminative features are used, beating the previous best single-classifier number of 82.3 % reported in [39]. Finally, when using SA + fBMMI features, the SR-CH method achieves an accuracy of over 85 %.

Accuracy Versus Size of Dictionary

One disadvantage of many exemplar-based methods is that as the number of training exemplars used to make a classification decision increases, the accuracy deteriorates significantly. For example, in the kNN method, this implies that the number of training examples from each class used during voting increases. Similarly, for SR methods, this is equivalent to the size of H growing. Parametric-based classification approaches such as GMMs do not suffer from a degradation in performance for increased training data size.

Figure 15.8 shows the classification error versus number of training-exemplars (i.e. size of H) for different classification methods. Note that the GMM method is trained with all of the training data, and is just shown here as a reference. In addition, since the feature vectors in H have dimension 120, and for our SR methods we assume H is over-complete, we only report results on SR methods when the number of examples in H is larger than 120.

First, observe that the error rates for the two purely exemplar-based methods, namely kNN and ABCS with no model term, increase exponentially as the size of H grows. However, the SR-CH exemplar-only methodology is much more robust

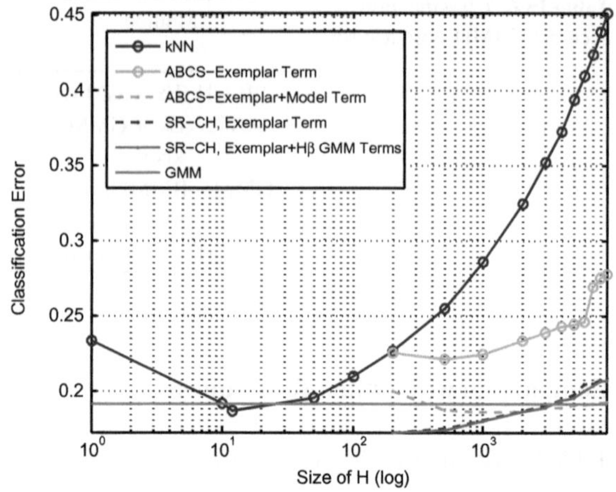

Fig. 15.8 Classification error vs. size of H

with respect to increased size of H, demonstrating the value of the convex hull regularization constraints. Including the extra GMM term into the SR-CH method improves the accuracy slightly. However, the SR-CH method still performs poorly compared to the ABCS technique which uses the GMM-based term. One explanation for this behavior is that GMM term for ABCS is capturing the probability of the data y given the GMM model, and thus the accuracy of the ABCS method eventually approaches the GMM accuracy. However, in SR-CH we capture the probability of $H\beta$ given the GMM. This is one drawback of SR-CH compared to ABCS for large H that we hope to address in the future.

15.6 Sparse Representation Features

In this section, we explore the use of a sparse representation exemplar-based technique [14] to create a new set of features while utilizing the benefits of HMMs to efficiently compare scores across frames. This is in contrast to previous exemplar-based methods which try to utilize the decision scores from the exemplar-based classifiers themselves to generate probabilities ([1, 2]). In our SR approach, given a test vector y and a set of exemplars h_i from the training set, which we put into a dictionary $H = [h_1; h_2 \ldots; h_n]$, we represent y as a linear combination of training examples by solving $y = H\beta$ subject to a spareness constraint on β. The feature $H\beta$ can be thought of as mapping test sample y back into the linear span of training examples in H. We will show that the frame classification accuracy is higher for the SR method[2] compared to a GMM, showing that not only does the $H\beta$ representation move test features closer to training, but also moves these features closer to the

[2] Using SRs to compute accuracy is described in [14].

correct class. Given these new set of $H\beta$ features, we train up an HMM on these features and perform recognition.

A speech signal is defined by a series of feature vectors, $Y = \{y^1, y^2 \dots y^n\}$, for example Mel-Scale Frequency Cepstral Coefficients (MFCCs). For every test sample $y^t \in Y$, we choose an appropriate H^t and then solve $y^t = H^t\beta^t$ to compute a β^t via ABCS. Then given this β^t, a corresponding $H^t\beta^t$ vector is formed. Thus a series of $H\beta$ vectors is created at each frame as $\{H^1\beta^1, H^2\beta^2 \dots H^n\beta^n\}$. The sparse representation features are created for both training and test. An HMM is then trained given this new set of features and recognition is performed in this new feature space.

15.6.1 Measure of Quality

We can measure how well y assigns itself to different classes in H by looking at the residual error between y and the $H\beta$ entries corresponding to a specific class [6]. Ideally, all nonzero entries of β should correspond to the entries in H with the same class as y and the residual error will be smallest within this class. More specifically, let us define a selector $\delta_i(\beta) \in \mathbb{R}^N$ as a vector whose entries are non-zero except for entries in β corresponding to class i. We then compute the residual error for class i as $\| y - H\delta_i(\beta) \|_2$. The best class for y will be the class with the smallest residual error. Mathematically, the best class i^* is defined as

$$i^* = \min_i \| y - H\delta_i(\beta) \|_2 . \tag{15.34}$$

15.6.2 Choices of Dictionary H

Success on the sparse representation features depends heavily on a good choice of H. Pooling together all training data from all classes into H will make the columns of H large (typically millions of frames), and will make solving for β intractable. Therefore, in this section we discussion various methodologies to select H from a large sample set. Recall that H is selected for each frame y, and then β is found using ABCS, in order to create an $H\beta$ feature for each frame.

- **Seeding H from Nearest Neighbors**: For each y, we find a neighborhood of closest points to y in the training set. These k neighbors become the entries of H. We refer the reader to [3] for a discussion on choosing the number of k neighbors for SRs. A set of $H\beta$ features is created for both training and test, but H is always seeded with data from training data. To avoid overtraining of $H\beta$ features on the training set, we require that only when creating $H\beta$ features on training, samples be selected from training that are of a different speaker than the speaker corresponding to frame y. While this kNN approach is computationally feasible on small-vocabulary tasks, using a kNN for large vocabulary tasks can be computationally expensive.

To address this, we discuss other choices for seeding H below, tailored to large vocabulary applications.

- **Using a Trigram Language Model**: Ideally only a small subset of Gaussians are typically evaluated at a given frame, and thus training data belonging to this small subset can be used to seed H. To determine these Gaussians at each frame, we decode the data using a trigram language model (LM), and find the best aligned Gaussian at each frame. For each Gaussian, we compute the 4 other closest Gaussians to this Gaussian. Here closeness is defined by finding Gaussian pairs which have the smallest Euclidean distance between their means. After we find the top five Gaussians at a specific frame, we seed H with the training data aligning to these top five Gaussians. Since this still typically amounts to thousands of training samples in H, we must sample this further. Our method for sampling is discussed in Sect. 15.6.3. We also compare seeding H using the top 10 Gaussians rather than top five.
- **Using a Unigram Language Model**: One problem with using a trigram LM is that this decode is actually the baseline system we are trying to improve upon. Therefore, seeding H with frames related to the top aligned Gaussian is essentially projecting y back down to the same Gaussian which initially identified it. Thus to increase variability between the Gaussians used to seed H and the best aligned Gaussian from the trigram LM decode, we explore using a unigram LM to find the best aligned Gaussian at each frame. Again, given the best aligned Gaussian, the four closest Gaussians to this are found and data from these five Gaussians is used to seed H.
- **Using no Language Model Information**: To further weaken the effect of the LM, we explore seeding H using only acoustic information. Namely, at each frame we find the top five scoring Gaussians. H is seeded with training data aligning to these Gaussians.
- **Enforcing Unique Phonemes**: Another problem with seeding H by finding the five closest Gaussians relative to the best aligned Gaussian is that all of these Gaussians could come from the same phoneme (i.e. phoneme "AA"). Therefore, we explore finding the five closest Gaussians relative to the best aligned such that the phoneme identities of these Gaussians are unique (i.e. "AA", "AE", "AW", etc.). H is then seeded by from frames aligning to these five Gaussians.
- **Using Gaussian Means**: The above approaches of seeding H use actual examples from the training set, which is computationally expensive. To address this, we investigate seeding H from Gaussian means. Namely, at each frame we use a trigram LM to find the best aligned Gaussian. Then we find the 499 closest Gaussians to this top Gaussian, and use the means from these 500 Gaussians to seed H.

15.6.3 Choice of Sampling

As discussed above, if we seed H using all training data belonging to specific Gaussians, this amounts to thousands of training examples in H. We explore two different approaches to sampling a subset of this data for seeding H.

- **Random Sampling**: For each gausssian we want to select training data from, we explore randomly sampling N training examples from the total set of training frames that aligned to this Gaussian. This process is repeated for each of the closest five Gaussians. We reduce the size of N as the "closeness" decreases. For example, for the closest 5 Gaussians, the number of data points N chosen from each Gaussian is 200, 100, 100, 50 and 50 respectively.
- **Sampling Based on Cosine Similarity**: While random sampling offers a relatively quick approach to select a subset of training examples, it does not guarantee that we select "good examples" from this Gaussian which actually are close to frame y. Alternatively, we explore splitting training points aligning to a Gaussian as being $1\sigma, 2\sigma$, etc. away from the mean of the Gaussian. Here σ is chosen to be the total number of training points aligned to this Gaussian, divided by number of samples N we want to sample from this Gaussian. Then within each σ set, we find the training point which has the closest cosine similarity to the test point y. This is repeated for all $1\sigma, 2\sigma$, etc. values. Again the number of samples taken from each Gaussian reduces as "closeness" decreases.

15.6.4 Experiments

The small vocabulary recognition experiments in this paper are conducted on the TIMIT phonetic corpus [16]. Similar to [40], acoustic models are trained on the training set, and results are reported on the core test set. The initial acoustic features are 13-dimensional MFCC features. The large vocabulary experiments are conducted on an English broadcast news transcription task [17]. The acoustic model is trained on 50 h of data from the 1996 and 1997 English Broadcast News Speech Corpora. Results are reported on 3 h of the EARS Dev-04f set. The initial acoustic features are 19-dimensional PLP features.

Both small and large vocabulary experiments utilize the following recipe for training acoustic models [40]. First, a set of CI HMMs are trained, either using information from the phonetic transcription (TIMIT) or from flat-start (broadcast news). The CI models are then used to bootstrap the training of a set of CD triphone models. In this step, at each frame, a series of consecutive frames surrounding this frame are joined together and a Linear Discriminative Analysis (LDA) transform is applied to project the feature vector down to 40 dimensions. Next, vocal tract length normalization (VTLN) and feature space Maximum Likelihood Linear Regression (fMLLR) are used to map the features into a canonical speaker space. Then, a set of discriminatively trained features and models are created using the boosted Maximum

Mutual Information (BMMI) criterion. Finally, the set of models is adapted using MLLR.

We create a set of $H\beta$ features from a set of fBMMI features. We choose this level as these features offer the highest frame accuracy relative to LDA, VTLN, or fMLLR features, allowing us to further improve on the accuracy with with the $H\beta$ features. A set of $H\beta$ features are created at each frame from the fBMMI features for both training and test. A new ML HMM is trained up from these new features and used for both training and test. Since $H\beta$ features create a linear combination of the discriminatively trained fBMMI features, we argue that some discrimination can be lost. Therefore, we explore applying another fBMMI transformation to the $H\beta$ features before applying model space discriminative training and MLLR.

In what follows we present results using $H\beta$ features on both small and large vocabulary tasks.

15.6.5 Sparsity Analysis

We first analyze the β coefficients obtained by solving $y = H\beta$ using ABCS [3]. For two randomly selected frames y, Fig. 15.9 shows the β coefficients corresponding to 200 entries in H for TIMIT and 500 entries for Broadcast News. Notice that for both datasets, the β entries are quite sparse, illustrating that only a few samples in H are used to characterize y. As [6] discusses, this sparsity can be thought of as a form of discrimination, as certain examples are selected as "good" in H while jointly assigning zero weights "bad" examples in H. We have seen advantages of the

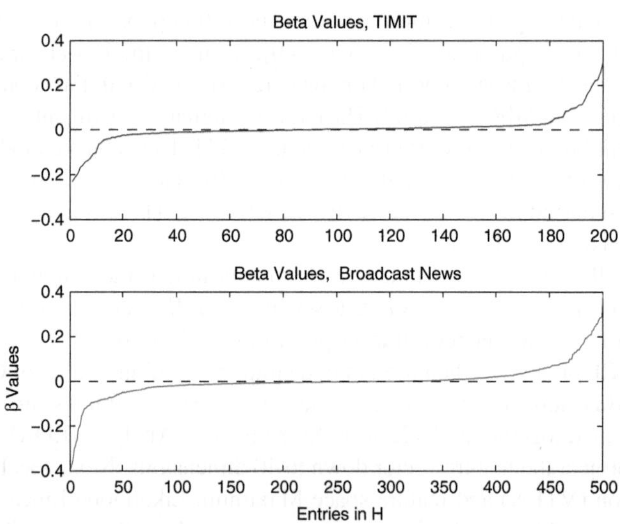

Fig. 15.9 β coefficients on TIMIT and broadcast news

Table 15.7 Frame accuracy on TIMIT testcore set

Classifier	frame Accuracy
GMM	70.4
Sparse representations	**71.7**

SR approach for classification, even on top of discriminatively trained y features, compared to a GMM [3]. We will also re-confirm this behavior in Sect. 15.6.6. The extra benefit of SRs on top of discriminatively trained fBMMI features, coupled with an exemplar-based nature of SRs, motivates us to further explore its behavior for recognition tasks.

15.6.6 TIMIT Results

Frame Accuracy

The success of $H\beta$ first relies on the fact that the β vectors give large support to correct classes and small support to incorrect classes (as demonstrated by Fig. 15.9) when computing $y = H\beta$ at each frame. Thus, the classification accuracy per frame, computed using (15.34), should ideally be high. Table 15.7 shows the frame accuracy for the GMM and SR methods.

Notice that the SR technique offers significant improvements over the GMM method, again confirming the benefit of exemplar-based classifiers.

Error Rate for $H\beta$ Features

Table 15.8 shows the recognition performance of $H\beta$ features on TIMIT. Due to the small vocabulary nature of TIMIT, we only explore seeding H from nearest neighbors. Notice that creating a set of $H\beta$ features in the fBMMI space offers a 0.7 % absolute improvement in PER. Given the small vocabulary nature of TIMIT, no gain was found applying another fBMMI transform to the baseline or $H\beta$ features. After applying BMMI and MLLR to both feature sets, the $H\beta$ features offer a 0.5 % improvement in PER over the baseline system. This shows that using exemplar-based SRs to produce $H\beta$ features not only moves test features closer to training, but also moves the feature vectors closer to the correct class, resulting in a decrease in PER.

Table 15.8 WER on TIMIT

Baseline system	PER	$H\beta$ System	PER
fBMMI	19.9	$H\beta$	**19.2**
+BMMI +MLLR	19.5	+BMMI +MLLR	**19.0**

15.6.7 Broadcast News Results

Selection of *H*

Table 15.9 shows the WER for the $H\beta$ features for different H choices discussed in Sect. 15.6.2. Note that the baseline fMMI system as a WER of 21.1 %. The following can be observed:

- There is little difference in WER when sampling is done randomly or using cosine similarity. For speed efficiencies, we use random sampling for H selection methods.
- There is little difference between using 5 and 10 Gaussians.
- Seeding H using nearest neighbors is worse than using the trigram LM. On broadcast news, we find that a kNN has lower frame-accuracy than a GMM, a result similarly observed in the literature for large vocabulary corpora [1]. This lower frame accuracy translates into a higher WER when H is seeded with nearest neighbors.
- Seeding H from unique Gaussians provides too much variability of phoneme classes into the $H\beta$ feature, also leading to a higher WER.
- Using a unigram LM to reduce the link between the Gaussians used to seed H and the best aligned Gaussian from the trigram LM decode offers a slight improvement in WER over the trigram LM.
- Utilizing no LM information results in a very high WER.
- Using Gaussian means to seed H reduces the computation to create $H\beta$ without a large increase in WER.

WER for $H\beta$ Features

Table 15.10 shows the performance of $H\beta$ features on the Broadcast News task. Creating a set of $H\beta$ features at the fBMMI space offers a WER of 21.1 % which is comparable to the baseline system. However, after applying an fBMMI transform to the $H\beta$ features we achieve a WER of 20.2 %, a 0.2 % absolute improvement when another fBMMI transform is applied to the original fBMMI features. Finally,

Table 15.9 WER of $H\beta$ features for different H

H selection method	WER
Trigram LM, random sampling, top 5 Gaussians	21.2
Trigram LM, cosine similarity sampling, top 5 Gaussians	21.3
Trigram LM, top 10 Gaussians	21.3
Nearest neighbor, 500	21.4
Trigram LM, 5 unique Gaussians	21.6
Unigram LM, top 5 Gaussians	**21.1**
No LM information, top 5 Gaussians	22.7
Gaussian means, top 500 Gaussians	21.4

Table 15.10 WER on broadcast news

Baseline system	WER	$H\beta$ system	WER
fBMMI	21.1	$H\beta$	21.1
+fBMMI	20.4	+fBMMI	**20.2**
+BMMI +MLLR	19.0	+BMMI +MLLR	**18.7**

after applying BMMI and MLLR to both feature sets, the $H\beta$ features offer a WER of 18.7 %, a 0.3 % absolute improvement in WER over the baseline system. This demonstrates again that using information about actual training examples to produce a set of features which are mapped closer to training and have a higher frame accuracy than GMMs improves accuracy for large vocabulary as well.

15.7 SR Phone Identification Features (S_{pif})

In this section, we review the use of SR for classification and use this framework to create our S_{pif} features. Let us, first, describe how we can use β to create a set of S_{pif} vectors. First, define matrix $H_{phnid} = [p_{1,1}, p_{1,2}, \ldots, p_{w,n_w}] \in \mathbb{R}^{r \times N}$, which has the same number of columns N as the original H, but a different number of rows r. Recall that each $x_{i,j} \in H$ has a corresponding class label i. We define each $p_{i,j} \in H_{phnid}$ corresponding to feature vector $x_{i,j} \in H$ to be a vector with zeros everywhere except at the index i corresponding to class of $x_{i,j}$. Figure 15.10 shows the H_{phnid} corresponding to H, where each $p_{i,j}$ becomes a phone identification vector with a value of 1 corresponding to the class of $x_{i,j}$. Here r, the dimension of each $p_{i,j}$, is equivalent to the total number of classes.

Once β is found by solving $y = H\beta$, we use this same β to select important classes within the new dictionary H_{phnid}. Specifically, let us define a new feature vector S_{pif}, as $S_{pif} = H_{phnid}\beta^2$, where each element of β is squared, i.e., $\beta^2 = \{\beta_i^2\}$. Notice that we are using β^2, as this is similar to the $\| \delta_i(\beta) \|_2$ classification rule given by (15.34). Each row i of the S_{pif} vector roughly represents the l_2 norm of β entries for class i.

A speech signal is defined by a series of feature vectors, $Y = \{y^1, y^2 \ldots y^n\}$, for example Mel-Scale Frequency Cepstral Coefficients (MFCCs). For every test sample $y^t \in Y$, we solve $y^t = H^t\beta^t$ to compute a β^t. Then given this β^t, a corresponding

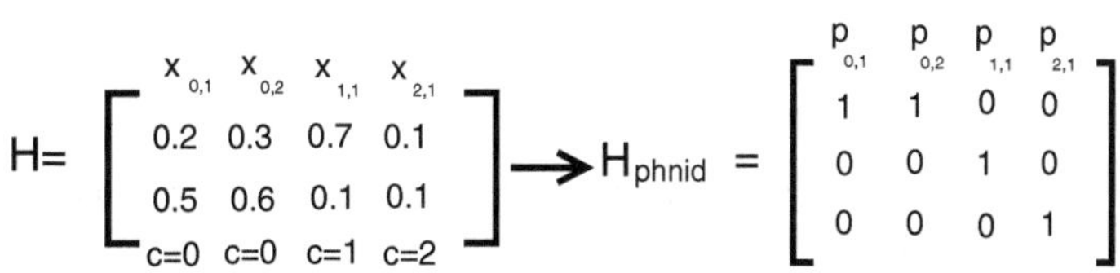

Fig. 15.10 H_{phnid} corresponding to H

S^t_{pif} vector is formed. Since β^t at each sample represents a weighting of entries in H^t that best represent test vector y^t, this makes it difficult to compare β^t values and the S^t_{pif} vectors across frames. Therefore, to ensure that the values can be compared across samples, the S^t_{pif} vectors are normalized at each sample. Thus, the new $\bar{S}^t_{pif}$ at sample t is computed as $\bar{S}^t_{pif} = \frac{S^t_{pif}}{\|S^t_{pif}\|_1}$. A series of S_{pif} vectors is created as $\{\bar{S}^1_{pif}, \bar{S}^2_{pif} \dots \bar{S}^n_{pif}\}$, and are used for recognition.

15.7.1 Construction of Dictionary H

Success of SRs depends on a good choice of H. In [14], various methods for seeding H from a large sample set were explored. Below we summarize the main techniques used in this work to select H.

Seeding H from Nearest Neighbors

For each y, we find a neighborhood of closest points to y from all examples in the training set. These k neighbors become the entries of H. While this approach works well on small-vocabulary tasks, it is computationally expensive for large data sets.

Using a Language Model

In speech recognition, when an utterance is scored using a set of HMMs (which have output distributions given by Gaussians), typically evaluating only a small subset of these Gaussians at a given frame allows for a large improvement in speed without a reduction in accuracy [41]. Using this fact, we use training data belonging to a small subset of Gaussians to seed H. To determine these Gaussians at each frame, we decode the data using a language model (LM), and find the best aligned Gaussian at each frame. For each Gaussian, we compute the four other closest Gaussians to this Gaussian. After we find the top five Gaussians at a specific frame, we seed H with the training data aligning to these top five Gaussians. We explore using both a trigram and unigram LMs to obtain the top Gaussians.

Using a Lattice

Seeding H as suggested above is similar to finding the best H at the frame level. However, the goal of speech recognition is to recognize words, and therefore we explore seeding H using information related to competing word hypotheses. Specifically, we create a lattice of competing word hypotheses and obtain the top Gaussians at each frame from the Gaussian alignments of the lattice. Gaussians to the best Gaussian are found and data from these five Gaussians is used to seed H.

15.7.2 Reducing Sharpness Estimation Error

As described in Sect. 15.7.1, for computational efficiency, S_{pif} features are created by first pre-selecting a small amount of data for dictionary H. This implies that only a few classes are present in H and only a few S_{pif} posteriors are non-zero, something we will define as feature sharpness. Feature sharpness by itself is advantageous—for example if we were able to correctly predict the right class at each frame and capture this in S_{pif} the WER would be close to zero. However, because we are limited by the amount of data that can be used to seed H, incorrect classes may have their probabilities boosted over correct classes, something we will refer to as sharpness estimation error. In this section, we explore various techniques to smooth out the sharp S_{pif} features and reduce estimation error.

Choice of Class Identification

The S_{pif} vectors are defined based on the class labels in H. We explore two choice of class labels in this paper. First, we explore using monophone class labels. Second, we investigate labeling classes in H by a set of context independent (CI) triphones. While using triphones increases the dimension of the S_{pif} vector, the elements in the vector are less sharp now since β values for a specific monophone are more likely to be distributed within the three different triphones of this monophone.

Posterior Combination

Another technique to reduce feature sharpness is to combine S_{pif} posteriors with posteriors coming from an HMM system, a technique which is often explored when posteriors are created using Neural Nets [17]. Specifically, let us define $h^j(y_t)$ as the output distribution for observation y_t and state j of an HMM system. In addition, define $S^j_{pif}(y_t)$ as the S_{pif} posterior corresponding to state j. Note that the number of S_{pif} posteriors could be less than the number of HMM states, so the same S_{pif} posterior could map to multiple HMM states. For example, the S_{pif} posterior corresponding to phone "aa" could map to HMM states "aa-b-0", "aa-m-0", etc. Given the HMM and S_{pif} posteriors, the final output distribution $b^j(y_t)$ is given by Eq. 15.35, where λ is a weight on the S_{pif} posterior stream, selected on a held-out set.

$$b^j(y_t) = h^j(y_t) + \lambda S^j_{pif}(y_t) \tag{15.35}$$

S_{pif} Feature Combination

As we will show in Sect. 15.7.5, S_{pif} features created using different methodologies to select H offer complementary information. For example, S_{pif} features created when H is seeded with a lattice have higher frame accuracy and incorporate more sequence information than when H is seeded using a unigram or trigram LM. However, S_{pif} features created from lattice information are much sharper compared to features created with a uni/trigram LM. Thus, we explore combining different S_{pif} features. If we denote S_{pif}^{tri}, S_{pif}^{uni} and S_{pif}^{lat} as being created from the three different H selection methodologies, we combine these features to produce a new S_{pif}^{comb} feature as given by Eq. 15.36. Weights $\{\alpha, \beta, \gamma\}$ are chosen on a held-out set with the constraint that $\alpha + \beta + \gamma = 1$.

$$S_{pif}^{comb} = \alpha S_{pif}^{tri} + \beta S_{pif}^{uni} + \gamma S_{pif}^{lat} \tag{15.36}$$

15.7.3 Experiments

The small vocabulary recognition experiments are conducted on TIMIT [16]. Similar to [14], acoustic models are trained on the training set, and results are reported on the core test set. The initial acoustic features are 13-dimensional MFCC features. The large vocabulary experiments are conducted on an English broadcast news transcription task [17]. The acoustic model is trained on 50 h of data from the 1996 and 1997 English Broadcast News Speech Corpora. Results are reported on 3 h of the EARS Dev-04f set. The initial acoustic features are 19-dimensional PLP features.

Both corpora utilize the following recipe for training. First, a set of CI HMMs are trained, either using information from the phonetic transcription (TIMIT) or from flat-start (Broadcast News). The CI models are then used to bootstrap the training of a set of CD triphone models. In this step, given an initial set of MFCC or PLP features, a set of LDA features are created. After the features are speaker adapted, a set of discriminatively trained features and models are created using the boosted Maximum Mutual Information (BMMI) criterion. Finally, models are adapted via MLLR.

On TIMIT, we explore creating S_{pif} features from both LDA and fBMMI features, while for Broadcast news, we only create S_{pif} features after the fBMMI stage. The initial LDA/fBMMI features are used for both y and H to solve $y = H\beta$ and crate S_{pif} features at each frame. In this work, we explore the ABCS method. Once series of S_{pif} vectors are created, an HMM is built on the training features.

15.7.4 TIMIT Results

Frame Accuracy

The success of S_{pif} first relies on the fact that the classification accuracy per frame, computed using Eq. 15.34, should ideally be high. Table 15.11 shows the classification accuracy for the GMM and SR methods,[3] for both LDA and fBMMI feature spaces. Notice that the SR technique offers significant improvements over the GMM method.

Recognition Results: Class Identification

Table 15.12 shows the phonetic error rate (PER) at the CD level for different class identification choices. Since only a kNN is used to seed H on TIMIT, we will call the feature S^{knn}_{pif}. We have also listed results for other CD-ML trained systems reported in the literature on TIMIT. Notice that smoothing out sharpness error of the S_{pif} features by using triphones rather than monophones results in a decrease in error rate. The S_{pif}-triphone features outperform the LDA features and also offer the best result of all methods on TIMIT at the CD level for ML trained systems.

We further explore S_{pif} features created after the fBMMI stage. Table 15.13 shows that the performance is now worse than the fBMMI system. Because the fBMMI features are already discriminative in nature and offer good class separability, S_{pif} features created in this space are too sharp, explaining the increase in PER.

Recognition Results: Posterior Combination

We explore reducing feature sharpness by combining S_{pif} posteriors with HMM posteriors, as shown in Table 15.14. We observe that on TIMIT, combining posteriors from two different feature streams has virtually no impact in recognition accuracy compared to the baseline fBMMI system, indicating there is little complementarity between the two systems. Because gains were not observed with posterior combination, further S_{pif} feature combination was not explored.

Table 15.11 Frame accuracy on TIMIT testcore set

Classifier	Frame Acc. (LDA)	Frame Acc. (fBMMI)
GMM	61.5	70.4
SR	**64.0**	**71.7**

[3] We have not included the accuracy of the HMM since this takes into account sequence information which both the GMM and SR methods do not.

Table 15.12 PER on TIMIT core test set—CD ML trained systems

System	PER (%)
S_{pif}^{knn} monophones, IBM CD HMM (this paper)	25.1
Monophone HTMs [42]	24.8
Baseline LDA features, IBM CD HMM	24.5
Heterogeneous measurements [43]	24.4
S_{pif}^{knn} triphones, IBM CD HMM (this paper)	**23.8**

Table 15.13 PER on TIMIT core test set—fMMI level

Features	PER
Baseline fBMMI features	19.4
S_{pif}^{knn} triphones	20.7

Table 15.14 PER on TIMIT core test set—posterior combination

Features	PER
Baseline fBMMI Features	19.4
S_{pif}^{knn}, Posterior Combination	19.4

15.7.5 Broadcast News

In this section we explore the S_{pif} features on Broadcast News.

Recognition Results: Choice of *H* and Class Identity

Table 15.15 shows the frame accuracy and WER on Broadcast news for different choice of H and class identity. We also quantify the sharpness estimation error between the different S_{pif} methods. We define "sharpness" of a S_{pif} vector by calculating the entropy from the non-zero probabilities of the feature. The sharper the S_{pif} feature, the lower the entropy. A very sharp S_{pif} feature that emphasizes the incorrect class for a frame will lead to a classification error. Therefore, we measure sharpness error by the average entropy of all misclassified S_{pif} frames. Please note that sharpness is only measured for monophone S_{pif} features. Using triphone S_{pif} smooths out class probabilities since the feature dimension is increased. However, it is difficult to quantifiably compare feature sharpness for the monophone and triphone S_{pif} features since the correct phone labels and dimensions are of the two features are different.

First, notice the trend between frame accuracy and entropy in Table 15.15. S_{pif}^{uni} features have a low frame accuracy and hence a low WER. While S_{pif}^{lat} features have a very high frame accuracy, they have a higher entropy on misclassified frames compared to S_{pif}^{tri} and S_{pif}^{uni}, and hence have a high WER. S_{pif}^{tri} features created from a trigram LM offer the best tradeoff between feature sharpness and accuracy, and achieve a WER close to the baseline. However, if feature sharpness is reduced by

Table 15.15 WER on broadcast news, class identification

Features	Frame Acc.	S_{pif} Entropy Error Frames	WER
Baseline fBMMI, ML training	–	–	19.4
S_{pif}^{tri} monophones	70.3	2.27	19.5
S_{pif}^{uni} monophones	68.3	2.23	29.0
S_{pif}^{lat} monophones	77.2	0.86	21.6
S_{pif}^{tri} triphones	–	–	19.8

using triphone S_{pif}^{tri} features, we see now on a word recognition task that the WER increases slightly.

Oracle Results of Reducing Estimation Error

We motivate the need for reducing sharpness error, with the following oracle experiment. Given the S_{pif}^{tri}-monophone features, x % of the frames which are misclassified are corrected to have a probability of 1 at the correct phone index and 0 elsewhere. Table 15.16 shows the results when 1 %, 3 %, and 5 % of the misclassified S_{pif} features are corrected. Notice that just by correcting a small % of misclassified features, the WER reduces significantly. This motivates us to explore different techniques to reduce S_{pif} sharpness in the next section.

Recognition Results: Posterior and S_{pif} Combination

In this section, we explore reducing sharpness through posterior and S_{pif} combination. Table 15.17 shows the baseline results for the fBMMI and S_{pif}-monphone features at 18.7 % and 19.5 % respectively. The frame accuracies and entropies of misclassified frames for various S_{pif} combination features are also listed. Note that the frame accuracy is only reported on the S_{pif} feature and does not include frame accuracy after posterior combination.

First, notice that through posterior combination, we reduce the WER by 0.5 % absolute from 18.7 % to 18.2 %, showing the complementarity between the fBMMI and S_{pif} feature spaces. Second, by doing additional S_{pif} feature combination, we

Table 15.16 WER on broadcast news, oracle results

Features	Frame accuracy	WER
S_{pif}^{tri} 0 % cheating	70.3	19.5
S_{pif}^{tri} 1 % cheating	71.4	19.4
S_{pif}^{tri} 3 % cheating	73.7	18.8
S_{pif}^{tri} 5 % cheating	76.1	17.6

Table 15.17 WER on broadcast news, posterior and S_{pif} combination

Features	Frame Acc.	S_{pif} Ent.	WER
Baseline fBMMI features, BMMI training + MLLR	–	–	18.7
S_{pif}^{tri} monophones	70.3	2.27	19.5
S_{pif}^{tri}, posterior combination	70.3	2.27	**18.2**
$\alpha S_{pif}^{tri} + \beta S_{pif}^{uni} + \gamma S_{pif}^{lat}$, posterior combination	76.3	2.29	**17.8**

are able to increase the frame accuracy from 70.3 % to 76.3 %, without a reduction in S_{pif} entropy as it increases slightly from 2.27 to 2.29. This results in a further decrease in WER of 0.4 % absolute from 18.2 % to 17.8 %, indicating the importance of reducing feature sharpness, particularly for misclassified S_{pif} frames.

15.8 Enhancing Exemplar-Based Posteriors for Speech Recognition Tasks

When errors occur in exemplar modeling, this results in wrong classes having their probabilities over-emphasized, something we will refer to as feature or posterior sharpness. In general, it can be argued that a more desired methodology for enhancing the posteriors is the one that simultaneously improves the frame accuracy and reduces the erratic sharpness across the frames. Given that through a NN transformation we have enhanced the posteriors by improving the frame error rate, we explore a new technique to smooth the posteriors. Specifically, we explore a technique similar to the tied mixture approach [20] where new posteriors are modeled as a tied mixture of the NN posteriors. Specifically, given feature o_t and a set of NN posterior scores $p(s_i|o_t)$ for all classes $i \in L$, we can estimate the posterior for state s_j as given by

$$p(s_j|o_t) = \sum_{i=1}^{L} p(s_i|o_t)p(s_j|o_t, s_i) \tag{15.37}$$

As in the tied mixture approach [20], a tying is invoked such that the term $p(s_j|o_t, s_i)$ for a given i is independent of o_t, which reduces (15.37) to

$$p(s_j|o_t) = \sum_{i=1}^{L} p(s_i|o_t)p(s_j|s_i) \tag{15.38}$$

where $p(s_j|s_i)$ is a set of mixing coefficients. Mixing NN posteriors from different classes helps to smooth over sharp posterior distributions [20].

In this section we look to learn a set of mixing coefficients $p(s_j|s_i)$ to mix state based posteriors from different states. More formally, we will refer to the $NN - S_{pif}$ posteriors $p(s_i|o_t)$ as a. If we assume there are L states, then the posterior probability $a_t(l)$ at time t for state l satisfies the following properties:

$$a_t(l) \geq 0 \quad \text{and} \quad \sum_{l=1}^{L} a_t(l) = 1 \tag{15.39}$$

Given state l and a set of $k = \{1, \ldots, L\}$ NN posteriors for this state l, we define a mixing coefficient $p(s_j|s_i)$ as $b(l, k)$, which satisfies the following properties:

$$b(l, k) \geq 0 \quad \text{and} \quad \sum_{k=1}^{L} b(l, k) = 1 \tag{15.40}$$

Our objective is to learn a set of mixing coefficients $b(l, k)$ via maximum likelihood. In this paper, we explore maximizing an objective function which linearly interpolates the original posteriors a, similar to the tied mixture approach [20]. Specifically, consider all frames aligned to a state l from $t = 1$ to T_l. We can define the mixed posterior for a specific frame t as

$$c_t(l) = \sum_{k=1}^{L} b(l, k)a_t(k) \tag{15.41}$$

It is easy to see that $c_t(l)$ satisfies (15.40) and is a posterior. The objective function of this posterior across all frames in the training data aligned to state l is given by

$$f_l(b) = \prod_{t=1}^{T_l} c_t(l) = \prod_{t=1}^{T_l} (\sum_{k=1}^{L} b(l, k)a_t(k)) \tag{15.42}$$

Because (15.42) is a polynomial with positive coefficients, the Baum-Welch update equation can be used to iteratively solve for $b(l, k)$ which maximizes the above objective function. The recursive update equation for $b(l, k)$ is given by

$$b(l, k) := \frac{b(l, k)\nabla_{b(l,k)} f_l(b)}{\sum_{j=1}^{L} b(l, j)\nabla_{b(l,j)} f_l(b)} \tag{15.43}$$

Here the gradient of the objective function $f_l(b)$ is

$$\nabla_{b(l,k)} f_l(b) = \sum_{t=1}^{T_l} f_l(b) \frac{a_t(k)}{\sum_{i=1}^{L} b(l, i)a_t(i)} \tag{15.44}$$

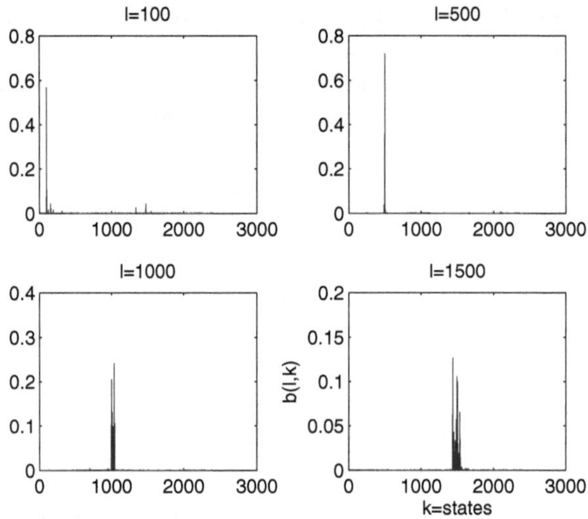

Fig. 15.11 Mixing coefficient examples

Substituting the gradient (15.44) into the update formula (15.43) yields the following update for $b(l, k)$

$$b(l, k) := \frac{1}{T_l} \sum_{t=1}^{T_l} \frac{b(l, k)a_t(k)}{\sum_{i=1}^{L} b(l, i)a_t(i)} \tag{15.45}$$

This equation shows that the mixing coefficients $b(l, k)$ learned for state l effectively take a linearly weighted average of posterior coefficients a over all training frames aligned to state l.

Note that (15.45) assumes an initial value of $b(l, k)$. We assume that the initial $b(l, k)$ is uniformly distributed as $1/L$ where L is the number of states. $b(l, k)$ is iteratively updated using (15.45) until the change in the objective function value between iterations is below a specified threshold.

Once $b(l, k)$ is learned, given state l, and the $NN - S_{pif}$ posteriors (denoted by a), a new posterior for state l is computed by taking a weighted average of the NN posteriors and mixing coefficients. This new posterior, denoted by $NN - S_{pif} - Post^{(l)}$ for state l is given by

$$NN - S_{pif} - Post^{(l)} = \sum_{k=1}^{L} b(l, k)a_t(k) \tag{15.46}$$

Figure 15.11 plots the mixing coefficients $b(l, k)$ for states $l = 100, 500, 1,000$, and $1,500$. We can observe that for all states, the non-zero mixing coefficients are clustered together, and thus come from context-dependent states which are similar to each other, for example states which map to the same monophone.

15.8.1 Results

The following experiments were conducted as described in Sect. 15.7.3.

Using S_{pif} Features As Output Probabilities

First, we explore the performance of S_{pif} posteriors when used as output probabilities directly in an HMM system. Table 15.18 shows that the performance of the S_{pif} posteriors is worse than the baseline GMM/HMM system trained on fBMMI features, illustrating the problem with deriving exemplar-based posterior features which are not learned through a discriminative process linked to WER. Furthermore, combining S_{pif} and GMM posteriors in tandem does not offer improvements over the baseline GMM/HMM system.

Enhancing Using Neural Networks

Second, we explore the performance of training a NN with S_{pif} features as input, and then again using the $NN - S_{pif}$ probabilities as output probabilities in an HMM system. Table 15.19 shows that the $NN - S_{pif}$ features offers a 1.3 % absolute reduction in WER over using S_{pif} features alone. This illustrates the importance of enhancing S_{pif} posteriors with a NN to create a set of posteriors better aligned the PER objective in speech. Furthermore, the PER of 19.0 % is better than the GMM/HMM system trained with fBMMI features [21], as well as a NN trained with fBMMI features [44]. This demonstrates the benefit of exemplar-based features over standard speech features (i.e. fBMMI).

Table 15.18 PER on TIMIT core test set, S_{pif} features

Features	PER
GMM/HMM fBMMI	**19.5**
S_{pif} posteriors	20.3
Tandem: S_{pif} + GMM	19.5

Table 15.19 PER on TIMIT core test set, NN enhancement

Features	PER
S_{pif}	20.3
$NN - S_{pif}$	**19.0**
GMM/HMM—fBMMI + BMMI + MLLR [21]	19.4
NN—fBMMI [44]	19.4

Table 15.20 PER on TIMIT Core Test set, posterior smoothing

Features	PER
$NN - S_{pif}$	19.0
$NN - S_{pif} - Post$	**18.7**

Smoothing with Posterior Modeling

Finally, we explore smoothing out $NN - S_{pif}$ posteriors through tied mixtures as discussed in this section. Again, mixed posteriors $NN - S_{pif} - Post$ are used as output probabilities in an HMM system. Table 15.20 shows that using posterior modeling, we can obtain a small improvement of 0.3 % absolute over the $NN - S_{pif}$ posteriors. illustrating the value of reducing posterior sharpness through tied mixture smoothing.

Error Analysis

Figure 15.12 shows the breakdown of error rates for the GMM/HMM, $NN - S_{pif}$ and $NN - S_{pif} - Post$ methods within six BPCs, namely vowels/semivowels, nasals, strong fricatives, weak fricatives, stops and closures/silence. Here the error rate was calculated by counting the number of insertions, deletions and substitutions that occur for all phonemes within a particular BPC. The $NN - S_{pif}$ method offers improvements over the GMM/HMM system in all classes except nasals and closures. Furthermore, we can see the gains with the $NN - S_{pif} - Post$ method are coming due to better modeling in the vowel, weak fricative and closure classes.

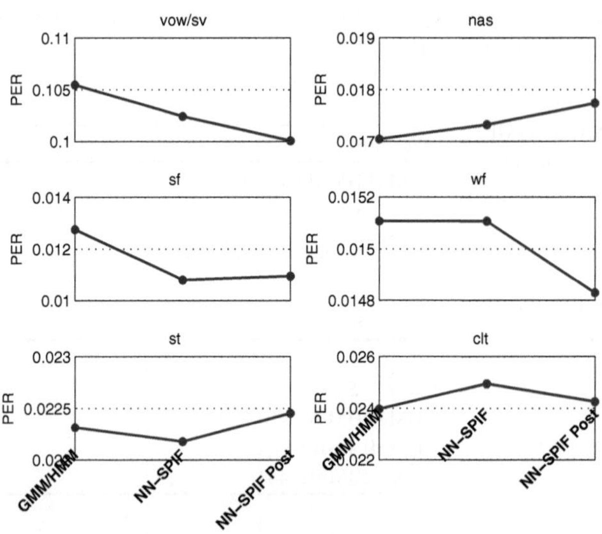

Fig. 15.12 Error rates within 6 BPCs for various methods

References

1. Deselaers T, Heigold G, Ney H (2007) Speech recognition with state-based nearest neighbour classifiers. In: Proceedings of the interspeech.
2. Gemmeke JF, Virtanen T (2010) Noise robust exemplar-based connected digit recognition. In: Proceedings of the ICASSP.
3. Sainath TN, Carmi A, Kanevsky D, Ramabhadran B (2010) Bayesian compressive sensing for phonetic classification. In: Proceedings of the ICASSP.
4. De Wachter M, Demuynck K, Van Compernolle D, Wambacq P (2003) Data driven example based continuous speech recognition. In: Proceedings of the european conference on speech communication and technology.
5. Tychonoff A, Arseny V (1977) Solution of ill-posed problems. Winston and Sons, Washington
6. Wright J, Yang A, Ganesh A, Sastry SS, Ma Y (2009) Robust face recognition via sparse representation. IEEE Trans Pattern Anal Mach Intell 31: 210–227
7. Carmi A, Gurfil P, Kanevsky D, Ramabhadran B (2009) ABCS: approximate bayesian compressive sensing. Technical Report Human Language Technologies, IBM
8. Sainath TN, Nahamoo D, Kanevsky D, Ramabhadrans B, Shah PM (2011) A convex hull approach to sparse representations for exemplar-based speech recognition. In: Proceedings of the ASRU.
9. Sainath T, Ramabhadran B, Olsen P, Kanevsky D, Nahamoo D (2011) A-Functions: a generalization of extended baum-welch transformations to convex optimization. In: Proceedings of the ICASSP.
10. Kanevsky D, Sainath TN, Ramabhadran B, Nahamoo D (2010) An analysis of sparseness and regularization in exemplar-based methods for speech classification. In: Proceedings of the interspeech.
11. Tibshirani R (1996) Regression shrinkage and selection via the lasso. J Roy Stat Soc Ser B (Methodol.) 58(1):267–288
12. Ji S, Xue Y, Carin L (2008) Bayesian compressive sensing. IEEE Trans Signal Process 56:2346–2356
13. Zou H, Hastie T (2005) Regularization and variable selection via the elastic net. J R Statist Soc B 67:301–320
14. Sainath TN, Ramabhadran B, Nahamoo D, Kanevsky D, Sethy A (2010) Exemplar-based sparse representation features for speech recognition. In: Proceedings of the interspeech.
15. Sainath TN, Nahamoo D, Ramabhadran B, Kanevsky D, Goel V, Shah PM (2011) Exemplar-based sparse representation phone identification features. In: Proceedings of the ICASSP.
16. Lamel L, Kassel R, Seneff S (1986) Speech database development: design and analysis of the acoustic-phonetic corpus. In: Proceedings of the DARPA speech recognition, workshop.
17. Kingsbury B (2009) Lattice-based optimization of sequence classification criteria for neural-network acoustic modeling In: Proceedings of the ICASSP.
18. De Wachter M, Matton M, Demuynck K, Wambacq P, Cools R, Van Compernolle D (2007) Template based continuous speech recognition. IEEE Trans Audio Speech Lang Process 15(4):1377–1390
19. Sainath TN, Ramabhadran B, Nahamoo D, Kanevsky D, Sethy A (2012) Enhancing exemplar-based posteriors for speech recognition tasks. In: Proceedings of the interspeech.
20. Bellegarda J, Nahamoo D (1990) Tied mixture continuous parameter modeling for speech recognition. IEEE Trans Acous Speech Signal Process 38(12):2033–2045
21. Sainath TN, Ramabhadran B, Picheny M, Nahamoo D, Kanevsky D (2011) Exemplar-based sparse representation features: From TIMIT to LVCSR. IEEE Trans Acous Speech and Signal Process 19(8):2598–2613
22. Candes EJ, Romberg J, Tao T (2006) Robust uncertainty principles: exact signal reconstruction from highly incomplete frequency information. IEEE Trans Inf Theory 52:489–509
23. Candes EJ (2006) Compressive sampling. Proceedings of the international congress of mathematicians, European Mathematical Society, Madrid, Spain

24. Gopalakrishnan PS, Kanevsky D, Nahamoo D, Nadas A (1991) An inequality for rational functions with applications to some statistical estimation problems. IEEE Trans. Information Theory 37(1): 107–113
25. Povey D (2003) Discriminative training for large vocabulary speech recognition. Ph.D. thesis, Cambridge University.
26. Sainath T, Ramabhadran B, Olsen P, Kanevsky D, Nahamoo D (2011) Convergence of line search a-function methods. In: Proceedings of the interspeech.
27. Kanevsky D (2005) Extended baum transformations for general functions, II", Technical Report, RC23645(W0506–120). Human Language Technologies, IBM
28. Carmi A, Gurfil P, Kanevsky D Ramabhadran B (2009) Extended compressed sensing: filtering inspired methods for sparse signal recovery and their nonlinear variants. Technical Report, RC24785, Human Language Technologies, IBM.
29. Carmi A, Gurfil P, Kanevsky D, Ramabhadran B (2009) ABCS: Approximate bayesian compressed sensing. Technical Report, RC24816, Human Language Technologies, IBM.
30. Carmi A, Gurfil P, Kanevsky D (April 2010) Methods for signal recovering using kalman filtering with embedded pseudo-measurement norms and quasi-norms. IEEE Trans Signal Process 58(4):2405–2409
31. Horesh L, Gurfil P, Ramabhadran B, Kanevsky D, Carmi A, Sainath TN (2010) Kalman filtering for compressed sensing. In: Proceedings of the information fusion, Edinburgh.
32. Ji S, Xue Y, Carin L (June 2008) Bayesian compressive sensing. IEEE Trans Signal Process 56:2346–2356
33. Efron B, Hassie B, Johnstone T, Tibshirani R (2004) Least angle regression. Ann Stat 32(2):407–451
34. Carmi A, Gurfil P (2009) Convex feasibility programming for compressed sensing. Technical Report, Technion
35. Mount D, Arya S (2006) ANN: A library for approximate nearest neighbor searching. Software available at http://www.cs.umd.edu/ mount/ANN/
36. Chang C, Lin C (2001) LIBSVM: A library for support vector machines. Software available at http://www.csie.ntu.edu.tw/ cjlin/libsvm
37. Kanevsky D (2004) Extended baum transformations for general functions. In: Proceedings of the ICASSP.
38. Povey D, Kanevsky D, Kingsbury B, Ramabhadran B, Saon G, Visweswariah K (2008) Boosted MMI for model and feature space discriminative training. In: Proceedings of the ICASSP.
39. Chang H, Glass J (2007) Hierarchical large-marging gaussian mixture models for phonetic classification. In: Proceedings of the ASRU.
40. Sainath TN, Ramabhadran B, Picheny M (2009) An exploration of large vocabulary tools for small vocabulary phonetic recognition. In: Proceedings of the ASRU.
41. Saon G, Zweig G, Kingsbury B, Mangu L, Chaudhari U (2003) An architecture for rapid decoding of large vocabulary conversational speech. In: Proceedings of the eurospeech.
42. Deng L, Yu D (2007) Use of differential cepstra as acoustic features in hidden trajectory modeling for phonetic recognition. In: Proceedings of the ICASSP.
43. Halberstat A, Glass J (1998) Heterogeneous measurements and multiple classifiers for speech recognition. In: Proceedings of the ICSLP.
44. Mohamad A, Sainath TN, Dahl G, Ramabhadrans B, Hinton GE, Picheny M (2011) Deep belief networks using discriminative features for phone recognition. In: Proceedings of the ICASSP.

MIX
Papier aus verantwortungsvollen Quellen
Paper from responsible sources
FSC® C105338

If you have any concerns about our products,
you can contact us on
ProductSafety@springernature.com

In case Publisher is established outside the EU,
the EU authorized representative is:
Springer Nature Customer Service Center GmbH
Europaplatz 3, 69115 Heidelberg, Germany

Printed by Libri Plureos GmbH
in Hamburg, Germany